OTHER GEOMETRY FORMULAS

Angle Sum in a Triangle:

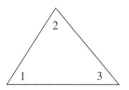

$$\angle 1 + \angle 2 + \angle 3 = 180°$$

Angle Sum in a Polygon:

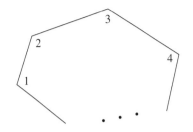

$$\angle 1 + \angle 2 + \angle 3 + ... + \angle n = (n-2) \cdot 180°$$

Pythagorean Theorem:

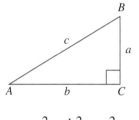

$$a^2 + b^2 = c^2$$

Special Right Triangles:

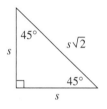

45-45 Right Triangle

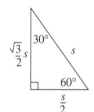

30-60 Right Triangle

Trigonometric Ratios:

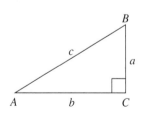

$$\tan A = \frac{a}{b} = \frac{\text{opposite side}}{\text{adjacent side}}$$

$$\sin A = \frac{a}{c} = \frac{\text{opposite side}}{\text{hypotenuse}}$$

$$\cos A = \frac{b}{c} = \frac{\text{adjacent side}}{\text{hypotenuse}}$$

Coordinate Geometry:

Slope of the line through $P(a,b)$ and $Q(c,d)$: $\qquad m = \dfrac{d-b}{c-a}$

Distance from $P(a,b)$ to $Q(c,d)$: $\qquad d = \sqrt{(c-a)^2 + (d-b)^2}$

Midpoint of segment $\overline{PQ}$: $\qquad M = \left(\dfrac{a+c}{2}, \dfrac{b+d}{2} \right)$

College
Geometry

College Geometry

A Problem-Solving Approach with Applications

Gary L. Musser

Oregon State University

Lynn E. Trimpe

Linn-Benton Community College

Prentice Hall

Upper Saddle River, New Jersey 07458

LIBRARY OF CONGRESS CATALOGING-IN-PUBLICATION DATA

Musser, Gary L.
　　College geometry : a problem-solving approach with applications /
　Gary L. Musser, Lynn E. Trimpe.
　　　　p.　　cm.
　　Includes index.
　　ISBN 0-02-385450-2
　　1. Geometry.　I. Trimpe, Lynn E.　II. Title.
　QA455.M87　1994
　516—dc20　　　　　　　　　　　　　　　　　　　　93-39365
　　　　　　　　　　　　　　　　　　　　　　　　　　　CIP

Editor: Robert Pirtle
Production Supervisor: Spectrum Publisher Services
Production Manager: Aliza Greenblatt
Text Designer: Robert Freese
Cover Designer: Robert Freese
Cover Illustration: Patricia Peticolas/Fundamental Photographs
Illustrations: Dartmouth Publishing, Inc.
Geometry Around Us and Applied Problems Illustrations: George Dick

This book was set in Times Roman, Souvenir, & Optima by Compset and was printed and bound by R.R. Donnelley & Sons Company. The Cover was printed by Lehigh Press.

© 1994 by Prentice-Hall, Inc.
Upper Saddle River, New Jersey 07458

Printed in the United States of America

10 9 8

ISBN　　0-02-385450-2

Prentice-Hall International (UK) Limited, London
Prentice-Hall of Australia Pty. Limited, Sydney
Prentice-Hall Canada Inc., Toronto
Prentice-Hall Hispanoamericana, S.A., Mexico
Prentice-Hall of India Private Limited, New Delhi
Prentice-Hall of Japan, Inc., Tokyo
Simon & Schuster Asia Pte. Ltd., Singapore
Editora Prentice-Hall do Brasil, Ltda., Rio de Janeiro

To my wife, Irene, to my son, Greg,
to my mother, Marge, and to my father, G.L.
—Gary

To my mother and father, Shirley and Howard,
and to my husband, Tim
—Lynn

Special Dedication to William F. Burger

This book is dedicated to Bill, Gary's friend and close associate at Oregon State University, who died prematurely from pancreatic cancer on March 23, 1991. Bill and Gary had taught together for over twelve years and coauthored *Mathematics for Elementary Teachers—A Contemporary Approach,* published by Macmillan.

Shortly after coming to Oregon State, Bill was the director of a research project funded by the National Science Foundation that studied how children learn geometry, especially with respect to the van Hiele levels. His pioneering work brought him international recognition in this area. Bill's passion for geometry led to the idea of writing this book. In the planning stage of this book and shortly before Bill passed away, Lynn was asked to join our team because of her expert teaching at the community college level and her valued contributions to the supplements for the Musser/Burger book for elementary teachers.

In addition to Bill's notable work in geometry, he received the OSU College of Science Carter Award for outstanding and inspirational teaching, he founded MAJIC, the Math Advising for Juniors Interested in College placement test for the state of Oregon, and he was a key participant in the OSU SMILE program, which encouraged science and mathematics achievement among Native American and Hispanic students in Oregon's rural schools. He spoke and published widely and was active in professional organizations. Most of all, he was revered by his many students at OSU.

In his personal life, Bill played oboe in a local woodwind quintet, was an active member in his church, loved opera, was a devoted fan of the Cleveland Indians since boyhood, played softball until he broke his ankle sliding into second base, enjoyed his push-button transmission Chrysler Newport with a square steering wheel, and treasured the time spent with his wife, Adrienne, and his precious daughter, Mary.

Bill was a very special person. It was a joy working with him. He had a feel for and love of geometry that is seldom seen, and his love of teaching and his students was inspiring. He is missed by all who were touched by him. Although this book would have no doubt been richer had he been here to work with us, we believe that its spirit would please him.

List of Supplements

Student Activity Manual and Study Guide

Instructor's Manual

Computerized Test Bank

Preface

This book has four main goals:

1. To help students become better problem solvers, especially in solving common application problems involving geometry
2. To help students learn many properties of geometric figures, to verify them using proofs, and to use them to solve applied problems
3. To expose students to the axiomatic method of synthetic Euclidean geometry at an appropriate level of sophistication
4. To provide students with other methods for solving problems in geometry, namely using coordinate geometry and transformation geometry.

To accomplish these goals, we have organized the text into three parts:

Part I—Problem Solving, Geometric Shapes, and Measurement
   (Chapters 1–3)
Part II—Formal Synthetic Euclidean Geometry (Chapters 4–7)
Part III—Alternate Approaches to Plane Geometry (Chapters 8–9).

Our rationale for this arrangement takes into account the students who will study this book. First, research on learning geometry suggests that the study of geometry should begin with informal experiences and gradually move toward formal proof. Second, most of these students will be using geometry to solve problems encountered in their future vocations as well as in subsequent course-work. Since the notions of geometric shapes and measurement geometry are particularly relevant to these students, we provide an early development of these ideas in an informal manner. In this way, many ideas that are proved in Part II are introduced informally and some are justified intuitively in Part I.

Part II provides the core of a standard Euclidean geometry course. We have postulated some results that could be theorems. This allows the students to get to the central results more quickly, giving them additional time to apply the results to other proofs and applied problems.

Part III opens students' eyes to the fact that there are other ways to prove results and solve problems in geometry. Students grow to appreciate the beauty and efficiency of using these alternate approaches.

FEATURES

The following features have been incorporated into the design of this book to enhance student learning.

Pedagogical

- Two-color format
- A liberal use of examples throughout
- Extensive use of figures, many in two colors, to enhance the concept development
- Boxes to highlight postulates, theorems, and important definitions
- Bold face type to highlight definitions
- Many theorems motivated by first considering specific examples
- A distinctive symbol to indicate the end of an example or a proof
- Problem sets that include various combinations of exercises, problems, applied problems, and proofs
- Answers for all odd-numbered exercises and problems
- Answers for odd-numbered proofs either outlined, or given in a paragraph or statement-reason format
- Problem-solving strategies and clues given in each chapter together with additional problems that utilize the highlighted strategy
- Writing for Understanding problems to give students an opportunity to deepen their understanding and to communicate in writing
- Chapter Reviews that require active student participation and self-assessment
- Chapter Tests that serve as a review of students' abilities to work with basic concepts, solve problems, and make proofs
- A brief Table of Contents and a second more complete one listing every section
- Common geometric formulas with figures, a symbol list, and table of conversions on the inside cover pages
- Topics Sections at the end of the book that provide some prerequisite material and some enrichment topics
- Appendix 1, Getting Started, to which students can refer when starting proofs
- Appendix 2, a quick reference of all the postulates, theorems, and corollaries

Motivational

- Chapter openers that set the scene for each chapter by presenting a historical tidbit
- Initial applied problems that motivate the material in each section and that have a solution given at the end of the section

· Geometry Around Us features at the end of each section that make students cognizant of examples of geometry in our world
· Vignettes about People in Geometry that acquaint students with some well-known geometers

SUPPLEMENTS

Student Activity Manual and Study Guide—This resource contains many hands-on activities correlated with chapters in the text to promote concept learning, solutions for every other odd-numbered problem, hints and written solutions for selected proofs, complete solutions to chapter tests, and additional practice tests with answers.

Instructor's Manual—This manual contains answers to all even-numbered exercises, problems, and applied problems, with outlines of all even-numbered proofs, overhead transparency masters, two chapter tests for each chapter and one for each topic section, and a listing of all questions that appear in the computerized test bank, and some problems for computer exploration.

Computerized Test Bank—All of the test items in the Instructor's Manual are available on disk, both in MS-DOS and in Macintosh™ formats.

ACKNOWLEDGMENTS

We wish to thank the following for their expert reviews of our manuscripts: Martin Brown, Jefferson Community College; John Longnecker, University of Northern Iowa; Jill McKenney, Lane Community College; Sue Nolen, Blue Mountain Community College; Bernadette Perham, Ball State University; Arlene Sego, Cuyahoga Community College; Darlene Whitkanack, Northern Illinois University. Also we owe a special thanks to the following colleagues who have used the preliminary version of the text and given us valuable feedback: Ricardo Bell, Candy Drury, Dale Green, Sue Nolen, Wally Reed, Sharon Rodecap, Gayle Smith, Gerry Swenson, and Betty Westfall; test bank author: Catherine Aune; answer checkers: Judy DeSzoeke, Dale Green, Crystal Gilliland, and Roger Maurer; Don Fineran for permission to use problems from the Oregon Vo-Tech Project; Marv Kirk and Roger Maurer for their contributions to problem sets; Tommy Bryan for his superb job of proofreading; Dave Metz for preparing the People in Geometry vignettes; and George Dick for his creative graphics in Geometry Around Us and Applied Problems.

We also thank the many students who have used our preliminary version of the text and given us encouragement through their success. Last we thank our production team, especially Kelly Ricci and Mercedes Jackson at Spectrum Publisher Services and Aliza Greenblatt at Macmillan, for bringing this book to print and our visionary editor, Bob Pirtle, for his smooth coordination of this project.

G.L.M.

L.E.T.

Brief Contents

Contents

PART **II**   **Formal Synthetic Euclidean Geometry**

PART III   Alternate Approaches to Plane Geometry

1

Problem Solving in Geometry

 George Polya (1887–1985) was a mathematician famous for his lifelong interest in and study of the process of problem solving. Born and educated in Hungary, Polya came to the United States in 1940. He is the author of numerous books and papers on problem solving, the most famous of which is *How to Solve It*. This book has been translated into 17 different languages, and more than 1,000,000 copies have been sold since it was first published in 1945.

Polya defined intelligence as the ability to solve problems and believed that "solving problems is human nature itself." Thus he felt strongly that a major goal of education should be the development of problem-solving skills. To that end, Polya devised what has become known as the four-step problem-solving process, which will be a focus of this chapter.

PROBLEM-SOLVING STRATEGIES

1. *Draw a Picture*
2. *Guess and Test*
3. *Use a Variable*
4. *Look for a Pattern*
5. *Make a Table*
6. *Solve a Simpler Problem*

One of the main goals of this book is to help you to become a better problem solver. This chapter introduces six strategies that will be useful in solving geometry problems. In addition, at the beginning of each of the remaining chapters, a new strategy is introduced. In this way, your ability to solve problems should grow much as the Problem-Solving Strategies boxes like the one to the left grow throughout the book.

INITIAL PROBLEM

Chinese checkers is a marble game played by 2 to 6 players on a board like the one shown below. This board has a starting pen (triangular region in one point of the star) with 4 rows of holes that holds 10 marbles for each player. Suppose that a more challenging game is desired and a similar board is constructed to have 6 rows (21 marbles) for each starting pen. How many holes, in all, would this expanded board have?

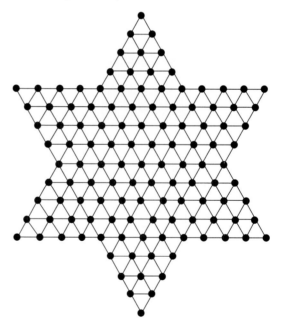

INTRODUCTION

Problem solving is considered by most people to be the main objective of any mathematics course. In a more general sense, problem solving is central to many disciplines, including business and engineering. This chapter provides an introduction to the study of problem solving with an emphasis on problems in geometry.

1.1
PROBLEM-SOLVING STRATEGIES

In our discussion of problem solving, we need to make clear what is meant by a problem. In this book a distinction is made between an "exercise" and a "problem." An **exercise** can be solved by applying a routine procedure. It may be similar to other exercises (or problems) that you have worked or have seen worked. A typical "word problem" in an algebra course is an exercise if you recognize the type of problem and can recall an appropriate procedure to apply.

On the other hand, a **problem** is nonroutine and unfamiliar. To solve a problem, you need to stop and think about how to attack it. You may need to try something completely new and different. You may get stuck several times and have to make new starts. The need for some kind of creative step on your part in solving the problem is what makes it different from an exercise.

George Polya was a mathematician whose name has come to be synonymous with problem solving. In an effort to encourage more students to "experience the tension and enjoy the triumph of discovery" that accompany solving a problem, he presented the following **four-step process** for problem solving.

STEP 1. Understand the Problem.
- Is it clear to you what is to be found?
- Do you understand the terminology used in the problem?
- Is there enough information?
- Is there irrelevant information?
- Are there any restrictions or special conditions to be considered?

STEP 2. Devise a Plan.
- How should the problem be approached?
- Does the problem appear similar to any others you have solved?
- What strategy might you use to solve the problem?

STEP 3. Carry Out the Plan.
- Apply the strategy or course of action chosen in Step 2 until a solution is found or you decide to try another strategy.

STEP 4. Look Back.

- · Is your solution correct?
- · Do you see another way to solve the problem?
- · Can your results be extended to a more general case?

In the four-step process, Step 2 is a critical one. Even if you thoroughly understand a problem (Step 1), you may not be able to progress further. On the other hand, once you have selected a workable strategy, it is usually not difficult to implement it (Step 3). Likewise, once a solution has been obtained, it is usually not difficult to verify whether that solution is correct (Step 4). Therefore, the purpose of this chapter is to focus on Step 2 in Polya's process and to present six general strategies that are frequently useful in solving geometric problems. Other strategies will be presented in subsequent chapters of the book, one strategy at the beginning of each chapter.

STRATEGY 1: DRAW A PICTURE

This strategy is a natural choice for solving problems in geometry because many of the problems of geometry are related to figures, shapes, and physical structures. Often drawing one or several pictures can help you to solve the problem or can help you to better understand the problem so that you can formulate a plan for solving it. The following clues may help you to identify situations where the Draw a Picture strategy might be useful.

CLUES

The Draw a Picture strategy may be appropriate when

- · A physical situation is involved.
- · Geometric figures or measurements are involved.
- · You want to gain a better understanding of the problem.
- · A visual representation of the problem is possible.

As you attempt to solve the following example problems, imagine solving the problems *without* looking at any pictures. Then try solving the problems with a picture to see if the picture helped in the process.

EXAMPLE 1.1   A large cube is formed by arranging 64 smaller cubes of the same size in a stack that measures 4 cubes by 4 cubes by 4 cubes. An open cardboard box is constructed in the shape of the larger cube, and the arrangement of small cubes just fits into the box. How many of the small cubes are *not* touching a side or the bottom of the box?

STEP 1. Understand the Problem.

There are no gaps between the small cubes and the sides of the box. Remember that the box has no top. We must determine how many of the 64 small cubes are in contact with neither a side nor the bottom of the box.

STEP 2. Devise a Plan.

Draw a picture of the cubes inside the box in order to better visualize their arrangement [Figure 1.1(a)]. The 4 by 4 by 4 cube has 64 small cubes in it, arranged in four layers.

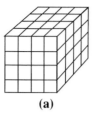 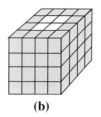

(a)　　　　　　　(b)　　　　　　　(c)

FIGURE 1.1

STEP 3. Carry Out the Plan.

Now we can start counting the small cubes that touch the sides and bottom of the box. Shading in those cubes that are in contact with a side or the bottom of the box may be helpful [Figure 1.1(b)]. Remove the central "core" of cubes [Figure 1.1(c)]. Some of these cubes touch only the bottom of the box. We can see that the top 3 layers of 4 cubes each, or 12 cubes in all, touch neither the sides nor the bottom of the box.

STEP 4. Look Back.

We also could have solved this problem by subtracting the number of cubes that *do* touch the sides and bottom of the box from 64. What would happen if the large cube measured 5 by 5 by 5? How many small cubes would touch neither the sides nor the bottom of that cube? Can this problem be generalized to an n by n by n cube? If so, how?

Additional Problems Where the Strategy "Draw a Picture" Is Useful

1. If the diagonals of a square are drawn in, how many triangles of all sizes are formed?

To Get You Started

Diagonal

FIGURE 1.2

   (a) Do you know what the diagonal of a square is? It is a line segment whose endpoints are two opposite corners of the square. One diagonal is shown in Figure 1.2. (A more general definition of a diagonal will be given later in the book.)

   (b) Remember that the triangles may be different sizes. Also, some of the triangles may overlap.

   (c) Although one picture may be helpful in formulating a solution to the problem, several pictures might make the solution even clearer.

2. A rectangular milk crate has spaces for 24 bottles in 4 rows and 6 columns. Can you put 18 bottles of milk into the crate so that each row and each

column of the crate have an even number of bottles in them? [HINT: You might consider the spaces that do *not* have bottles in them, rather than the 18 that do.]

STRATEGY 2: GUESS AND TEST

Also known as "Trial and Error," the Guess and Test strategy is an extremely useful method for solving many problems. Guess and Test is often the very first strategy employed by experienced mathematicians and novice problem solvers alike. Even if it does not lead immediately to a solution, the Guess and Test strategy helps you to get a "feeling" for a problem and may suggest other strategies that could be used to solve the problem.

The Guess and Test method does not necessarily imply random guessing. In fact, often the conditions of the problem will limit the possible guesses. After the first few guesses, you may make some observations that will further limit your guesses. Rather than random Guess and Test, what is more often utilized is a form of systematic Guess and Test. With systematic Guess and Test there is some order to the guesses and a means of recording which guesses have been tested. The following clues may help you to identify situations where the Guess and Test strategy may be useful.

CLUES

The Guess and Test strategy may be appropriate when

· There is a limited number of possible solutions to test.
· You want to gain a better understanding of the problem.
· You have a good idea what the solution is.
· You can systematically try possible solutions.
· Your choices have been narrowed down by first using other strategies.
· There is no obvious strategy to try.

EXAMPLE 1.2   Five friends were sitting on one side of a table. Gary sat next to Bill. Mike sat next to Tom. Howard sat in the third seat from Bill. Gary sat in the third seat from Mike. Who sat on the side of Tom opposite from Mike?

STEP 1. Understand the Problem.

The five seats are arranged in a straight line. We are to assign the five people to the five seats and determine who, in addition to Mike, is next to Tom. The order of the persons may vary and still meet the conditions of the problem. For example, Gary-Bill or Bill-Gary would satisfy the condition "Gary sat next to Bill."

STEP 2. Devise a Plan.

Draw a picture such as the one in Figure 1.3(a) to represent the seats and use the initial letter of each man's name to represent each of the five men. Test various arrangements of the men using the conditions in the problem to narrow down the guesses.

__ __ __ __ __

(a)

FIGURE 1.3

STEP 3. Carry Out the Plan.

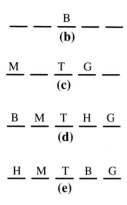

(b)

(c)

(d)

(e)

FIGURE 1.3

We might start by testing some possibilities for the middle seat. Remember that the problem stated that Howard sat in the third seat away from Bill. Let's try placing Bill in the middle seat [Figure 1.3(b)]. Notice that this makes it impossible for Howard to be placed three seats away from Bill. The same problem arises if Howard is seated in the middle. Likewise, because Gary must be placed in the third seat from Mike, neither Gary nor Mike can be in the middle seat. So Tom must be the man in the middle seat.

Now we will try some arrangements with Tom in the middle seat. In Figure 1.3(c), Gary is in the third seat from Mike as required, but Mike is not next to Tom. In Figure 1.3(d), Mike is now next to Tom and Howard and Bill are placed so that Howard is in the third seat from Bill. But Gary is not seated next to Bill. We can fix that by switching Howard and Bill [Figure 1.3(e)]. Now this arrangement meets all of the stated conditions. We can see that Bill must be seated on the other side of Tom.

STEP 4. Look Back.

Is there any other arrangement of the men that works? One other arrangement that satisfies the conditions of the problem is G B T M H. This arrangement still places Bill on the other side of Tom. Are there other solutions? Why or why not?

Additional Problems Where the Strategy "Guess and Test" Is Useful

1. Is it possible to divide the shape in Figure 1.4 into four parts so that the four parts are the same size as each other and have the same shape as the original figure?

To Get You Started

(a) The four parts must be exact duplicates of each other, and they must all fit together in the larger shape with no gaps or overlaps. You might verify that the smaller pieces are exact duplicates of each other by cutting them out and fitting them on top of each other.

(b) Draw four copies of the original shape and see if you can put them together to form a larger version of the same shape.

2. Six identical coins are arranged as shown in Figure 1.5. Change the arrangement of coins on the left into the arrangement of coins on the right by moving only two of the coins.

FIGURE 1.4

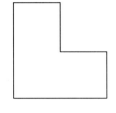

FIGURE 1.5

STRATEGY 3: USE A VARIABLE

The Use a Variable strategy is appropriate when relationships between unknown quantities are stated, when an equation can be written to model a problem situation, or when a general formula is desired. In applying this strategy you will be able to utilize your skills from algebra. The following clues may help you to identify situations where the Use a Variable strategy may be useful.

CLUES

The Use a Variable strategy may be appropriate when

· A phrase similar to "for any number" is present or implied.
· A problem suggests an equation.
· A proof or a general solution is desired.
· There are a large number of cases.
· A proof is required in a problem involving numbers.
· An unknown quantity is related to known quantities.
· There are infinitely many numbers involved.
· You are trying to develop a general formula.

EXAMPLE 1.3   The measure of the largest angle of a triangle is three times the measure of the smallest angle. The measure of the third angle of the triangle is 40° more than the measure of the smallest angle. What are the measures of the angles in the triangle? (Use the fact that the sum of the measures of the angles in a triangle is 180°. This fact will be developed later in this book.)

STEP 1. Understand the Problem.

The relationships between the measures of the angles can be described by introducing a variable. If we use x to represent the measure of the smallest angle, we have

$$x = \text{the measure of the smallest angle}$$
$$3x = \text{the measure of the largest angle}$$
$$x + 40° = \text{the measure of the third angle (Figure 1.6)}$$

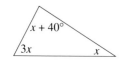

FIGURE 1.6

We must use the relationships we have described to determine the number of degrees in each angle of the triangle.

STEP 2. Devise a Plan.

We can use what we know about the angles of a triangle to relate the expressions x, $3x$, and $x + 40°$. Because the sum of the measures of the angles in a triangle is 180°, we can write

$$x + 3x + (x + 40°) = 180°$$

STEP 3. Carry Out the Plan.

Now we can solve the equation to find the measures of the angles.

$$3x + x + (x + 40°) = 180°$$
$$5x + 40° = 180°$$
$$5x = 140°$$
$$x = 28°$$

This means that $3x = 84°$ and $x + 40° = 68°$. Therefore, the largest angle measures 84°, the smallest measures 28°, and the other angle measures 68°.

STEP 4. Look Back.

Verify that the sum of the measures of the angles is 180°.

$$\text{Check:} \qquad 28° + 84° + 68° = 180° \checkmark$$

Could we have solved the problem another way by using x to represent the measure of one of the other two angles?

Additional Problems Where the Strategy "Use a Variable" Is Useful

1. Find a formula for the number of diagonals that can be drawn from one vertex of a polygon with a given number of sides. For example, the pentagon in Figure 1.7 is a polygon with 5 sides. In that example, two different diagonals can be drawn from vertex A.

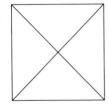

FIGURE 1.7

To Get You Started

    (a) You need to know that a polygon, in simple terms, is a geometric figure formed by enclosing a region with straight line segments. (A more formal definition of polygon will be given later in the book.) The line segments are the **sides** of the polygon, and the endpoints of the line segments are the **vertices** of the polygon.

    (b) Draw several other examples of polygons and count the number of diagonals that can be drawn from one vertex. Do you see a relationship between the number of sides in the polygon and the number of diagonals from a vertex? If you let n be the number of sides of the polygon, what expression using n represents the number of diagonals from one vertex?

2. A new football field under construction will have a total length (including end zones) that is 40 feet more than twice its width. The area of the football field must be calculated in order to determine the amount of grass seed necessary to seed the field. Find the area of the field if its perimeter will be 1040 feet. [NOTE: Recall that the area of a rectangle is $A = lw$ and the perimeter is $P = 2l + 2w$.]

Answers to Additional Problems for Strategies 1–3

Draw a Picture

1. There are four large triangles and four smaller ones (Figure 1.8). Thus, there are eight triangles in all.

2. Many solutions are possible, one of which is shown below. X's mark the six positions in the crate that are *empty* (Figure 1.9).

FIGURE 1.8

O	O	O	O	O	O
X	X	O	O	O	O
X	O	X	O	O	O
O	X	X	O	O	O

FIGURE 1.9

Guess and Test
1.

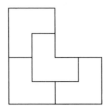

2. One solution follows

Use a Variable
1. The number of diagonals from one vertex is $n - 3$.
2. $w = 160$ ft, $l = 360$ ft, $A = 57,600$ ft^2

GEOMETRY AROUND US

Of the polygons that appear in nature, such as squares and triangles, one of the most frequently seen is the regular hexagon, a polygon with six sides all the same length and six angles all having the same measure. The most familiar examples are probably the snowflake and cells of a honeycomb. It is interesting to note that although it is said of snowflakes that "no two are alike," every such crystal can be outlined with a regular hexagon. Also, although the cells of a honeycomb are hexagonal, they are cylindrical tubes when first made by the bees. The pressure of the tubes against one another results in their hexagonal shape.

PROBLEM SET 1.1

EXERCISES/PROBLEMS

1. How many equilateral triangles of all sizes are there in the 3 by 3 by 3 equilateral triangle shown here? (An **equilateral triangle** is a triangle with three sides of the same length. Equilateral triangles will be discussed later in the book.)

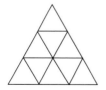

2. A **tetromino** is a shape made up of 4 squares where two squares are joined along an entire side.

Not a Tetromino   Tetromino

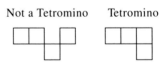

Two tetromino shapes are the same if they can be matched by turning or flipping. For example, the following two tetrominos are the same as the tetromino shown at right above.

How many different tetromino shapes are possible?

3. A **pentomino** is made up of 5 squares where two squares are joined along an entire side. Two pentominos are the same if they can be matched by turning or flipping (see problem 2). How many different pentomino shapes are possible?

4. Find 3 different pentomino shapes that fit together to form a 3 by 5 rectangle.

5. Find 5 different pentomino shapes that fit together to form a 5 by 5 square.

6. (a) Determine if exactly 6 squares of the 3 by 3 grid shown can be shaded so that no 3 shaded squares line up horizontally, vertically, or diagonally.

   (b) Determine if exactly 12 squares of a 4 by 4 grid can be shaded so that no 4 shaded squares line up horizontally, vertically, or diagonally.

7. Determine if a circular pizza can be cut into 11 pieces with 4 straight cuts. [HINT: The pieces need not be the same size or shape.]

8. An organization has access to 10 rectangular tables that will each seat 4 people on a side and one person on each end. For an awards banquet the tables are to be lined up end to end to form one long table. How many people can be seated at this long table? How many people could be seated at a long table made from 25 of the rectangular tables?

9. A man has a 10-meter by 10-meter rectangular garden that he wants to fence. How many fence posts will he need if each corner has a post and posts are spaced 1 meter apart along the sides?

10. Toothpicks are arranged to form the following figure.
   (a) Remove exactly 6 toothpicks and leave 2

squares. (There should be no "leftover" toothpicks.)

   (b) Remove exactly 5 toothpicks and leave 3 squares. (Again, there should be no "leftover" toothpicks.)

11. Use the toothpick arrangement in problem 10 to do the following.
   (a) Remove exactly 7 toothpicks to leave 2 squares.
   (b) Remove 2 toothpicks to leave 6 squares.

12. Greg has 1002 meters of fencing. He wants to fence a rectangular region that is four times as long as it is wide. One of the longer sides is bordered by a river, so that side will not be fenced. What will be the dimensions of the region?

13. A child has 10 blocks with heights of 1 cm, 2 cm, 3 cm, 4 cm, 5 cm, 6 cm, 7 cm, 8 cm, 9 cm, and 10 cm. The child wants to use only these 10 blocks to build two separate towers that are exactly the same height. Can this be done? If so, how? If not, why not?

14. Trace the following figure and cut it into five pieces along the indicated lines. Rearrange the pieces to form a square. You must use all five pieces and have no gaps or overlaps, and you cannot turn the pieces over.

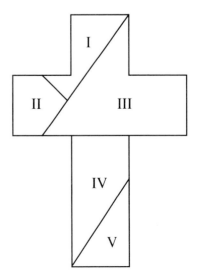

15. Trace the following figure and cut out the pieces. Rearrange the pieces to form (a) a rectangle that is not a square and (b) a triangle. This dissection puzzle is called the **Chinese Tangram Puzzle** and has been popular for over two hundred years.

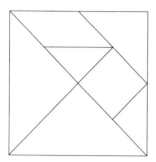

16. Six coins are arranged in a triangular shape as shown. Rearrange the coins to form a circular shape by moving only two coins. To "move" a coin, you must slide it along the surface, without disturbing any other coins, and place it in a new position so that it is touching another coin.

17. (a) Arrange nine toothpicks to form exactly five equilateral triangles not necessarily of the same size. Toothpicks may not be bent or broken.

   (b) Arrange six toothpicks to form exactly four equilateral triangles of the same size. Toothpicks may not be bent or broken.

APPLICATIONS

18. A standard tennis court is shown next. How many rectangles of all sizes are there on a tennis court? [NOTE: Treat the line under the net as one of the lines of the court.]

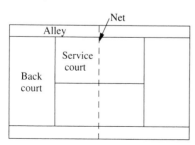

19. A famous problem, posed by puzzler Henry Dudenay, presented the following situation. Suppose that houses are located at points *A*, *B*, and *C*. We want to connect each house to water, electricity, and gas located at points *W*, *G*, and *E* without any of the pipes/wires crossing each other. [NOTE: This problem is related to a branch of mathematics called graph theory.]

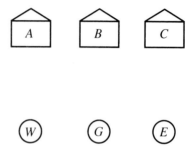

(a) Try making all of the connections. Can it be done? If so, how?

(b) Suppose that the owner of one of the houses, say *B*, is willing to let the pipe for one of his neighbors' connections pass through his house. Then can all of the connections be made? If so, how?

20. A water main for a street is being laid using mechanical joint pipe. The pipe comes in 18-foot and 20-foot sections, and a designer determines that the water main would require 14 fewer sections of 20-foot pipe than if 18-foot sections were used. Find the total length of the water main.

21. A man living in the Northwest wants to mail a Christmas tree to a friend living in the South. He plans to bundle the tree tightly and pack it in a long box with a square end. He was told by the mailing service that the length plus the girth of his package cannot exceed 130 inches. If the Christmas tree is 5'10" tall, what is the maximum width and height for his package? [NOTE: The **girth** is the distance around the package as shown.]

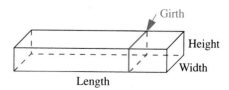

22. A problem that challenged mathematicians for many years concerns the coloring of maps. That is, what is the minimum number of colors necessary to color any map? In 1976 it was finally proved, with the aid of a computer, that no map requires more than four colors. Determine the smallest number of colors necessary to color each map shown. [NOTE: Two "countries" that share only one point can be colored the same color, but if they have more than one point in common, they must be colored differently.]

(a)

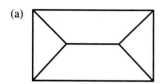

(b)

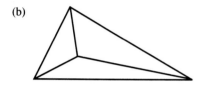

(c)

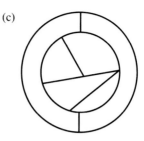

(d)

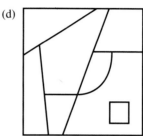

(e)

1.2

MORE PROBLEM-SOLVING STRATEGIES

In this section we will continue to examine strategies useful in solving problems in geometry. The three strategies presented in this section are often used together in solving a given problem. In fact, at times it may be difficult to solve a problem using only one of these strategies and not employing any others. The same is true for many of the strategies that will be discussed in this book. It is probably more common to use a combination of strategies in solving a problem than to use one particular strategy. For example, frequently the first step in solving a problem is to "play with" the problem using the Guess and Test strategy or the Draw a Picture strategy. After some time is spent on a problem in this way, we might decide to use a variable to write an equation or begin to look for a pattern.

STRATEGY 4: LOOK FOR A PATTERN

Much of mathematics consists of observing and describing patterns. Patterns can be observed in sequences of numbers, relationships between numbers, and in geometric figures. The Look for a Pattern method is often used in conjunction with other problem-solving strategies, such as Make a Table or Solve a Simpler

Problem, which will be discussed soon. The following clues may help you to identify situations where the Look for a Pattern strategy may be useful.

CLUES

The Look for a Pattern strategy may be appropriate when

· A list of data is given.
· A sequence of numbers or figures is involved.
· Listing special cases helps you to deal with a complex problem.
· You are asked to make a prediction or a generalization.
· Information can be expressed and viewed in an organized manner, such as in a table.

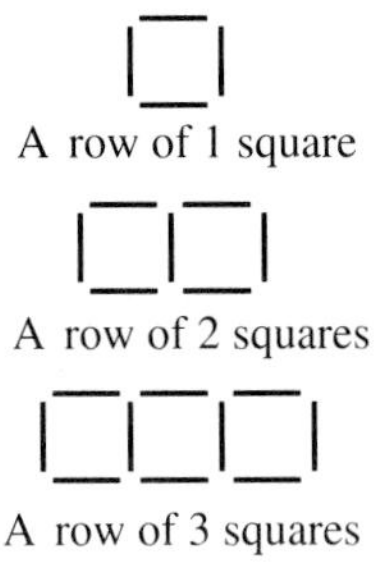

EXAMPLE 1.4   Toothpicks can be arranged to form rows of squares as shown in Figure 1.10. How many toothpicks are required to form a row of 100 squares in this manner?

A row of 1 square

A row of 2 squares

A row of 3 squares

FIGURE 1.10

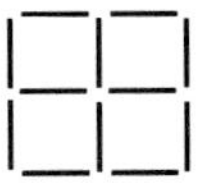

FIGURE 1.11

STEP 1.   Understand the Problem.

The toothpicks cannot be broken. The toothpicks must be arranged in a straight line of squares, not in a rectangular array of squares. For example, the arrangement in Figure 1.11 would not be acceptable. We must determine the number of toothpicks to form a row of 100 squares.

STEP 2.   Devise a Plan.

We might try a few more rows of squares, say 4 squares and then 5 squares (Figure 1.12). Then we could examine the results to see if there appears to be a relationship between the number of squares in a row and the number of toothpicks used.

STEP 3.   Carry Out the Plan.

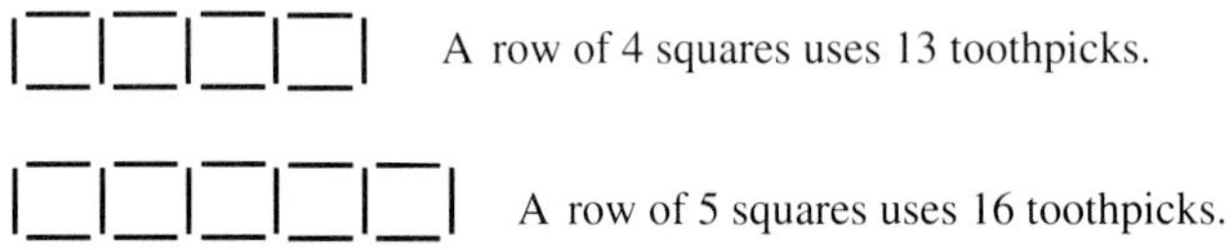

A row of 4 squares uses 13 toothpicks.

A row of 5 squares uses 16 toothpicks.

FIGURE 1.12

So far, from Figures 1.10 and 1.12, we have the following:

1 square	requires	4 toothpicks
2 squares	require	7 toothpicks
3 squares	require	10 toothpicks
4 squares	require	13 toothpicks
5 squares	require	16 toothpicks

It appears that for each additional square 3 more toothpicks are required. Therefore, we can say that the row with 100 squares will require 3 more toothpicks than the row with 99 squares. But how many toothpicks are required for the row with 99 squares? We would prefer not to count so many toothpicks. Looking at the pictured rows of squares we might also observe a visual pattern. A row of 2 squares consists of 2 sets of 3 toothpicks plus a single toothpick necessary to complete the last square as shown in Figure 1.13.

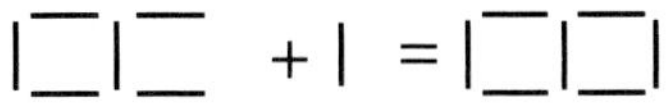

$$|\_|\_ \ + | \ = |\_|\_|$$

FIGURE 1.13

Similarly, a row of 3 squares consists of 3 sets of 3 toothpicks plus a single toothpick and a row of 4 squares consists of 4 sets of 3 toothpicks plus a single toothpick, as shown in Figure 1.14.

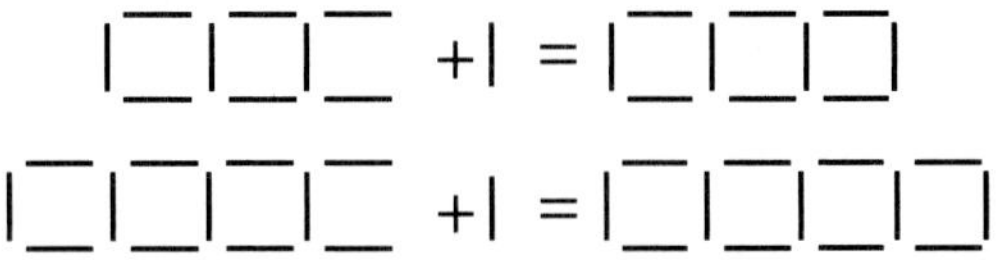

$$|\_|\_|\_ \ +| \ = |\_|\_|\_|$$

$$|\_|\_|\_|\_ \ +| \ = |\_|\_|\_|\_|$$

FIGURE 1.14

It appears to be the case that with each added square one additional set of 3 toothpicks is needed. This observation is consistent with the numerical pattern we observed earlier. So a row of 5 squares would require 5 sets of 3 toothpicks plus one single toothpick, or a total of 5(3) + 1 = 16 toothpicks.

This means that a row of 100 squares would require 100 sets of 3 toothpicks plus one more toothpick to complete the last square. So the total number of toothpicks necessary for a row of 100 squares would be 100(3) + 1 = 301. Other visual patterns may be observed for this problem, but they should all lead to the same final answer, namely 301 toothpicks.

STEP 4. Look Back.

How many toothpicks would be required for other rows of squares? We could use a variable to generalize the pattern we have observed. Let n be the number of squares in a row. Then the number of toothpicks required can be calculated as in Step 3. The row would require n sets of 3 toothpicks plus one more toothpick, so the total number of toothpicks required would be $n(3) + 1 = 3n + 1$. This formula can be used to determine the number of toothpicks required to form a row of any size. For example, a row of 200 squares would require 3(200) + 1, or 601 toothpicks.

In the preceding toothpick problem, we observed five different examples of rows of toothpicks, noticed a pattern, and then concluded that this pattern would hold for a row of any length. When we look at specific examples and draw a general conclusion from the pattern observed, we are using what is called **inductive reasoning**.

We often use inductive reasoning to draw conclusions. For example, scientists reason inductively when they perform experiments, observe the results, and form a hypothesis to explain those results. We may observe the sequence of numbers 5, 7, 11, 13, 17, 19, 23, . . ., and notice a pattern. Namely, the differences between successive terms alternate between 2 and 4, that is, $7 - 5 = 2$, $11 - 7 = 4$, $13 - 11 = 2$, and so on. Reasoning inductively, we may predict that the next number in this sequence will be 2 more than 23, or 25.

In the situations involving inductive reasoning described so far, specific examples are observed and a general conclusion is drawn. This transition from the specific to the general is the essence of inductive reasoning and distinguishes it from deductive reasoning, which is used extensively in geometry. In **deductive reasoning**, some general statements are accepted as true and then specific conclusions are drawn from those statements. Both inductive and deductive reasoning are used in geometry. Deductive reasoning will be a major focus of our study in Chapter 4.

One pitfall of inductive reasoning must be considered. Based on our observations of a few examples, we conclude that the pattern we notice will continue and that it is the *only* correct one. This assumption may not be warranted. For example, in the case of the numerical sequence 5, 7, 11, 13, 17, 19, 23, . . ., it may seem reasonable to assume that the next term will be 25. However, it is just as reasonable to predict that the next term will be 29 because all numbers listed are prime numbers (counting numbers that have only two factors) and the next prime number after 23 is 29. Thus, in this example, more than one valid conclusion is possible. Due to this difficulty, often a combination of inductive and deductive reasoning is used to justify a conclusion.

Additional Problems Where the Strategy "Look for a Pattern" Is Useful

1. Examine the sequence of rectangles with subdivisions and a shading as shown in Figure 1.15. Draw the figure that would come next in the sequence.

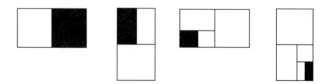

FIGURE 1.15

To Get You Started

   (a) Compare consecutive figures and consider the size and position of the shaded region in each of them.
   (b) What do you observe about the orientation of the large rectangle?
   (c) Try focusing on only one attribute at a time and follow it through the sequence of figures from left to right.

2. The rectangular numbers are the whole numbers that represent the arrays of dots in Figure 1.16.

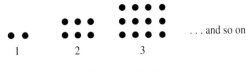

FIGURE 1.16

Notice that there are 6 dots in the second rectangular number. Find the number of dots in the *n*th rectangular number.

STRATEGY 5: MAKE A TABLE

It is often helpful to make a list or a table when searching for a pattern. For instance, in Example 1.4 the number of toothpicks for rows of different sizes was calculated and recorded. This data could have been neatly summarized in a table such as the one shown next.

Number of Squares in a Row	Number of Toothpicks Required
1	4
2	7
3	10
4	13
5	16

Having data arranged in a table such as the one shown may make patterns more easily recognized. For example, we might observe that adding 2 to each number in the right column yields a column of numbers that are multiples of 3. You will often see tables combined with the Look for a Pattern strategy.

A table can also provide a convenient means for systematically recording guesses when applying the Guess and Test strategy. A table is a good way to organize data as it is generated, summarize information given in a problem, or list possible values of variables. The following clues may help you to identify situations where the Make a Table strategy might be useful.

CLUES

The Make a Table strategy may be appropriate when

· Information can easily be organized and presented.
· Data will be generated.
· You want to list the results obtained by using Guess and Test.
· You are asked "in how many ways" something can be done.
· You are trying to learn about a collection of numbers or figures generated by a rule or formula.

EXAMPLE 1.5   A man wishes to enclose a rectangular area of 100 square feet for his dog (Figure 1.17). He is considering rectangles with *whole number* dimensions. What dimensions would require the least amount of fencing?

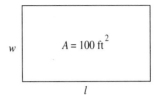

w $A = 100 \text{ ft}^2$ l

FIGURE 1.17

STEP 1. Understand the Problem.

The rectangle must have an area of exactly 100 square feet, and its dimensions must be whole numbers. In order for the amount of fencing to be as small as possible, the perimeter of the rectangle must be as small as possible. Recall that the area of a rectangle is $A = l \times w$ and that the perimeter of a rectangle is $P = 2l + 2w$, or $P = 2(l + w)$. We must determine the length and width that give the smallest perimeter.

STEP 2. Devise a Plan.

What lengths and widths give an area of 100 square feet? Make a table summarizing the possible dimensions and showing the perimeter in each case.

STEP 3. Carry Out the Plan.

Possible lengths and widths are summarized in the following table.

Width	Length	Area $= l \times w$	Perimeter $= 2(l + w)$
1	100	100	202
2	50	100	104
4	25	100	58
5	20	100	50
10	10	100	40

Because 1, 2, 4, 5, 10, 20, 25, 50, and 100 are the only whole number factors of 100, the table shown contains all possible dimensions for the rectangle. Therefore the minimum perimeter is 40 feet, and it occurs when the rectangle measures 10 ft by 10 ft. Notice that the rectangle with the minimum perimeter in this case is a square. A square is a special type of rectangle.

STEP 4. Look Back.

Will a square always give the minimum perimeter? Try rectangles with different areas. Would it make a difference if only three sides of the enclosure were fenced and the fourth side was against a garage?

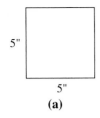

5"

5"

(a)

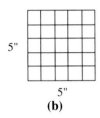

5"

5"

(b)

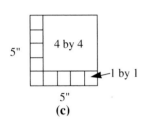

5"

4 by 4

—1 by 1

5"

(c)

FIGURE 1.18

Additional Problems Where the Strategy "Make a Table" Is Useful

1. A square piece of plywood measures 5″ by 5″ [Figure 1.18(a)]. In how many ways can it be cut into smaller pieces such that each of the smaller pieces is a square with whole-number dimensions?

To Get You Started

    (a) Here you might want to use both the Draw a Picture and Make a Table strategies.

    (b) There are many solutions. One possibility is that the square could be cut into 25 squares 1 inch on a side [Figure 1.18(b)].

    (c) The squares that result need not have the same dimensions. For example, some of the pieces might be 1 by 1 squares, and one piece might be a 4 by 4 square [Figure 1.18(c)] .

    (d) What dimensions are possible for the smaller squares? You might list those possibilities as headings in a table.

2. A dartboard consisting of 4 circles is shown in Figure 1.19. Three colors—red, yellow, and black—are available to repaint the board. If the bull's-eye of the dartboard is to be red, and no two bordering circles can be the same color, how many different color schemes are possible?

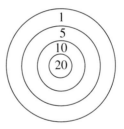

FIGURE 1.19

STRATEGY 6: SOLVE A SIMPLER PROBLEM

If the numbers in a problem are very large, very small, or difficult to work with, it is often helpful to solve a simpler version of the same problem. For example, large numbers may be replaced with smaller, more manageable numbers, or numbers involving many decimal places may be replaced with whole numbers. Once a technique is found for solving the simpler problem, the same technique can be used to solve the original problem.

    For instance, in Example 1.4 the number of toothpicks necessary to form 100 squares was desired. Rather than solve the problem with 100 squares, several simpler problems were studied. Rows consisting of 3, 4, and 5 squares were examined, and a pattern was observed that could be extended to the case of 100 squares.

    The Solve a Simpler Problem strategy can also be utilized when a figure is complex. We can look at simpler versions of the same figure and see if a pattern emerges that will lead us to a solution of the original problem. The following clues may help you to identify situations where the Solve a Simpler Problem strategy might be useful.

CLUES

The Solve a Simpler Problem strategy may be appropriate when

· The problem involves complicated computations.
· The problem involves very large or very small numbers.
· A direct solution is too complex.
· You want to gain a better understanding of the problem.
· The problem involves a large array or diagram.

EXAMPLE 1.6   There are 20 people at a party. If each person shakes hands with each other person at the party, how many handshakes will there be?

STEP 1. Understand the Problem.

Each person must shake hands with every other person exactly once. If person *A* shakes hands with person *B*, then person *B* does not shake hands with person *A* again. We must count the number of handshakes that occur and avoid duplication.

FIGURE 1.20

STEP 2. Devise a Plan.

We might represent the people as points on a circle and draw a line segment to represent each handshake. For example, the line segment joining *A* and *B* represents persons *A* and *B* shaking hands (Figure 1.20).

STEP 3. Carry Out the Plan.

Figure 1.20 shows all the handshakes for *A*. If the line segments are drawn for the remaining handshakes, it will be very difficult to count all of them. However, simpler problems, or parties with fewer people, can be solved more easily. Figure 1.21 shows the results for parties with 2, 3, and 4 people.

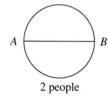

2 people

1 handshake

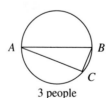

3 people

2 handshakes for *A* and 1 additional one for *B*.

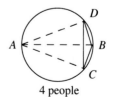

4 people

3 handshakes for *A*, 2 additional ones for *B*, and 1 additional one for *C*.

FIGURE 1.21

We can summarize the results in a table and look for a pattern.

Number of People	Number of Handshakes
2	1
3	$2 + 1 = 3$
4	$3 + 2 + 1 = 6$
5	$4 + 3 + 2 + 1 = 10$

Based on the pattern in the table, it appears that the number of handshakes for 6 people would be $5 + 4 + 3 + 2 + 1 = 15$. Similarly, for any number of people, say n, the number of handshakes would be

$$(n - 1) + (n - 2) + \cdots + 2 + 1$$

So, for 20 people, the number of handshakes would be

$$19 + 18 + 17 + \cdots + 3 + 2 + 1$$

for a total of 190 handshakes.

Another pattern could be used to shorten the computation. Notice that if we pair terms as shown next, the sum of each pair of terms is 20.

$$19 + 18 + 17 + 16 + \cdots + 4 + 3 + 2 + 1$$

$$19 + 1 = 20, 18 + 2 = 20, \cdots, 11 + 9 = 20$$

There would be a total of 9 pairs of terms, where each pair adds up to 20, plus the "leftover" middle term of 10. So we could say that the total number of handshakes would be $9(20) + 10 = 180 + 10 = 190$.

STEP 4. Look Back.

Is there another way in which this problem can be solved? Also, if n represents the number of people at the party, can a general formula be found for the number of handshakes?

Additional Problems Where the Strategy "Solve a Simpler Problem" Is Useful

1. How many squares of all sizes are there on an 8 by 8 checkerboard (Figure 1.22)?

To Get You Started

    (a) You might start by drawing a smaller 2 by 2 or 3 by 3 checkerboard and counting the number of squares of all sizes formed. Several checkerboards of each size may be easier to use than a single checkerboard.

    (b) Listing the number of squares of various sizes in a table may help you to better keep track of the number of squares. For example, in the case of

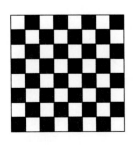

FIGURE 1.22

the 3 by 3 checkerboard you would want to list the number of different 1 by 1, 2 by 2, and 3 by 3 squares on the board.

(c) Can you see a pattern in the number of squares of each size for each checkerboard?

2. A schoolroom has 13 different desks and 13 different chairs. Suppose you want to arrange the desks and chairs so that each desk has a chair with it. How many such arrangements are possible? (Assume that desk number 1 paired with chair number 1 is a different arrangement from desk number 1 paired with chair number 2, etc.)

Solutions to Additional Problems for Strategies 4–6

Look for a Pattern

1.

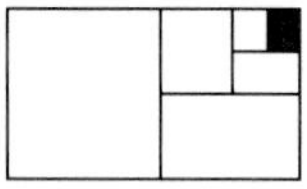

2. $n(n + 1)$

Make a Table

1. There are 10 possibilities, as shown in the following table.

4 by 4	3 by 3	2 by 2	1 by 1
1			9
	1	3	4
	1	2	8
	1	1	12
	1		16
		4	9
		3	13
		2	17
		1	21
			25

2. There are 8 possibilities, as shown in the following table.

20	10	5	1
Red	Yellow	Red	Black
Red	Yellow	Red	Yellow
Red	Yellow	Black	Yellow
Red	Yellow	Black	Red
Red	Black	Red	Black
Red	Black	Red	Yellow
Red	Black	Yellow	Black
Red	Black	Yellow	Red

Solve a Simpler Problem
1. $1 + 4 + 9 + 16 + 25 + 36 + 49 + 64 = 204$
2. $13(12)(11)(10) \cdots (1) = 6{,}227{,}020{,}800$

GEOMETRY AROUND US

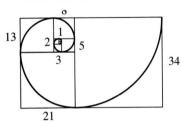

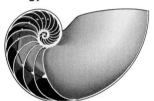

As the shell of a chambered nautilus grows, it creates a new wall behind its body every few months, forming a spiral array of chambers. A nautilus spiral formed in this way can be approximated by a sequence of squares whose side lengths follow what is known as the Fibonacci sequence: 1, 1, 2, 3, 5, 8, 13, ...

PROBLEM SET 1.2

EXERCISES/PROBLEMS

1. The x's in half of each figure can be counted in two ways.

x	x	x
x	x	x

$$1 + 2 = \frac{1}{2}(2 \times 3)$$

x	x	x	x
x	x	x	x
x	x	x	x

$$1 + 2 + 3 = \frac{1}{2}(3 \times 4)$$

(a) Draw a similar figure to find $1 + 2 + 3 + 4$.
(b) Express the sum $1 + 2 + 3 + 4$ in a similar way. Check to see that the sum is correct.
(c) Look for a pattern in the sums. Use it to find the sum of the whole numbers from 1 to 50 and from 1 to 75.
(d) Generalize your findings in part (c) by writing a formula for

$$1 + 2 + 3 + \cdots + (n - 1) + n$$

2. Find the number of toothpicks required to form each of the patterns below and then complete the table which follows.

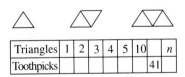

Triangles	1	2	3	4	5	10		n
Toothpicks						41		

3. The **triangular numbers** are the whole numbers that represent the following shapes

. . . and so on.

Find the number of dots in the nth triangular number.

4. The **pentagonal numbers** are the whole numbers that represent the following shapes

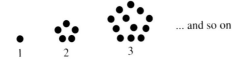

... and so on

Find the number of dots in the *n*th pentagonal number.

5. If *A* belongs with *B*, then *X* belongs with *Y*. Which of (i), (ii) or (iii) is the best choice for *Y*?

(a)   *A*   *B*   *X*

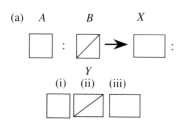

(b)   *A*   *B*   *X*

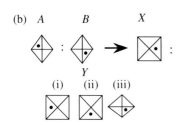

(c)   *A*   *B*   *X*

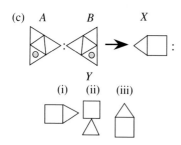

(d)   *A*   *B*   *X*

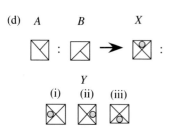

6. Use inductive reasoning to draw the next figure in each of the following sequences of figures.

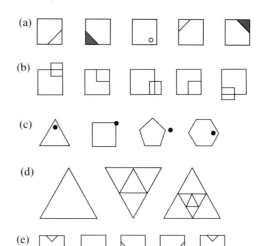

7. In each of the following problems, the first five terms of a sequence are given. Use inductive reasoning to determine the 6th, 7th, and 12th terms of each sequence. Then write a formula for the *n*th term.
   (a) 1, 4, 9, 16, 25, . . .
   (b) −2, −4, −6, −8, −10, . . .
   (c) 1, 3, 9, 27, 81, . . .
   (d) 3, 5, 7, 9, 11, . . .

8. Six cities *A*, *B*, *C*, *D*, *E*, and *F* are located as shown next. If bikepaths were constructed so that each pair of cities would be connected by a direct path, how many paths would be required?

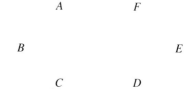

9. Julie walks to school at point *B* from her house at point *A*, a distance of 6 blocks. For variety, she likes to try different routes each day. How many different paths of exactly 6 blocks can she take? One path is shown.

10. A well-known sequence of numbers is shown next.

1, 1, 2, 3, 5, 8, 13, 21, 34, . . .

This sequence is called the **Fibonacci sequence** and is named for a 10th-century Italian mathematician whose pen name was Fibonacci and who posed a problem involving this sequence. This sequence arises in the shapes of nature, art, and architecture, as well as in music and probability. Write the next five terms of this sequence.

11. Determine the missing numbers in each of the following Fibonacci-type sequences.
(a) 2, 5, 7, 12, — , — , — , . . .
(b) 1, — , 7, — , 20, — , . . .
(c) 3, — , — , 19, — , . . .
(d) 10, — , — , — , — , 95, . . .

12. How many triangles of all sizes and shapes are in the picture?

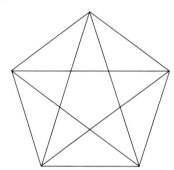

13. How many triangles of all sizes are contained in this figure?

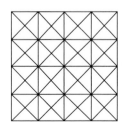

14. How many squares will be in the 10th set of squares in the following sequence? In the *n*th set?

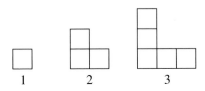

1   2   3

15. How many squares will be in the 10th set of squares in the following sequence? In the *n*th set?

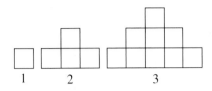

1   2   3

16. How many blocks will be in the 10th set of cubes in the following sequence? In the *n*th set?

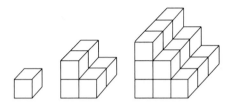

17. How many cubes are in the 10th set of cubes in this sequence?

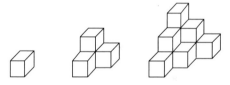

18. A 3 by 3 by 3 cube was painted on all faces and then cut apart into twenty-seven 1 by 1 by 1 cubes.
(a) How many 1 by 1 by 1 cubes had no faces painted? One face painted? Two faces painted? Three faces painted? Four or more faces painted?
(b) Answer the same questions for a 4 by 4 by 4 cube and for a 5 by 5 by 5 cube.
(c) Answer the same questions for an *n* by *n* by *n* cube, where *n* is any whole number greater than 1.

19. (a) If the following pattern is continued, how many letters will be in the "M" column?

```
            X   . . .
        Y   X   . . .
    Z   Y   X   . . .
        Y   X   . . .
            X   . . .
```

(b) Which column will contain 45 letters?

20. Two corner squares are cut from an 8 by 8 checkerboard to obtain the following figure. Can this new board be exactly covered by 31 dominoes which measure 2 by 1? [NOTE: You may not cut any of the dominoes.]

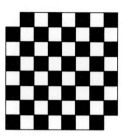

21. What fraction of the large square region is shaded? Assume that the pattern of shading continues forever.

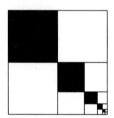

22. If the terms of the Fibonacci sequence (see problem 10) are squared and then added as shown next, an interesting pattern emerges.

$$
\begin{aligned}
1^2 + 1^2 &= 2 \\
1^2 + 1^2 + 2^2 &= 6 \\
1^2 + 1^2 + 2^2 + 3^2 &= 15 \\
1^2 + 1^2 + 2^2 + 3^2 + 5^2 &= 40
\end{aligned}
$$

Use this pattern to predict the following sum without performing the addition.

$$1^2 + 1^2 + 2^2 + 3^2 + \cdots + 144^2$$

Check your answer using a calculator.

APPLICATION

23. A certain type of gutter comes in 6-ft, 8-ft, and 10-ft sections. How many different lengths of gutter can be formed using three sections of guttering?

Solution to Initial Problem

We can look at simpler versions of the problem and see if we can observe a pattern. Suppose that a board had only two or three rows of holes in each starting pen as shown below.

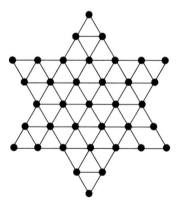

The total number of holes in all six pens is $6(1 + 2) = 6(3) = 18$.
The number of holes in the inner hexagonal region is

$$3 + 4 + 5 + 4 + 3 = 2(3) + 2(4) + 5$$

We can summarize and extend this information in the following table.

Number of Rows in a Starting Pen	Number of Holes in Starting Pens	Number of Holes in the Central Hexagon	Total Number of Holes
2	$6(1+2)=18$	$2(3)+2(4)+5=19$	$18+19=37$
3	$6(1+2+3)=36$	$2(4)+2(5)+2(6)+7=37$	$36+37=73$
4	$6(1+2+3+4)=60$	$2(5)+2(6)+2(7)+2(8)+9=61$	$60+61=121$
n	$6(1+2+\cdots+n)$	$2(n+1)+2(n+2)+\cdots+2(2n)+2n+1$	The sum of the numbers in columns 2 and 3

Substituting 6 for n in the first row we see that a game with 6 rows of holes (or 21 holes) in a starting pen has the following total number of holes:

$$T = 6(1+2+3+4+5+6)+2(7)+2(8)+2(9)+2(10)+2(11)+2(12)+13$$
$$= 253 \text{ holes}$$

Writing for Understanding

1. Find a problem that you want to spend time solving, making sure that it *is* a problem for you, not an exercise. Then begin to solve the problem, writing down all of your steps and writing down your insights into the problem as they occur to you. Also, write down the things you say to yourself as you solve the problem. Focus on your approach to the problem, the strategies you use, and how you are feeling about your success with the problem. The goal here is for you to learn about your own problem-solving process, not necessarily to find the solution to the problem at this time.

2. Select six problems from this chapter's problem set, one for each of the strategies. Identify the clue that led you to select the strategy you used and explain how that clue was useful.

3. Select three problems from this chapter's problem set. For each of the three problems, use the Look Back step to construct two new problems that are extensions or minor modifications of the original problem.

4. Revisit a real-world problem you have solved. Explain how Polya's four-step process could be applied to solve that problem.

CHAPTER REVIEW

Following is a list of key vocabulary, notation, and ideas for this chapter. Mentally review these items and, where appropriate, write down the meaning of each term. Then restudy the material that you are unsure of before proceeding to take the chapter test.

PEOPLE IN GEOMETRY

Euclid of Alexandra (circa 300 B.C.) has been called the "father of geometry." He authored many works, but he is most famous for *The Elements*. Although many of the results of plane geometry were already known, Euclid's unique contribution was the use of definitions, postulates, and axioms with statements to be proved, called propositions. Most of the theorems and much of the development contained in a typical high school geometry course today are taken from Euclid's *Elements*.

Euclid founded the first school of mathematics in Alexandria, Greece. As one story goes, a student who had learned the first theorem asked Euclid, "But what shall I get by learning this?" Euclid called his slave and said "Give him three pence, since he must make gain out of what he learns." As another story goes, a king asked Euclid if there were not an easier way to learn geometry than by studying *The Elements*. Euclid replied by saying, "There is no royal road to geometry."

CHAPTER 1 TEST

1. List the six strategies for problem solving you have learned in this chapter.

2. Given the following two "clues," which problem-solving strategy should you attempt?
 (a) A phrase similar to "for any number" is present or implied.
 (b) You are trying to develop a general formula.

3. Move just 4 toothpicks to make 3 squares.

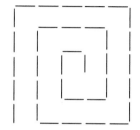

4. Twelve toothpicks form six equilateral triangles. Move four toothpicks to leave three equilateral triangles.

5. Two angles of a triangle have the same measure. The measure of the third angle is five less than three times the measure of each of the other angles. Find the measures of the angles in the triangle. (Use the fact that the sum of the measures of the angles in a triangle is 180°.)

6. In the following figure, the big square represents a farm and the shaded square represents a house and yard. The farmer wishes to retire and remain in the house. She also wants to divide the rest of the farm into pieces having the same size and shape for her five sons. How can she do this?

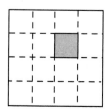

7. Find the maximum number of pieces a round cake can be cut into using
 (a) 3 straight cuts   (b) 4 straight cuts
 (c) 5 straight cuts
 Assume all cuts are perpendicular to the bottom of the cake.

8. How many squares will be in the 10th set of squares in the following sequence? In the *n*th set?

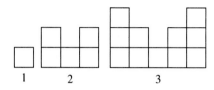

9. Use inductive reasoning to write the 10th term in each sequence below. Then write a formula for the *n*th term.
 (a) 4, 13, 22, 31, 40, . . .
 (b) 3, 5, 9, 17, 33, . . .

10. A large cube is made up of 1 by 1 by 1 cubes. Find the total number of cubes of all sizes in a large cube having dimensions:
 (a) 2 by 2 by 2   (b) 3 by 3 by 3
 (c) 4 by 4 by 4   (d) *n* by *n* by *n*

11. Trace the following figure and cut it into five pieces along the indicated lines. Rearrange the pieces to form a square. You must use all five pieces and have no gaps or overlaps, and you cannot turn the pieces over.

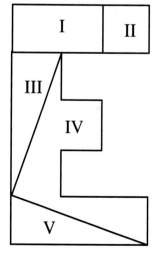

Geometric Shapes and Measurement

Geometers used informal reasoning until approximately 300 B.C. when Euclid wrote his famous set of books called *The Elements*. In *The Elements*, Euclid made two very important contributions to the way we study mathematics. First, he organized existing results in a manner that linked them together. In this way, later results can be verified or proved using earlier results. Second, Euclid introduced us to an axiomatic system that was built upon common notions, definitions, postulates and axioms, and theorems. Later mathematicians replaced some of Euclid's definitions with undefined terms to form what is now known as an axiomatic system.

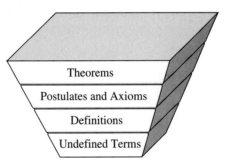

Elements of an Axiomatic System

PROBLEM-
SOLVING
STRATEGIES

1. Draw a Picture
2. Guess and Test
3. Use a Variable
4. Look for a Pattern
5. Make a Table
6. Solve a Simpler Problem
7. *Look for a Formula*

STRATEGY 7: LOOK FOR A FORMULA

INITIAL PROBLEM

Find the number of squares of all sizes in the 10 by 10 grid of squares shown. [NOTE: To find the total number of squares, you must count the number of different 1 by 1 squares, 2 by 2 squares, etc.]

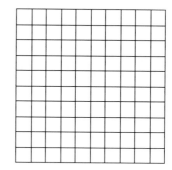

CLUES

The Look for a Formula strategy may be appropriate when

· A problem suggests a pattern that can be generalized.
· Ideas such as distance, area, volume, or other measurable attributes are involved.
· Applications in science, business, and so on are involved.
· A problem involves counting a number of possibilities.

A solution for the initial problem above is on page 86.

INTRODUCTION

Initially, geometry was used to solve problems involving measurements on the earth ("geo" means earth and "metry" means measure). Although the earth is essentially spherical in shape, when we view the earth in small regions it appears to be flat (Figure 2.1).

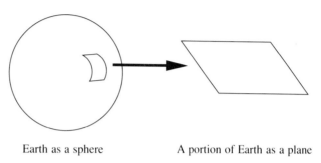

Earth as a sphere   A portion of Earth as a plane

FIGURE 2.1

Thus, in many practical situations involving measurement, a small portion of the surface of the earth is assumed to be flat and considered to be part of a plane. This is one of the reasons why plane geometry has been a main focus of school mathematics for the past several centuries.

In Chapter 1, we solved problems related to geometry by reasoning informally. In this chapter, we begin to formalize the study of geometry in the spirit of Euclid. A modern **axiomatic system** begins with a small set of undefined terms and builds through the addition of definitions and postulates to the point where many rich mathematical theorems can be proved (Figure 2.2).

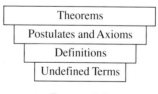

FIGURE 2.2

In Section 2.1 we introduce some undefined terms and postulates, and many definitions of plane geometry. Then we study common two-dimensional geometric shapes in Section 2.2. In Section 2.3 we look at angle measure in polygons and at how polygons can be arranged to cover a plane. In Section 2.4 we examine three-dimensional shapes. Finally, in Section 2.5 we introduce dimensional analysis, a useful technique for solving problems involving units of measure.

2.1
UNDEFINED TERMS, DEFINITIONS, POSTULATES, SEGMENTS, AND ANGLES

Applied Problem

A **compass card**, or **compass rose**, is marked with the 32 equally spaced points of the compass and is used in navigation. An example with some of the directions labelled is shown. If a ship sailing WNW changes its course to NE, it has turned through an angle of how many degrees?

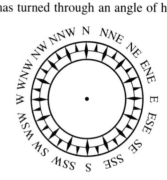

ELEMENTS OF AN AXIOMATIC SYSTEM

Definitions, as they appear in dictionaries, are circular. That is, if you continue to look up key words in a definition, you will likely revisit one of the words you have found already. By contrast, in an axiomatic system, we agree to leave some terms undefined and then build definitions of other terms and postulates from those undefined terms.

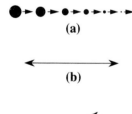

(a)

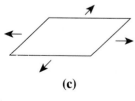

(b)

Undefined terms can be *described* but cannot be given precise definitions using simpler known terms. Three undefined terms in geometry are point, line, and plane. We think of a **point** as a circular dot that is shrunk until it has no size [Figure 2.3(a)]. By a **line**, we can think of a wire stretched as tightly as possible (hence "straight"), of infinite length (denoted by arrowheads), and having no thickness [Figure 2.3(b)]. Finally, a **plane** can be thought of as a sheet of paper with no thickness, stretched tightly, and extending infinitely in all directions [Figure 2.3(c)]. These descriptions provide a way for us to visualize points, lines, and planes, but they are *not* definitions. We *define* **space** to be the set of all points. Any collection of points is called a **geometric figure**. In particular, lines and planes are composed of points, and hence are geometric figures.

Postulates are statements that we *assume* to be true. Postulates state relationships among defined and undefined terms. The purpose of stating postulates is to establish some first principles upon which a geometry is based. This technique of developing a system, such as geometry, by beginning with a few first principles was perhaps the most significant achievement of Euclid. After

(c)

FIGURE 2.3

introducing postulates and some definitions, new results may be deduced. Results that are *deduced* from undefined terms, definitions, postulates, and/or results that follow from them are called **theorems**.

RELATIONSHIPS AMONG POINTS, LINES, AND PLANES

Now we begin to examine some of the postulates of geometry that will be assumed in this book.

Postulate 2.1

Every line contains at least two distinct points.

FIGURE 2.4

When a line contains a point, we say that the point is "on" the line. Lines are named by any two of their points together with a double arrow over the letters. For example, $\overleftrightarrow{AB}$ (or $\overleftrightarrow{BA}$) is shown in Figure 2.4. A line also may be named by a single lowercase letter. For example, $\overleftrightarrow{AB}$ in Figure 2.4 can be called line *l*. Although the preceding postulate states that any line contains at least two points, in fact we will see that a line contains infinitely many points. This result follows from a postulate to be stated soon.

As you may remember from your experience with graphing lines in algebra, two points determine a unique line. The next postulate formalizes this idea.

Postulate 2.2

Two points are contained in one and only one line.

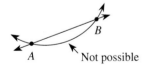

FIGURE 2.5

Postulate 2.2 suggests that all lines are straight because otherwise more than one line could be drawn through two points (Figure 2.5). Also, because we can imagine two points arbitrarily far apart, this postulate suggests that lines are infinite in length.

The next postulate suggests a connection between the "straightness" of lines and the "flatness" of planes (Figure 2.6). That is, just as lines are "straight," planes are "flat."

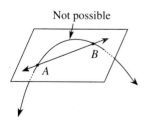

FIGURE 2.6

Postulate 2.3

If two points are in a plane, then the line containing these points is also in the plane.

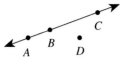

FIGURE 2.7

Points are said to be **collinear** if there is a line containing all of the points and **noncollinear** if there is no line that contains all of the points. In Figure 2.7, A, B, and C are collinear. Also A, B, and D are noncollinear since they do not lie on one line. Just as two points determine a unique line, three points determine a unique plane. This fact is stated in the next postulate.

Postulate 2.4

Three noncollinear points are contained in one and only one plane, and every plane contains at least three noncollinear points.

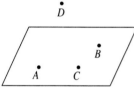

FIGURE 2.8

Postulate 2.4 guarantees that a plane consists of more than a single line.

Points are called **coplanar** if there is a plane containing all of the points and are said to be **noncoplanar** if there is no plane that contains all of the points. In Figure 2.8, for example, D is above the plane containing A, B, and C, so points A, B, and C are coplanar whereas A, B, C, and D are noncoplanar.

The next postulate makes a similar statement about space.

Postulate 2.5

In space, there exist at least four points that are not all coplanar.

Postulate 2.5 assures us that space is not just a single plane; that is, space is three-dimensional.

LINE SEGMENTS AND THEIR MEASURE

The next postulate involves the real number line and is the one that will allow us to draw many conclusions about plane geometry as well as allow us to find lengths of line segments. For a discussion of real numbers, see Topic 2 at the back of the book.

Postulate 2.6

The Ruler Postulate

Every line can be made into an exact copy of the real number line using a 1-1 correspondence.

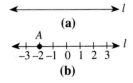

FIGURE 2.9

Figure 2.9(a) shows a line l, and Figure 2.9(b) shows l labelled as a real number line using the Ruler Postulate. The real number associated with a point on a line is called the **coordinate** of that point. For example, in Figure 2.9(b) the coordinate of point A is -2. Only integers are labelled in Figure 2.9(b) because it is impossible to label all the real numbers. Because the set of real numbers is infinite, Postulate 2.6 implies that every line has infinitely many points. Also, because the real number line is infinite in length, this postulate implies that *all* lines are infinite in length.

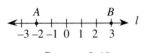

FIGURE 2.10

Distances between points are measured using the coordinates of the points. The **distance from A to B** is defined to be the nonnegative difference between their coordinates. In Figure 2.10, the distance from A to B is $3 - (-2) = 5$.

The Ruler Postulate allows us to *define* a line segment as well as to find its length. If A and B are two points on a line l, then the **line segment** determined by A and B, written $\overline{AB}$ or $\overline{BA}$, is points A and B together with all points on the line whose coordinates are between those of A and B (Figure 2.11). Points A and B are called the **endpoints** of the segment $\overline{AB}$. The **length of the line segment** $\overline{AB}$, written as AB, is the distance from A to B. In Figure 2.10, $AB = 5$.

FIGURE 2.11

Two segments are said to be **congruent** if they have the same length. When the segments $\overline{AB}$ and $\overline{CD}$ are congruent, we write $\overline{AB} \cong \overline{CD}$, which is read "$\overline{AB}$ **is congruent to** $\overline{CD}$," or we write $AB = CD$ to mean that the distance from A to B equals the distance from C to D. Congruent line segments are indicated by marks as illustrated in Figure 2.12(a). If more than one pair of segments is congruent, double or triple marks are used [Figure 2.12(b)].

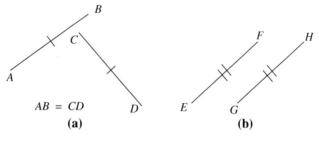

$$AB = CD$$

(a)   **(b)**

FIGURE 2.12

FIGURE 2.13

A portion of a line that has one endpoint and extends indefinitely in the one direction is called a **ray** (Figure 2.13). More precisely, if points A and B are on a line l, then ray $\overrightarrow{AB}$ consists of the points A and B and all points X on l whose coordinates satisfy one of the following: Either (1) the coordinate of X is between the coordinates A and B, or (2) the coordinate of B is between the coordinates of A and X.

ANGLES AND THEIR MEASURE

In addition to measuring lengths, it is important to be able to measure regions, such as plots of land or walls of a building. Such regions are often bounded by shapes composed of line segments (Figure 2.14).

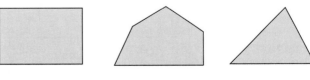

FIGURE 2.14

The sides of these shapes meet to form corners of various sizes. Two line segments or rays meeting at a common endpoint form an **angle**. The common endpoint is called the **vertex** (plural **vertices**) of the angle and the segments or rays are called the **sides** of the angle. Angles are usually named using the angle symbol "$\angle$" and three points: a point on one side, then the vertex, then a point on the other side. Figure 2.15(a) shows $\angle ABC$ (or $\angle CBA$) and $\angle DEF$ (or $\angle FED$).

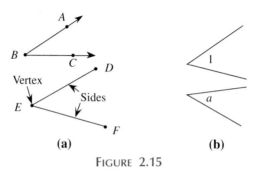

(a)   **(b)**

FIGURE 2.15

Points B and E are the vertices of $\angle ABC$ and $\angle DEF$, respectively. Rays $\overrightarrow{BA}$ and $\overrightarrow{BC}$ are the sides of $\angle ABC$, and $\overrightarrow{ED}$ and $\overrightarrow{EF}$ are the sides of $\angle DEF$. When the meaning is clear, an angle may be named using the letter at its vertex. For example, $\angle ABC$ may be called $\angle B$. Angles also are named by numbers or lowercase letters as illustrated in Figure 2.15(b).

Angles are usually measured in terms of degrees or radians. In this book, we will use degree measure. If two rays coincide, such as $\overrightarrow{AB}$ and $\overrightarrow{AC}$ in Figure 2.16, the angle they form has measure 0°. If $\overrightarrow{AC}$ is rotated counterclockwise one complete revolution as shown in Figure 2.17 until it coincides with $\overrightarrow{AB}$ again, we say the measure of this angle is 360°.

The measure of an angle that is 1/360 of a complete revolution is one **degree**, written 1°. Degrees are represented by the small, raised circle "°". Angles can be measured using a **protractor**, such as the one shown in Figure 2.18, together with the following postulate.

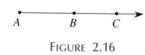

FIGURE 2.16

FIGURE 2.17

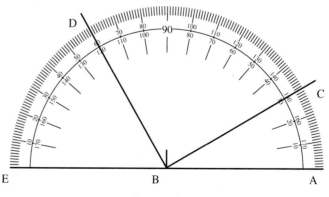

FIGURE 2.18

Postulate 2.7

The Protractor Postulate

If one ray of an angle is placed at 0° on a protractor and the vertex is placed at the midpoint of the bottom edge, then there is a 1-1 correspondence between all other rays that can serve as the second side of the angle and the real numbers between 0° and 180° inclusive, as indicated by a protractor.

If the vertex of an angle is placed at the center of the bottom of the protractor and one ray of the angle is placed at 0° at either end of the protractor, then the **measure of the angle** is given by the number that falls on the other ray. The inner numbers on the protractor are used for angles measured counterclockwise from the right, and the outer numbers are used for angles measured clockwise from the left. For example, in Figure 2.18 the measure of $\angle ABC$ is 30°, written as m($\angle ABC$) = 30°, or $\angle ABC$ = 30° when the context is clear. The measure of $\angle ABD$ is 120°, and the measure of $\angle EBC$ is 150°. Because the measure of $\angle ABC$ is less than 90°, it is an example of what is called an acute angle. On the other hand, $\angle ABD$, whose measure is greater than 90° but less than 180°, is an example of an obtuse angle. In fact, there are five special classes of angles; they are defined next and illustrated in Figure 2.19.

Definition

Type of Angle	Measure
Acute	Between 0° and 90°
Right	90°
Obtuse	Between 90° and 180°
Straight	180°
Reflex	Between 180° and 360°

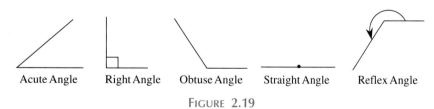

Acute Angle   Right Angle   Obtuse Angle   Straight Angle   Reflex Angle

FIGURE 2.19

Notice that in Figure 2.19 a right angle is indicated by the "square corner" placed in the right angle. Two lines that form a right angle are said to be **perpendicular**. If lines l and m are perpendicular, we write $l \perp m$.

Two angles that have the same measure are said to be **congruent**. Congruent angles are designated using one or more small arcs. In Figure 2.20, $\angle ABC$ is congruent to $\angle DEF$. When two angles are congruent, as in this case, we write

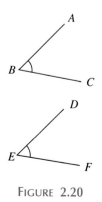

FIGURE 2.20

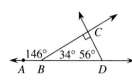

FIGURE 2.21

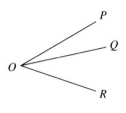

FIGURE 2.22

$\angle ABC \cong \angle DEF$ or $\angle B \cong \angle E$. When the context is clear, we also may write $\angle B = \angle E$ to indicate that the two angles have the same measure.

Two angles whose sum is 90° are said to be **complementary**. Two angles whose sum is 180° are said to be **supplementary**. For example, in Figure 2.21, $\angle CBD$ and $\angle BDC$ are complementary because $34° + 56° = 90°$, and $\angle ABC$ and $\angle CBD$ are supplementary because $146° + 34° = 180°$.

Angles $\angle ABC$ and $\angle CBD$ in Figure 2.21 are a pair of adjacent angles. In general, two angles $\angle ABC$ and $\angle CBD$ are said to be **adjacent** if they have a common vertex (B) and a common side ($\overline{BC}$) and if $\overline{BC}$ is between $\overline{BA}$ and $\overline{BD}$. In other words, the sum of the measures of $\angle ABC$ and $\angle CBD$ must equal the measure of $\angle ABD$. In Figure 2.22, $\angle POQ$ and $\angle QOR$ are adjacent angles. The angles $\angle POR$ and $\angle QOR$ are *not* adjacent angles because the common side $\overline{OR}$ is *not* between $\overline{OP}$ and $\overline{OQ}$.

EXAMPLE 2.1   In the Figure 2.23, $\overline{CG} \perp \overline{GE}$, $\angle AGB = 77°$, $\angle CGD = 54°$, and $\angle FGC = 157°$. Using this information, answer the following questions.

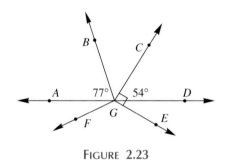

FIGURE 2.23

(a) What are the measures of $\angle DGE$, $\angle BGC$, $\angle AGF$, and $\angle FGE$?

(b) What is one pair of complementary angles?

(c) What are four pairs of supplementary angles?

(d) What are four pairs of adjacent angles?

SOLUTION

(a)   (i) Find $\angle DGE$: Because $\overline{CG} \perp \overline{GE}$, we know $\angle CGE = 90°$. In addition, $\angle CGD + \angle DGE = \angle CGE$. So we know that $54° + \angle DGE = 90°$, or $\angle DGE = 90° - 54° = 36°$.

(ii) Find $\angle BGC$: Because $\angle AGD$ is a straight angle, we know that

$$\angle AGB + \angle BGC + \angle CGD = 180°$$

so,

$$77° + \angle BGC + 54° = 180°$$
$$\angle BGC = 180° - 131°$$
$$\angle BGC = 49°$$

(iii) Find $\angle AGF$: Because $\angle FGC$ was given as $157°$, we know that

$$\angle AGF + \angle AGB + \angle BGC = 157°$$

so,

$$\angle AGF + 77° + 49° = 157°$$
$$\angle AGF = 157° - 126°$$
$$\angle AGF = 31°$$

(iv) Find $\angle FGE$: Because the sum of the measures of all six angles is $360°$, we can find the measure of the remaining angle by subtracting from $360°$.

$$\angle FGE = 360° - 77° - 49° - 54° - 36° - 31° = 113°.$$

(b) Complementary angles are angles the sum of whose measures is $90°$. Because $\angle CGE = 90°$, $\angle CGD$ and $\angle DGE$ are complementary.

(c) Supplementary angles are angles the sum of whose measures is $180°$. Because $\angle AGD = 180°$, we can say that the following pairs of angles are supplementary: $\angle AGB$ and $\angle BGD$, $\angle AGC$ and $\angle CGD$, $\angle AGF$ and $\angle FGD$, and $\angle AGE$ and $\angle EGD$.

(d) Adjacent angles share a common side and vertex, but do not overlap. Four examples of adjacent angles in the figure are $\angle AGB$ and $\angle BGC$, $\angle BGC$ and $\angle CGD$, $\angle AGE$ and $\angle EGC$, and $\angle EGF$ and $\angle FGC$. There are other pairs that also could have been listed. [NOTE: All of the pairs of angles in part (c) are also adjacent angles.]

Solution to Applied Problem

Because there are thirty-two equally spaced directions on the compass card shown, the points are $360°/32 = 11.25°$ apart. Therefore, the angle from WNW to N is $6(11.25°) = 67.5°$, and the angle from N to NE is $4(11.25) = 45°$. Thus, the ship has turned through an angle of $67.5° + 45° = 112.5°$.

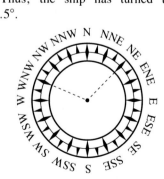

GEOMETRY AROUND US

Cameras and artists' canvasses are often set upon some form of tripod. One explanation for the use of a tripod is that three points determine one and only one plane. This means that a three-legged stand will be more stable than one with four or more legs even if the floor is not perfectly level. In fact, when dairy barns had dirt floors, milking stools were three-legged for this reason.

PROBLEM SET 2.1

EXERCISES/PROBLEMS

Exercises 1–6 refer to the following figure.

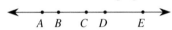

1. How many different line segments appear in the figure? Name them.

2. How many different line segments have *B* as one endpoint? Name them.

3. How many different rays appear in the figure? Name them.

4. How many different rays have *C* as their initial point? Name them.

5. Name the line $\overleftrightarrow{AE}$ in eight other ways.

6. Name the ray $\overrightarrow{EA}$ in three other ways.

7. Points *P*, *Q*, *R*, and *S* are located on line *l*. Their corresponding coordinates are $P = -3.78$, $Q = -1.35$, $R = 0.56$, and $S = 2.87$. Find each of the following distances.
   (a) *PR*   (b) *RQ*   (c) *PS*   (d) *QS*

$$\begin{array}{ccccc} & P & Q & R & S \\ \hline -4 & -3 & -2 & -1 & 0 & 1 & 2 & 3 & 4 \end{array} \; l$$

8. Points *A*, *B*, *C*, and *D* are located on line *l*. Their corresponding coordinates are $A = -5\sqrt{2}$, $B = -\sqrt{2}$, $C = \sqrt{2}$, and $D = 4\sqrt{2}$. Find each of the following distances.
   (a) *AC*   (b) *BD*   (c) *BC*   (d) *AD*

$$\begin{array}{ccccc} A & & B & C & D \\ \hline -8 & -6 & -4 & -2 & 0 & 2 & 4 & 6 & 8 \end{array} \; l$$

9. (a) Three different angles are pictured in the following figure. Name each of them in two different ways.
   (b) Name two adjacent angles in the figure.

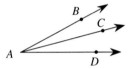

10. (a) Three different angles in the following figure have *Q* as a vertex. Name each of them in two different ways.
    (b) Name two pairs of adjacent angles in the figure.

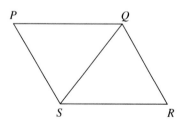

11. (a) How many different angles less than 180° are shown in the following figure?
    (b) How many of them are obtuse?
    (c) How many of them are acute?

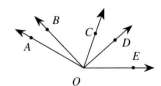

12. (a) How many different angles are shown in the following figure, where *T*, *S*, and *W* are collinear?

(b) How many of them are obtuse?

(c) How many pairs of angles are supplementary? Name them.

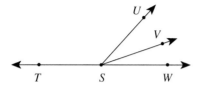

13. Use your protractor to measure the following angles. Classify each of them as acute, right, or obtuse. If necessary, extend the sides of an angle using a straightedge so that you can measure the angle more accurately.

(a)

(b)

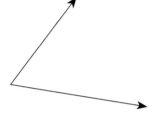

(c)

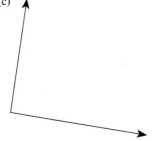

14. Use your protractor to measure the following angles. Classify each of them as acute, right, or obtuse. If necessary, extend the sides of an angle.

(a)

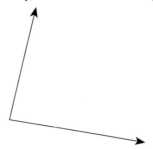

(b)

(c)

15. In the figure, ∠*BFC* = 55°, ∠*AFD* = 150°, and ∠*BFE* = 120°. Determine the measures of ∠*AFB* and ∠*CFD*. [NOTE: Do *not* measure the angles with your protractor.]

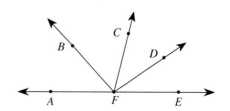

16. In the following figure $\overline{AO}$ is perpendicular to $\overline{CO}$. If $\angle AOD = 150°$ and $\angle BOD = 72°$, determine the measures of $\angle AOB$ and $\angle BOC$. [NOTE: Do *not* measure the angles with your protractor.]

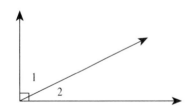

17. In the following figure, the measure of $\angle 1$ is 9° less than half the measure of $\angle 2$. Find the measures of $\angle 1$ and $\angle 2$.

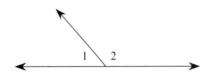

18. In the following figure, the measure of $\angle 1$ is 14° less than twice the measure of $\angle 2$. Find the measures of $\angle 1$ and $\angle 2$. Round your answers to the nearest hundredth of a degree.

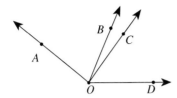

19. If $\angle A$ and $\angle B$ are complementary but $\angle A$ is four times as large as $\angle B$, find the measure of $\angle A$.

20. The measure of $\angle X$ is 9° more than twice the measure of $\angle Y$. If $\angle X$ and $\angle Y$ are supplementary angles, find the measure of $\angle X$.

21. Two lines drawn in a plane separate the plane into three different regions if the lines are parallel.

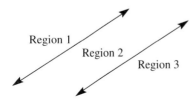

If the lines are intersecting, then they will divide the plane into four regions.

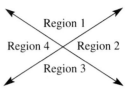

Thus, the greatest number of regions into which two lines can divide a plane is four. Determine the greatest number of regions into which a plane can be divided by three lines, four lines, five lines, and ten lines. Generalize to n lines.

22. If three points are located in a plane, there is one line that can be drawn through them if the points are collinear.

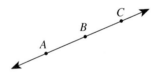

If the three points are noncollinear, then there are three lines that can be drawn through pairs of points.

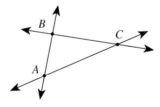

Therefore, for three points, three is the greatest number of lines that can be drawn through pairs of points. Determine the greatest number of lines that can be drawn for four points, five points, and six points. Generalize to n points.

APPLICATIONS

In surveying, navigation, and other applications of angle measure, angles must often be measured more precisely than to the nearest degree. Then angles may be measured to the nearest tenth or hundredth of a degree or in minutes and seconds, where 60 minutes = 1 degree, and 60 seconds = 1 minute. Minutes are denoted by a single mark (e.g., 45′) and seconds are denoted by a double mark (e.g., 45″). So an angle measure might be written as 32°15′40″.

23. Convert each of the following angle measures to degrees and decimal fractions of a degree. For example, $13°30′ = 13\frac{30}{60} = 13.5°$. Round to the nearest thousandth of a degree. [NOTE: Many scientific calculators will perform these conversions. Consult your manual.]
(a) 41°24′   (b) 84°15′
(c) 25°30′15″   (d) 123°34′48″

24. Convert each of the following angle measures in degrees to degrees and minutes. Round to the nearest minute.
(a) 31.6°   (b) 95.75°
(c) 102.16°   (d) 58.029°

25. In surveying, directions are usually specified in terms of bearings. The **bearing** of a line segment is the acute angle that the line segment makes with a north-south line. The angle and the quadrant are specified. For example, the bearing of $\overline{AB}$ is N 20° E. We also could say that the bearing of $\overline{BA}$ is S 20° W. (In this case, imagine placing B at the origin rather than A.)

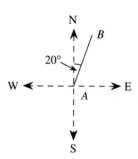

Sketch line segments with the following bearings.
(a) N 17° W   (b) S 48° E
(c) N 78° E   (d) S 65° W

26. The **azimuth** of a line segment is the angle that the line segment makes with a north-south line. It is measured from 0° to 360° in a clockwise direction from the north. For example, the azimuth of $\overline{AB}$ below is 250°.

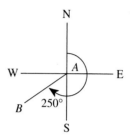

Sketch line segments with the following azimuths.
(a) 115°   (b) 225°   (c) 72°   (d) 329°25′

27. Convert the following bearings to azimuths.
(a) N 40° E   (b) N 40° W
(c) N 82°17′ W   (d) S 24°35′ E

28. Convert the following azimuths to bearings.
(a) 163°   (b) 241°
(c) 123°48′   (d) 329°25′

29. Determine the measure of $\angle ABC$ in the following figure. The bearings of $\overline{AB}$ and $\overline{BC}$ are indicated.

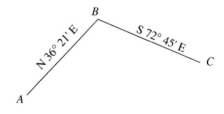

30. Given the bearings of $\overline{AB}$ and $\overline{BC}$ as shown next, determine the measure $\angle ABC$. [NOTE: Be careful with the order of the points here. The bearing of $\overline{AB}$ is not the same as the bearing of $\overline{BA}$. See problem 25.]

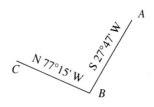

2.2
POLYGONS AND CIRCLES

Applied Problem

A box-end wrench has a six-point opening at one end. What is the name of the polygonal opening, and why would this shape be used?

TRIANGLES

Many geometric figures made up of line segments can be described by the lengths of their segments or by the measures of the angles formed. The simplest of these figures is the triangle. Given three *noncollinear* points A, B, and C, the **triangle** determined by these three points, written $\triangle ABC$, is formed by the **sides** $\overline{AB}$, $\overline{AC}$, and $\overline{BC}$ (Figure 2.24). A triangle has three angles and three sides.

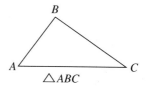

$\triangle ABC$

FIGURE 2.24

Triangles can be classified by both the lengths of their sides and the measures of their angles. Many special triangles are defined next, and Figure 2.25 shows examples of each of them.

Definition

Types of Triangles	Definition
Equilateral	All three sides are congruent.
Isosceles	At least two sides are congruent.
Scalene	No two sides are congruent.
Acute	All three angles are less than 90°.
Right	One angle is a right angle.
Obtuse	One angle is an obtuse angle.
Equiangular	All three angles are congruent.

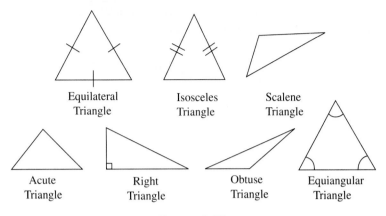

FIGURE 2.25

In an isosceles triangle, the angles opposite the congruent sides are called **base angles**, and the side between these two angles is usually called the **base** of the triangle. In Figure 2.26(a), the base angles are $\angle A$ and $\angle C$ and the base is $\overline{AC}$. The angle between the two congruent sides is often called the **vertex angle**. In Figure 2.26(a), $\angle B$ is the vertex angle.

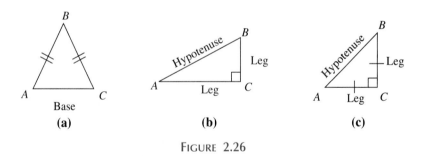

FIGURE 2.26

Notice that because an isosceles triangle has *at least* two sides congruent, any equilateral triangle also can be classified as an isosceles triangle. In a right triangle, the side opposite the right angle is called the **hypotenuse** and the other two, shorter sides are called the **legs** of the triangle. In Figure 2.26(b), $\overline{AC}$ and $\overline{BC}$ are legs and $\overline{AB}$ is the hypotenuse of right triangle $\triangle ABC$.

Generally, more than one of the terms just defined are used to describe a triangle. For example, a triangle with two congruent sides and a right angle is called an isosceles right triangle. Figure 2.26(c) shows right isosceles triangle $\triangle ABC$. A triangle with one right angle but with no two sides congruent is called a right scalene triangle [Figure 2.26(b)]. We will show in Chapter 4 that every equilateral triangle is also equiangular and vice versa.

SIMPLE CLOSED CURVES

Triangles are part of a more general class of geometric figures called simple closed curves. A **simple closed curve** is a figure that lies in a plane and can be traced so that the starting and ending points are the same and no part of the curve is crossed or retraced (Figure 2.27).

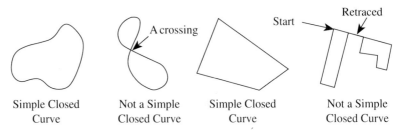

Simple Closed   Not a Simple   Simple Closed   Not a Simple
Curve   Closed Curve   Curve   Closed Curve

FIGURE 2.27

A **circle** is a simple closed curve that consists of the set of all points equidistant from a given point, called the **center** of the circle. A circle is named by its center. Circle O is pictured in Figure 2.28(a).

(a)   (b)

FIGURE 2.28

A line segment joining the center with a point on the circle is called a **radius** of the circle; its length is called **the radius** of the circle. A line segment containing the center and with endpoints on the circle is called a **diameter** of the circle; its length is called **the diameter** of the circle. In Figure 2.28(b), r is a radius and d is a diameter. Notice that the words radius and diameter are used to refer to line segments as well as to the lengths of those line segments. So we might say that in Figure 2.28(b) $\overline{PO}$ is a radius of the circle. We also could say that the radius of circle O is 6 inches and the diameter is 12.

POLYGONS

As can be seen in Figure 2.27, a simple closed curve need not be "curved." For example, a triangle is a simple closed curve composed of three line segments. In general, a **polygon** is a simple closed curve composed of line segments. Figure 2.29 shows some examples of polygons.

Polygons are usually named by the number of sides they contain. Some names of common polygons are listed next.

Definition

Type of Polygon	Number of Sides
Triangle	3
Quadrilateral	4
Pentagon	5
Hexagon	6
Octagon	8
Decagon	10
n-gon	n

By these definitions, in Figure 2.29, *ABCD* is a quadrilateral, *EFGHIJKL* is an octagon, and *PQRSTU* is a hexagon. Notice that in a polygon, the number of vertices is the same as the number of sides.

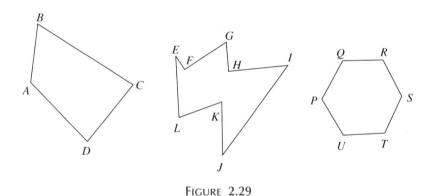

FIGURE 2.29

QUADRILATERALS

Many quadrilaterals are named by the lengths of their sides, the measures of their angles, or some other attribute such as parallelism. Two lines are said to be **parallel** if they are in the same plane and do not intersect. Two line segments are parallel if the lines containing the segments are parallel. Parallel lines are identified using one or more arrowheads in the middle of the lines. In Figure 2.30, line *l* is parallel to line *m*, and we write *l* ∥ *m*. Several types of quadrilaterals are described in the next table, and Figure 2.31 displays examples of each type.

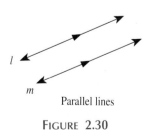

Parallel lines

FIGURE 2.30

Definition

Type of Quadrilateral	Definition
Square	All sides are congruent and all angles are right angles.
Rectangle	All angles are right angles.
Rhombus	All sides are congruent.
Parallelogram	Opposite sides are parallel.
Trapezoid	Exactly one pair of sides is parallel.
Isosceles trapezoid	Nonparallel sides are congruent.
Kite	Two pairs of adjacent sides are congruent and nonoverlapping.

The parallel sides in a trapezoid are called **bases** of the trapezoid. The sides that are not parallel are called the **legs** of the trapezoid. Thus, in each trapezoid in Figure 2.31, the bases of the trapezoid are indicated by the parallel line marks and the other two sides are the legs.

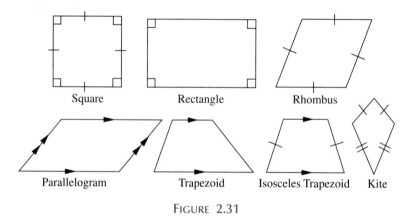

FIGURE 2.31

You may have noticed that there is some overlap in the definitions. For example, every square is also a rectangle because a square has four right angles. Several other such relationships will be explored in the problem set and in Chapter 5.

SYMMETRY

We have seen that polygons can be described in terms of the number of sides they have or in terms of attributes of their sides and angles. Also symmetries of various types can be used to describe polygons.

There are two common types of symmetry a geometric figure may possess: reflection symmetry and rotation symmetry. A figure has **reflection symmetry** if there is a line along which the figure may be folded so that one half of the figure matches exactly with the other half. The fold line is called the **line of symmetry** or the **axis of symmetry**.

The isosceles triangle shown in Figure 2.32(a) has reflection symmetry because it can be folded along the axis of symmetry and the two portions of the triangle would match exactly. A figure can have more than one axis of symmetry, as the fold lines for the square in Figure 2.32(b) demonstrate. The triangle in Figure 2.32(c) has no line of symmetry. Hence it has no reflection symmetry.

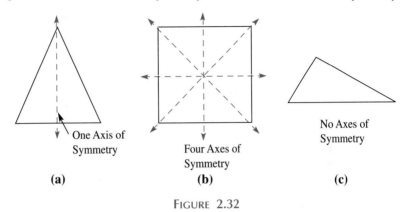

(a)
One Axis of Symmetry

(b)
Four Axes of Symmetry

(c)
No Axes of Symmetry

FIGURE 2.32

Because an isosceles triangle has reflection symmetry, we can establish a relationship between its base angles. When the triangle in Figure 2.32(a) is folded along the axes of symmetry, ∠1 will match ∠2. This means that these two angles have the same measure. We usually state this result as "base angles of an isosceles triangle are congruent."

An equilateral triangle is a special type of isosceles triangle, so it also has base angles congruent. But an equilateral triangle has three axes of symmetry, one through each vertex. Thus, any two angles of the equilateral triangle will have the same measure. This means that an equilateral triangle is also equiangular. We discuss these angle relationships more formally in Chapter 4.

A geometric figure has **rotation symmetry** if the figure can be rotated about a point less than a full turn, so that the image is identical to the original figure. The point is called the **center of rotation symmetry**. The equilateral triangle in Figure 2.33 has rotation symmetry because it can be rotated through an angle of 120° (one third of a full turn) and the image will match exactly with the original triangle. The equilateral triangle also has a rotation symmetry of 240°.

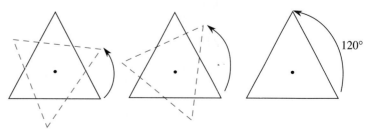

120°

FIGURE 2.33

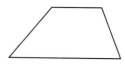

FIGURE 2.34

The trapezoid in Figure 2.34 does not have rotation symmetry because it must be rotated through one full turn of 360° before the image will match the original figure.

We can test figures for both reflection and rotation symmetries. To test for reflection symmetry, try folding the figure in half. To test for rotation symmetry, trace a copy of the figure and try to superimpose the tracing on the original figure by rotating it about a point.

Solution to Applied Problem

The opening in the wrench has six sides, so it is a hexagon. The wrench has six rotational symmetries and fits a hexagonal bolt in any of six positions, each 60° apart.

GEOMETRY AROUND US

An excellent example of a pentagon is the Pentagon in Arlington, VA, home to the United States Department of Defense. This building has the largest ground area of any office building in the world. Each of the outer sides of the Pentagon measures 921 feet, and it covers an area of 1,263,240 ft². Each day 29,000 people work in the building and travel along its 17 miles of hallways.

PROBLEM SET 2.2

EXERCISES/PROBLEMS

1. Several triangles are shown with congruent sides, congruent angles, and right angles as indicated.
   (a) Which triangles are isosceles?
   (b) Which triangles are equilateral?
   (c) Which triangles are scalene?
   (d) Which triangles are right triangles?
   (e) Which triangles are obtuse?

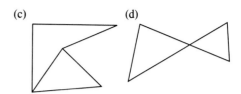

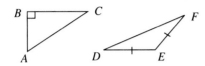

2. Several triangles are shown with congruent sides, congruent angles, and right angles as indicated.
   (a) Which triangles are scalene?
   (b) Which triangles are isosceles?
   (c) Which triangles are right triangles?
   (d) Which triangles are equiangular?
   (e) Which triangles are obtuse?

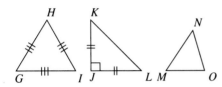

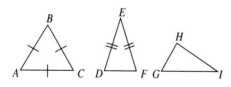

3. Which of the following shapes are polygons? If one is not, explain why.

(a)   (b)

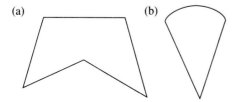

(c)   (d)

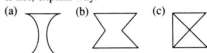

4. Which of the following shapes are polygons? If one is not, explain why.
   (a)   (b)   (c)

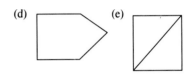

   (d)   (e)

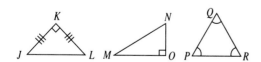

5. Find the specified shapes in the following figure. Assume that angles that appear to be right angles *are* right angles, segments that appear to be parallel *are* parallel, and segments that appear to be congruent *are* congruent.

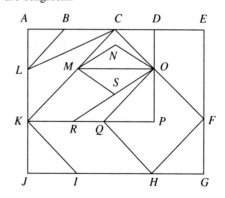

   (a) A square
   (b) A rectangle that is not a square
   (c) A parallelogram that is not a rectangle
   (d) An isosceles right triangle
   (e) An isosceles triangle with no right angles
   (f) A rhombus that is not a square
   (g) A kite that is not a rhombus
   (h) A scalene triangle with no right angles
   (i) A right scalene triangle
   (j) A trapezoid that is not isosceles
   (k) An isosceles trapezoid

6. Find the specified shapes in the following figure. Assume that angles that appear to be right angles *are* right angles, segments that appear to be parallel *are* parallel, and segments that appear to be congruent *are* congruent.

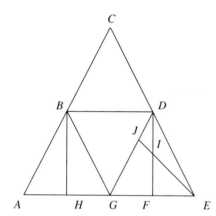

(a) A rectangle
(b) A kite
(c) An isosceles triangle with no right angle
(d) An equilateral triangle
(e) A right isosceles triangle
(f) A right scalene triangle
(g) A parallelogram that is not a rectangle
(h) An isosceles trapezoid
(i) An obtuse scalene triangle
(j) A trapezoid that is not isosceles
(k) A pentagon

7. The following scalene triangle is formed from 14 toothpicks. We can describe this triangle as a 3-5-6 triangle.

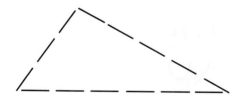

Use toothpicks to answer each of the following questions.

(a) How many different isosceles triangles can be formed using 20 toothpicks? [NOTE: You should have no gaps or overlapping toothpicks, nor can toothpicks be broken.]
(b) How many different scalene triangles can be formed using 20 toothpicks?

(c) How many different equilateral triangles can be formed using 20 toothpicks?

8. (a) How many different isosceles triangles can be formed using 24 toothpicks? (See the directions for the preceding problem.)
(b) How many different scalene triangles can be formed using 24 toothpicks?
(c) How many different equilateral triangles can be formed using 24 toothpicks?

9. Fold a rectangular piece of paper on the dotted line as shown in each figure below. Then make cuts in the paper as indicated. Sketch what you think the shape will be when the paper is unfolded. Then unfold your paper to check your picture.

(a)

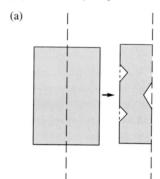

(b)

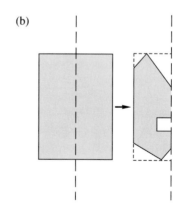

(c)

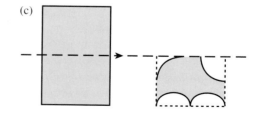

10. Fold a rectangular piece of paper on the dotted line as shown in each figure below. Then make cuts in the paper as indicated. Sketch what you think the shape will be when the paper is unfolded. Then unfold your paper to check your picture.

(a)

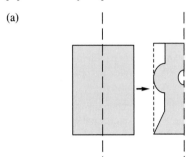

(b)

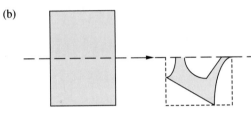

(c)

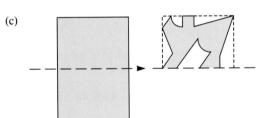

11. Each shape shown next was obtained by folding a rectangular piece of paper in half vertically (lengthwise) and then making appropriate cuts in the paper. For each figure, draw the folded paper and show the cuts that must be made to make the figure. Try folding and cutting a piece of paper to check your answer.

(a)

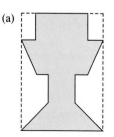

(b)

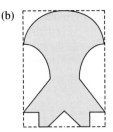

(c)

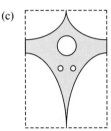

12. Each shape shown next was obtained by folding a rectangular piece of paper in half horizontally and then making appropriate cuts in the paper. For each figure, draw the folded paper and show the cuts that must be made to make the figure. Try folding and cutting a piece of paper to check your answer.

(a)

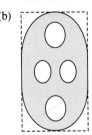

(b)

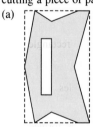

(c)

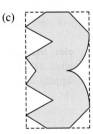

13. Consider the square lattice shown here.

(a) How many different triangles can be drawn that have $\overline{AB}$ as one side?

(b) How many of these are isosceles?

(c) How many are right triangles?

(d) How many are acute?

(e) How many are obtuse? [HINT: Use your answers to parts (a), (c), and (d).]

14. Consider the square lattice shown here.

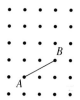

(a) How many different triangles can be drawn that have $\overline{AB}$ as one side?

(b) How many of these are isosceles?

(c) How many are right triangles?

(d) How many are acute?

(e) How many are obtuse? [HINT: Use your answers to parts (a), (c), and (d).]

15. Given the square lattice shown, draw quadrilaterals having $\overline{AB}$ as a side.

(a) How many different parallelograms are possible?

(b) How many rectangles are possible?

(c) How many rhombuses?

(d) How many squares?

16. Given the triangular lattice shown, draw quadrilaterals having $\overline{AB}$ as a side.

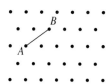

(a) How many different parallelograms are possible?

(b) How many rectangles are possible?

(c) How many rhombuses?

(d) How many trapezoids?

17. Consider the equally spaced points on the following circle.

(a) How many different triangles can be drawn using three of the points of the circle as vertices?

(b) How many of the triangles are isosceles?

(c) How many are right triangles?

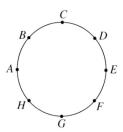

18. Consider the equally spaced points on the following circle.

(a) How many different quadrilaterals can be drawn using four of the points on the circle as vertices?

(b) How many of the quadrilaterals are rectangles?

(c) How many are kites?

(d) How many are trapezoids?

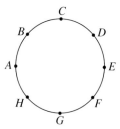

19. Given a square and a circle, draw an example, if possible, where they intersect in exactly the number of points given.

(a) No points   (b) One point

(c) Two points   (d) Three points

20. Given a triangle and a circle, draw an example, if possible, where they intersect in exactly the number of points given.
    (a) No points   (b) One point
    (c) Two points   (d) Three points

21. Draw two copies of the hexagon shown.
    (a) Divide one hexagon into three identical parts so that each part is a rhombus.
    (b) Divide the second hexagon into six identical kites.

22. Draw two copies of the hexagon shown.
    (a) Divide one hexagon into four identical trapezoids.
    (b) Divide the second hexagon into eight identical polygons.

23. (a) Does the rectangle shown have reflection symmetry? If so, how many axes of symmetry does it have?
    (b) Does the rectangle have rotation symmetry? If so, how many different rotation symmetries does it have?

24. (a) Does the rhombus shown have reflection symmetry? If so, how many axes of symmetry does it have?
    (b) Does the rhombus have rotation symmetry? If so, how many different rotation symmetries does it have?

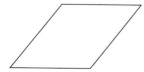

25. (a) Does the isosceles trapezoid shown have reflection symmetry? If so, how many axes of symmetry does it have?
    (b) Does the isosceles trapezoid have rotation symmetry? If so, how many different rotation symmetries does it have?

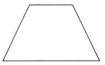

26. (a) Does the right isosceles triangle shown have reflection symmetry? If so, how many axes of symmetry does it have?
    (b) Does the right isosceles triangle have rotation symmetry? If so, how many different rotation symmetries does it have?

27. (a) Draw the lines of symmetry in the regular n-gons in (i)–(iii). How many does each have?
    (b) How many lines of symmetry does a regular n-gon have?

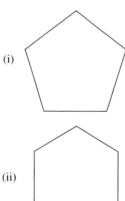
(i)

(ii)

(iii)

28. (a) Trace the regular *n*-gons in problem 27 to find all of their rotation symmetries for a regular pentagon, a regular hexagon, and a regular octagon.
(b) How many rotation symmetries does a regular *n*-gon have?

APPLICATION

29. Bingo is played on a 5 by 5 grid of squares in which certain squares must be covered in order to win, usually five squares in a row vertically, horizontally, or diagonally.

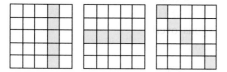

To make the games more interesting, other patterns on the grid are often chosen to be winners. Several examples are shown. For each one, tell whether the pattern has reflection symmetry, rotation symmetry, or both.

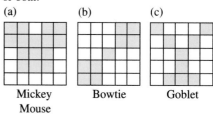

(a)   (b)   (c)

Mickey   Bowtie   Goblet
Mouse

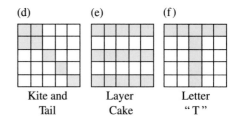

(d)   (e)   (f)

Kite and   Layer   Letter
Tail   Cake   "T"

2.3
ANGLE MEASURE IN POLYGONS
AND TESSELLATIONS

Applied Problem

A decorative floor tiling pattern is shown. It utilizes two types of polygons to cover the floor without gaps or overlaps. Explain why this arrangement of tiles is possible.

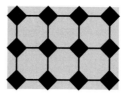

TESSELLATIONS

A **polygonal region** is a polygon together with the portion of the plane that is enclosed by the polygon (Figure 2.35).

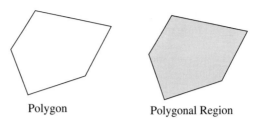

Polygon   Polygonal Region

FIGURE 2.35

Polygonal regions can be arranged to cover a plane just as ceramic tiles or carpet squares can be arranged to cover a countertop or a floor. Such an arrangement is called a **tessellation** or a **tiling** if (1) the entire plane can be covered without gaps, and (2) no two polygons overlap; that is, polygons share only common sides. Figure 2.36 shows some partial tessellations of the plane.

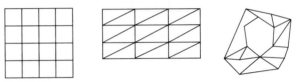

FIGURE 2.36

ANGLE MEASURE IN A TRIANGLE

A tessellation composed of triangles may be used to develop a useful relationship among the angle measures of a triangle. Figure 2.37(a) shows a triangle that can be used to tessellate the plane. That is, identical copies of the triangle can be arranged to cover the plane [Figure 2.37(b)].

(a)

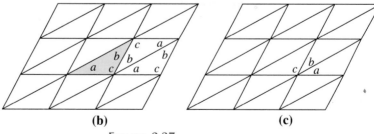

(b)   (c)

FIGURE 2.37

Notice that the sum of the angles in each of the triangles is $a + b + c$. However, because of the arrangement of triangles, we can see that the angles labelled c, b, and a in Figure 2.37(c) form a straight angle. Hence $c + b + a = 180°$. But c, b, and a are the angle measures in any one of the triangles. Thus, the sum of the angle measures in each of the triangles is 180°. This relationship holds true for any triangle, and we state it in the next theorem. We will prove this theorem formally in Chapter 5.

Theorem 2.1

Angle Measure in a Triangle

The sum of the measures of the angles in a triangle is 180°.

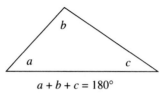

$$a + b + c = 180°$$

This theorem can be used to find missing angle measures in many geometric figures.

EXAMPLE 2.2   Find the measures of the missing angles in each of the following figures. Congruent angles and right angles are indicated.

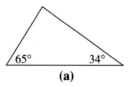

(a)

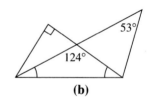

(b)

SOLUTION

(a) Let x = measure of the missing angle [Figure 2.38(a)]. We know that

$$x + 65° + 34° = 180°$$

Therefore, $x + 99° = 180°$, or $x = 81°$.

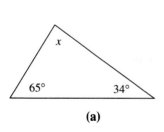

(a)

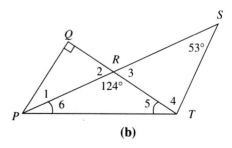

(b)

FIGURE 2.38

(b) First number the missing angles and label the vertices [Figure 2.38(b)]. Consider $\triangle PRT$. Because the sum of the angle measures is $180°$ and $\angle 5 = \angle 6$, we can say

$$\angle 5 + \angle 6 + 124° = 180°$$
$$\angle 5 + \angle 6 = 56°$$
$$2(\angle 6) = 56°$$
$$\angle 6 = 28°$$

Hence, $\angle 5$ and $\angle 6$ both have measure $28°$.

We also might notice that $\angle QRT$ is a straight angle. Therefore, $\angle 2 + 124° = 180°$, so $\angle 2 = 56°$. Likewise, $\angle 3 = 56°$. Now we know the measures of two of the three angles in $\triangle PQR$. Thus we can say

$$\angle 1 + \angle 2 + 90° = 180°$$
$$\angle 1 + 56° + 90° = 180°$$
$$\angle 1 = 34°$$

In a similar way we can use $\triangle RST$ to find the measure of $\angle 4$ because we know the measures of two angles in the triangle. Therefore, we have found the following:

$$\angle 1 = 34°, \angle 2 = 56°, \angle 3 = 56°, \angle 4 = 71°, \angle 5 = 28°, \angle 6 = 28° \quad \bullet$$

VERTEX ANGLES IN A POLYGON

The angles in a polygon are called its **vertex angles**. For example, in Figure 2.39(a), the vertex angles in the pentagon are $\angle V$, $\angle W$, $\angle X$, $\angle Y$, and $\angle Z$. Line segments joining nonadjacent vertices in a polygon are called **diagonals**. In Figure 2.39(b), $\overline{WZ}$ and $\overline{WY}$ are two of the diagonals of $VWXYZ$. The sum of the measures of the vertex angles of a polygon can be found by subdividing the polygon into triangles using diagonals.

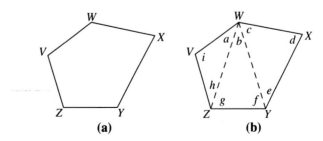

(a)   **(b)**

FIGURE 2.39

EXAMPLE 2.3   Find the sum of the measures of the vertex angles of a pentagon.

SOLUTION   Draw a pentagon and then divide it into triangles by drawing in all of the diagonals from any *one* vertex [Figure 2.39(b)]. Because the sum of the angle measures in a triangle is 180°, in Figure 2.39(b) we have $a + h + i = 180°$, $b + g + f = 180°$, and $c + d + e = 180°$. Thus, the sum of the vertex angles in pentagon $VWXYZ$ is $3(180°) = 540°$.   •

The technique used in Example 2.3 can be generalized to find the sum of the measures of the vertex angles in any polygon. In the pentagon, three triangles were formed by drawing diagonals from one vertex. Analogously, in a polygon with n sides, $n - 2$ triangles will be formed. This leads to the next theorem.

Theorem 2.2

Angle Measure in a Polygon

The sum of the measures of the vertex angles in a polygon with n sides is $(n - 2)180°$.

REGULAR POLYGONS

Regular polygons are polygons in which all of the sides are congruent and all of the angles are congruent. Squares and equilateral triangles are examples of regular polygons. Three other regular polygons are shown in Figure 2.40.

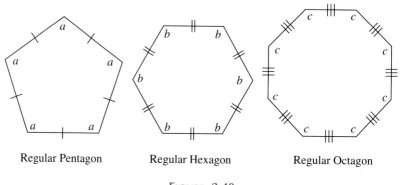

Regular Pentagon   Regular Hexagon   Regular Octagon

FIGURE 2.40

EXAMPLE 2.4   Find the measure of a vertex angle of a regular pentagon (Figure 2.40).

SOLUTION   The sum of the angle measures in a regular pentagon is 540°, as shown in Example 2.3. Because the vertex angles in a regular polygon are congruent and there are five such angles, the measure of *each* vertex angle is $540°/5 = 108°$.   •

The technique illustrated in Examples 2.3 and 2.4 can be generalized to any regular n-gon. The next theorem summarizes this more general result. It will be verified in the problem set.

Theorem 2.3

Vertex Angle Measure in a Regular Polygon

The measure of a vertex angle in a regular n-gon is $\dfrac{(n - 2)180°}{n}$.

Using Theorem 2.3, we can calculate the measure of a vertex angle in any regular polygon. Table 2.1 contains a list of several regular n-gons together with the measure of their vertex angles.

n	Measure of a Vertex Angle in a Regular n-gon
3	$(3 - 2)180°/3 = 60°$
4	$(4 - 2)180°/4 = 90°$
5	$(5 - 2)180°/5 = 108°$
6	$(6 - 2)180°/6 = 120°$
8	$(8 - 2)180°/8 = 135°$
10	$(10 - 2)180°/10 = 144°$

TABLE 2.1

Notice that as the number of sides in the regular polygon increases, the measure of one of the polygon's vertex angles also increases.

Theorem 2.3 can be used to determine the number of sides in a regular polygon if we know the measure of a vertex angle. The next example demonstrates this.

EXAMPLE 2.5   A vertex angle of a regular polygon measures 157.5°. How many sides does the polygon have?

SOLUTION   We know that the measure of the vertex angle will be $\dfrac{(n - 2)180°}{n}$, where n is the number of sides of the polygon. In this case, $\dfrac{(n - 2)180°}{n} = 157.5°$. We can find n by solving this equation.

$$\frac{(n - 2)180°}{n} = 157.5$$

$$(n - 2)180° = 157.5n$$

$$180n - 360 = 157.5n$$

$$-360 = -22.5n$$

$$\frac{-360}{-22.5} = n$$

$$n = 16$$

Therefore, the polygon has 16 sides.

A **regular tessellation** is a tessellation composed of regular polygons that are all the same size and shape. Figure 2.41 shows three regular tessellations.

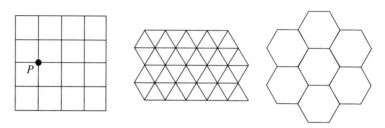

FIGURE 2.41

To form a regular tessellation, the polygonal regions must fit together exactly around each vertex. In other words, the sum of the vertex angle measures must equal 360°. For example, in the case of the tessellation of squares in Figure 2.41, four 90° angles fit together around vertex P and add up to 360°. This tessellation is possible because 4(90°) = 360°; that is, 90° is a factor of 360°. If the number of degrees in the vertex angle of a regular polygon is a factor of 360°, then a regular tessellation can be formed using that regular polygon. Using Table 2.1, it can be seen that 60°, 90°, and 120° are measures of vertex angles of regular polygons that are also factors of 360°. Also, when $n > 6$, the vertex angle measure is between 120° and 180°, hence it will not divide evenly into 360°. Thus, the only possible regular tessellations are composed of equilateral triangles, squares, and regular hexagons, which are the regular tessellations shown in Figure 2.41.

Solution to Applied Problem

The tessellation pictured is composed of squares and regular octagons. Each vertex angle of a square measures 90° and each vertex angle of a regular octagon measures 135°. When two octagons and one square are placed together at a common vertex, the sum of their vertex angles is 2(135°) + 90 = 360°. Thus, the region surrounding the common vertex is covered completely, without any overlap. This tiling composed of regular polygons of different types is an example of what is called a semiregular tessellation.

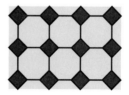

GEOMETRY AROUND US

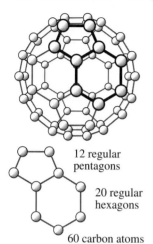

Buckyballs are currently a focus of study among physicists and chemists. **Buckyballs** are composed of regular pentagons and hexagons and resemble soccer balls or symmetrical spheres of chicken wire when depicted using a chemist's ball-and-stick models. More formally, they are known as buckminsterfullerenes, named for their resemblance to the geodesic domes of Buckminster Fuller. Their unique properties suggest many new applications including new lubricants, drugs, fuels, batteries, and high strength materials.

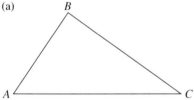

12 regular pentagons

20 regular hexagons

60 carbon atoms

PROBLEM SET 2.3

EXERCISES/PROBLEMS

1. Use your protractor to measure each vertex angle in each polygon shown. If necessary, extend the sides of the angles in order to measure them. Then find the sum of the measures of the vertex angles in each polygon. What *should* the angle sum be in each case?

(a)

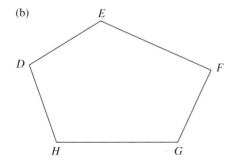

(b)

2. Use your protractor to measure each vertex angle in each polygon shown. If necessary, extend the sides of the angles in order to measure them. Then find the sum of the measures of the vertex angles in each polygon. What *should* the angle sum be in each case?

(a)

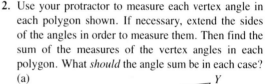

(b)

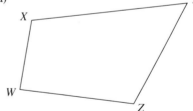

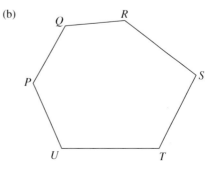

3. Draw examples, if possible, of each of the following types of triangles. If a triangle is not possible, explain why.
   (a) Acute scalene triangle
   (b) Obtuse scalene triangle
   (c) Right equilateral triangle
   (d) Scalene equiangular triangle

4. Draw examples, if possible, of each of the following types of triangles. If a triangle is not possible, explain why.
   (a) Right scalene triangle
   (b) Right isosceles triangle
   (c) Obtuse isosceles triangle
   (d) Acute right triangle

5. The measures of two angles in a triangle are given. Find the measure of the missing angle.
   (a) 90°, 60°   (b) 120°, 40°
   (c) 85°, 33°   (d) 79°, 67°

6. The measures of $\angle A$, $\angle B$, and $\angle C$ are given. Can a triangle $\triangle ABC$ be made with the given angle measures? Explain.
   (a) $\angle A = 36°$, $\angle B = 78°$, $\angle C = 66°$
   (b) $\angle A = 124°$, $\angle B = 56°$, $\angle C = 20°$
   (c) $\angle A = 90°$, $\angle B = 74°$, $\angle C = 18°$

7. Find the measures of the missing angles in the quadrilaterals shown.
   (a)
   (b)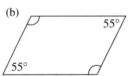
   (c)
   (d)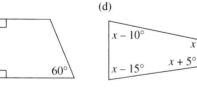

8. Find the measures of the missing angles in the quadrilaterals shown.
   (a)
   (b)

(c)
(d)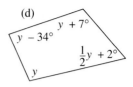

9. On a square lattice, draw a tessellation using each of the following triangles. It may be helpful to cut out a copy of each triangle.
   (a)   (b)

10. On a square lattice, draw a tessellation using each of the following quadrilaterals. It may be helpful to cut out a copy of each quadrilateral.
    (a)   (b)   (c)

11. For the following regular n-gons, give the measure of a vertex angle.
    (a) 12-gon   (b) 16-gon   (c) 10-gon
    (d) 20-gon   (e) 18-gon   (f) 36-gon

12. Measures of a vertex angle of regular polygons are given. Determine how many sides each polygon has.
    (a) 150°   (b) 156°   (c) 174°   (d) 178°

13. A regular n-gon must have n congruent angles and n congruent sides. Show that having n congruent angles is not sufficient for a polygon to be regular by sketching examples of
    (a) An equiangular quadrilateral that is not regular
    (b) An equiangular pentagon that is not regular

14. Show that having n congruent sides is not sufficient for a polygon to be regular by sketching examples of
    (a) An equilateral quadrilateral that is not regular
    (b) An equilateral hexagon that is not regular

15. Suppose that the sum of the measures of the vertex angles of a polygon is 1620°. How many sides does the polygon have?

16. What is the sum of the measures of the vertex angles of a 35-sided polygon?

17. Calculate the measure of each lettered angle. Congruent angles and right angles are indicated.

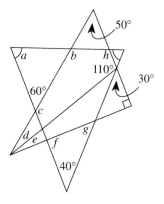

18. Calculate the measure of each lettered angle. Congruent angles and right angles are indicated.

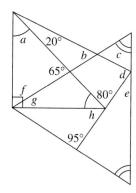

19. In the five-pointed star shown, what is the sum of the angle measures at A, B, C, D, and E? Assume that the pentagon inside the star is regular.

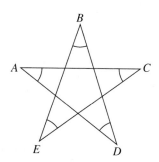

20. Complete the following table. Let V represent the number of vertices, D the number of diagonals from each vertex, and T the total number of diagonals in the polygon.

Polygon	V	D	T
Triangle			
Quadrilateral			
Pentagon	5	2	5
Hexagon			
Octagon			
n-gon			

21. Verify that the measure of a vertex angle in a regular polygon is $(n - 2)180°/n$ (Theorem 2.2), by completing the given table as follows: (i) Sketch regular polygons with 4, 5, 6, and more sides, and (ii) divide each polygon into triangles by drawing all possible diagonals from one vertex. Use inductive reasoning to develop the formula given in Theorem 2.2.

Number of Sides and Angles in a Polygon	Number of Triangles into which the Polygon can be Divided	Sum of Angle Measures in the Polygon	Measure of One Angle in the Polygon
3	1	180°	180°/3 = 60°
4	2	2 (180°) = 360°	360°/4 = 90°
5			
6			
8			
10			
n			

22. Use the fact that the sum of the measures of the vertex angles in a quadrilateral is 360° to explain why *any* quadrilateral, not just a square or a rectangle, can be used to tessellate the plane.

23. Draw a large copy of △*ABC* on scratch paper and cut it out.

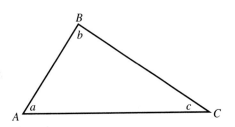

Fold down vertex *B* so that it lies on $\overline{AC}$ and so that the fold line is parallel to $\overline{AC}$.

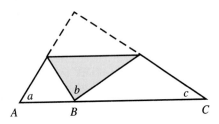

Now fold vertices *A* and *C* into point *B*.

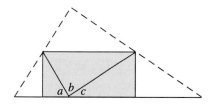

(a) What does the resulting figure tell you about the measures of ∠*A*, ∠*B*, and ∠*C*? Explain.
(b) What kind of polygon is the folded shape and what is the length of its base?
(c) Try this same procedure with two other types of triangles. Are the results the same?

APPLICATIONS

24. A forester determines that the **angle of elevation** (the acute angle between the horizontal and the line of sight) to the top of a fir tree is 42°15′, as shown next. Find the measure of ∠*CBA* in degrees, minutes, and seconds.

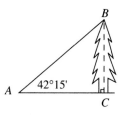

25. From a plane at point *A* in the following picture, the **angle of depression** (the acute angle between the horizontal and the line of sight) to a control tower at point *C* is 20°35′50″, as shown. Find the measure of ∠*ACB* in degrees, minutes, and seconds.

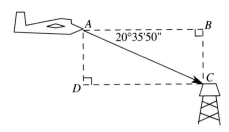

26. In surveying, a **traverse** is the measurement of a polygonal region, made in a series of legs, or line segments. The end of one leg is the start of the next. So the end of the final leg should connect with the beginning of the first leg. When this happens, the traverse is said to "close." The interior angles of a certain traverse are given next. Determine whether each traverse closes (that is, forms a closed polygon). [HINT: You might begin by drawing each figure by using a protractor.]
(a) ∠*A* = 97.8°, ∠*B* = 61.3°, ∠*C* = 115.5°, and ∠*D* = 82.6°.
(b) ∠*A* = 116°15′, ∠*B* = 89°45′, ∠*C* = 103°30″, ∠*D* = 128°45′, and ∠*E* = 101°45′.

27. The traditional way to check the accuracy of a closed traverse is to measure the interior angles turned at each point and compare the sum of their measures to (*n* − 2)180°.
(a) A survey party traversed a boundary of a proposed timber sale that contained five sides. The interior angles formed were 141°32′, 78°17′, 63°25′, 97°10′, and 159°36′. Determine the error, if any, in this traverse.
(b) A closed traverse with six sides has interior angles of 109°28′, 92°15′, 136°46′, 112°53′, 145°35′, and 121°8′. Determine the error, if any, in this traverse.

2.4
THREE-DIMENSIONAL SHAPES

Applied Problem

A restaurant packages take-out food in cardboard containers like the one shown. The box shapes are convenient in that when assembled they stack inside one another. They also can be unfolded and laid flat. Ignoring the flaps to hold the cover shut, what is the shape of the box when unfolded?

POLYHEDRA

Thus far we have been studying figures and regions on the plane; that is, figures that are 0-dimensional (points), 1-dimensional (such as lines), and 2-dimensional (such as triangles). Now we begin a study of 3-dimensional shapes.

A **polyhedron** (plural **polyhedra**) is a three-dimensional shape composed of polygonal regions, any two of which have at most a common side. In addition, it is an enclosed, connected finite portion of space without holes. Figure 2.42 shows three examples of polyhedra.

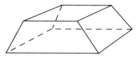

FIGURE 2.42

The polygonal regions of a polyhedron are called **faces**, the common line segments are called **edges**, and a point where edges meet is called a **vertex** (Figure 2.43).

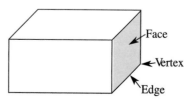

FIGURE 2.43

PRISMS

Two common types of polyhedra we will study are prisms and pyramids. A **prism** is a polyhedron with two opposite faces, called **bases**, that are identical polygonal regions in parallel planes; the other faces are called **lateral faces**. Prisms are named by the shape of their bases (Figure 2.44).

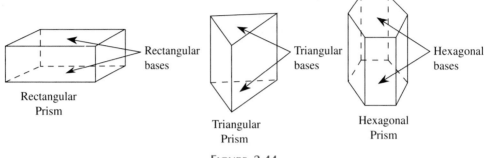

FIGURE 2.44

Prisms whose lateral faces are rectangular regions are called **right prisms**; otherwise they are called **oblique prisms** (Figure 2.45).

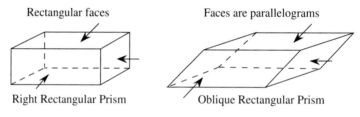

FIGURE 2.45

PYRAMIDS

A **pyramid** is a polyhedron consisting of a polygonal region for its **base** and triangular regions as **lateral faces**. A pyramid is formed when a point, its **apex**, not in the plane of the base is connected by line segments to the vertices of the base. The name of a pyramid is determined by the shape of its base. The perpendicular distance from the apex to the base of a pyramid is called the **height** of the pyramid (Figure 2.46).

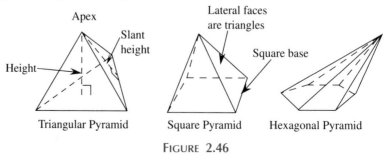

FIGURE 2.46

When the base of a pyramid is a regular polygon and its lateral faces are *isosceles* triangles, the pyramid is called a **right regular pyramid**; otherwise it is an **oblique regular pyramid**. The **slant height** of a right regular pyramid is the height of any of its faces. The triangular and square pyramids in Figure 2.46 are right regular pyramids and the hexagonal pyramid is oblique.

EXAMPLE 2.6   Name each of the polyhedra shown in Figure 2.47. Give as complete a description as possible.

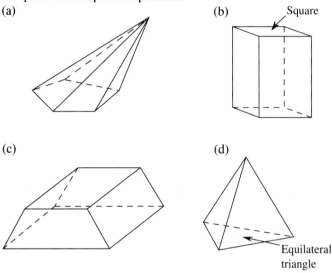

(a)

(b) Square

(c)

(d) Equilateral triangle

FIGURE 2.47

SOLUTION
(a) Oblique pentagonal pyramid
(b) Right square prism
(c) Right trapezoidal prism
(d) Right regular triangular pyramid, or right equilateral triangular pyramid
    [NOTE: In (c) the base of the prism is not the "bottom" of the prism.]   •

REGULAR POLYHEDRA

Polyhedra that have regular polygons as faces form a group of especially interesting figures. The Greeks studied the regular polyhedra and imbued them with special significance. A **regular polyhedron** is a polyhedron in which all faces are regular polygons of exactly the same size and shape. The Greeks discovered early on that only five such regular polyhedra exist. Those five are shown on the next page in Figure 2.48.

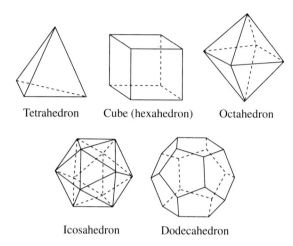

Tetrahedron   Cube (hexahedron)   Octahedron

Icosahedron   Dodecahedron

FIGURE 2.48

The following table summarizes information about the regular polyhedra.

Polyhedron	Shape of Face	Number of Faces, F	Number of Vetrices, V	Number of Edges, E
Tetrahedron	Triangle	4	4	6
Hexahedron	Square	6	8	12
Octahedron	Triangle	8	6	12
Dodecahedron	Pentagon	12	20	30
Icosahedron	Triangle	20	12	30

Examine the last three columns in the table and see if you notice a pattern in the numbers. The mathematician Leonard Euler noticed a pattern in the relationship among F, V, and E. He observed that in every case $F + V = E + 2$. This relationship has come to be known as **Euler's formula**, and it holds for *all* polyhedra, not just for regular polyhedra.

EXAMPLE 2.7   At first glance, it might appear that a soccer ball is an example of a regular polyhedron because its faces are all regular polygons (Figure 2.49). Why is a soccer ball not a regular polyhedron?

SOLUTION   The faces of a soccer ball are regular pentagons (the black faces) and regular hexagons (the white faces). Thus, it is not a regular polyhedron because its faces are not identical.   •

FIGURE 2.49

If the faces of a soccer ball were all identical regular pentagons, the ball would take on the shape of a dodecahedron and would not be sufficiently "round." If the faces were all identical regular hexagons, the ball would be a flat surface, because as we saw in Section 2.3, regular hexagons tessellate the plane.

The arrangement of faces on the soccer ball does make it an example of a semiregular polyhedron. A **semiregular polyhedron** has several different regular polygons as faces, but the same arrangement of polygons appears at each vertex.

CYLINDER

There are common three-dimensional curved shapes analogous to the prism and pyramid. A **circular cylinder** is the shape formed by two identical circular regions in parallel planes together with the surface formed by line segments joining corresponding points of the two circles (Figure 2.50).

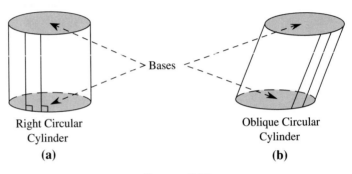

Right Circular
Cylinder
(a)

Oblique Circular
Cylinder
(b)

FIGURE 2.50

The two circular regions are called the **bases** of a cylinder. When the line segments joining corresponding points of the two bases are perpendicular to the planes containing the bases, the cylinder is called a **right circular cylinder** [Figure 2.50(a)]; otherwise it is an **oblique circular cylinder** [Figure 2.50(b)].

CONES

A **circular cone** is the shape formed by a circular region, called the **base**, together with the surface formed when the **apex**, a point not in the same plane as the circle, is joined by line segments to every point on the circle (Figure 2.51).

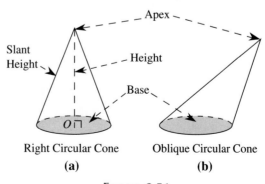

Right Circular Cone
(a)

Oblique Circular Cone
(b)

FIGURE 2.51

A **right circular cone** is a circular cone with its apex on the line that is perpendicular to the base at its center [Figure 2.51(a)]; otherwise circular cones are **oblique** [Figure 2.51(b)]. In the right circular cone, the perpendicular distance from the apex to the center of the base is called the **height** of the cone. For a right circular cone, the distance from the apex to a point on the edge of the base is called the **slant height**.

SPHERES

A **sphere** is the set of all points in space that are a fixed distance from a given point (the **center**) of the sphere. A line segment whose endpoints are the center and a point on the sphere is called **a radius** of the sphere; the length of any radius is called **the radius** of the sphere. A line segment that contains the center of the sphere and whose endpoints are on the sphere is called **a diameter** of the sphere; the length of any diameter is called **the diameter** of the sphere (Figure 2.52).

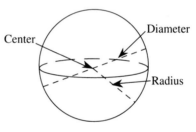

FIGURE 2.52

Solution to Applied Problem

The shape of the container is called a **frustum**. The top and bottom are rectangles and the lateral faces are all isosceles trapezoids. So the shape of the box when unfolded (ignoring flaps for securing the sides and top) is as shown, where the rectangle for the top has been decomposed into two rectangles.

GEOMETRY AROUND US

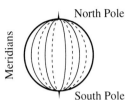

Meridians

North Pole

South Pole

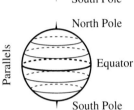

Parallels

North Pole

Equator

South Pole

Every cross section of a sphere is a circle. Cross sections that have as their center the center of the sphere are called **great circles**. **Meridians** on a globe are all great circles. They are used to measure longitude. **Parallels** on a globe, except for the equator, are not great circles because their centers are not the center of the earth. Parallels are used to measure latitude.

PROBLEM SET 2.4

EXERCISES/PROBLEMS

1. Which dark circle is behind the others in the figure? Look at the figure for one minute before answering.

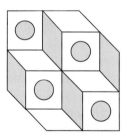

2. Is the cube in the inside corner of the larger figure or is it on the outside corner of another cube?

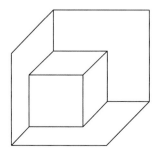

3. Pictured is a stack of cubes. Also given are the top view, the front view, and the right side view. (Assume that the only hidden cubes are ones that support a pictured cube.)

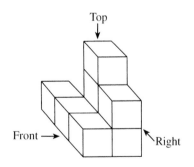

Top

Front →

Right

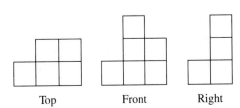

Top   Front   Right

Draw the three views of each of the following stacks of cubes.

(a)

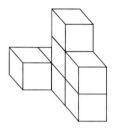

(b)

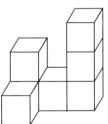

(c)

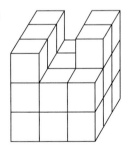

(d)

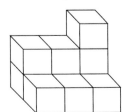

4. In the following figure, the drawing on the left shows a shape. The drawing on the right tells you how many cubes are in each stack forming the shape. It is called a **base design**.

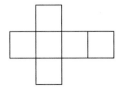

2	2	3
1	1	1

(a) Which is the correct base design for the following shape?

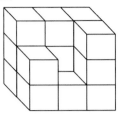

(i)

3	2	3
2	1	2

(ii)

3	3	3
2	1	3

(iii)

3	3	3
3	2	1

(b) Make a base design for the next shape.

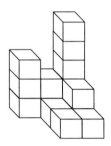

(c) Draw a cube picture for this base design.

3	3	2	1
1	1	2	

5. Recall that a pentomino is a geometric shape composed of 5 connected squares, where two sides are joined along an entire side. (See problems 2 and 3 in Section 1.1). Which of the 12 different pentominos can be folded to form an open-topped cube?

6. A **hexomino** is a geometric shape composed of 6 connected squares, where two squares are joined along an entire side. There are 35 different hexominos. Which of them can be folded to form a cube?

7. A cube when unfolded might look like the following.

Such an arrangement of faces of a polyhedron is called a **net**. Which of the following nets can be folded to form a tetrahedron?

(a)   (b)   (c)

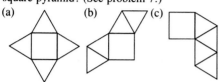

8. Which of the following nets can be folded to form a square pyramid? (See problem 7.)

(a)   (b)   (c)

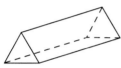

9. Draw a net for the octahedron.

10. Draw a net for the triangular prism shown.

11. Which of the three cubes represents a different view of the cube on the left? All faces have different letters.

(i)   (ii)   (iii)

12. All the faces of the cube on the left have different figures on them. Which of the three other cubes represents a different view of that cube?

(i)   (ii)   (iii)

13. Answer the following for the given prism.

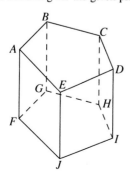

(a) Name the bases of the prism.
(b) Name the lateral faces of the prism.
(c) Name the faces that are hidden from view.
(d) Name the prism by type.

14. Answer the following for the given prism.

(a) Name the bases of the prism.
(b) Name the lateral faces of the prism.
(c) Name the faces that are hidden from view.
(d) Name the prism by type.

15. Name the following pyramids according to type.

(a)   (b)   (c)

16. Name the following prisms by type.

(a)   (b)   (c)

17. Identify these figures by name.

(a)   (b)   (c)

Base is   Base is   All cross-
a circle.   a circle.   sections
                                  are circles.

18. Which of the following are prisms? pyramids? neither?

(a)   (b)   (c)

19. Given are several prisms. Use these to complete the following table to see if Euler's formula holds for prisms.

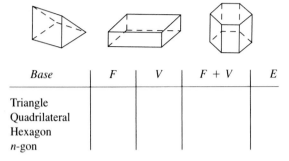

Base	F	V	F + V	E
Triangle				
Quadrilateral				
Hexagon				
n-gon				

20. Given are several pyramids. Use these to complete the following table to see if Euler's formula holds for pyramids.

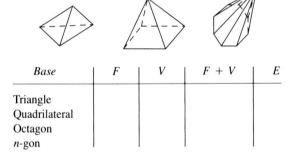

Base	F	V	F + V	E
Triangle				
Quadrilateral				
Octagon				
n-gon				

21. Given are three arbitrary three-dimensional shapes. Use these to complete the following table to see if Euler's formula holds for these shapes.

(i)   (ii)   (iii)

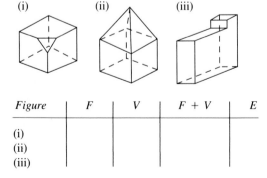

Figure	F	V	F + V	E
(i)				
(ii)				
(iii)				

22. An **antiprism** is a polyhedron that resembles a prism except that its lateral faces are triangles. Three antiprisms are shown. Notice how the bases are rotated images of each other. Use these antiprisms to complete the following table to see if Euler's formula holds for these shapes.

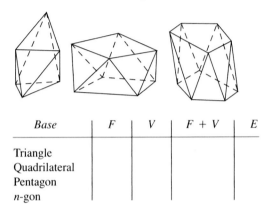

Base	F	V	F + V	E
Triangle				
Quadrilateral				
Pentagon				
n-gon				

23. A polyhedron has 18 edges and 8 faces. How many vertices does it have? Sketch a polyhedron that satisfies these conditions, and name the polyhedron you drew.

24. A polyhedron has 10 edges and 6 vertices. How many faces does it have? Sketch a polyhedron that satisfies these conditions, and name the polyhedron you drew.

25. Describe the cross sections formed when a plane cuts the following cylinders as shown.

(a)   (b)

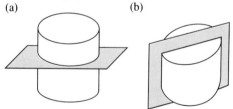

26. Describe the cross sections formed when a plane cuts the following right circular cones as shown.

(a)   (b)

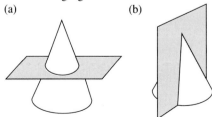

27. If the cube illustrated is cut by a plane midway between opposite faces and the front portion is placed against a mirror, the entire cube appears to be formed. The cutting plane is called a **plane of symmetry** and the figure is said to have **reflection symmetry**.

(a) How many planes of symmetry of this type are there for a cube?

(b) A plane passing through pairs of opposite edges is also a plane of symmetry. How many planes of symmetry of this type are there in a cube?

(c) How many planes of symmetry of all types are there for a cube?

28. The line connecting centers of opposite faces of a cube is an **axis of rotational symmetry**, because the cube can be turned about the axis and appears to be in the same position as it was initially. In fact, the cube can be turned about the axis four times as it returns to its initial position. This axis of symmetry is said to have order 4. How many axes of symmetry of order 4 are there in a cube?

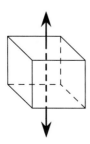

29. How many planes of symmetry do the following figures have? (Models may help.)
(a) Regular tetrahedron
(b) Right square pyramid

(c) Right regular pentagonal prism
(d) Right circular cylinder

30. Find the axes of symmetry for the following figures. Indicate the order of each type. (Models may help.)

(a) Regular tetrahedron
(b) Right regular pentagonal prism

31. Imagine a regular tetrahedron cut with a plane through the midpoints of three edges as shown. Describe the shape of each intersection.

(a)   (b)

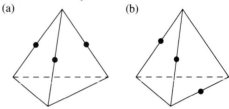

32. Imagine a cube cut with a plane through the midpoints of three edges as shown. Describe the shape of each intersection.

(a)   (b)

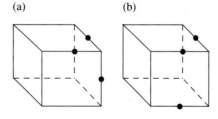

(c)   (d)

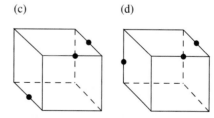

33. Show how to slice a cube with four cuts to make a regular tetrahedron. (Use a model—for example, slice some clay or dough.)

34. A regular tetrahedron is attached to a face of a square pyramid with equilateral faces where the faces of the tetrahedron and the pyramid are identical triangles. What is the fewest number of faces possible for the resulting polyhedron? (Use a model—it will suggest a surprising answer.)

2.5
DIMENSIONAL ANALYSIS

Applied Problem

David is planning a summer motorcycle trip around the United States. He drew his route on a map and estimated its length to be about 265 cm. The scale on his map is 1 cm = 39 km. His motorcycle averages about 50 miles per gallon. If he figures gasoline will cost about $1.20 per gallon, how much will the gasoline for his trip cost him?

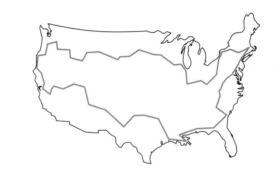

CONVERTING USING DIMENSIONAL ANALYSIS

Various units of measure and systems of measure are available for measuring the length of a line segment, size of an angle, or area of a polygonal region. Conversions within these systems or between two systems are frequently necessary. **Dimensional analysis** is a problem-solving strategy that can be used to make conversions. This strategy may be appropriate when units of measure are involved, the problem involves physical quantities, or conversions are required.

In the technique of dimensional analysis, fractions equivalent to 1 that include units of measure are utilized. These fractions are called **unit ratios**. For example, because 1 foot = 12 inches, we can write unit ratios of $\frac{1 \text{ foot}}{12 \text{ inches}}$ or $\frac{12 \text{ inches}}{1 \text{ foot}}$. In each case the numerator equals the denominator, so each fraction is equivalent to 1. Other examples of unit ratios are $\frac{60 \text{ min}}{1 \text{ hr}}$, $\frac{1 \text{ hr}}{60 \text{ min}}$, $\frac{4 \text{ qt}}{1 \text{ gal}}$, and $\frac{1 \text{ kg}}{1000 \text{ g}}$.

To perform a conversion, a unit ratio is treated as a fraction and multiplied times a measurement or another unit fraction. The unit ratio must be carefully chosen so that the desired units are obtained.

EXAMPLE 2.8   Convert a length of 9 feet to inches.

SOLUTION   First we write the given measurement as a fraction: $9 \text{ ft} = \dfrac{9 \text{ ft}}{1}$. If
we multiply this fraction by any other fraction equivalent to 1, we will obtain a
new fraction equivalent to the original one. Two unit ratios that might be used
are $\dfrac{12 \text{ in.}}{1 \text{ ft}}$ and $\dfrac{1 \text{ ft}}{12 \text{ in.}}$ because they both relate feet to inches. [NOTE: Inch is
abbreviated using a period "in." to distinguish it from "in" as in "in the
picture."] However, using the first ratio will allow us to "cancel" units as
follows:

$$9 \text{ ft} = \frac{9\cancel{\text{ft}}}{1} \cdot \frac{12 \text{ in.}}{\cancel{\text{ft}}} = \frac{9(12)}{1(1)} \text{ in.} = 108 \text{ in.}$$

This means that a length of 9 ft is the same as 108 in.   •

Notice that because we wished to convert from feet to inches and the original
fraction had ft in the numerator, we chose a unit ratio, $\dfrac{12 \text{ in.}}{1 \text{ ft}}$, having 1 ft in the
denominator. In this way the units of feet cancelled and we were left with the
units we desired, namely inches.

The method of dimensional analysis is especially helpful if several
conversions must be made. In that case, unit ratios may be chained together in
a multiplication process. The next example demonstrates such a conversion
within the metric system.

EXAMPLE 2.9   A rectangular garden plot measures 345 cm on one side. What
is this length in kilometers?

SOLUTION   Because 100 cm = 1 m and 1000 m = 1 km, we can write

$$345 \text{ cm} = \frac{345 \cancel{\text{cm}}}{1} \cdot \frac{1 \cancel{\text{m}}}{100 \cancel{\text{cm}}} \cdot \frac{1 \text{ km}}{1000 \cancel{\text{m}}} = \frac{345(1)(1)}{1(100)(1000)} \text{ km} = 0.00345 \text{ km} \bullet$$

Notice again that we chose to use $\dfrac{1 \text{ m}}{100 \text{ cm}}$ rather than the unit ratio $\dfrac{100 \text{ cm}}{1 \text{ m}}$ in
order for the appropriate units to "cancel."

Dimensional analysis works equally well when we are required to convert
between the English and metric systems. Although converting from feet to
inches or cm to km without using dimensional analysis is fairly straightforward,
it may be more difficult to decide whether to multiply or divide when converting
between systems or when working with unfamiliar units. At those times,
dimensional analysis is an excellent choice of strategy.

To convert between the English and metric units for length, only one
relationship between the two systems is really necessary: 1 inch equals exactly
2.54 centimeters. All other conversions can be performed using this relationship.

EXAMPLE 2.10   The radius of the earth is approximately 6380 km. What is the radius of the earth in miles?

SOLUTION   We will use the relationship 1 in. = 2.54 cm and will multiply several unit ratios together. First we convert km to cm.

$$6380 \text{ km} = \frac{6380 \text{ km}}{1} \cdot \frac{1000 \text{ m}}{1 \text{ km}} \cdot \frac{100 \text{ cm}}{1 \text{ m}} = 638{,}000{,}000 \text{ cm}$$

Next we convert from metric to English units.

$$6380 \text{ km} = \frac{6380 \text{ km}}{1} \cdot \frac{1000 \text{ m}}{1 \text{ km}} \cdot \frac{100 \text{ cm}}{1 \text{ m}} \cdot \frac{1 \text{ in.}}{2.54 \text{ cm}} \approx 251{,}000{,}000 \text{ in.}$$

Finally, we convert from inches to miles.

$$6380 \text{ km} = \frac{6380 \text{ km}}{1} \cdot \frac{1000 \text{ m}}{1 \text{ km}} \cdot \frac{100 \text{ cm}}{1 \text{ m}} \cdot \frac{1 \text{ in.}}{2.54 \text{ cm}} \cdot \frac{1 \text{ ft}}{12 \text{ in.}} \cdot \frac{1 \text{ mi}}{5280 \text{ ft}}$$

$$= \frac{6380(1000)(100)(1)(1)(1)}{1(1)(1)(2.54)(12)(5280)} \text{ mi} \approx 3964 \text{ mi}$$

So, the radius of the earth is about 3964 mi.

The same steps as those used in Example 2.10 can be used to derive a new conversion ratio that allows us to convert directly from mi to km or km to mi.

EXAMPLE 2.11   Find the number of miles in a kilometer; that is, 1 km is the same as ? mi.

SOLUTION

$$1 \text{ km} = \frac{1 \text{ km}}{1} \cdot \frac{1000 \text{ m}}{1 \text{ km}} \cdot \frac{100 \text{ cm}}{1 \text{ m}} \cdot \frac{1 \text{ in.}}{2.54 \text{ cm}} \cdot \frac{1 \text{ ft}}{12 \text{ in.}} \cdot \frac{1 \text{ mi}}{5280 \text{ ft}}$$

$$= \frac{1(1000)(100)(1)(1)(1)}{1(1)(1)(2.54)(12)(5280)} \text{ mi} \approx 0.6214 \text{ mi}$$

So, for future reference, we have these *approximate* conversion ratios:

$$\frac{1 \text{ km}}{0.6214 \text{ mi}} \quad \text{and} \quad \frac{0.6214 \text{ mi}}{1 \text{ km}}$$

CONVERTING RATES USING DIMENSIONAL ANALYSIS

Different units of measure are often mixed to form some kind of **rate**, such as 35 mi/gal, 88 ft/sec, or 1.49 dollars/liter. Dimensional analysis can be helpful in converting from one rate to another equivalent rate, as shown next.

EXAMPLE 2.12   A well produces water at a rate of 30 gallons per minute. Express this rate in liters per day.

SOLUTION   We must convert the rate gal/min to liters/day. This means we must multiply by unit fractions to convert gallons to liters and minutes to days. We can perform either of these conversions first or do them simultaneously. We need a conversion ratio that relates gallons and liters. We'll use the fact that 1 liter = 1.057 quarts and that 1 gallon = 4 quarts.

We can think of the rate 30 gal/min as the ratio $\dfrac{30 \text{ gal}}{1 \text{ min}}$. That is, 30 gal of water is produced for every 1 min of time that passes. To convert from gallons per minute to liters per minute, we can use the following sequence of fractions:

$$\frac{30 \text{ gal}}{1 \text{ min}} = \frac{30 \text{ gal}}{1 \text{ min}} \cdot \frac{4 \text{ qt}}{1 \text{ gal}} \cdot \frac{1 \text{ L}}{1.057 \text{ qt}} \approx \frac{113.5 \text{ L}}{1 \text{ min}}$$

If these fractions are multiplied, the result is in L/min. Now we must convert from minutes to days.

$$\frac{30 \text{ gal}}{1 \text{ min}} = \frac{30 \text{ gal}}{1 \text{ min}} \cdot \frac{4 \text{ qt}}{1 \text{ gal}} \cdot \frac{1 \text{ L}}{1.057 \text{ qt}} \cdot \frac{60 \text{ min}}{1 \text{ hr}} \cdot \frac{24 \text{ hr}}{1 \text{ day}}$$

$$= \frac{30(4)(1)(60)(24)}{1(1)(1.057)(1)(1)} \frac{\text{L}}{\text{day}}$$

$$\approx 163{,}500 \text{ L/day}$$

Notice that we chose the unit ratio $\dfrac{60 \text{ min}}{1 \text{ hr}}$ because the original rate, $\dfrac{30 \text{ gal}}{1 \text{ min}}$, had minutes in the denominator.

The next and last two examples demonstrate how the technique of dimensional analysis can be utilized in more general problem-solving situations.

PROBLEM SOLVING USING DIMENSIONAL ANALYSIS

EXAMPLE 2.13   The sun is 93,000,000 miles from the earth and light travels at a speed of approximately 186,000 miles/second. How long, in minutes, does it take for light from the sun to reach the earth?

SOLUTION   We can use the rate 186,000 miles/second as $\dfrac{186{,}000 \text{ mi}}{1 \text{ sec}}$ or as $\dfrac{1 \text{ sec}}{186{,}000 \text{ mi}}$. We will multiply the latter by the given distance of 93,000,000 mi so that the miles will cancel and we will be left with a unit of time.

$$\frac{(93{,}000{,}000) \text{ mi}}{1} \cdot \frac{1 \text{ sec}}{(186{,}000) \text{ mi}} \cdot \frac{1 \text{ min}}{60 \text{ sec}} = \frac{(93{,}000{,}000)(1)(1)}{(186{,}000)(60)} \text{ min} = 8\tfrac{1}{3} \text{ min}$$

So light from the sun hits the earth in about $8\tfrac{1}{3}$ minutes.

EXAMPLE 2.14   A do-it-yourselfer is building a backyard barbecue. The project requires 525 bricks and they must be moved to the building site in a wheelbarrow that will carry a maximum of 150 lbs of bricks. If the bricks weigh $2\tfrac{1}{4}$ tons per thousand, how many trips with the wheelbarrow will be necessary?

SOLUTION   We start with the 525 bricks and multiply by ratios of the appropriate form to convert to wheelbarrow loads.

$$525 \text{ bricks} = \frac{525 \cancel{\text{ bricks}}}{1} \cdot \frac{2.25 \cancel{\text{ tons}}}{1000 \cancel{\text{ bricks}}} \cdot \frac{2000 \cancel{\text{ lb}}}{1 \cancel{\text{ ton}}} \cdot \frac{1 \text{ wheelbarrow load}}{150 \cancel{\text{ lb}}}$$

$$= \frac{525(2.25)(2000)(1)}{1(1000)(1)(150)} \text{ loads}$$

$$= 15.75 \text{ loads}$$

So it will take 16 trips with the wheelbarrow to move all of the bricks.

Solution to Applied Problem

We can use dimensional analysis to solve this problem if we chain the conversions required. First, we determine the actual distance to be traveled according to the scale and his measurements.

$$265 \text{ cm} = \frac{265 \cancel{\text{ cm}}}{1} \cdot \frac{39 \text{ km}}{1 \cancel{\text{ cm}}}$$

To convert from km to mi, we could use the fact that 1 in. = 2.54 cm. However, we can obtain an approximate solution with fewer calculations if we use 1 km ≈ 0.6214 mi from Example 2.9.

$$\frac{265 \cancel{\text{ cm}}}{1} \cdot \frac{39 \cancel{\text{ km}}}{1 \cancel{\text{ cm}}} \cdot \frac{0.6214 \text{ mi}}{1 \cancel{\text{ km}}}$$

Finally, we can find the total cost by multiplying by two more unit ratios to find the number of gallons of gas needed and the cost for that amount of gas.

$$\frac{265 \cancel{\text{ cm}}}{1} \cdot \frac{39 \cancel{\text{ km}}}{1 \cancel{\text{ cm}}} \cdot \frac{0.6214 \cancel{\text{ mi}}}{1 \cancel{\text{ km}}} \cdot \frac{10 \cancel{\text{ gal}}}{50 \cancel{\text{ mi}}} \cdot \frac{\$1.20}{1 \cancel{\text{ gal}}} \approx \$154.13$$

GEOMETRY AROUND US

The United States remains an island in a metric world. In 1993, the only nonmetric countries in the world other than the United States were Burma and Liberia. Actually, now our units of measure are defined in terms of the metric units. For example, in 1959 one inch was defined to be exactly 2.54 centimeters. Even though many industries in the United States, including the automobile and pharmaceutical industries, already use metric tools and measures, the change to everyday use of the metric system has been very gradual. In 1988, President Reagan signed a bill that required all government agencies to be metric by 1992.

PROBLEM SET 2.5

EXERCISES/PROBLEMS

1. Give the appropriate unit ratios to convert each of the following measures.
   - (a) inches to feet
   - (b) miles to feet
   - (c) pounds to ounces
   - (d) quarts to gallons

2. Give the appropriate unit ratios to convert each of the following measures.
   - (a) meters to centimeters
   - (b) millimeters to centimeters
   - (c) kilograms to grams
   - (d) milliliters to liters

3. Give the appropriate unit ratios to convert each of the following measures.
   - (a) 17 hours to minutes
   - (b) 360 seconds to minutes
   - (c) 720 inches to yards
   - (d) 1440 man-hours to man-days

4. Give the appropriate unit ratios to convert each of the following measures.
   - (a) 45 degrees to minutes
   - (b) 0.4 meters to millimeters
   - (c) 60 hours to days
   - (d) 15¢/foot to cents/inch

5. Using dimensional analysis, perform the following conversions.
   - (a) 4.5 lb to oz
   - (b) 3744 min to deg
   - (c) 25 mi to in.
   - (d) 10,500 mL to kL

6. Using dimensional analysis, perform the following conversions.
   - (a) 58 mg to g
   - (b) 0.7 m to km
   - (c) 3 tons to oz
   - (d) 2400 sec to days

7. Using dimensional analysis, perform the following conversions.
   - (a) 45 mi/hr to ft/sec
   - (b) $200/day to dollars/hour
   - (c) 0.3 in./year to ft/century
   - (d) 16 km/L to m/mL

8. Using dimensional analysis, perform the following conversions.
   - (a) 40 mi/gal to mi/qt
   - (b) 64 in./sec to mi/hr
   - (c) 50 kg/m to g/cm
   - (d) $1.99/L to cents/mL

9. Using the fact that 1 in. = 2.54 cm, perform the following conversions.
   - (a) 6-in. snowfall to cm
   - (b) 100-yd football field to m
   - (c) 420-ft home run to m
   - (d) 1-km racetrack to mi

10. Using the fact that 1 in. = 2.54 cm, perform the following conversions.
    - (a) 26.2-mi marathon to m
    - (b) 1-m ruler to in.
    - (c) 195-mm tire width to in.
    - (d) 8-ft ceiling to cm

11. Using the fact that 1 in. = 2.54 cm and 1 kg = 2.205 lb, perform the following conversions.
    - (a) 100 ft/sec to m/sec
    - (b) 0.4 kg/m to lb/yd

12. Using the fact that 1 in. = 2.54 cm, perform the following conversions.
    - (a) 48¢/ft to cents/cm
    - (b) 60 km/hr to ft/min

APPLICATIONS

13. The speed limit on some U.S. highways is 65 mph. If metric speed limit signs are posted, what will they read?

14. A seamstress has agreed to sew four bridesmaids' dresses for an upcoming wedding. Each dress requires $1\frac{5}{8}$ yds of lace that sells for 89¢ per inch. Use dimensional analysis to find the total cost for the lace.

15. Mary Decker has run a mile on an indoor track in 4 min 17.6 sec. What was her speed in km/hr?

16. Bamboo can grow as much as 35.4 inches per day. Express this rate in km/hr.

17. A car burns oil at an average rate of 1 quart per 2500 miles. At what rate, in mL/m, does the car burn oil?

18. A foreign car's gas tank holds 50 L of gasoline. What will it cost to fill the tank if regular gas is selling for $1.09 a gallon?

19. Human hair typically grows about one-half inch in four weeks. At what rate, in mph, does hair grow?

20. A mason laying brick for a wall 8'9'' tall allows $2\frac{5}{8}''$ for each "course," or row, of bricks, including $\frac{1}{2}''$ for the mortar joint. Use dimensional analysis to determine how many courses of brick must be laid.

21. We see lightning before we hear thunder because light travels faster than sound. If sound travels at about 1000 km/hr through air at sea level at 15°C and light travels at 186,282 mi/sec, how many times faster is the speed of light than the speed of sound? Use dimensional analysis.

22. A DC-9, which burns approximately 7000 lb of fuel per hour, makes a daily round trip between St. Louis and San Francisco. The trip from St. Louis to San Francisco takes 4 hr 15 min and the return trip takes 3 hr 45 min because of prevailing winds. How many tons of fuel are burned by the DC-9 in a 30-day month?

23. Light travels 186,282 miles per second. Use dimensional analysis to answer each of the following questions.
    (a) Based on a 365-day year, how far in miles will light travel in one year? This unit of distance is called a **light year**.
    (b) If a star in Andromeda is 76 light years away from earth, how many miles will light from the star travel on its way back to earth?
    (c) The planet Jupiter is about 480,000,000 miles from the sun. How long, in hours, does it take for light to travel from the sun to Jupiter?

Solution to Initial Problem

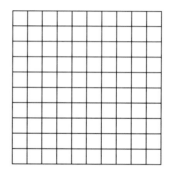

One way to keep count of the various overlapping squares is to count the upper right-hand corners of all of the squares of a particular size. The following table lists the number of various sized squares in the 10 by 10 grid.

Length of Side	10	9	8	7	6	5	4	3	2	1
Number of Squares	1	4	9	16	25	36	49	64	81	100

The total number of squares is 385. Notice that the number of squares is always a square number and that placing a dot on the upper right-hand corner of all of the squares of a particular size results in a square array of dots. Notice also that the formula for the number of squares is n^2.

Additional Problems Where the Strategy "Look for a Formula" Is Useful

1. Jack's beanstalk increases in height by $\frac{1}{2}$ the first day, $\frac{1}{3}$ the second day, $\frac{1}{4}$ the third day, and so on. What is the smallest number of days it would take to become at least 100 times as tall as its original height?

2. How many different (nonzero) angles are formed in a fan of rays like the one pictured next, but one having 100 rays?

Answers for the additional problems can be found in the Answers section near the end of the book.

Writing for Understanding

1. Describe as many applications of angle measure in other fields as you can. For example, angles are important in sports such as golf, where the flight of the ball is related to the angle at which it is hit as well as to the angle of the clubface.

2. Describe the types of symmetry you see in the room around you. Be sure to consider both types of symmetry and two-dimensional as well as three-dimensional shapes.

3. **Fermi problems** are problems to be solved by estimation without seeking exact numbers from references. Describe how to use estimation and your current knowledge, including dimensional analysis, to solve the following Fermi problem:

    Suppose that all the newspapers sold in the United States in one year were stacked in one pile. Estimate the height of that stack in miles. Explain your reasoning at each step.

4. "Geometry Around Us" at the end of Section 2.3 discusses the metric system, which is formally referred to as the Système International (SI). Write up a report contrasting this system with our customary system, which uses feet, pounds, and so on. Make a case for the retention of our customary system or for converting to the metric system. Be sure to support your arguments.

CHAPTER REVIEW

Following is a list of key vocabulary, notation, and ideas for this chapter. Mentally review these items and, where appropriate, write down the meaning of each term. Then restudy the material that you are unsure of before proceeding to take the chapter test.

Section 2.1—Undefined Terms, Definitions, Postulates, Segments, and Angles

Vocabulary/Notation

Axiomatic system 33
Undefined terms 34
Point 34
Line ($\overleftrightarrow{AB}$) 34
Plane 34
Space 34
Geometric figure 34
Postulate 34
Theorem 35
Collinear 36
Noncollinear 36

Coplanar 36
Noncoplanar 36
Ruler Postulate 36
Coordinate 36
Distance from A to B 37
Line segment ($\overline{AB}$) 37
Endpoint 37
Length of a line segment 37
Congruent segments 37
Ray ($\overrightarrow{AB}$) 37
Angle ($\angle$) 38

Vertex (vertices) 38
Sides (of an angle) 38
Degree 38
Protractor 38
Protractor Postulate 39
Measure of an angle 39
Acute angle 39
Right angle 39

Obtuse angle 39
Straight angle 39
Reflex angle 39
Perpendicular lines ($\perp$) 39
Congruent angles 39
Complementary angles 40
Supplementary angles 40
Adjacent angles 40

Main Ideas/Results

1. A modern axiomatic system consists of undefined terms, definitions, postulates, and theorems.

2. The Ruler Postulate

3. The Protractor Postulate

Section 2.2—Polygons and Circles

Vocabulary/Notation

Triangle ($\triangle ABC$) 46
Sides of a triangle 46
Equilateral triangle 46
Isosceles triangle 46
Scalene triangle 46
Acute triangle 46
Right triangle 46
Obtuse triangle 46
Equiangular triangle 46
Base angles (of an isoceles
    triangle) 47
Vertex angle 47
Base (of an isosceles triangle) 47
Hypotenuse 47
Leg 47
Simple closed curve 47
Circle 48
Center (of a circle) 48
Radius (of a circle) 48
Diameter (of a circle) 48
Polygon 48

Quadrilateral 49
Pentagon 49
Hexagon 49
Octagon 49
Decagon 49
n-gon 49
Parallel lines (segments) ($\parallel$) 49
Square 50
Rectangle 50
Rhombus 50
Parallelogram 50
Trapezoid 50
Base of a trapezoid 50
Legs of a trapezoid 50
Isosceles trapezoid 50
Kite 50
Reflection symmetry 50
Line of symmetry 51
Axis of symmetry 51
Rotation symmetry 51
Center of rotation symmetry 51

Main Ideas/Results

1. The study of simple closed curves includes the study of circles, and polygons (which includes triangles and quadrilaterals).

2. Symmetry is useful in describing figures.

Section 2.3—Tessellations and Regular Polygons

Vocabulary/Notation

Polygonal region 58
Tessellation (tiling) 59
Vertex angles (in a polygon) 61

Diagonal of a polygon 61
Regular polygon 62
Regular tessellation 64

Main Ideas/Results

1. Tessellations can be used to discover relationships in geometric figures.
2. The sum of the measures of the angles in an n-gon (and hence the measure of any angle in a regular polygon) is determined by the number of its sides.

Section 2.4—Three-Dimensional Shapes

Vocabulary/Notation

Polyhedron (polyhedra) 69
Face (of a polyhedron) 69
Edge (of a polyhedron) 69
Vertex (of a polyhedron) 69
Prism 70
Base (of a prism) 70
Lateral face 70
Right prism 70
Oblique prism 70
Pyramid 70
Base (of a pyramid) 70
Lateral face (of a pyramid) 70
Apex (of a pyramid) 70
Height (of a pyramid) 70
Right regular pyramid 71
Oblique regular pyramid 71
Slant height 71
Regular polyhedron 71

Euler's formula 72
Semiregular polyhedron 73
Circular cylinder 73
Base (of a cylinder) 73
Right circular cylinder 73
Oblique circular cylinder 73
Circular cone 73
Base (of a cone) 73
Apex (of a cone) 73
Right circular cone 74
Oblique circular cone 74
Height (of a cone) 74
Slant height (of a cone) 74
Sphere 74
Center (of a sphere) 74
Radius (of a sphere) 74
Diameter (of a sphere) 74

Main Idea/Result

1. Important three-dimensional shapes to be studied are prisms, pyramids, cylinders, cones, and spheres, as well as regular polyhedra.

Section 2.5—Dimensional Analysis

Vocabulary/Notation

Dimensional analysis 80
Unit ratio 80

Rate 82

Main Idea/Result

1. Dimensional analysis is used to convert between various systems of measurement and to solve problems.

PEOPLE IN GEOMETRY

Leonhard Euler (1707–1783) was one of the most prolific of all mathematicians. He published 530 books and papers during his lifetime and left much unpublished work at the time of his death. From 1771 on, he was totally blind, yet his mathematical discoveries continued. He would work mentally, then dictate to assistants, sometimes using a large chalkboard on which to write the formulas for them. His writings are a model of clear exposition, and much of our modern mathematical notation has been influenced by his style. For instance, the modern use of the symbol π is due to Euler. In geometry, he is best known for the Euler line of a triangle and the formula $V - E + F = 2$, which relates the number of vertices, edges, and faces of any simple closed polyhedron.

CHAPTER 2 TEST

TRUE-FALSE

Mark as true any statement that is always true. Mark as false any statement that is never true or that is not necessarily true. Be able to justify your answers.

1. Any three points lie in one and only one plane.

2. If two adjacent angles are supplementary, then their noncommon sides form a straight line.

3. An isosceles triangle has reflection symmetry but not rotation symmetry.

4. The vertex angle of a hexagon measures 120°.

5. The radius of a circle is twice its diameter.

6. An equiangular triangle must be an acute triangle.

7. A circle is the set of all points in a plane that are the same distance from a fixed point.

8. A parallelogram with four congruent sides is a rhombus.

9. A triangle is an example of a simple closed curve.

10. A right circular cylinder is a polyhedron.

EXERCISES/PROBLEMS

11. All squares are kites, but not all kites are squares. What additional conditions must be satisfied for a kite to be a square?

12. Find the measures of the lettered angles. Congruent angles are marked.

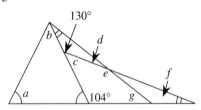

13. Find the measure of a vertex angle for a regular 24-gon.

14. Find the specified shapes in the following figure. Assume that angles which appear to be right angles are right angles, segments which appear to be parallel are parallel, and segments which appear to be congruent are congruent.

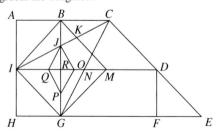

(a) Three squares

(b) A rectangle that is not a square

(c) A parallelogram that is not a rectangle

(d) Seven congruent right isosceles triangles

(e) An isosceles triangle not congruent to those in (d)

(f) A rhombus that is not a square

(g) A kite that is not a rhombus

(h) A scalene triangle with no right angles

(i) A right scalene triangle

(j) A trapezoid that is not isosceles

(k) An isosceles trapezoid

15. In the following figure, find one example of the following angles:

(a) supplementary   (b) complementary

(c) right   (d) adjacent

(e) acute   (f) obtuse

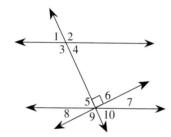

16. Determine the number of rotation and reflection symmetries for each of the following figures.

(a)

(b)

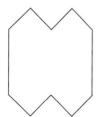

(c)

17. Form a tessellation of the plane using the quadrilateral *ABCD*.

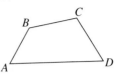

18. Given the triangular lattice shown, draw triangles having $\overline{AB}$ as a side.

(a) How many different triangles are possible?

(b) How many are isosceles triangles?

(c) How many are obtuse triangles?

(d) How many are right triangles?

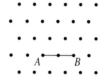

19. One millisecond is 1/1000 second. Use dimensional analysis to convert $3\frac{1}{4}$ hours to milliseconds.

20. Use your protractor and a ruler to draw an example of each of the following.

(a) A quadrilateral with three angles that each measure 100° and with no two sides congruent.

(b) A hexagon that is equiangular but not regular.

APPLICATIONS

21. Use dimensional analysis to convert the rate 6 ounces per cubic foot to grams per liter.

22. A traverse with five sides has interior angles with the following measures: 121°45′, 118°25′, 87°32′, 98°34′, and 113°54′. Does the traverse close? If not, determine the error in the traverse.

23. A long-playing record album turns at a rate of $33\frac{1}{3}$ revolutions per minute (rpm). Joan puts an album on the stereo and listens for only 45 seconds. In that time, how many revolutions did the record make?

24. The bed of a gravel truck has the shape shown below. Describe that shape as completely as possible.

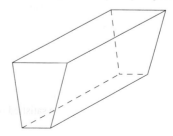

CHAPTER

3

Perimeter, Area, and Volume

Archimedes (287–212 B.C.) is considered to have been the greatest mathematician in antiquity. One of his discoveries was that by using appropriately dimensioned cross sections, a solid cone and solid sphere placed two units from the fulcrum of a lever would balance a solid cylinder placed one unit from the fulcrum. Because the volume of the cylinder and the cone were known, he then found the formula for the volume of the sphere by using a law of physics.

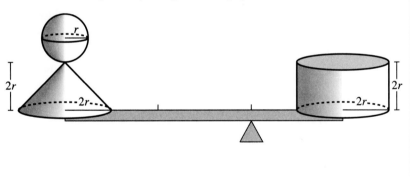

STRATEGY 8: USE A MODEL

The Use a Model strategy is useful in problems involving figures. A model may be as simple as a paper, wooden, or plastic shape, or as complicated as a carefully constructed replica that an architect or engineer might use.

INITIAL PROBLEM

Describe a solid shape that will fill each of the holes in this template as well as pass through each hole.

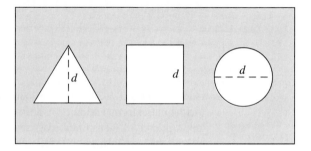

CLUES

The Use a Model strategy may be appropriate when

· Physical objects can be used to represent the ideas involved.
· A drawing is either too complex or inadequate to provide insight into the problem.
· A problem involves three-dimensional objects.

A solution for the initial problem above is on page 157.

INTRODUCTION

Knowledge of geometry is required in problems that involve perimeter, area, and volume. Perimeter involves measurement in one dimension. Area and volume, which are found by using linear measure, are important ideas in two and three dimensions, respectively. Sections 3.1 and 3.2 develop the notions of perimeter and area as applied to polygons and circles. In Section 3.3 the concept of area is used to develop an important relationship between the sides of a right triangle. Then, in Sections 3.4 and 3.5, ideas involving linear and area measure are applied in three dimensions to find the surface area and volume of common three-dimensional shapes.

3.1
PERIMETER, CIRCUMFERENCE, AND AREA
OF RECTANGLES AND TRIANGLES

Applied Problem

A family plans to invest in new carpet for a living room. The floor plan and dimensions shown include a brick hearth and a hardwood entry, which will not be carpeted. If the carpet costs $22.50/yd^2$ installed and 5% extra area is allowed for cutting, how much will it cost to carpet the room?

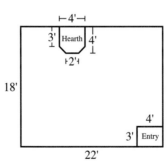

PERIMETER OF A POLYGON

The **perimeter** of a polygon is the sum of the lengths of its sides. "Peri" means around and "meter" means measure; hence perimeter literally means the measure around. Perimeter formulas can be developed for some common quadrilaterals. A square and a rhombus both have four congruent sides. If one side is of length s, then the perimeter of each of them can be represented by $4s$ (Figure 3.1).

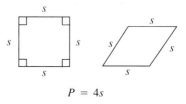

$$P = 4s$$

FIGURE 3.1

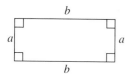

In rectangles and parallelograms, pairs of opposite sides are congruent. This will be verified formally in Chapter 5. Thus, if the lengths of their sides are a and b, then the perimeter of a rectangle or a parallelogram is $2a + 2b$. A similar formula can be used to find the perimeter of a kite (Figure 3.2). These results are summarized in the following theorem. [NOTE: A **theorem** is a result that is proved or derived from other known facts or results.]

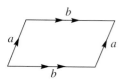

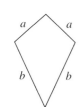

$$P = 2a + 2b$$

FIGURE 3.2

Theorem 3.1

Perimeters of Common Quadrilaterals	
Figure	Perimeter
Square with sides of length s	$4s$
Rectangle with sides of lengths a and b	$2a + 2b$
Parallelogram with sides of lengths a and b	$2a + 2b$
Rhombus with sides of length s	$4s$
Kite with sides of lengths a and b	$2a + 2b$

EXAMPLE 3.1   Find the perimeters of the following (Figure 3.3):

(a) A triangle whose sides have lengths 5 cm, 7 cm, and 9 cm

(b) A square with sides of length 8 ft

(c) A rectangle with one side of length 9 mm and another side of length 5 mm

(d) A rhombus one of whose sides has length 7 in.

(e) A parallelogram with sides of lengths 7.3 cm and 9.4 cm

(f) A kite whose shorter sides are $2\frac{2}{3}$ yd and whose longer sides are $6\frac{1}{5}$ yd

(g) A trapezoid whose bases have lengths 13.5 ft and 7.9 ft and whose other sides have lengths 4.7 ft and 8.3 ft

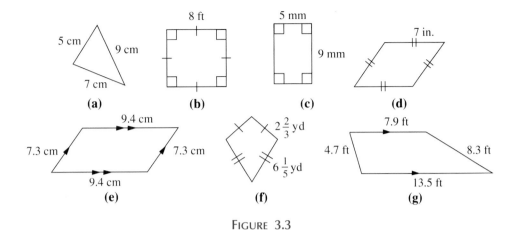

FIGURE 3.3

SOLUTION

(a) $P = 5$ cm $+ 7$ cm $+ 9$ cm $= 21$ cm

(b) $P = 4 \times 8$ ft $= 32$ ft

(c) $P = 2(9$ mm$) + 2(5$ mm$) = 28$ mm

(d) $P = 4 \times 7$ in. $= 28$ in.

(e) $P = 2(7.3$ cm$) + 2(9.4$ cm$) = 33.4$ cm

(f) $P = 2(2\frac{2}{3}$ yd$) + 2(6\frac{1}{5}$ yd$) = 5\frac{1}{3}$ yd $+ 12\frac{2}{5}$ yd $= 17\frac{11}{15}$ yd

(g) $P = 13.5$ ft $+ 7.9$ ft $+ 4.7$ ft $+ 8.3$ ft $= 34.4$ ft

CIRCUMFERENCE OF A CIRCLE

The perimeter of a circle, that is, the distance around the circle, is called its **circumference**. Since ancient times geometers have known that the ratio of the circumference to the diameter in any circle is a constant. In about 225 B.C., Archimedes determined that this ratio was between $3\frac{10}{71}$ and $3\frac{1}{7}$. That is, if C represents the circumference and d represents the diameter of any circle, then $\frac{C}{d}$ can be *approximated* by $3\frac{1}{7}$, or 22/7. In decimal form, $\frac{C}{d}$ is *approximately* 3.14.

The ratio $\frac{C}{d}$ has come to be represented by the Greek letter π (**pi**) which has been determined to be an irrational number, that is, a nonterminating, nonrepeating decimal. An approximation of π more precise than 3.14 is 3.14159265. By 1989, the value of π had been calculated to more than one billion places with the aid of a computer.

The definition of π as $\dfrac{C}{d}$ allows us to calculate the circumference of a circle.

That is, because $\dfrac{C}{d} = \pi$, we can write $C = \pi d = \pi(2r) = 2\pi r$.

Postulate 3.1

Circumference of a Circle

The circumference of a circle is the product of π and the diameter of the circle.

$$C = \pi d = 2\pi r$$

EXAMPLE 3.2   Find the perimeter of each geometric figure in Figure 3.4. The figure on top of the square in Figure 3.4(b) is a semicircle.

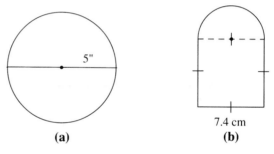

(a)   (b)

FIGURE 3.4

SOLUTION

(a) $C = 2\pi r = 2\pi(5'') = 10\pi$ inches, or approximately 31.42 inches. [NOTE: The exact answer is 10π, whereas 31.42 is a decimal approximation.]

(b) The figure consists of a square topped by a semicircle, or half of a circle. The distance around this figure is the sum of the lengths of three sides of the square plus half of the circumference of a circle, or

$$P = 3(7.4 \text{ cm}) + \tfrac{1}{2}\pi(7.4 \text{ cm}) \approx 33.82 \text{ cm}$$ •

AREA

Tessellations can be used to determine the area of the region enclosed by a simple closed curve. For convenience, we use a square tessellation, where one of the small squares is taken to be our unit square [Figure 3.5(a)]. By counting the full squares and adjusting for the partial squares, the area of the region enclosed by the simple closed curve in Figure 3.5(a) can be estimated to be about 6 square units. If smaller squares are used in the tessellation, such as four per unit square as in Figure 3.5(b), then a better estimate of the area may be possible.

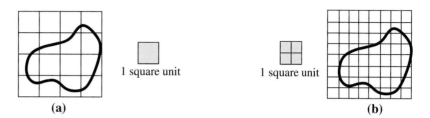

FIGURE 3.5

This discussion leads to our first area postulate.

Postulate 3.2

Area of a Region Enclosed by a Simple Closed Curve

(a) For every simple closed curve and unit square, there is a positive real number that gives the number of unit squares (and parts of unit squares) that exactly tessellate the region enclosed by the simple closed curve.

(b) The area of a region enclosed by a simple closed curve is the sum of the areas of the smaller regions into which the region can be subdivided.

The real number described in Postulate 3.2(a) is commonly called the **area** of the region enclosed by the simple closed curve. For example, if a region has area 2.3 square units, this means that the region can be covered exactly by an equivalent of 2.3 unit squares.

AREA OF A RECTANGLE

Postulate 3.2 leads to some convenient formulas when it is applied to certain polygonal regions. For example, suppose that a 3 cm by 5 cm rectangle is covered by a square tessellation where the squares are 1 cm by 1 cm (Figure 3.6).

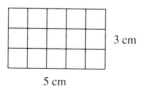

3 cm

5 cm

FIGURE 3.6

Three rows of 5 squares, or exactly 15 square centimeters, cover the rectangular region. We say that the area of the rectangular region is 15 square centimeters, which is also written as 15 cm². When the length and width of a rectangle are whole numbers, as in Figure 3.6, the area of the rectangular region (or area of the rectangle, for short) is the product of its length and its width. Moreover, this relationship holds when the length and width of the rectangle are not necessarily whole numbers, but may be any real numbers.

Definition

Area of a Rectangle

The area of a rectangle is the product of its length and its width.

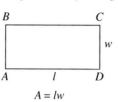

$$A = lw$$

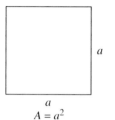

$$A = a^2$$

FIGURE 3.7

A square is a special rectangle where the length and width are the same. So, in the case of a square with side of length a, its area is $a \cdot a$, or a^2 (Figure 3.7).

AREA OF A TRIANGLE

Before finding the area of an arbitrary triangle, we will find the area of a right triangle. But first we will assume the following postulate. It essentially states that two triangles that are exact duplicates of each other have the same area.

Postulate 3.3

Any two triangles whose corresponding angles and sides are congruent have the same area.

Using Postulates 3.2 and 3.3, we can find the area of right triangle $\triangle ABC$ in Figure 3.8 by making a copy of $\triangle ABC$ and arranging the two triangles to form a quadrilateral. Because the angle sum in a triangle is 180°, it can be shown that all four angles of the quadrilateral are right angles. Hence $AC'BC$ is a rectangle (Figure 3.8).

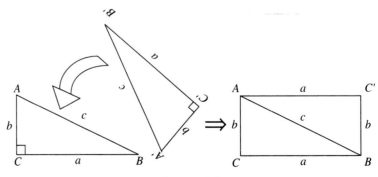

FIGURE 3.8

According to Postulates 3.2 and 3.3, because two identical right triangles make up rectangle $AC'BC$, each of their areas must be half the area of $AC'BC$. This leads to the following theorem.

Theorem 3.2

Area of a Right Triangle

The area of a right triangle is half the product of the lengths of its legs.

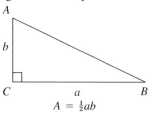

$$A = \tfrac{1}{2}ab$$

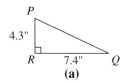

(a)

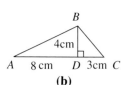

(b)

FIGURE 3.9

EXAMPLE 3.3   Determine the area of the following:

(a) $\triangle PQR$ [Figure 3.9(a)]

(b) $\triangle ABC$ where $\triangle ABD$ and $\triangle BCD$ are right triangles [Figure 3.9(b)]

SOLUTION

(a) Because $\triangle PQR$ is a right triangle, $A = \tfrac{1}{2}(7.4'')(4.3'') = 15.91$ in^2.

(b) Both $\triangle ABD$ and $\triangle BCD$ are right triangles. Thus, the area of $\triangle ABD$ is $\tfrac{1}{2}(8$ cm$)(4$ cm$)$, or 16 cm^2, and the area of $\triangle BCD$ is $\tfrac{1}{2}(3$ cm$)(4$ cm$)$, or 6 cm^2. Then, by Postulate 3.2(b), the area of $\triangle ABC$ is 16 cm^2 + 6 cm^2, or 22 cm^2.

A line segment from a vertex of a triangle that is perpendicular to the opposite side of the triangle is called an **altitude** of the triangle (Figure 3.10). The *length* of an altitude is also referred to as the altitude or **height** to a particular side. The side to which the altitude is drawn is called the **base** of the triangle associated with that altitude. So in Figure 3.9, $\overline{BD}$ is the altitude to base $\overline{AC}$. In Example 3.3(b), the sum of the two areas could have been written as $\tfrac{1}{2}(8)(4) + \tfrac{1}{2}(3)(4)$, or $\tfrac{1}{2}(8 + 3)(4)$. This suggests the next theorem whose verification is left for the problem set.

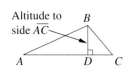

FIGURE 3.10

Theorem 3.3

Area of a Triangle

The area of a triangle is half the product of the length of one side and the height to that side.

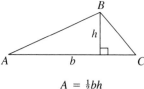

$$A = \tfrac{1}{2}bh$$

EXAMPLE 3.4   Calculate the areas of the triangles in Figure 3.11.

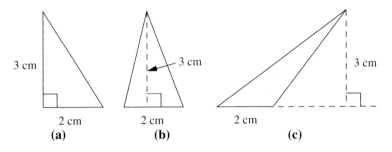

FIGURE 3.11

SOLUTION   Although the shapes of the three triangles are very different, each of the triangles has a base of 2 cm with a height of 3 cm. Thus, all three triangles will have exactly the same area, namely

$$A = \tfrac{1}{2}bh = \tfrac{1}{2}(2 \text{ cm})(3 \text{ cm}) = 3 \text{ cm}^2$$ •

It should be emphasized that any side can serve as "the base" of a triangle when the area is to be determined. The base need not be the "bottom" side of a triangle.

EXAMPLE 3.5   Calculate the area of $\triangle XYZ$ in Figure 3.12.

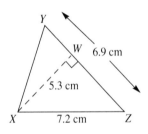

FIGURE 3.12

SOLUTION   We will use $\overline{YZ}$ as the base, where $\overline{XW}$ is the altitude to that base. Thus, $A = \tfrac{1}{2}bh = \tfrac{1}{2}(6.9 \text{ cm})(5.3 \text{ cm}) \approx 18.3 \text{ cm}^2$.   •

Any triangle has three different altitudes, or heights, one to each side (Figure 3.13). The area of a triangle does not depend on which side and corresponding height are chosen. They all yield the same area.

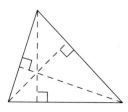

FIGURE 3.13

CONVERTING UNITS OF AREA USING DIMENSIONAL ANALYSIS

In Section 2.5 we used dimensional analysis to convert from one unit of measure to another. We can use this technique to convert from one unit of area measure to another as shown in the next example.

EXAMPLE 3.6   Convert 10 yd² to ft².

SOLUTION   First notice that we are converting square units to square units. It makes no sense to convert *square* yards to *linear* feet, for example.

One square yard, or 1 yd², can be represented as a square that measures 1 yd by 1 yd [Figure 3.14(a)]. Since 1 yd = 3 ft, we can also view the 1 yd by 1 yd square as a 3 ft by 3 ft square [Figure 3.14(b)].

Thus

$$1 \text{ yd}^2 = (1 \text{ yd})(1 \text{ yd}) = (3 \text{ ft})(3 \text{ ft}) = 9 \text{ ft}^2$$

Notice that in Figure 3.14(b), there are 9 square feet in the square yard. So, to convert 10 yd² to ft², we can write

$$10 \text{ yd}^2 = \frac{10 \text{ yd}^2}{1} \cdot \frac{3 \cdot 3 \text{ ft}^2}{1 \cdot 1 \text{ yd}^2} = \frac{10(3)(3)}{1(1)(1)}\text{ft}^2 = 90 \text{ ft}^2$$

In Example 3.6, because 1 yd² had two dimensions, we used the conversion factor 1 yd = 3 ft *twice* to make the conversion.

We can also use dimensional analysis to convert between English and metric units of area as shown in the next example.

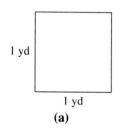

1 yd

1 yd

(a)

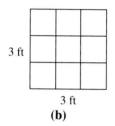

3 ft

3 ft

(b)

FIGURE 3.14

EXAMPLE 3.7   An Olympic-sized swimming pool of uniform depth measures 50 m by 21 m. The bottom of the pool is to be treated with a sealer, and one gallon of the sealer covers 300 ft² of surface. How many gallons of sealer will be required for the job? (Recall: 1 in. = 2.54 cm.)

SOLUTION   First we must determine the area of the bottom of the pool in square feet. The area can be calculated and the conversion performed at the same time as follows.

$$A = lw$$

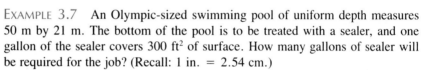

$$= \frac{50 \text{ m} \cdot 21 \text{ m}}{1} \cdot \frac{100 \cdot 100 \text{ cm}^2}{1 \cdot 1 \text{ m}^2} \cdot \frac{1 \cdot 1 \text{ in}^2}{(2.54)(2.54) \text{ cm}^2} \cdot \frac{1 \cdot 1 \text{ ft}^2}{(12)(12) \text{ in}^2}$$

$$= \frac{50(21)(100)(100)(1)(1)(1)(1)}{1(1)(1)(2.54)(2.54)(12)(12)}\text{ft}^2 \approx 11{,}302 \text{ ft}^2$$

Now we can use dimensional analysis again to determine the number of gallons of sealer needed.

$$\frac{11302 \text{ ft}^2}{1} \cdot \frac{1 \text{ gal}}{300 \text{ ft}^2} = \frac{11302 \ (1)}{1(300)} \text{ gal} \approx 38 \text{ gal}$$

Therefore, approximately 38 gallons of sealer would be needed.

Solution to Applied Problem

The total area of the room is $18'(22') = 396$ ft². The areas that will not be carpeted consist of (i) the entry: $3'(4') = 12$ ft², and (ii) the hearth, which has been divided here into two rectangles and two triangles:

$$3'(4') + 2'(1') + 2(\tfrac{1}{2})(1')(1') = 15 \text{ ft}^2$$

Thus, the area to be carpeted is $396 \text{ ft}^2 - 12 \text{ ft}^2 - 27 \text{ ft}^2 = 369 \text{ ft}^2$. Taking into account the additional 5%, $369 \text{ ft}^2 + (0.05)(369 \text{ ft}^2) = 387.45 \text{ ft}^2$ must be ordered. The total cost at \$22.50 per square yard is

$$\frac{387.45 \text{ ft}^2}{1} \cdot \frac{1 \text{ yd}^2}{9 \text{ ft}^2} \cdot \frac{\$22.50}{1 \text{ yd}^2} = \frac{(387.45)(1)(\$22.50)}{1(9)(1)} \approx \$968.63$$

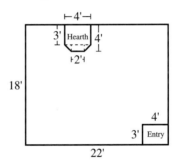

GEOMETRY AROUND US

On June 14, 1991, in its first public display, the world's largest flag, a 411-foot-long, 210-foot-wide, seven-ton version of Old Glory, was displayed on the Mall in Washington, DC. The flag is big enough that two space shuttles could be wrapped in it, 250 bushels of corn could be harvested from under it, and a football field and most of the bleachers could be covered by it.

PROBLEM SET 3.1

EXERCISES/PROBLEMS

1. By counting squares and parts of squares, find the area of each figure shown on the following square lattices. Express your answer in square units, where 1 square unit is the region shown.

☐ = 1 square unit

2. By counting squares and parts of squares, find the area of each figure shown on the following square lattices.

☐ = 1 square unit

(a)   (b)   (c)

(a)   (b)   (c)

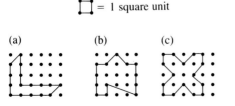

3. On the square lattice shown sketch examples of six different triangles each of which has an area of 2 square units.

4. The area of the rectangle *ABCD* is 40 in². What is the area of the unshaded region? Explain.

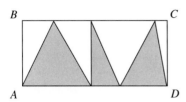

5. In Section 1.2 toothpicks were arranged to form rows of squares and equilateral triangles. If one toothpick represents one unit, determine the perimeter if *n* polygons of the following types were arranged in a row.

(a) *n* equilateral triangles:

(b) *n* squares:

(c) *n* pentagons:

(d) *n* hexagons:

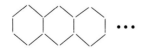

(e) *n* octagons:

(f) *n* *m*-sided polygons

6. Suppose that a large square was formed by arranging toothpicks as shown.

(a) Find the perimeter and area of this square.

(b) Suppose now that the toothpicks in the corners are rearranged to form a small square one toothpick per side in every corner. Find the perimeter and area for this new figure.

(c) If the toothpicks are rearranged again to form squares with two toothpicks on a side in each corner, find the perimeter and area of the resulting figure.

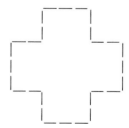

(d) If the squares in the corners have 3 toothpicks per side, find the perimeter and area of the figure.

(e) What pattern do you observe? Suppose the original square had *m* toothpicks on a side and the small squares in each corner had *n* toothpicks on a side. Write a formula for the perimeter and a formula for the area of the figure.

7. Find the area and perimeter of each rectangle.

(a)   (b)

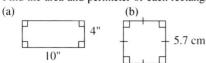

4"   5.7 cm

10"

8. Find the area and perimeter of each rectangle.

(a)   (b)

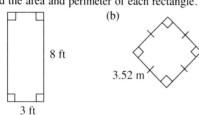

8 ft

3.52 m

3 ft

9. Determine the perimeter and area of each of the following right triangles.

(a)

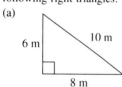

6 m   10 m

8 m

(b)

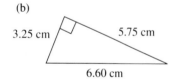

3.25 cm   5.75 cm

6.60 cm

(c)

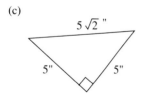

5 √2 "

5"   5"

10. Determine the perimeter and area of each of the following triangles.

(a)

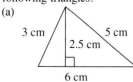

3 cm   5 cm

2.5 cm

6 cm

(b)

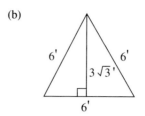

6'   6'

3√3'

6'

(c)

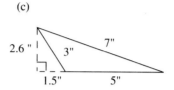

2.6 "   7"

3"

1.5"   5"

11. Calculate the circumference of each circle shown. Round your answers to two decimal places.

(a)   (b)

4"   10.8 m

12. For each circle, the circumference is given. Find the radius and diameter for each circle. Round to two decimal places for approximate answers.

(a)   (b)

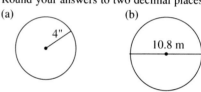

$C = 12\pi$ ft $C = 15.87$ cm

$d = ?$ $d = ?$

$r = ?$ $r = ?$

13. Determine the perimeter of each figure shown.

(a)   (b)   11.2 m

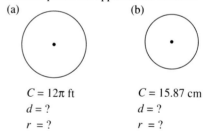

8.7 m

7"   4"

24.6 m

14. Determine the perimeter of each figure shown.

(a)   (b) 16"   9"

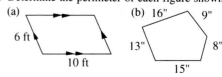

6 ft

10 ft

13"   8"

15"

15. Determine the perimeter of each regular polygon shown.

(a)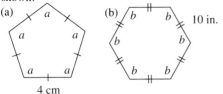

4 cm

(b)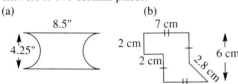

10 in.

(c) Write a formula for the perimeter of a regular polygon with n sides, each of length s.

16. Determine the perimeter of each figure shown. The ends of the figure in (a) are semicircles. Round your answers to two decimal places.

(a)

8.5"

4.25"

(b)

7 cm

2 cm

2 cm

2.8 cm

6 cm

17. Use dimensional analysis to perform each of the following conversions. Where answers are approximate, round to two decimal places.
(a) 50 in² to ft²
(b) 2,000 m² to cm²
(c) 3.5 yd² to m²

18. Use dimensional analysis to perform each of the following conversions. Where answers are approximate, round to two decimal places.
(a) 0.25 mi² to ft²
(b) 400 m² to km²
(c) 248 cm² to yd²

19. Find the area of the rectangle shown in (a) square feet and (b) square meters. Round your answers to two decimal places.

4'8"

8'3"

20. Find the area of the triangle shown in (a) square meters and (b) square inches. Round your answers to two decimal places.

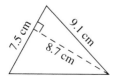

7.5 cm

9.1 cm

8.7 cm

21. The length of a rectangle is 6 inches less than twice its width. If the perimeter of the rectangle is 72 inches, find the dimensions of the rectangle.

22. The width of a rectangle is 18 inches less than its length. If the area of the rectangle is 1440 in², find the dimensions of the rectangle.

23. What happens to the perimeter of a rectangle if its length and width are both multiplied by 2? by 3? by 10? by n?

24. What happens to the area of a rectangle if its length and width are both multiplied by 2? by 3? by 10? by n?

25. Suppose that two rectangles both have a perimeter of 40 m. Must they also have the same area? Try several examples to test your conjecture.

26. Suppose that two rectangles both have areas of 144 cm². Must they also have the same perimeter? Try several examples to test your conjecture.

27. For the following triangle, measure the length of each side in millimeters and then measure the height to each of those sides. Record your results in a table like the one shown. Use each pair to calculate the area of the triangle. Did you get approximately the same area in each case?

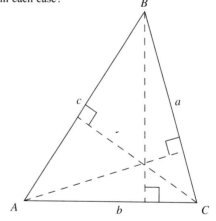

	side	height	area
$a =$			
$b =$			
$c =$			

28. For the following triangle, measure the length of each side in millimeters and then measure the height to each of those sides. Record your results in a table such as the one shown. Use each pair to calculate the area of the triangle. Did you get approximately the same area in each case?

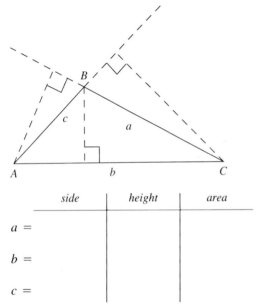

side	height	area
$a =$		
$b =$		
$c =$		

29. Recall that a tetromino is a geometric shape composed of 4 connected squares, where two squares are joined along an entire side. (See problem 2 in Section 1.1.)
(a) Which tetrominos have the maximum perimeter?
(b) Which tetrominos have the minimum perimeter?

30. A pentomino is a geometric shape composed of 5 connected squares, where two squares are joined along an entire side. (See problem 3 in Section 1.1.)
(a) Which pentominos have the maximum perimeter?
(b) Which pentominos have the minimum perimeter?
(c) Generalize your results from problems 29, 30(a), and 30(b) for a shape formed of 8 squares.

31. Imagine that a cable is wrapped tightly around the Earth at the equator. Then suppose that the cable is cut, an additional 30 ft are added to the cable, and that this longer cable is lifted so that it is equidistant from the equator around the Earth. Because the cable is now a little longer than the circumference of the Earth, there will be a gap between the cable and the surface of the Earth. How large is the gap? Would you be able to crawl under the cable? Walk under it? [NOTE: The equatorial diameter of the Earth is about 7926 mi.]

32. The two coins shown have radii of 3 cm and 1 cm.

The smaller coin rolls along the circumference of the fixed larger coin until the smaller coin returns to its original position. How many revolutions has the small coin made?

33. Plutarch (circa A.D. 100) pointed out that only two rectangles with whole-number dimensions that have perimeters numerically equal to their areas are possible. Find the length and width of each of these two rectangles.

34. Heron of Alexandria, an early Greek mathematician, provided another method for finding the area of a triangle. Heron's formula can be used to find the area of a triangle if the lengths of the three sides are known. **Heron's Formula** gives the area of a triangle as $A = \sqrt{s(s - a)(s - b)(s - c)}$, where a, b, and c are the lengths of the three sides and s is half the perimeter, namely $s = \dfrac{a + b + c}{2}$. The number s is called the **semiperimeter** of the triangle. Use this formula to find the areas of triangles with the following side lengths. For answers that are not exact, round to two decimal places.
(a) 5 cm, 12 cm, 13 cm
(b) 4 m, 5 m, 6 m
(c) 4 ft, 5 ft, 8 ft

35. Use the method in Example 3.3(b) to verify Theorem 3.2 by answering the following questions about $\triangle ABC$.

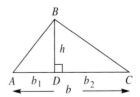

(a) In △ABC, b is the length of $\overline{AC}$ and h is the height to that side. If $AD = b_1$ and $DC = b_2$, what is the area of right triangle △ABD? (Use Theorem 3.2.)

(b) What is the area of right triangle △CBD?

(c) Express the area of △ABC as the sum of the areas of △ABD and △CBD.

(d) Simplify your answer in (c) by factoring out common factors. What does the sum $b_1 + b_2$ represent? Substitute that expression for $b_1 + b_2$ in your answer.

(e) Using your answer in (d), what is the area of △ABC? Does this result agree with Theorem 3.3?

APPLICATIONS

36. One of the world's largest synchrotrons, or atom smashers, is at the Fermi National Accelerator Laboratory in Illinois. The accelerator is in a circular underground tunnel 2 km in diameter. Find the distance that atomic particles travel in one pass through the tunnel. Round your answer to the nearest meter.

37. You may have heard that a basketball hoop is large enough for two basketballs to go through the ring side by side. The inside diameter of a basketball rim is 18″. The circumference of a basketball must be no less than 29.5″ and no greater than 30″. Is the hoop wide enough to accommodate two basketballs at the same time?

38. According to the directions on a 20-lb bag of fertilizer/weed killer, one bag will cover 5000 ft² of lawn. If a house, driveway, and lot have dimensions as shown in the following diagram, will one bag be enough? Assume that all of the *unshaded* region is to be fertilized.

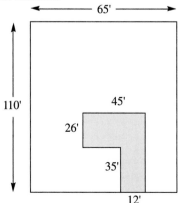

39. The floor plan of a room in a house is shown. The room will be carpeted, and quarter-round molding will be laid around the edges of the room.

(a) Determine the length of quarter-round that will be required. Allow for the 7′ patio doors and the two 3′ wide doors in the figure. Round your answer to the nearest foot.

(b) Determine the number of square yards of carpet which will be needed. Round your answer to the nearest square yard.

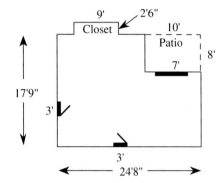

40. A mason is laying tile on three walls surrounding a shower. Two of the walls to be tiled are 4 feet wide, and one wall is 3 feet wide. If tile will extend 7 feet up each wall, approximately how many 4″ by 4″ tiles will be needed for the job?

41. A dining room measures 12′10″ by 18′6″ and has 9′ ceilings. One long and one short wall of the room will be papered. Each wall to be papered has a standard 3′ by 5′ window. If the wallpaper costs $15.99/roll and one roll of wallpaper has a useable yield of 32 ft², how many rolls of wallpaper will be required? How much will paper for the room cost? [NOTE: Only whole rolls of wallpaper can be purchased.]

42. Calculate the length of the circular orbit of the space shuttle if it travels 350 miles above the surface of the Earth. How many times larger is the orbit of the shuttle than the circumference of the Earth? (Use 7920 miles as the approximate diameter of the Earth.)

43. A photographer plans to use a 10-ft strip of molding to make square frames for two photographs. The side of one frame will be three times as long as a side of the smaller frame. If the entire 10-ft strip of molding will be used and 10% of the strip is allowed for cuts and waste, how large can the two picture frames be?

44. There are about 5 billion people on Earth. If they all lined up and joined hands, how many times would the line of people wrap around the equator? Assume that each person takes up about 2 yards of space with arms extended and that the radius of the Earth is about 3960 miles.

45. A car tire has a diameter of 22 in. When the car is travelling at 30 mph, how many revolutions per minute is the tire turning? Round your answer to the nearest tenth of a revolution. [HINT: Dimensional analysis may be helpful here.]

46. A coil of cable consists of 42 loops. The inside diameter of the coiled cable is 16″, and the outside diameter is 30″. The total length of the cable can be calculated using the average circumference as follows:

average circumference = π(average diameter)
length = (average circumference)
× (number of loops)

Determine the total length of the cable, in feet.

47. A track is designed as shown, with semicircular ends and inner dimensions as given.

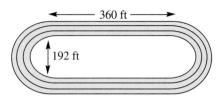

(a) Calculate the distance (to the nearest hundredth of a foot) covered in one lap around the innermost lane. Then express your answer in miles, rounded to two decimal places.

(b) Suppose that each lane is 3 feet wide. How much further does a runner travel in one lap around the second lane than in a lap around the first lane (innermost lane)? Round your answer to the nearest hundredth of a foot.

(c) How much further does a runner travel in one lap around the track in the third lane than in a lap around the second lane? Round your answer to the nearest hundredth of a foot.

(d) If a race were to be run and starting points in each lane staggered so that all runners would travel exactly the same distance, how much of a head start should runners in the outer lanes be given?

3.2
MORE AREA FORMULAS

Applied Problem

A single 4″ water pipe is to be replaced by smaller 2″ pipes. How many 2″ pipes will be required to carry at least as much water as the 4″ pipe?

AREA OF A PARALLELOGRAM

Now that we can use Theorem 3.3 to find the area of any triangle, we can use Postulate 3.2 to find the areas of other common polygons.

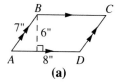

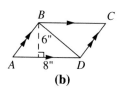

FIGURE 3.15

EXAMPLE 3.8   Find the area of the parallelogram in Figure 3.15(a).

SOLUTION   If diagonal $\overline{BD}$ is drawn [Figure 3.15(b)], two triangles having the same size and shape are formed. (We prove this in Chapter 5.) Thus the area of $ABCD$ is twice the area of $\triangle ABD$. Because the height of $\triangle ABD$ is 6″ and the base has measure 8″, its area is $\frac{1}{2}(6″)(8″) = 24$ in². Because $\triangle ABD$ and $\triangle CDB$ are identical triangles, each with area 24 in², by Postulate 3.2 the area of the parallelogram is $2(24) = 48$ in².   •

Notice that we could have found the area in Example 3.8 by multiplying 6″ (the height of one of the triangles) by 8″ (the length of the base of one of the triangles).

In Chapter 5 we prove that the perpendicular distance between two parallel lines is a constant. In the case of a parallelogram, this distance is called the **height** with respect to these parallel sides. As is the case with triangles, a height is associated with a particular side, called the corresponding **base** of the parallelogram.

Example 3.8 leads to the following general result. The verification is left for the problem set.

Theorem 3.4

Area of a Parallelogram

The area of a parallelogram is equal to the product of the length of one of its sides and the height to that side.

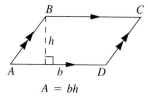

$$A = bh$$

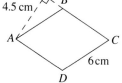

FIGURE 3.16

In Chapter 5 we will prove that a rhombus is a special type of parallelogram. Thus, the area of a rhombus can be found using the formula in Theorem 3.4.

EXAMPLE 3.9   Find the area of the rhombus $ABCD$ in Figure 3.16.

SOLUTION   Any side of the rhombus may be used as the base. We will use $\overline{BC}$ since the height to that side is given. A rhombus has four congruent sides, so $BC = CD = 6$ cm. Thus, the area of $ABCD$ is given by

$$A = bh = (6 \text{ cm})(4.5 \text{ cm}) = 27 \text{ cm}^2 \qquad •$$

AREA OF A TRAPEZOID

The same technique used to derive the formula for the area of a parallelogram can be used to find a formula for the area of a trapezoid.

EXAMPLE 3.10   Find the area of the trapezoid in Figure 3.17(a).

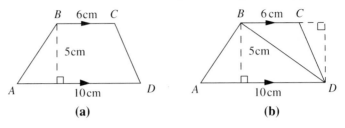

FIGURE 3.17

SOLUTION   If diagonal $\overline{BD}$ is drawn [Figure 3.17(b)], two triangles are formed. Thus, the area of *ABCD* is the sum of the areas of the two triangles. Because the height of both triangles is 5 cm and the bases have lengths 10 cm and 6 cm, the area of the trapezoid is

$$\tfrac{1}{2}(10 \text{ cm})(5 \text{ cm}) + \tfrac{1}{2}(6 \text{ cm})(5 \text{ cm}) = 25 \text{ cm}^2 + 15 \text{ cm}^2 = 40 \text{ cm}^2 \qquad •$$

Notice that $\tfrac{1}{2}(10)(5) + \tfrac{1}{2}(5)(6)$ can be rewritten as $\tfrac{1}{2}(5)(10 + 6)$. More generally, we have the following result whose verification is left for the problem set.

Theorem 3.5

Area of a Trapezoid

The area of a trapezoid is equal to half the product of the height and the sum of the lengths of its two bases.

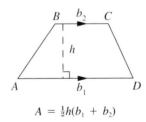

$$A = \tfrac{1}{2}h(b_1 + b_2)$$

EXAMPLE 3.11   Find the area of the trapezoid in Figure 3.18(a) in two different ways.

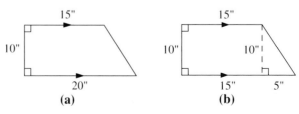

FIGURE 3.18

SOLUTION

(i) Using Theorem 3.4, we have

$$A = \tfrac{1}{2}h(b_1 + b_2) = \tfrac{1}{2}(10'')(15'' + 20'') = 175 \text{ in}^2$$

(ii) Subdividing the trapezoid into a rectangle and a right triangle as in Figure 3.18(b), we have

$$A = 15''(10'') + \tfrac{1}{2}(5'')(10'') = 150 \text{ in}^2 + 25 \text{ in}^2 = 175 \text{ in}^2 \qquad \bullet$$

AREA OF A REGULAR POLYGON

It can be shown that there is a point in the plane that is equidistant from all vertices of a regular polygon and from all the sides of the polygon. This point is called the **center** of the polygon. In Figure 3.19(a), point O is the center of the regular hexagon.

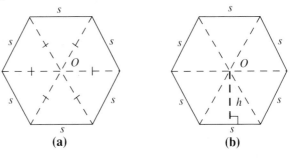

(a)   (b)

FIGURE 3.19

The area of any regular polygon can be found by conveniently subdividing the polygon into triangles as suggested by the regular hexagon in Figure 3.19(a). The area of each of the six triangles is $\tfrac{1}{2}sh$, where h is the perpendicular distance from the center of the polygon to any side [Figure 3.19(b)]. Thus, the area of the hexagon is $6(\tfrac{1}{2}sh)$. This equation can be rewritten as $\tfrac{1}{2}h(6s)$, which is equivalent to $\tfrac{1}{2}hP$, where $P = 6s$ is the perimeter of the hexagon. This technique can be generalized to any *regular* polygon. This is the content of the next theorem.

Theorem 3.6

Area of a Regular Polygon

The area of a regular polygon is equal to half the product of the perimeter of the polygon and the perpendicular distance from its center to one of its sides.

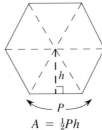

$$A = \tfrac{1}{2}Ph$$

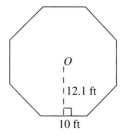

O

12.1 ft

10 ft

FIGURE 3.20

EXAMPLE 3.12   A bandstand in a city park is to be constructed in the shape of a regular octagon (Figure 3.20). To estimate the lumber needed, the area of the platform must be calculated. Using the dimensions given in the figure where point *O* is the center, determine the area of the octagon.

SOLUTION   Using Theorem 3.6, we have

$$A = \tfrac{1}{2}Ph = \tfrac{1}{2}(8 \cdot 10 \text{ ft})(12.1 \text{ ft}) = 484 \text{ ft}^2$$

AREA OF A CIRCLE

Now imagine a series of regular polygons where the number of sides increases to approach a circle (Figure 3.21).

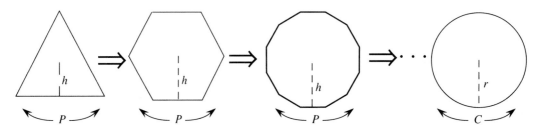

FIGURE 3.21

The h in Theorem 3.6 becomes the radius, r, of the circle and the perimeter P becomes the circumference C of the circle. Thus, by substituting r for h and C for P in the area formula in Theorem 3.6, the area of a circle of radius r is equal to $\tfrac{1}{2}rC = \tfrac{1}{2}r(2\pi r) = \pi r^2$.

Theorem 3.7

Area of a Circle

The area of a circle is equal to the product of π and the square of the radius.

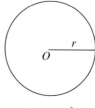

$$A = \pi r^2$$

EXAMPLE 3.13   Figure 3.22 is composed of one large and four small semicircles. Find the area enclosed by these five semicircles.

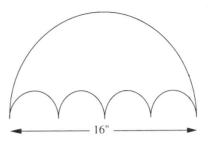

FIGURE 3.22

SOLUTION   The diameter D of the larger semicircle is 16 inches, so its radius R is 8 inches. The diameter d of each small semicircle is $16/4 = 4$ inches, so the radius r is 2 inches. The area of the large semicircle is

$$A_L = \tfrac{1}{2}\pi R^2 = \tfrac{1}{2}\pi(8'')^2 = 32\pi \text{ in}^2$$

The area cut out by the four small semicircles is

$$A_S = 4 \cdot \tfrac{1}{2}\pi(r)^2 = 2\pi(r)^2 = 2\pi(2'')^2 = 8\pi \text{ in}^2$$

Thus, the total area is given by

$$A_{\text{Total}} = A_L - A_S = 32\pi \text{ in}^2 - 8\pi \text{ in}^2 = 24\pi \text{ in}^2$$

The exact area of the shaded region is 24π in². For many practical applications, a decimal approximation is more appropriate. In this case,

$$A_{\text{Total}} = 24\pi \text{ in}^2 \approx 75.4 \text{ in}^2 \qquad \text{(to one decimal place)} \qquad \bullet$$

Solution to Applied Problem

To replace a 4″ pipe with 2″ pipes, the total cross-sectional area of the 2″ pipes must be at least as great as the cross-sectional area of the 4″ pipe. A 4″ pipe has cross-sectional area of $\pi(2^2) = 4\pi$ in². A 2″ pipe has cross-sectional area $\pi(1^2) = \pi$ in². Hence, four 2″ pipes will have the same cross-sectional area as one 4″ pipe.

4 in.

GEOMETRY AROUND US

CARS PRODUCED
PER YEAR

United States
7,800,000

Germany
3,800,000

The pictograph to the left compares the number of cars produced in a year by two countries. This graph is deceptive, however. Using the numbers given, the production of cars in the United States appears to be about twice that of Germany. However, the car symbol for the United States appears to be about *four* times as large. The reason is that both the length and the height of the car were doubled which results in the area being quadrupled.

To present an accurate two-dimensional comparison using the car symbols where one represents n times as many cars as the other, either of the following is correct: (a) One of the dimensions of the first car should be displayed as n times the corresponding dimension of the second car and the other dimensions of the two cars should be exactly the same, or (b) both dimensions of the first car should be $\sqrt{n}$ times those of the second car.

PROBLEM SET 3.2

EXERCISES/PROBLEMS

1. Find the area of each figure.

(a)
7 cm
4 cm
5 cm
5 cm
7 cm

(b)
5" 5"
5"
5" 4"

(c)
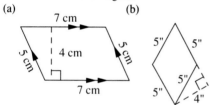
3.1 cm
4.5 cm
5.6 cm
6.4 cm
8 cm
4.8 cm

(d)
10"

(c)
$6\sqrt{2}$ cm

(d)
11.5 mm
14.6 mm
18 mm

2. Find the area of each figure.

(a)
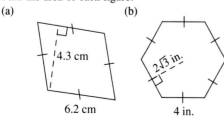
4.3 cm
6.2 cm

(b)
$2\sqrt{3}$ in.
4 in.

3. Complete the following table, where information is given about a circle with radius r, diameter d, circumference C, and area A. Round your answers to two decimal places.

	r	d	C	A
(a)	12			
(b)			45π	
(c)		$\sqrt{6}$		
(d)				92.8

4. Complete the following table, where information is given about a parallelogram with base b, height h, and area A. Round your answers to two decimal places.

	b	h	A
(a)		6	40
(b)	$3\sqrt{2}$	$2\sqrt{3}$	
(c)	13.6		74.9

5. For the parallelogram shown, measure the length of each side in millimeters. Then measure the height to each side. Record your results in a table like the one shown. Use each pair of measurements to calculate the area of the parallelogram. Were your results equal or nearly equal?

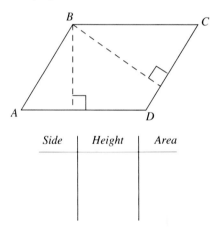

Side	Height	Area

6. Apply the directions for problem 5 to the following parallelogram.

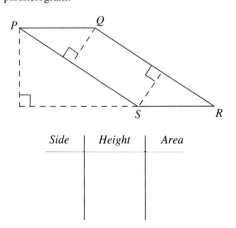

Side	Height	Area

7. Find the total area of each of the figures shown. $\triangle ABC$ is equilateral and all curves are semicircles. Round your answers to two decimal places.

(a)   (b)

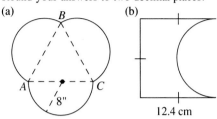

12.4 cm

8. Find the total area of each of the figures shown. In (b), the two regular octagons have the same center, $OP = 5.3$ cm, and $PQ = 1$ cm.

(a)

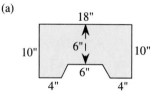

(b)

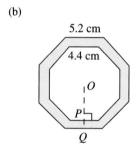

9. The area of a trapezoid is 69 in². Its bases have lengths 9 in. and 14 in. Find the height of the trapezoid.

10. A trapezoid has a height of 5 cm and one base which measures 8 cm. If the area of the trapezoid is 28 cm², find the length of the other base.

11. If the radius of a circle is doubled, what is the area of the resulting circle? If the radius is multiplied by 5, what is the new area? What happens if the radius is multiplied by n?

12. The base and height of a parallelogram are both multiplied by 3. What is the area of the resulting parallelogram? If both the base and the height are multiplied by 7, what is the new area? What happens if the base is multiplied by m and the height is multiplied by n?

13. Determine the area of the shaded region in each of the following figures. Round your answers to two decimal places.

(a)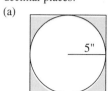

(b)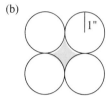

14. Find the area of the shaded machined piece shown, which is made by cutting a small semicircle out of a larger semicircle. The radius of the larger semicircle is twice the radius of the smaller. Round your answer to the nearest square centimeter.

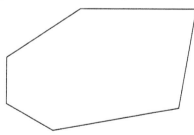

10.5 cm

15. Determine the approximate area of the figure shown by measuring appropriate dimensions in millimeters. Round your answer to the nearest square millimeter. [HINT: Try subdividing the shape into smaller regions. Heron's formula may be useful.]

16. Determine the approximate area of the figure shown by measuring appropriate dimensions in millimeters. Round your answer to the nearest square millimeter.

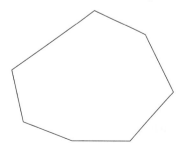

17. In the following figure, each petal inside the square is formed by the intersection of two semicircles. Find the area of the shaded region. Give the *exact* answer.

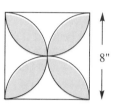

8"

18. The circles in the figure have radii of 6", 4", 4", and 2". Which is larger, the shaded area inside the large circle or the shaded area outside the large circle?

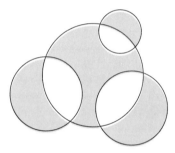

19. Which has the larger area, a square with a perimeter of 20 inches or a circle with a circumference of 20 inches?

20. Concentric circles are circles that have the same center. Two concentric circles have radii of 10 cm and 14 cm.
(a) Which is larger, the area inside the smaller circle or the area between the two circles?
(b) What would the radius of the larger circle have to be in order for the area of the inner circle to equal the area between the two circles?

21. Review Example 3.8. Use the method of that example to verify Theorem 3.4 by answering the following questions about parallelogram *ABCD*.

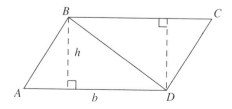

(a) In *ABCD*, *b* is the length of base $\overline{AD}$ and *h* is the height to that base. What is the area of △*ABD*?

(b) How does △*ABD* compare with △*CDB*? What is the area of △*ABD*? of △*CDB*?

(c) Express the area of parallelogram *ABCD* in terms of the areas of △*ABD* and △*CDB*.

(d) Considering your answer in (c), how can you find the area of a parallelogram? Does your result agree with Theorem 3.3?

22. Review Example 3.10. Use the method of that example to verify Theorem 3.4 by answering the following questions about trapezoid *ABCD*.

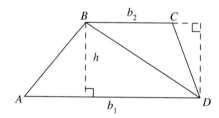

(a) In *ABCD*, b_1 and b_2 are the lengths of its bases, and *h* is its height. What is the area of △*ABD*?

(b) What is the height of △*BCD*? What is the area of △*BCD*?

(c) Express the area of trapezoid *ABCD* in terms of the areas of △*ABD* and △*BCD*.

(d) Simplify your answer in (c) by factoring out common factors.

(e) Considering your answer in (d), how can you find the area of a trapezoid? Does your answer agree with Theorem 3.4?

APPLICATIONS

23. A take-and-bake pizza store offers three sizes of circular pizzas: small (10″ diameter), medium (12″ diameter), and large (16″ diameter). A special combination pizza comes "with everything" and costs $6.25 for a small, $9.49 for a medium, and $13.99 for a large size. Which size pizza is the best buy?

24. A circular table has a diameter of 48″. The table can be enlarged by inserting two rectangular leaves, each

18″ wide and 48″ long. What is the area of the expanded table? Round your answer to the nearest tenth of a square *foot*.

25. A soybean farmer uses an irrigation system in which 8 sections of 165-ft pipe are connected end to end, and the resulting length of sprinkler pipe travels in a circle around a well as shown in the next figure.

(a) How many acres of the square field are irrigated? (One **acre** is 43,560 ft².)

(b) Wheat is usually planted in the corners of the field, which are not irrigated. How many acres of wheat are planted in this field?

26. Two 1-inch pipes empty into one larger pipe. What is the minimum diameter of the larger pipe that can accommodate the two smaller pipes?

27. A rectangular metal plate 8.4″ by 10.2″ has four holes drilled in it. Two holes have a diameter of $\frac{3}{4}$″ and two holes have a diameter of $\frac{1}{2}$″. If the plate material weighs 7.5 lbs/ft², find the weight of the plate. Round your answer to the nearest tenth of an *ounce*.

28. An indoor running track has the dimensions shown. The ends of the track are semicircles. If the track is to be resurfaced, how many square feet of material will be required? Round your answer to the nearest square foot.

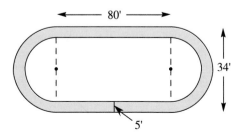

29. A goat is tied to one corner of a square pen 20 ft on a side.
   (a) If the goat's tether is 10 ft long, what percentage of the grass in the pen can the goat reach?
   (b) What should the length of the tether be so that the goat can reach half of the grass? Round your answer to the nearest tenth of a foot.

30. An archery target consists of 5 concentric circles as shown.

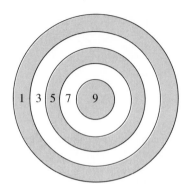

Points for each region on the target are indicated. The radius of the bull's-eye is 4.5″, and each ring surrounding the bull's-eye is 4.625″ wide.
   (a) What is the area of the bull's-eye?
   (b) What is the area of the 7-point region?
   (c) What is the area of the 5-point region?
   (d) What is the area of the entire target?
   (e) The probability of hitting the bull's-eye on a random shot that hits the target is

$$P(\text{bull's-eye}) = \frac{\text{area of bull's-eye}}{\text{area of entire target}}$$

   Calculate the probability of hitting the bull's-eye. Express your answer as a percent. This gives the percentage of the time in which a shot that hits the target at random would be expected to hit the bull's-eye.
   (f) What is the probability that a random shot hits the 7-point region? The 5-point region? The 3-point region? The 1-point region?
   (g) Based on your calculations in part (f), determine if the points assigned to the bull's-eye and rings are fair.

31. A paraglider wants to land in the unshaded region in the square field illustrated because the shaded regions

(four quarter circles) are briar patches. If the paraglider hits the field randomly due to unexpected wind currents, what is the probability that she misses the briar patches? Use the fact that the probability of landing in the unshaded region is the area of the unshaded region divided by the total area of the field.

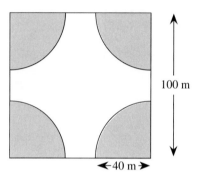

32. A clock face dart board is shown. Suppose that the dart board is 16 inches in diameter and has other dimensions as follows:

   no-score ring: 1 inch wide
   double-score ring: 0.5 inch wide
   triple-score ring: 0.5 inch wide
   25-point ring: 0.25 inch wide
   50-point ring (bull's-eye): 1 inch in diameter

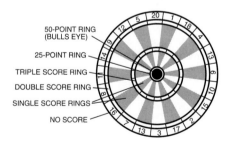

   (a) What is the area of the entire dart board?
   (b) What is the area of the no-score ring?
   (c) What is the area of the double-score ring?
   (d) What is the probability that a dart hitting the target at random lands in the no-score ring?
   (e) What is the probability that a dart hitting the target at random lands in the double-score ring?

3.3
THE PYTHAGOREAN THEOREM AND RIGHT TRIANGLES

Applied Problem

A building has the dimensions shown. If the rafters have a 2-ft overhang, determine the length of a rafter and the number of squares of shingles required to roof the building. [NOTE: A square of shingles covers 100 ft².]

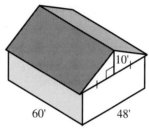

THE PYTHAGOREAN THEOREM

The Pythagorean Theorem, which relates the lengths of the legs and the hypotenuse of a right triangle, is perhaps the most famous of all theorems in geometry. It is named for the Greek mathematician, Pythagoras (circa 500 B.C.), who presented a proof of the relationship. However, there is no question that the results of the theorem were known and used for hundreds of years before his time.

Theorem 3.8

Pythagorean Theorem

The sum of the squares of the lengths of the legs of a right triangle is equal to the square of the length of the hypotenuse.

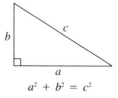

$$a^2 + b^2 = c^2$$

Theorem 3.8 can be verified using areas of triangles and squares as shown next.

EXAMPLE 3.14   Show that the Pythagorean Theorem holds for right triangle $\triangle ABC$ [Figure 3.23(a)].

SOLUTION   Arrange four copies of $\triangle ABC$ to form a large square as shown in Figure 3.23(b).

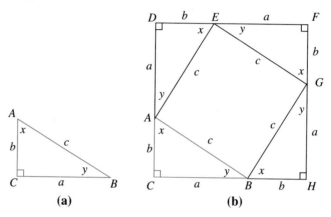

FIGURE 3.23

Because $\triangle ABC$ has a right angle at C, $x + y = 90°$. Because $\angle CBH$ is a straight angle and $x + y = 90°$, we can conclude that $\angle ABG$ is a right angle. Similarly, $\angle BGE$, $\angle AEG$, and $\angle EAB$ are right angles. Thus, $AEGB$ is a square.

Now we will compute (a) the area of square $CDFH$, and (b) the area of square $AEGB$ together with the areas of the four triangles:

$$\text{area of } CDFH = (a + b)^2 = a^2 + 2ab + b^2$$
$$\text{area of } AEGB + 4(\text{area } \triangle ABC) = c^2 + 4(\tfrac{1}{2}ab) = c^2 + 2ab$$

The area of the larger square must be the same as the area of the smaller square plus the areas of the four right triangles. Therefore, $a^2 + 2ab + b^2 = c^2 + 2ab$. Subtracting $2ab$ from both sides, we have $a^2 + b^2 = c^2$.   •

Many geometric figures as well as real-life designs and constructions include right angles, providing opportunities to apply the Pythagorean Theorem.

EXAMPLE 3.15   Find the lengths of the missing sides in each triangle in Figure 3.24.

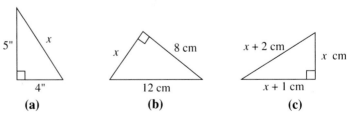

FIGURE 3.24

SOLUTION

(a) In the triangle in Figure 3.24(a), the hypotenuse is unknown. However, by the Pythagorean Theorem, we have

$$4^2 + 5^2 = x^2$$
$$16 + 25 = x^2$$
$$41 = x^2$$
$$x = \sqrt{41} \text{ in.} \qquad \text{(exact answer), or}$$
$$x \approx 6.4 \text{ in.}$$

[NOTE: Because the length of a side cannot be negative, we chose the positive square root of 41.]

(b) In the triangle in Figure 3.24(b), one of the legs is unknown. So, we have

$$x^2 + 8^2 = 12^2$$
$$x^2 + 64 = 144$$
$$x^2 = 80$$
$$x = \sqrt{80} \text{ cm} = 4\sqrt{5} \text{ cm} \qquad \text{(exact answer)}$$
$$x \approx 8.9 \text{ cm} \qquad \text{(to the nearest tenth)}$$

(c) In the triangle in Figure 3.24(c), none of the lengths of the sides is known but a relationship between them is given. To find all three sides, we will use the Pythagorean Theorem.

$$x^2 + (x + 1)^2 = (x + 2)^2$$
$$x^2 + x^2 + 2x + 1 = x^2 + 4x + 4$$
$$x^2 - 2x - 3 = 0$$
$$(x - 3)(x + 1) = 0$$
$$x = 3 \text{ cm} \qquad \text{or} \qquad x = -1 \text{ cm}$$

Because the length of a side cannot be negative, x must be 3 cm. Hence, $x + 1 = 4$ cm and $x + 2 = 5$ cm.   •

The Pythagorean Theorem can be used to calculate areas of geometric figures as illustrated by the next example.

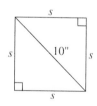

FIGURE 3.25

EXAMPLE 3.16   Find the area of a square that has a diagonal 10 inches long (Figure 3.25).

SOLUTION   The diagonal and the sides of the square form two right triangles (Figure 3.25). To find the length of a side of the square, we will use the Pythagorean Theorem.

$$s^2 + s^2 = 10^2$$
$$2s^2 = 100$$
$$s^2 = 50$$

Hence, the area of the square is $A = s^2 = 50$ in^2.   •

TEST FOR RIGHT TRIANGLES

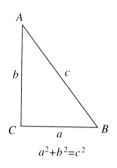

$a^2+b^2=c^2$

FIGURE 3.26

The Pythagorean Theorem states that in a right triangle with legs of lengths a and b and hypotenuse of length c we have $a^2 + b^2 = c^2$. On the other hand, sometimes we need to determine whether a triangle is in fact a right triangle. There is a test based on the Pythagorean Theorem that indicates whether or not a triangle contains a right angle. This test depends on the relationship among the three sides of a triangle. If $\triangle ABC$ has sides of length a, b, and c as in Figure 3.26 and if $a^2 + b^2 = c^2$, then $\triangle ABC$ has a right angle at C. If $a^2 + b^2 \neq c^2$, then $\triangle ABC$ is *not* a right triangle. A verification of this test will be provided in Chapter 4.

EXAMPLE 3.17   Lengths of sides of two triangles are given. Determine if either triangle is a right triangle.
(a) 11, 15, 19   (b) 7, 16.8, 18.2

SOLUTION
(a) Let $a = 11$, $b = 15$, and $c = 19$, the longest side. Then $a^2 + b^2 = 11^2 + 15^2 = 346$ and $c^2 = 19^2 = 361$.
Because $a^2 + b^2 \neq c^2$, the triangle is not a right triangle.
(b) $a^2 + b^2 = 7^2 + (16.8)^2 = 331.24$ and $c^2 = (18.2)^2 = 331.24$. Because $a^2 + b^2 = c^2$, a triangle with sides of lengths 7, 16.8, and 18.2 is a right triangle.   •

SPECIAL RIGHT TRIANGLES

Two special right triangles that appear frequently in geometric applications will be discussed next. First consider an equilateral triangle. Recall that an equilateral triangle is also equiangular [Figure 3.27(a)].

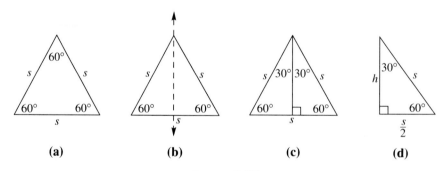

FIGURE 3.27

Also, in Chapter 2 we saw that an equilateral triangle has three lines of symmetry. If the triangle is folded on a line of symmetry as shown in Figure 3.27(b), two identical triangles are formed. Because the angles on either side of the line of symmetry must be congruent, the angles in each triangle measure 30°, 60°, and 90° [Figure 3.27(c)]. Thus, each triangle that is formed is called a

30-60 right triangle or a **30-60-90 triangle**. Because the base of the equilateral triangle is bisected, that is, cut in half, by the fold line as shown in Figure 3.27(c), each 30-60 right triangle has a base of $\frac{s}{2}$. Using the Pythagorean Theorem, we can determine the height, h, of any 30-60 right triangle in terms of s as follows [Figure 3.27(d)].

$$h^2 + \left(\frac{s}{2}\right)^2 = s^2 \Leftrightarrow h^2 + \frac{s^2}{4} = s^2 \Leftrightarrow h^2 = \frac{3s^2}{4}$$

Therefore, $h = \sqrt{\frac{3s^2}{4}} = \frac{s\sqrt{3}}{2}$. This discussion is summarized in the next theorem.

Theorem 3.9

30-60 Right Triangle

In a 30-60 right triangle the length of the leg opposite the 30° angle is half the length of the hypotenuse. The length of the leg opposite the 60° angle is $\frac{\sqrt{3}}{2}$ times the length of the hypotenuse.

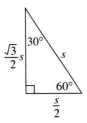

The fact that the relationships stated in Theorem 3.9 are true for any 30-60 right triangle means that if we know the length of any one side of such a triangle, we can find the lengths of the other two.

EXAMPLE 3.18   Find the lengths of the missing sides for each of the 30-60 right triangles in Figure 3.28.

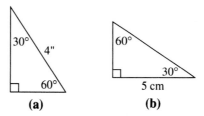

(a)   (b)

FIGURE 3.28

SOLUTION

(a) Using the notation in Theorem 3.9, $s = 4''$, so $\dfrac{s}{2} = 2''$ and $\dfrac{\sqrt{3}}{2}s = \dfrac{\sqrt{3}}{2}(4'') = 2\sqrt{3}$ in. Therefore, the leg opposite the 30° angle is 2″, and the leg opposite the 60° angle is $2\sqrt{3}$ in.

(b) In this case, $\dfrac{\sqrt{3}}{2}s = 5$ cm, so $s = 5\left(\dfrac{2}{\sqrt{3}}\right) = \dfrac{10}{\sqrt{3}} = \dfrac{10\sqrt{3}}{3}$ cm, and $\dfrac{s}{2} = \dfrac{5\sqrt{3}}{3}$ cm. Thus, the leg opposite the 30° angle has length $\dfrac{5\sqrt{3}}{3}$ cm and the hypotenuse has length $\dfrac{10\sqrt{3}}{3}$ cm.   •

A similar relationship exists between the lengths of the sides of the right triangle formed by folding a square along a line of symmetry containing one of its diagonals [Figure 3.29(a)].

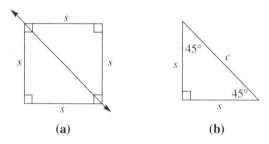

(a)   (b)

FIGURE 3.29

Because the angles formed on either side of the fold line must be congruent and because they also are complementary, each of these acute angles must measure 45°. Therefore, each of the triangles formed is called a **45-45 right triangle** or **45-45-90 triangle** [Figure 3.29(b)]. Notice that each triangle formed is also a right isosceles triangle.

We can apply the Pythagorean Theorem to determine the length of the hypotenuse, c, in a 45-45 right triangle in terms of the length of one of the sides, s, as follows [Figure 3.29(b)]:

$$s^2 + s^2 = c^2 \quad \text{so} \quad c^2 = 2s^2 \quad \text{or} \quad c = s\sqrt{2}$$

This result is summarized in our next theorem.

Theorem 3.10

45-45 Right Triangle

In a 45-45 right triangle, the length of the hypotenuse is $\sqrt{2}$ times the length of a leg.

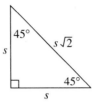

As was true for the 30-60 right triangle, this theorem allows us to find the missing lengths of two sides of a 45-45 right triangle if we know the length of any one side.

EXAMPLE 3.19   Find the area, to the nearest square inch, of parallelogram *ABCD* in Figure 3.30(a).

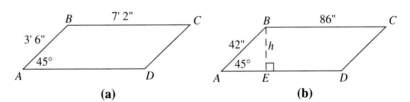

(a)   (b)

FIGURE 3.30

SOLUTION   Convert the dimensions to inches and draw in the height to the longer side [Figure 3.30(b)]. $\triangle ABE$ is a 45-45 right triangle so we have

$$h \sqrt{2} = 42 \qquad \text{or} \qquad h = \frac{42}{\sqrt{2}} \text{ in.}$$

Therefore, the area of *ABCD* is $A = bh = (86)\left(\dfrac{42}{\sqrt{2}}\right) \approx 2554 \text{ in}^2.$

We have shown that if a triangle is a 30-60 right triangle or a 45-45 right triangle and if we know the length of one side of the triangle, then we can determine the lengths of the other two sides. We will generalize this technique in Chapter 6 so that given the measure of one angle of *any* size and the length of one side of a right triangle, we can find the lengths of the other two sides.

Solution to Applied Problem

The following figure is an end view of the roof portion of the house.

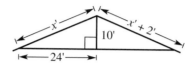

Because a rafter forms a right triangle with the end of the house, we can apply the Pythagorean Theorem to find the length of a rafter, $x + 2$. We have

$$24^2 + 10^2 = x^2$$
$$676 = x^2 \quad \text{or} \quad x = 26'$$

Therefore, the length of the rafter is $x + 2 = 28$ ft. Each half of the roof is a rectangle measuring 60' by 28'. Hence the total area of the roof is given by

$$A = 2(28' \times 60') = 3360 \text{ ft}^2$$

The number of 100 ft^2 in 3360 ft^2 is 33.60. Thus the roof requires 33.6 squares of shingles.

GEOMETRY AROUND US

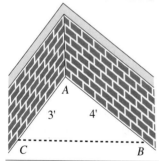

Carpenters, bricklayers, and other builders frequently make use of the Test for Right Triangles to lay out a square corner. A distance of 3 feet may be marked off on one wall and a distance of 4 feet marked off on the adjacent wall as shown. If the corner at point A is to be a right angle, then the distance from B to C must be exactly 5 feet since $3^2 + 4^2 = 5^2$. This 3-4-5 triangle is probably the most commonly used right triangle, but any multiple such as 6-8-10 could be used. Other common right triangle ratios, such as 5-12-13 or 8-15-17, can also be used to "square a corner."

PROBLEM SET 3.3

EXERCISES/PROBLEMS

1. Find the length of the missing side in each right triangle.

   (a)   (b)   (c)

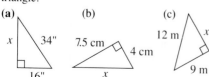

2. Find the length of the missing sides in each right triangle.

   (a)   (b)   (c)

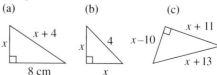

3. Find the missing lengths in the following figures. Round your answers to the nearest tenth.

(a)   (b)

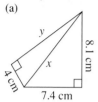

(b)

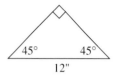

4. Calculate the area of the quadrilateral using the dimensions shown. Round your answer to the nearest tenth of a square centimeter. [HINT: Heron's formula may be useful here.]

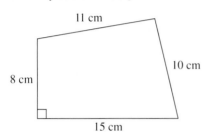

11 cm

10 cm

8 cm

15 cm

5. Find the areas of the polygons with dimensions as shown.

(a)   (b)

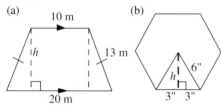

10 m

13 m

h

20 m

6"

h

3" 3"

6. Complete the following table, where information is given about a right triangle with legs a and b, hypotenuse c, perimeter P, and area A. Round your answers to two decimal places.

	a	b	c	P	A
(a)		4.81	7.65		
(b)	10.55				78.23
(c)	$\sqrt{5}$	$\sqrt{15}$			
(d)	x	x		60	

7. Which of the following numbers could be the lengths of sides of a right triangle?
(a) 7, 24, 25   (b) 12, 24, 26
(c) 28, 21, 35

8. Which of the following numbers could be the lengths of sides of a right triangle?
(a) 11, 60, 61   (b) 8, 9, 15
(c) 10, 22, 26

9. Find the missing lengths in each triangle.

(a)   (b)

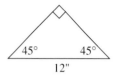

10. Find the missing lengths in each triangle.

(a)   (b)

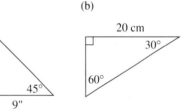

11. Determine the height of trapezoid $ABCD$. What is the area of the trapezoid, to the nearest tenth of a square centimeter?

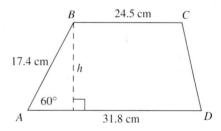

B 24.5 cm C

17.4 cm

h

60°

A 31.8 cm D

12. Determine the lengths of the legs of isosceles trapezoid $PQRS$. Calculate the perimeter and area of the trapezoid.

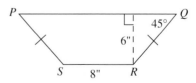

P Q

45°

6"

S 8" R

13. Find the area of parallelogram $ABCD$ given $AB = 12$ cm, $AD = 20$ cm, and $\angle A = 30°$.

14. Find the area of isosceles trapezoid $ABCD$ if the bases measure 10″ and 16″ and two base angles each measure 60°. Give an exact answer.

15. Find the area of a rhombus with a side of length 9.4 m and a vertex angle that measures 45°. Round your answer to two decimal places.

16. Find the area of trapezoid *PQRS* if $\angle P = 60°$, $\angle S = 45°$, $PS = 16$ in., and the height is 5 in. Round your answer to two decimal places.

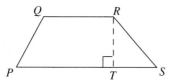

17. The area of a rectangle is 24 in². A diagonal of the rectangle measures exactly $\sqrt{73}$ in. Find the length and width of the rectangle.

18. In the figure shown, the areas of the two large circles are both 25π cm². The area of $\triangle OPQ$ is 6 cm². Find the area of circle *P*.

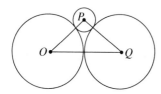

19. In the triangle shown, $\angle P = 45°$, $\angle Q = 105°$, and $QR = 12$ inches. Find the lengths *PQ* and *PR*.

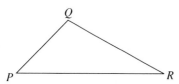

20. The altitude of an equilateral triangle cuts the base in half. Use this fact to find a formula for the area of the following equilateral triangle.

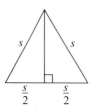

21. A triangle has sides of lengths *a*, *b*, and *c*, where $a < b < c$. Consider the relationship between a^2, b^2, and c^2 for each of the following questions.

(a) When will the triangle be a right triangle?
(b) When will the triangle be an acute triangle?
(c) When will the triangle be an obtuse triangle?

22. Given are the lengths of sides of a triangle. Indicate whether each triangle is a right triangle. If not, specify whether it is acute or obtuse. [HINT: See the previous problem.]
(a) 70, 54, 90   (b) 63, 16, 65
(c) 24, 48, 52   (d) 27, 36, 45
(e) 48, 46, 50   (f) 9, 40, 46

23. The proof of the Pythagorean Theorem that was presented in this section depended upon a figure containing squares and right triangles. Another famous proof of the Pythagorean Theorem also uses squares and right triangles. It is attributed to a 12th-century Hindu mathematician, Bhaskara, and it is said that he simply wrote the word "Behold" above the figure, believing that the proof was evident from the drawing. Use algebra and the areas of the right triangles and squares in the figure to verify the Pythagorean Theorem.

BEHOLD!

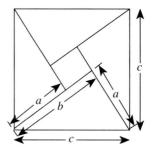

24. A **Pythagorean triple** is a set of three positive integers such that $a^2 + b^2 = c^2$. For example, (3, 4, 5) is a Pythagorean triple. Other Pythagorean triples can be generated by the following process:

(i) Choose any two positive integers, *u* and *v*, such that $u > v$.
(ii) Let $a = 2uv$, $b = u^2 - v^2$, and $c = u^2 + v^2$.

(a) Generate *a*, *b*, and *c* for several pairs of values for *u* and *v*. Then verify that (a, b, c) forms a Pythagorean triple in each case.
(b) Using algebra, show why the expressions given for *a*, *b*, and *c* will always satisfy the equation $a^2 + b^2 = c^2$.

APPLICATIONS

25. An angle-iron bracket is to be made in the shape shown. How long must the third piece be to complete the construction?

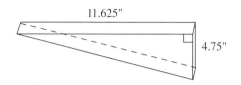

26. A baseball diamond is a square that measures 90 feet on a side. How far must the catcher throw the ball from home plate to second base to pick off a runner?

27. A 16-foot ladder will be used to paint a house. If the foot of the ladder must be placed at least 4 feet away from the house to avoid flowers and shrubs, what is the highest point on the house that the ladder will reach? Round your answer to the nearest tenth of a foot.

28. Two boats leave a dock at the same time. One heads due east at 20 mph, and one heads due south at 15 mph. How far apart are the two boats after 45 minutes have elapsed?

29. As an airplane takes off from a runway, it makes an angle of 30° with the ground. When it has travelled 1100 feet at this take-off angle, what is its altitude?

30. A conveyor belt is to be set up to move bundles of shingles up to a roof. If the edge of the roof is 10 feet above the ground and the conveyor can make no more than a 45° angle with the ground, how long a conveyor belt will be needed? How far away from the house will the bottom of the conveyor be placed? Round your answers to the nearest foot.

31. A faucet is located at one corner of a yard that measures 50 ft by 65 ft. How long a garden hose will be necessary to be able to water plants in any part of the yard? Round your answer to the nearest foot.

32. A rectangular field measures 325 m by 240 m. How much shorter would it be to walk diagonally across the field than to walk around two sides of the field? Round your answer to the nearest meter.

33. To determine a distance across the lake shown, the indicated measurements were taken. To the nearest foot, what is the distance *PR*?

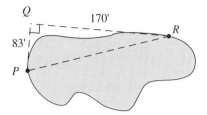

34. In the figure shown, the center of one pulley is 11′8″ above and 18′3″ to the right of the center of the other pulley. Find the distance between their centers to the nearest inch.

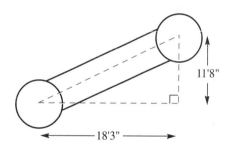

35. Square plugs are often used to check the diameter of a hole. What must the length of the side of the plug be to test a hole with diameter 3.16 cm? Round down to the nearest 0.01 cm.

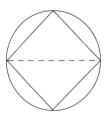

36. A 19″ TV screen measures 19″ diagonally across the rectangular face of the screen. If the screen is about 4″ wider than it is high, find the dimensions of the screen. Round your answer to the nearest tenth of an inch.

37. A 40-foot antenna is to be held in place by four guy wires that will be attached to the antenna halfway up and anchored to the flat roof at points 15 feet from the base of the antenna. If you allow 20 feet of wire for fastening the wires to the antenna and to the roof, how many feet of wire will be required? Round your answer to the nearest foot.

38. The roof of the house shown is known to have a **pitch** of "4 in 12." That is, the roof rises 4 feet for each 12 feet of span. Two dimensions on the end of the house have been measured. Use them to calculate the gable height x of the roof and the length of the rafter y. (Assume an 18″ overhang.)

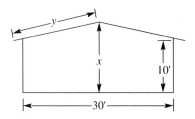

39. A pattern for a square block of a quilt is shown. The quilt will be made of 80 blocks 9″ on a side like the one shown, and four colors will be used. Fabric to be used for the quilt is 36″ wide. How many inches of fabric of each color must be purchased to construct the quilt? Ignore waste.

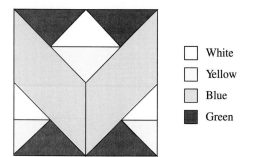

☐ White

☐ Yellow

☐ Blue

☐ Green

40. A log 30″ in diameter is to be milled into a single piece of lumber with a square end.
(a) What are the dimensions of the largest piece that can be cut from the log? [NOTE: It will be shown in Chapter 7 that the diagonals of the square must pass through the center of the circle.]

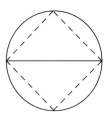

(b) What percent of the log is utilized in the piece of lumber?

41. A rectangular packing box measures 24″ by 18″ by 15″ as shown. Will a tennis racket 32″ long fit into this box? Explain.

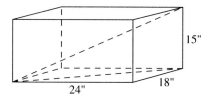

42. Three pipes are stacked lengthwise on a flat-bed truck as shown. If each pipe has a diameter of 60 inches, how high is the stack, to the nearest inch?

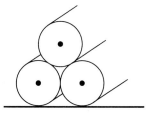

43. A circular tube, or conduit, is needed to enclose four wires, each with a diameter of 1/4 inch. Determine the smallest diameter conduit which will contain the cables. Round your answer to the nearest hundredth of an inch.

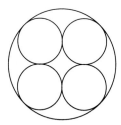

3.4
SURFACE AREA

Applied Problem

A hopper designed to store wood chips and empty them into trucks has the shape shown. How many square feet of metal were used in its construction, including material for the top?

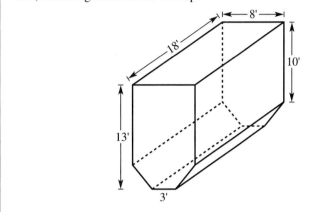

In Sections 3.1 and 3.2 we discussed areas of regions in the plane. Now we will extend that idea to shapes in three dimensions.

SURFACE AREA OF A RIGHT PRISM

The **surface area of a polyhedron** is the sum of the areas of its lateral faces and bases. SA is an abbreviation for "surface area."

EXAMPLE 3.20   Find the surface area of the right rectangular prism in Figure 3.31.

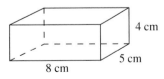

FIGURE 3.31

SOLUTION   The area of each base is $8 \cdot 5 = 40$ cm². Two of the lateral faces each have an area of $4 \cdot 5 = 20$ cm² and two have area $8 \cdot 4 = 32$ cm², so the surface area is $2 \cdot 40 + 2 \cdot 20 + 2 \cdot 32 = 2(40 + 20 + 32) = 184$ cm².   •

Notice that the surface area of the right rectangular prism in Example 3.20 could be expressed as follows:

$$SA = 2(8 \cdot 5 + 4 \cdot 5 + 4 \cdot 8)$$
$$= 2(8 \cdot 5) + 2(4 \cdot 5 + 4 \cdot 8)$$
$$= 2(8 \cdot 5) + (2 \cdot 5 + 2 \cdot 8)4$$
$$= 2(\text{area of the base}) + (\text{perimeter of the base})(\text{height})$$

This result holds for any right prism; that is, the base need not be a rectangle.

Theorem 3.11

Surface Area of a Right Prism

The surface area of a right prism is the sum of twice the area of its base plus the product of the perimeter of the base and the height of the prism.

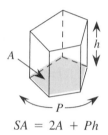

$$SA = 2A + Ph$$

SURFACE AREA OF A RIGHT PYRAMID

The surface area of a pyramid is the sum of the areas of its base and lateral faces.

EXAMPLE 3.21   Find the surface area of a right regular triangular pyramid if the sides of the base are 6 cm and the slant height is 8 cm [Figure 3.32(a)].

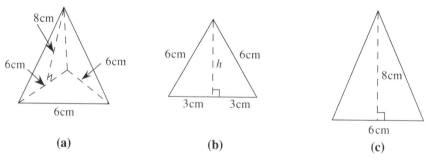

(a)   **(b)**   **(c)**

FIGURE 3.32

SOLUTION   First, the pyramid is a right *regular triangular* pyramid. Therefore, its base is an equilateral triangle [Figure 3.32(b)]. By the Pythagorean Theorem, $h^2 + 3^2 = 6^2$, or $h^2 = 36 - 9 = 27$. Thus, $h = \sqrt{27} = 3\sqrt{3}$ cm and the area of the base is $\frac{1}{2}bh = \frac{1}{2}(6)(3\sqrt{3}) = 9\sqrt{3}$. [NOTE: The height of the base could also be found by observing that the base is divided into two 30-60 right triangles.] The three faces are identical isosceles triangles [Figure 3.32(c)]. The area of one face is $\frac{1}{2}(6)(8) = 24$ cm². Thus, the area of all three faces is $3 \cdot 24 = 72$ cm². Finally, the surface area, the sum of the areas of all faces, is $9\sqrt{3} + 72$ cm², or approximately 87.6 cm² to one decimal place. •

The surface area of the pyramid in Example 3.21 can be expressed as follows:

$$SA = 9\sqrt{3} + 3[\tfrac{1}{2}(6 \cdot 8)]$$
$$= 9\sqrt{3} + \tfrac{1}{2}(3 \cdot 6)8$$
$$= \text{area of the base} + \tfrac{1}{2}(\text{perimeter of the base})(\text{slant height})$$

This result holds for any right regular pyramid, regardless of the shape of the base. The surface area of a pyramid with a base that is not regular is found by adding the areas of the base and the lateral faces.

Theorem 3.12
___

Surface Area of a Right Regular Pyramid

The surface area of a right pyramid is the sum of the area of its base plus half the product of the perimeter of the base and the slant height.

$$SA = A + \tfrac{1}{2}Pl$$

SURFACE AREA OF RIGHT CYLINDER

The surface area of a cylinder can be found by adapting the formula for the surface area of a prism as suggested by Figure 3.33.

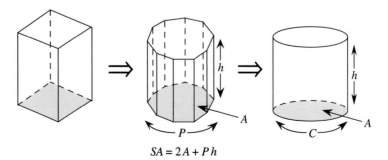

$$SA = 2A + Ph$$

FIGURE 3.33

In the case of a right circular cylinder whose base has radius r, A represents the area of the base, namely πr^2, and P represents the perimeter (or C the circumference), namely $2\pi r$. This leads to the following theorem.

Theorem 3.13

Surface Area of a Right Circular Cylinder

The surface area of a right circular cylinder is twice the area of its base plus the circumference of its base times its height.

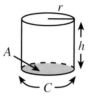

$$SA = 2A + Ch = 2\pi r^2 + 2\pi rh$$

EXAMPLE 3.22   Find the surface area of the right circular cylinder whose height is 7 cm and whose base has diameter 12 cm.

SOLUTION $SA = 2\pi(6^2) + 2\pi(6)(7) = 72\pi + 84\pi = 156\pi$ cm^2.   •

SURFACE AREA OF A RIGHT CIRCULAR CONE

As in the case of the cylinder, the surface area of a cone may be found by adapting the formula for the surface area of a pyramid as suggested in Figure 3.34.

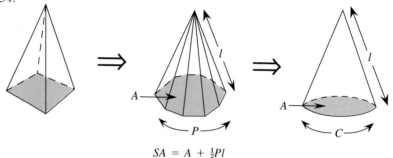

$$SA = A + \tfrac{1}{2}Pl$$

FIGURE 3.34

In the case of a right circular cone where r is the radius of the base, A represents the area of the base, namely πr^2, and C represents the circumference of the base, $2\pi r$. Also, the slant height is represented by l. By substituting for A and P in the formula $A + \tfrac{1}{2}Pl$ we can arrive at another formulation for surface area of a right circular cone.

$$SA = A + \tfrac{1}{2}Cl = \pi r^2 + \tfrac{1}{2}(2\pi rl) = \pi r(r + l)$$

This discussion leads to our next result.

Theorem 3.14

Surface Area of a Right Circular Cone

The surface area of a right circular cone is the sum of the area of its base and half the circumference of its base times its slant height.

$$SA = A + \tfrac{1}{2}Cl = \pi r(r + l)$$

FIGURE 3.35

EXAMPLE 3.23   Find the surface area of the right circular cone whose height is 8 cm and whose base has radius 6 cm (Figure 3.35).

SOLUTION   First we must find the slant height, l. Here

$$l^2 = 6^2 + 8^2 = 36 + 64 = 100 \quad \text{or} \quad l = 10 \text{ cm}$$

[NOTE: We also could have observed that $6 = 2 \cdot 3$ and $8 = 2 \cdot 4$. Thus, $l = 2 \cdot 5 = 10$ because it is the hypotenuse of a right triangle whose sides are multiples of the sides of a 3-4-5 right triangle.]
Therefore,

$$SA = \pi(6^2) + \tfrac{1}{2}(2\pi)(6)(10) = 36\pi + 60\pi = 96\pi \text{ cm}^2 \qquad \bullet$$

SURFACE AREA OF A SPHERE

The derivation of the formula for the surface area of a sphere is quite complex. However, the following observation due to Archimedes simplifies the task of remembering the formula.

Consider the smallest right circular cylinder containing a sphere of radius r [Figure 3.36(a)]. Because the sphere has radius r, the base of the cylinder has radius r and the height of the cylinder is $2r$ [Figure 3.36(b)]. Therefore, the surface area of the *cylinder* is as follows:

$$SA = 2A + Ch = 2\pi r^2 + 2\pi r(2r) = 6\pi r^2$$

Archimedes' observation was that *both* the surface area and the volume of a sphere are two-thirds the respective surface area and volume of the smallest right circular cylinder containing the sphere. Thus, the surface area of the sphere of radius r can be found using the previous equation as follows:

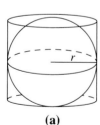

(a)

(b)

FIGURE 3.36

$$SA_{\text{Sphere}} = \tfrac{2}{3} SA_{\text{Cylinder}} = \tfrac{2}{3} (6\pi r^2) = 4\pi r^2$$

Theorem 3.15

Surface Area of a Sphere

The surface area of a sphere is 4π times the square of its radius.

$$SA = 4\pi r$$

Solution to Applied Problem

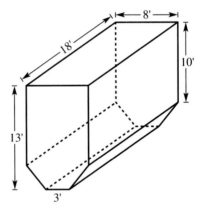

We need to find the surface area of this right hexagonal prism. Using the Pythagorean Theorem to solve for x in the following figure, we find that

$$3^2 + (2.5)^2 = x^2 \qquad \text{or} \qquad x \approx 3.91 \text{ ft}$$

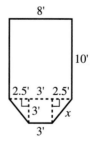

Thus, the area of the six rectangular faces is given by $A = Ph$, or $A = (8' + 10' + 3.91' + 3' + 3.91' + 10')18'$, which equals 698.76 ft². The area of the two bases is $2[8'(10') + \frac{1}{2}(3')(3' + 8')] = 193$ ft². Thus, the surface area is approximately

$$193\text{ft}^2 + 698.76\text{ft}^2 = 891.76 \text{ ft}^2$$

GEOMETRY AROUND US

Crystals grow naturally in the shape of polyhedra. In fact, some take on the shapes of the regular polyhedra. In table salt (sodium chloride) we can see cubic crystals with the naked eye. Sodium chlorate crystals occur as tetrahedra. The octahedral shape of most diamond crystals facilitates the cutting of the gem to take advantage of its light-reflective properties.

PROBLEM SET 3.4

EXERCISES/PROBLEMS

1. Find the surface area of each of the following right prisms.

(a)

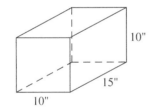

10"
15"
10"

(b)

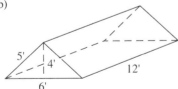

5'
4'
12'
6'

(c)

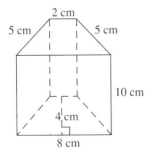

2 cm
5 cm   5 cm
10 cm
4 cm
8 cm

2. Find the surface area of each of the following right square pyramids.

(a)

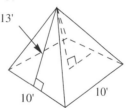

13'
10'   10'

(b)

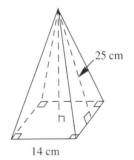

25 cm
14 cm

(c)

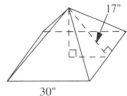

17"
30"

3. Find the surface area of each of the following cans to the nearest square centimeter.
   (a) Coffee can
       $r = 7.6$ cm
       $h = 16.3$ cm
   (b) Soup can
       $r = 3.3$ cm
       $h = 10$ cm
   (c) Juice can
       $r = 5.3$ cm
       $h = 17.7$ cm
   (d) Shortening can
       $r = 6.5$ cm
       $h = 14.7$ cm

4. Find the surface area of spheres having the following radii or diameters. Round to the nearest whole unit.
   (a) $r = 6$ in.
   (b) $r = 2.3$ in.
   (c) $d = 24$ m
   (d) $d = 6.7$ km

5. Find the surface area of the following regular polyhedra to the nearest cm².
   (a) Tetrahedron with edge length of 5 cm
   (b) Octahedron with edge length of $4\sqrt{3}$ cm
   (c) Icosahedron with edge length of 12 cm

6. Find the surface area of each right prism with the given features.
   (a) The bases are equilateral triangles with sides of length 8 ft, and the height of the prism is 10 ft.
   (b) The bases are trapezoids with parallel sides of lengths 7 cm and 9 cm perpendicular to one side of length 6 cm. The height of the prism is 12 cm.
   (c) The base is a right triangle with legs of length 5 and 12 and the height of the prism is 20.

7. Find the surface area of each cone to the nearest whole unit.
   (a)

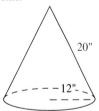

20"
12"

   (b)

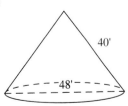

40'
48'

   (c)

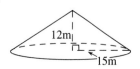

12m
15m

8. Find the surface area of each right pyramid with the following features. Round to the nearest whole unit.
   (a) Its height is 14″ and the base is a regular hexagon with sides of length 12″.
   (b) Its base is a 10″ by 18″ rectangle and its height is 12″.

9. Find the total surface area to the nearest whole unit.
   (a)

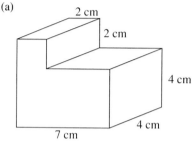

2 cm
2 cm
4 cm
4 cm
7 cm

   (b)

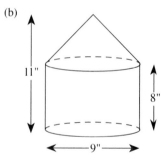

11"
8"
9"

10. Find the total surface area of the following solid shapes.
    (a)

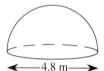

4.8 m

    (b)

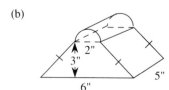

2"
3"
6"
5"

11. Find the total surface area of the hollowed out hemisphere shown.

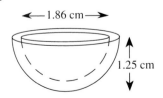

1.86 cm
1.25 cm

12. Find the total surface area of the hollowed out cylinder shown.

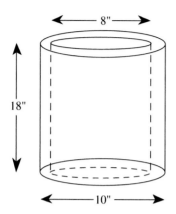

8"

18"

10"

13. Find the diameter, to the nearest millimeter, of a sphere with surface area 215.8 cm².

14. A right circular cylinder has a total surface area of 112π. If the height of the cylinder is 10, find the diameter of the base.

15. The radius of one sphere is 4 cm and the radius of a second sphere is 12 cm. How do the surface areas of the two spheres compare?

16. A spherical balloon has a diameter of 8 feet. By how much should the radius of the balloon be increased if the surface area of the balloon is to be doubled?

17. The top of a rectangular box has an area of 96 square inches. Its side has area 72 square inches and its end has area 48 square inches. What are the dimensions of the box?

18. The surface area of a right circular cylinder is 1200 cm². If the height of the cylinder is twice its diameter, find the dimensions of the cylinder. Round to two decimal places.

19. Suppose you have 36 unit cubes like the one shown below.

Those cubes could be arranged to form right rectangular prisms of various sizes. For example, the cubes could be arranged in 9 rows of 4 cubes as shown. The surface area of that prism would be 98 square units.

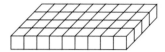

(a) Sketch an arrangement of the cubes in the shape of a right rectangular prism that has a total surface area of exactly 96 square units.

(b) Sketch an arrangement of the cubes in the shape of a right rectangular prism that has a total surface area of exactly 80 square units.

(c) What arrangement of the cubes gives a right rectangular prism with the smallest possible total surface area? What is this minimum surface area?

(d) What arrangement of the cubes gives a right rectangular prism with the largest possible total surface area? What is this maximum surface area?

20. Thirty unit cubes are stacked in square layers to form a tower. The bottom layer measures 4 cubes × 4 cubes, the next layer 3 cubes × 3 cubes, the next layer 2 cubes × 2 cubes, and the top layer is a single cube.

(a) Determine the total surface area of the tower of cubes.

(b) Suppose that the number of cubes and the height of the tower were increased so that the bottom layer of cubes measured 8 cubes × 8 cubes. What would be the total surface area of this tower?

(c) What would be the total surface area of the tower if the bottom layer measured 20 cubes × 20 cubes?

APPLICATIONS

21. A box without a top is to be made from sheet metal. If the box will measure 5'6" by 2'10" by 1'8" high, how many square feet of sheet metal will be needed? Round your answer to the nearest tenth of a square foot.

22. The inside of a cylindrical tank used to hold resin must be coated with a corrosion preventative. The tank measures 7' in diameter and 5' in height. One gallon of corrosion preventative will cover 10 ft². How many gallons will be needed to coat the tank? Assume that the sides, top, and bottom of the tank will be treated.

23. A room measures 4 m by 7 m and the ceiling is 3 m high. A liter of paint covers 20 m². How many liters of paint will it take to paint all but the floor of the room?

24. A golf ball has a diameter of 1.68 inches. A box to hold three golf balls will be made out of cardboard. What is the least amount of cardboard that will be needed, assuming the centers of the balls will be collinear when in the box? Ignore flaps to be folded in on the ends and round your answer to the nearest tenth of a square inch.

25. A machined piece is a right regular hexagonal prism with a hole 1/2 inch in diameter drilled through it. Find the total surface area, rounding to the nearest tenth of a square inch.

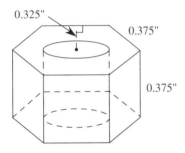

26. A gutter is to be formed from a length of tin bent so that the cross section is a semicircle 4 inches in diameter. If a total length of 210 feet is needed and an extra 3/4 inch at each edge is required to finish the edge, calculate the number of square feet of tin needed. Round to the nearest square foot.

27. The cylindrical storage tank with hemispherical ends as shown must be painted. If the paint costs $21.95 per gallon and one gallon of paint covers 200 ft², what will the paint cost? Round your answer to the nearest cent.

28. Assuming that the earth is a sphere with an equatorial diameter of 12,760 kilometers, what is the surface area of the earth?

29. A room measures 4 m wide by 4 m high by 8 m long. At point *P* in the center of the front and 0.5 m from the floor, is a VCR. At a point *Q* in the center of the back wall and 0.5 m down from the ceiling, is a stereo speaker. The speaker is to be connected to the VCR. The speaker wire must be in contact with a wall, the ceiling, or the floor at all times. What is the shortest length of wire that can be used? [HINT: Draw a two-dimensional picture.]

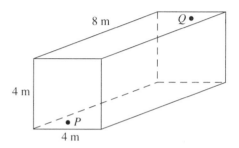

30. A water tower in the shape of a right circular cylinder has a stairway to the top which winds around the outside of the tower. The tower is 50 ft in diameter and 40 ft tall, and the stairs wrap around the cylinder exactly twice. How long, to the nearest foot, is the stairway? [HINT: See problem 29.]

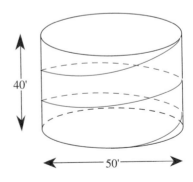

31. A barber pole consists of a cylinder of radius 10 cm on which 1 red, 1 white, and 1 blue helix, each of equal width are painted. The cylinder is 1 m high. If each stripe makes a constant angle of 60° with the vertical axis of the cylinder, how much surface area is covered by the red stripe?

3.5
VOLUME

Applied Problem

At a steel mill a thick slab of steel 1 m by 1 m by 15 cm will be rolled into a long strip 1 m wide and 0.5 cm thick. How long will the rolled strip be?

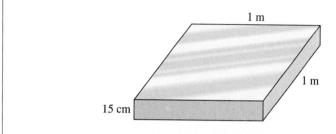

VOLUME

Volume is the three-dimensional analog of area. To find the volume of a shape, a unit cube must be designated. Then we must determine how many unit cubes completely fill a given three-dimensional shape (Figure 3.37). The following postulate for volume is analogous to the Area Postulate.

1 cubic unit

FIGURE 3.37

Postulate 3.4

Volume Postulate

(a) For every polyhedron and unit cube, there is a real number that gives the number of unit cubes (and parts of unit cubes) which exactly fill the region enclosed by the polyhedron.

(b) The volume of the region enclosed by a polyhedron is the sum of the volumes of the smaller regions into which the region can be subdivided.

The real number described in Postulate 3.4 is called the **volume** of the region enclosed by the polyhedron. For the sake of simplicity, we usually say "the volume of the polyhedron" to mean the volume of the region enclosed by the polyhedron. The next example shows how to find the volume of a right rectangular prism.

EXAMPLE 3.24   Using a unit cube of 1 cm by 1 cm by 1 cm, find the volume of a right rectangular prism 3 cm by 4 cm by 2 cm by counting the number of unit cubes that will fill the prism exactly (Figure 3.38).

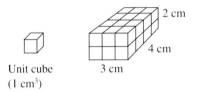

Unit cube
(1 cm³)

FIGURE 3.38

SOLUTION   As can be seen in Figure 3.38, the prism would hold two layers of cubes that have dimensions 3 cm by 4 cm. Because there are $3 \times 4 = 12$ cubes in each layer, there are $3 \times 4 \times 2 = 24$ unit cubes in all. Thus the volume is 24 cm³. This result may be viewed as the product of the area of the base of the prism (3×4) and its height (2).   •

VOLUME OF A PRISM

As in the case of finding the area of a rectangle, the volume of a prism having real number dimensions is found using the technique of Example 3.24.

Definition

Volume of a Prism

The volume of a prism is the product of the area of its base and its height.

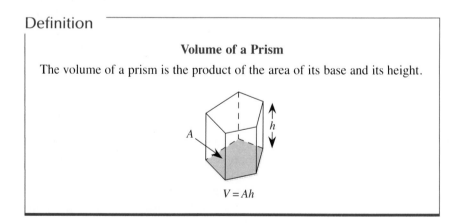

$$V = Ah$$

The following two results are immediate consequences of this definition of volume.

Theorem 3.16

Volume of a Right Rectangular Prism

The volume of a right rectangular prism is the product of its length, width, and height.

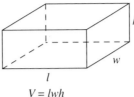

$$V = lwh$$

Corollary 3.17

Volume of a Cube

The volume of a cube is the cube of the length of its side.

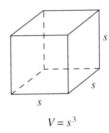

$$V = s^3$$

[NOTE: A **corollary** is a theorem that is a direct result or a special case of a preceding theorem. In this case, a cube is a special type of right rectangular prism.]

VOLUME OF A PYRAMID

The volume of a pyramid may be inferred from Figure 3.39, where three pyramids identical to the one on the left are shown to be filling a right prism completely.

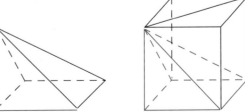

FIGURE 3.39

Even though Figure 3.39 pictures a square pyramid, the ratio of three pyramids to one prism holds in general. That is, the volume of a pyramid is one third the volume of the corresponding prism determined by the base and height of the pyramid.

Theorem 3.18

Volume of a Pyramid

The volume of a pyramid is one third the product of the area of its base and its height.

$$V = \tfrac{1}{3}Ah$$

EXAMPLE 3.25   Find the volumes of the polyhedra with dimensions shown in Figure 3.40.

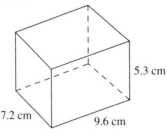

Right Rectangular Prism
(a)

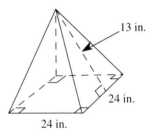

Right Square Pyramid
(b)

FIGURE 3.40

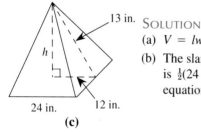

SOLUTION

(a) $V = lwh = 7.2 \text{ cm} \times 9.6 \text{ cm} \times 5.3 \text{ cm} = 366.336 \text{ cm}^3$

(b) The slant height is 13 in. The distance from the center of the base to an edge is $\tfrac{1}{2}(24 \text{ in.}) = 12$ in. [Figure 3.40(c)]. Thus, the height, h, satisfies the equation $h^2 + 12^2 = 13^2$, or $h = 5$ in. The volume, then, is given by

$$V = \tfrac{1}{3}Ah = \tfrac{1}{3}(24^2)(5) = 960 \text{ in}^3$$

VOLUMES OF CYLINDERS AND CONES

The formulas for the volumes of cylinders and cones are obtained in a manner similar to that used to derive the formulas for their surface areas (Figure 3.41).

In the case of the circular cylinder with a base of radius r, A represents the area of the base, or πr^2. The height of the cylinder is represented by h. Figure 3.41 suggests the following theorems.

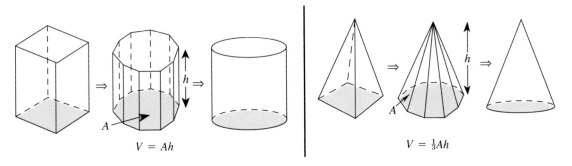

FIGURE 3.41

Theorem 3.19

Volume of a Right Circular Cylinder

The volume of a cylinder is the product of the area of its base and its height.

$$V = Ah = \pi r^2 h$$

Theorem 3.20

Volume of a Right Circular Cone

The volume of a circular cone is one third the product of the area of its base and its height.

$$V = \tfrac{1}{3}Ah = \tfrac{1}{3}\pi r^2 h$$

EXAMPLE 3.26   Find the volumes of the right circular cylinder and right circular cone with dimensions given in Figure 3.42.

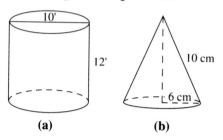

(a)   (b)

FIGURE 3.42

SOLUTION

(a) $V = \pi r^2 h = \pi (5^2)(12) = 300\pi$ ft^3

(b) $V = \frac{1}{3}\pi r^2 h = \frac{1}{3}\pi (6^2)\sqrt{10^2 - 6^2} = \frac{1}{3}\pi (6^2)(8) = 96\pi$ cm^3 •

Although formulas were given for the volumes of right prisms, pyramids, cylinders, and cones, the same formulas hold for similar oblique shapes.

VOLUME OF A SPHERE

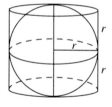

FIGURE 3.43

It was stated in Section 3.4 that Archimedes observed that both the surface area and volume of a sphere are two thirds of the respective surface area and volume of the smallest right circular cylinder containing the sphere (Figure 3.43). The sphere has radius r, so the base of the cylinder has radius r and the height of the cylinder is $2r$. Therefore the volume of the cylinder is

$$V = Ah = \pi r^2 (2r) = 2\pi r^3$$

Thus, the volume of the sphere of radius r can be found as

$$V_{\text{Sphere}} = \tfrac{2}{3} V_{\text{Cylinder}} = \tfrac{2}{3}(2\pi r^3) = \tfrac{4}{3}\pi r^3$$

We summarize this result in the following theorem.

Theorem 3.21

Volume of a Sphere

The volume of a sphere is $\frac{4}{3}\pi$ times the cube of its radius.

$$V = \tfrac{4}{3}\pi r^3$$

EXAMPLE 3.27   Find the radius of the sphere for which the surface area and volume are the same number.

SOLUTION   Let r be the radius of the sphere. Then $4\pi r^2 = \frac{4}{3}\pi r^3$. Next solve for r.

$$4\pi r^2 = \tfrac{4}{3}\pi r^3$$
$$12\pi r^2 = 4\pi r^3$$
$$4\pi r^3 - 12\pi r^2 = 0$$
$$4\pi r^2(r - 3) = 0$$

Therefore, $r = 0$ or $r = 3$. Because $r = 0$ produces a sphere consisting of a single point, the answer is $r = 3$.

As a check, a sphere of radius 3 has surface area $4\pi(3^2) = 36\pi$ and volume $\frac{4}{3}\pi(3^3) = 36\pi$. Of course, the dimensions are square units for the surface area and cubic units for the volume.

The following table summarizes the surface area and volume formulas we have studied. Connections formed by observing the similarities and differences in the various formulas will help you remember them more easily.

Geometric Shape	Surface Area	Volume
Right prism	$SA = 2A + Ph$	$V = Ah$
Right circular cylinder	$SA = 2A + Ch$	$V = Ah$
	$= 2\pi r(r + h)$	$= \pi r^2 h$
Right regular pyramid	$SA = A + \tfrac{1}{2}Pl$	$V = \tfrac{1}{3}Ah$
Right circular cone	$SA = A + \tfrac{1}{2}Cl$	$V = \tfrac{1}{3}Ah$
	$= \pi r(r + l)$	$= \tfrac{1}{3}\pi r^2 h$
Sphere	$SA = 4\pi r^2$	$V = \tfrac{4}{3}\pi r^3$

TABLE 3.1

USING DIMENSIONAL ANALYSIS TO CONVERT UNITS OF VOLUME

In Section 3.1 we used dimensional analysis to convert units of area measure. The next example shows how dimensional analysis can be used to convert between units of volume measure.

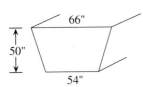

FIGURE 3.44

EXAMPLE 3.28   Calculate the volume of dirt that must be removed from a 200-foot-long trench with dimensions shown in Figure 3.44. Find the volume in cubic inches first, and then convert to cubic yards.

SOLUTION   The trench is a right prism with a base in the shape of a trapezoid and a height equal to the length of the trench, namely 200', or 2400". Thus we have

$$V = Ah$$
$$= \tfrac{1}{2}[(50")(66" + 54")](2400")$$
$$= 7,200,000 \text{ in}^3$$

To convert from cubic inches to cubic feet, we use the fact that $1 \text{ ft}^3 = (1 \text{ ft})(1 \text{ ft})(1 \text{ ft}) = (12 \text{ in.})(12 \text{ in.})(12 \text{ in.}) = 1728 \text{ in}^3$. Hence,

$$V = \frac{7,200,000 \cancel{\text{in}^3}}{1} \cdot \frac{1 \cdot 1 \cdot 1 \cancel{\text{ft}^3}}{12 \cdot 12 \cdot 12 \cancel{\text{in}^3}} \cdot \frac{1 \cdot 1 \cdot 1 \text{ yd}^3}{3 \cdot 3 \cdot 3 \cancel{\text{ft}^3}} \approx 154 \text{ yd}^3$$

So, approximately 154 yd³ of dirt will be removed from the trench.   •

In the previous example we calculated the volume first and then performed the conversion. However, we could have converted to yards first and then found the volume in cubic yards. Also, because we were working with cubic units, or three dimensions, we used each conversion factor *three* times.

Solution to Applied Problem

The volume of the 1 m by 1 m by 0.5 m strip will be the same as the volume of the 1 m by 1 m by 15 cm slab. The volume of the slab in cubic meters is

$$V = lwh = (1 \text{ m})(1 \text{ m})(0.15 \text{ m}) = 0.15 \text{ m}^3$$

Because the volume of the new, longer strip will also be 0.15 m³, we can find the length, *l*, of the new strip.

$$0.15 \text{ m}^3 = l(1 \text{ m})(0.005 \text{ m})$$
$$l = \frac{0.15 \text{ m}^3}{0.005 \text{ m}^2} = 30 \text{ m}$$

Therefore, the new strip will be 30 m long.

GEOMETRY AROUND US

The sphere is one of the most commonly occurring shapes in nature. Frog eggs, tomatoes, oranges, the Earth, and the moon are a very few examples. The regular occurrence of the spherical shape is not coincidental. One explanation for the frequency of the sphere's appearance is that for a given surface area, the sphere encloses the greatest volume. In other words, a sphere requires the least amount of natural material to surround a given volume.

PROBLEM SET 3.5

EXERCISES/PROBLEMS

1. Find the volume of each of the following right prisms.

(a)

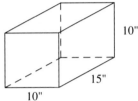

(b)

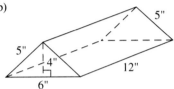

(c)

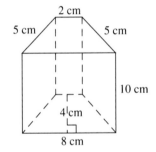

2. Find the volume of each of the following right pyramids.

(a)

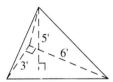

(b)

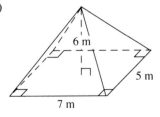

(c)

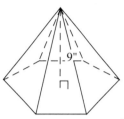

Area of the base is 38 in.²

3. Find the volume of each of the following cans to the nearest cubic centimeter.

(a) Coffee can
   $r = 7.6$ cm
   $h = 16.3$ cm

(b) Soup can
   $r = 3.3$ cm
   $h = 10$ cm

(c) Juice can
   $r = 5.3$ cm
   $h = 17.7$ cm

(d) Shortening can
   $r = 6.5$ cm
   $h = 14.7$ cm

4. Find the volume of each right circular cone to the nearest whole unit.

(a)

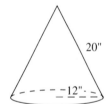

(b)

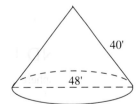

(c)

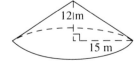

5. Shown below are base designs for stacks of unit cubes (see problem set in Section 2.4). For each one, determine the volume and total surface area (including the bottom) of the stack described.

(a)

3	4	2
1	1	3

(b)

(c)

3	3	3
1	2	3
1	2	3

6. Shown below are base designs for stacks of unit cubes (see problem set in Section 2.4). For each one, determine the volume and total surface area (including the bottom) of the stack described.

(a)

1	2	4	3

(b)

2	3
1	5

(c)

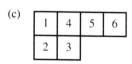

7. Find the volume of the following spheres to the nearest whole unit.
   (a) $r = 6$ in.   (b) $r = 2.3$ in.
   (c) $d = 24$ m   (d) $d = 6.7$ km

8. Find the volume of each right prism with the given features.
   (a) The bases are equilateral triangles with sides of length 8 ft, and the height of the prism is 10 ft.
   (b) The bases are trapezoids with bases of lengths 7 cm and 9 cm perpendicular to one side of length 6 cm, and the height of the prism is 12 cm.
   (c) The base is a right triangle with legs of length 5 in. and 12 in., and the height of the prism is 20 in.

9. Sketched below is a pattern (or net) for a three-dimensional figure. Use the dimensions given to calculate the volume and the total surface area of the three-dimensional figure formed by the pattern. Round your answers to the nearest cm³ or cm².

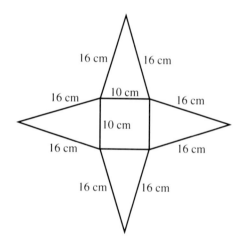

10. Sketched below is a pattern (or net) for a three-dimensional figure. Use the dimensions given to calculate the volume and the total surface area of the three-dimensional figure formed by the pattern. Round your answers to the nearest cm³ or cm².

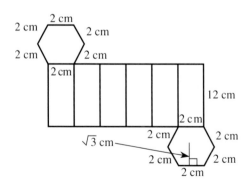

11. Use dimensional analysis to perform each of the following conversions. Round to two decimal places for approximate answers.
   (a) 400 in³ to ft³   (b) 1.2 m³ to cm³
   (c) 0.4 ft³ to mm³

12. Use dimensional analysis to perform each of the following conversions. Round to two decimal places for approximate answers.
   (a) 115 yd³ to ft³   (b) 42 mm³ to cm³
   (c) 2.3 m³ to yd³

13. Suppose that all the dimensions of a square prism are doubled. How would the volume change?

14. (a) How does the volume of a right circular cylinder change if its radius is doubled?
   (b) How does the volume of a right circular cylinder change if its height is doubled?

15. (a) If the side of one square is three times as long as the side of a second square, how do their areas compare?
   (b) If the side of a cube is three times as long as the side of a second cube, how do their volumes compare?
   (c) If all the dimensions of a rectangular box are doubled, what happens to its volume?

16. How does the volume of a sphere change if its radius is doubled?

17. A right rectangular prism has a volume of 324 cubic units. One edge has measure twice that of a second edge and nine times that of the third edge. What are the dimensions of the prism?

18. A right circular cone has a volume of 140 in³. The height of the cone is the same length as the diameter of the base. Find the dimensions of the cone to the nearest tenth of an inch.

19. A rectangular piece of 8½″ by 11″ paper can be rolled into a cylinder in two different directions. If there is no overlapping, which cylinder has the greater volume, the one with the long side of the rectangle as its height, or the one with the short side of the rectangle as its height?

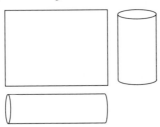

20. Given are three boxes.

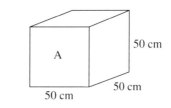

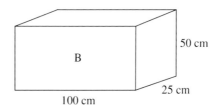

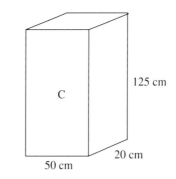

   (a) Find the volume of each box.
   (b) Find the surface area of each box.
   (c) Do boxes with the same volume always have the same surface area?
   (d) Which box used the least amount of cardboard?

APPLICATIONS

21. The Pyramid Arena in Memphis, Tennessee, has a square base that measures approximately 544 feet on a side. The pyramid is 321 feet high.
   (a) Calculate its volume.
   (b) Calculate its lateral surface area.

22. A soft-drink cup is in the shape of a right circular cone with capacity 250 milliliters. The radius of the circular base is 5 centimeters. How deep is the cup?

23. A 4-inch-thick concrete slab is being poured for a circular patio 10 feet in diameter. Concrete costs $50 per cubic yard. Find the cost of the concrete. (Assume that you cannot purchase a fraction of a cubic yard of concrete.)

24. The directions on the back of a cake mix box call for two round cake pans, each 9″ in diameter, or one rectangular 9″ by 13″ pan. Which method will yield the taller cake? (Assume the two 9″ round cakes are *not* stacked on top of one another.)

25. A waffle cone has a diameter of 9 cm and a height of 15 cm. How much frozen yogurt, to the nearest cubic centimeter, will the cone hold if filled so that the yogurt is level with the top of the cone?

26. Noah's ark was reported to have been 300 cubits long, 50 cubits wide, and 30 cubits high. Use a rectangular prism with no top as an approximation to the shape of the ark. What is the capacity of the ark in cubic meters? [HINT: Use dimensional analysis—a **cubit** equals 21 inches and 1 inch equals 2.54 cm.]

27. (a) If one inch of rain fell over one acre, how many cubic inches of water would that be? How many cubic feet?
    (b) If one cubic foot of water weighs approximately 62 pounds, what is the weight, in tons, of one inch of rain over one acre of ground?
    (c) The weight of one gallon of water is about 8.3 pounds. A rainfall of one inch over one acre of ground means about how many gallons of water?

28. A father and his son working together can cut 48 ft³ of firewood per hour.
    (a) If they work an 8-hour day and are able to sell all the wood they cut at $100 per cord, how much money can they earn? A **cord** is defined as 4′ by 4′ by 8′.
    (b) If they split the money evenly, at what hourly rate should the father pay his son?
    (c) If the delivery truck can hold 100 ft³, how many trips would it take to deliver all the wood cut in a day?
    (d) If they sell their wood for $85 per truckload, what price are they getting per cord?

29. The volume of an object with an irregular shape can be determined by measuring the volume of water it displaces.
    (a) A rock placed in an aquarium measuring $2\frac{1}{2}$ feet long by 1 foot wide causes the water level to rise 1/4 inch. What is the volume of the rock?
    (b) With the rock in place, the water level in the aquarium is one half inch from the top. The owner wants to add to the aquarium 200 solid marbles, each with a diameter of 1.5 cm. Will the addition of these marbles cause the water in the aquarium to overflow?

30. The **density** of a substance is the ratio of its mass to its volume.

$$\text{density} = \frac{\text{mass}}{\text{volume}}$$

Density is usually expressed in terms of grams per cubic centimeter (g/cm³). For example, the density of copper is 8.94 g/cm³.
    (a) Express the density of copper in kg/dm³.
    (b) A chunk of oak firewood has a mass of about 2.85 kg and has a volume of 4100 cm³. Determine the density of oak in g/cm³, rounding to the nearest thousandth.
    (c) A piece of iron weighs 45 ounces and has a volume of 10 in³. Determine the density of iron in g/cm³, rounding to the nearest tenth.

31. A sculpture made of iron has the shape of a right square prism topped by a sphere. The metal in each part of the sculpture is 2 mm thick. If the outside dimensions are as shown and the density of iron is 7.87 g/cm³, calculate the approximate mass of the sculpture in kilograms.

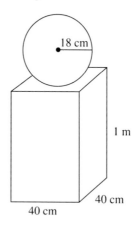

32. While rummaging in his great aunt's attic, Bernard found a small figurine that he believed to be made out of silver. To test his guess he looked up the density of silver in his chemistry book and found it to be 10.5 g/cm³. He found that the figure had a mass of 149 g. To determine its volume he dropped it into a cylindrical glass of water. If the diameter of the glass was 6 cm and the figurine was pure silver, by how much should the water level in the glass rise?

33. A hollow rubber ball has a circumference of 22 cm and is made of rubber which is 0.6 cm thick. Find the volume of rubber in the ball, rounding to the nearest cm^3.

34. A standard tennis ball can is a cylinder that holds three tennis balls.
  (a) Which is greater, the circumference of the can or its height?
  (b) If the radius of a tennis ball is 3.5 cm, what percentage of the can is occupied by air?

35. A tank full of a liquid is in the shape of a sphere 6 ft in diameter. If 200 gallons of liquid are pumped out of the tank, how many gallons remain in the tank? (Recall that 1 ft^3 ≈ 7.48 gal.)

36. A cylindrical steel pipe has an inside diameter of 2″ and an outside diameter of 2.5″. If the pipe is 15′ long, how many cubic inches of steel does the pipe contain? If steel weighs 490 lb/ft^3, what is the weight of the pipe? What volume of water will the pipe hold when full? Round each of your answers to the nearest whole unit.

37. The first three steps of a 10-step staircase are shown.

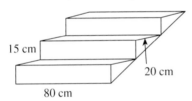

  (a) Find the amount of concrete, to the nearest cm^3, needed to make the exposed portion of the staircase.
  (b) Find the amount of carpet needed to cover the fronts, tops, and sides of the concrete steps, to the nearest cm^2.

38. (a) How many square meters of tile are needed to tile the sides and bottom of the swimming pool illustrated?
  (b) How much water does the pool hold?

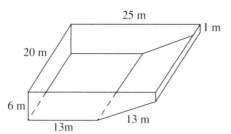

39. Find the volume, to the nearest cubic inch, of the simplified I-beam whose cross section is shown and whose length is 25′.

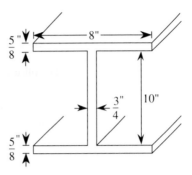

40. The **displacement** of a cylinder in a piston engine is the volume it displaces when it travels from its lowest point to its highest point. The **bore** of the cylinder is its diameter, and the **stroke** is the height the piston travels. Find the displacement of one cylinder if it has a bore of 92 mm and a stroke of 67 mm. Express your answer in cubic centimeters. If the engine has 4 cylinders, what is the total displacement, to the nearest cubic centimeter?

41. In water treatment applications, a unit of volume called the acre foot is sometimes used. One **acre foot** of water is the volume of water that would cover an acre to a depth of one foot. Determine the number of gallons in one acre foot. (Use the fact that 1 acre is 43,560 ft^2 and 1 ft^3 is 7.48 gal.)

42. At a water treatment plant a cylindrical digester 25 ft in diameter and 20 ft deep is to be emptied into two rectangular drying beds. If the beds are 60 ft long by 20 ft wide and will be filled to a depth of 15 inches, by how many inches will the level in the digester be lowered?

43. A 3/8-inch drill is used to cut a hole through a cube of metal 2″ on a side. How much metal is removed (to the nearest hundredth of an inch)? What percentage of the original block of metal remains?

44. A manufacturing company is shipping ceramic figurines in 9″ by 6″ by 8″ boxes. How many of these boxes can be packed in a 2′ by 3′ by 5′ shipping crate?

45. Plywood glue is stored in a tank that has the shape shown. How many cubic feet of glue will this tank hold? Round your answer to the nearest whole number.

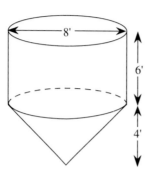

46. Lumber is measured in board feet. A **board foot** is the volume of a square piece of wood measuring 1 foot long, 1 foot wide, and 1 inch thick. A surfaced "two by two" actually measures 1.5 inches by 3.5 inches, a "two by six" measures 1.5 by 5.5, and so on. Plywood is sold in *exact* dimensions and is differentiated by thickness. Find the number of board feet in the following pieces of lumber.
    (a) 6-foot-long two by four
    (b) 10-foot two by eight
    (c) 4-foot by 8-foot sheet of $\frac{3}{4}$-inch plywood
    (d) 4-foot by 6-foot sheet of $\frac{5}{8}$-inch plywood

47. A farmer's irrigation system sprays water in a circle with a radius of 1320 ft at the rate of 1000 gal/min. At this rate, how long, to the nearest hour, must the sprinkler run to distribute the equivalent of 3/4 inch of rainfall over the area?

48. A bucket has the shape of a truncated circular cone, or a **frustum**. The formula for the volume of a frustum is

$$V = \tfrac{1}{3}h(B_L + B_S + \sqrt{B_L B_S})$$

where h is the height of the frustum, B_L is the area of the larger base, and B_S is the area of the smaller base. (This formula will be verified in Chapter 6.)

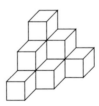

If the diameter of the top of the bucket is 12″, the diameter of the bottom is 9″, and the bucket is 10″ tall, calculate the volume of the bucket in gallons. Use the fact that 1 gal ≈ 231 in³.

49. Two designs for an oil storage tank are being considered: spherical and cylindrical. The two tanks would have the same capacity and would each have an inside diameter of 60 feet.
    (a) What would be the height of the cylindrical tank?
    (b) If one cubic foot holds 7.5 gallons of oil, what is the capacity of each tank in gallons?
    (c) Which of the two designs has the smallest surface area and would thus require less material in its construction?

50. Suppose that you have ten separate unit cubes. The total volume is 10 cubic units and the surface area is 60 square units. If you arrange the cubes as shown, the volume is still 10 cubic units, but the surface area is only 36 square units. For each of the following problems, assume that all cubes are stacked to form a single shape sharing complete faces (no loose cubes allowed).

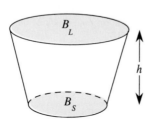

    (a) How can you arrange 10 cubes to get a surface area of 34 square units? Sketch your answer.
    (b) What is the greatest possible surface area you can obtain with 10 cubes? Sketch.
    (c) How can you arrange the 10 cubes to get the smallest possible surface area? What is this area?
    (d) Answer the questions in parts (a), (b), and (c) for 27 and 64 cubes.
    (e) What arrangement has the greatest surface area for a given number of cubes?
    (f) What arrangement seems to have the least surface area for a given number of cubes?
    (g) Biologists have found that an animal's surface area is an important factor in its regulation of body temperature. Why do you think desert animals such as snakes are long and thin? Why do furry animals curl up in a ball to hibernate?

Solution to Initial Problem

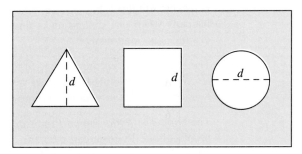

A cylinder whose base has diameter d and whose height is d will pass through the square and circle exactly [Figure 3.45(a)]. We will modify a model of this cylinder to get a shape with a triangular cross section.

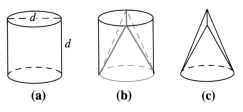

(a)   (b)   (c)

FIGURE 3.45

If we slice the cylinder along the heavy lines in Figure 3.45(b), the resulting model will pass through the triangular hole exactly. Figure 3.45(c) shows the resulting model that will pass through all three holes exactly.

Additional Problems Where the Strategy "Use a Model" Is Useful

1. A store is promoting a sale on cola, and a display is being set up at the end of an aisle. One hundred forty-four six packs have been arranged in a 12 by 12 square on the floor to form the first layer of a square pyramid. What is the greatest number of six packs that can be stacked on top of this to form a pyramid?

2. If a cube is sliced with a plane parallel to one of the faces, the cross section that is revealed will be a square. Describe how to slice a cube with a plane to obtain each of the following cross sections. [NOTE: Only *one* cut is allowed in each case.]
   (a) A rectangle   (b) An isosceles triangle
   (c) A rhombus other than a square

Writing for Understanding

1. Describe how you would solve the following Fermi problem (see Chapter 2, "Writing for Understanding" for a description of a Fermi problem):
   Suppose that two shoes for each resident of your state were lined up toe to heel. About how long in meters would this line of shoes be? What would be the dimensions of the smallest cubical box that could contain all of the shoes? Explain your reasoning.

2. Discuss how a change in the dimensions of a figure affects the area, perimeter, volume, and surface area where appropriate. For example, what happens when one dimension is doubled? Include a rationale and specific examples in your answer.

3. The Pythagorean Theorem is one of the most famous theorems in mathematics. Why do you think that it is so well-known and what makes it so useful? Include examples in your discussion.

4. The book *The Pythagorean Proposition* contains over one hundred different proofs of the Pythagorean theorem. Find this book in your library. Select three of your favorite proofs and explain why you prefer them.

CHAPTER REVIEW

Following is a list of key vocabulary, notation, and ideas for this chapter. Mentally review these items and, where appropriate, write down the meaning of each term. Then restudy the material that you are unsure of before proceeding to take the chapter test.

Section 3.1—Perimeter, Circumference, and Area of Rectangles and Triangles

Vocabulary/Notation

Perimeter 95	Altitude 101
Circumference 97	Height (of a triangle) 101
Area 99	Base (of a triangle) 101

Main Ideas/Results

1. The Area Postulate for simple closed curves
2. Formulas can be derived for the perimeter of common polygons and the circumference of circles.
3. Area formulas can be derived for rectangles and triangles.
4. Dimensional analysis can be used to convert between units of area measure and to solve problems involving areas.

Section 3.2—More Area Formulas

Vocabulary/Notation

Height (of a parallelogram) 111	Center (of a regular polygon) 113
Base (of a parallelogram) 111	

Main Idea/Result

1. Area formulas can be derived for parallelograms (and rhombuses), trapezoids, regular polygons, and circles.

Section 3.3—The Pythagorean Theorem and Right Triangles

Vocabulary/Notation

30-60 right triangles 125   45-45 right triangles 126

Main Ideas/Results

1. The Pythagorean Theorem can be derived using areas of triangles and squares.
2. There is a test based on the lengths of the sides of a triangle to determine if a triangle is a right triangle.
3. There are special relationships among the lengths of the sides of 30-60 and 45-45 right triangles.

Section 3.4—Surface Area

Vocabulary/Notation

Surface area 133

Main Idea/Result

1. Surface area formulas can be derived for certain prisms, pyramids, cylinders, cones, and spheres.

Section 3.5—Volume

Vocabulary/Notation

Volume 143

Main Ideas/Results

1. The Volume Postulate
2. Volume formulas can be derived for certain prisms, pyramids, cylinders, cones, and spheres.
3. Dimensional analysis can be used to convert between units of volume measure and to solve problems involving volumes.

PEOPLE IN GEOMETRY

Pythagoras (circa 580–500 B.C.) was a Greek mathematician who founded a mystical school in southern Italy. The members of the school, known as the Pythagoreans, were a brotherhood with secret religious rites. Their motto was "all is number." They studied the properties of numbers, believing that whole numbers were the basis of the structure of the universe. The Pythagoreans were the first to discover irrational numbers, numbers that cannot be expressed as the quotient of two whole numbers. This discovery was deeply disturbing for the Pythagoreans, for it invalidated some of their mathematical proofs and challenged the assumption that whole numbers were the ultimate reality. Of course, Pythagoras is most famous for the theorem named after him. Although he likely did not discover the Pythagorean Theorem, he may have been the first to prove it.

CHAPTER 3 TEST

TRUE-FALSE

Mark as true any statement that is always true. Mark as false any statement that is never true or that is not necessarily true. Be able to justify your answers.

1. The formula for the area of a rectangle is $A = 2l + 2w$.

2. If two sides of a triangle are known, the third side can always be found using the Pythagorean Theorem.

3. To find the area of a trapezoid, you need the height of the trapezoid and the lengths of each of the parallel sides.

4. If two rectangles have equal perimeters, then their areas are equal.

5. If the formula for the volume of a three-dimensional shape is $\frac{1}{3}Ah$, then the shape is a cone.

6. If a sphere and the base of a cylinder have the same radius and the height of the cylinder is the same as the diameter of its base, then their surface areas are the same.

7. Every altitude of a triangle has the same length.

8. There are infinitely many great circles of a sphere.

9. A triangle with sides measuring 6.75 cm, 17.55 cm, and 16.2 cm is a right triangle.

10. If the circumference of a circle is doubled, then the area of the circle is also doubled.

EXERCISES/PROBLEMS

11. Express the formula for the area of a circle in terms of its
    (a) Diameter
    (b) Circumference

12. An isosceles right triangle has an area of 32 cm². Find the lengths of its legs.

13. (a) Find the volume of the right square pyramid shown.

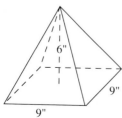

(b) What happens to the volume if the length of a side of the square is doubled?

(c) What happens to the volume if the length of a side of the square is tripled?

(d) Find the surface area of the pyramid.

14. Develop a formula for the area of a regular hexagon with side length x by partitioning it into two trapezoids.

15. Complete the following table for a sphere:

Diameter	Radius	SA	Circumference	V
(a)			56π	
(b)		100π		
(c)				972π
(d) 6				

16. Find the area and perimeter of the following figure where the curved piece is a semicircle. Round to two decimal places.

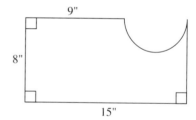

17. A new car owner needs to convert the following measurements:

    Weight: 3000 lbs to kg
    Length: 4000 mm to ft
    Tire pressure: 27 lbs/in² to g/cm²

    Use dimensional analysis to perform each of these conversions given 1 in. = 2.54 cm and 1 kg = 2.2 lb.

18. Find the area of the figure shown on the square lattice.

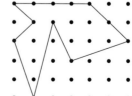

19. Find the surface area and volume of the solid figure shown, rounding to the nearest whole number.

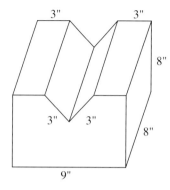

APPLICATIONS

20. A piece of rubber tubing is in the shape of a hollowed-out cylinder. If the outside radius is 24 inches, the inside radius is 20 inches, and the length of the tube is 10 feet, find the total surface area of the tube.

21. Alice and Bob took a 20-mile bicycle ride together. Alice rode a touring bicycle and Bob rode his mountain bike. If the outside diameter of the tires on Alice's bike is 27 inches and on Bob's bike is 26 inches, how many more revolutions did Bob's wheels make on the trip than Alice's wheels?

22. An Oregon wheat silo is in the shape of a right circular cylinder with a hemisphere on the top. If the radius of the cylinder is 10 feet and the cylinder is 70 feet tall, what volume of wheat can be stored inside?

23. The Great Wall of China is about 1500 miles long. The cross section of the wall is a trapezoid 25 feet high, 25 feet wide at the bottom, and 15 feet wide at the top. How many cubic yards of material make up the wall?

24. A boy flying a kite has 100 feet of string out when the kite lodges in a tree. He reels in 20 feet of the string as he walks 25 feet closer to the tree. How high in the tree is his kite stuck? Round your answer to the nearest foot.

4

Reasoning and Triangle Congruence

EUCLID—THE FATHER OF GEOMETRY

Euclid of Alexandria (circa 300 B.C.) has been called the "father of geometry." Euclid's major contribution to mathematics came when he organized much of the geometry of his day into one book called *The Elements. The Elements* consisted of thirteen "books," five on plane geometry, three on solid geometry, and five on geometric explanations of the mathematics now studied in algebra. Because much of the geometry he organized continues to be the heart of high school geometry courses, it is often titled Euclidean Geometry. His most famous original contribution to the book was a proof of the Pythagorean Theorem. The diagram he used to prove it, sometimes called the "windmill," is displayed below.

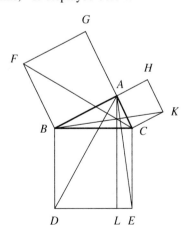

PROBLEM-SOLVING STRATEGIES

1. Draw a Picture
2. Guess and Test
3. Use a Variable
4. Look for a Pattern
5. Make a Table
6. Solve a Simpler Problem
7. Look for a Formula
8. Use a Model
9. *Identify Subgoals*

STRATEGY 9: IDENTIFY SUBGOALS

Often a problem can be solved more easily by breaking it down into smaller problems. Then the original problem is solved by solving the various parts, called **subgoals**, and assembling those solutions in a meaningful sequence.

INITIAL PROBLEM

The following figure has a regular octagon adjacent to a regular pentagon. Find the measure of ∠*ABC*.

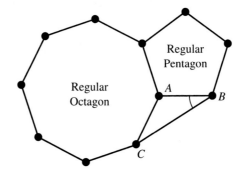

CLUES

The Identify Subgoals strategy may be appropriate when

· A problem can be broken down into a series of simpler problems.
· The statement of the problem is very long and complex.
· You can say "If I only knew . . . , then I could solve the problem."
· There is a simple, intermediate step that would be useful.
· There is other information that you wished the problem contained.

A solution for the initial problem above is on page 207.

INTRODUCTION

In Chapters 1–3, results were often derived from known results by reasoning informally. In this chapter we begin to reason more formally by constructing proofs. These proofs are applied to the study of triangle congruence. The triangle congruence results are then used to draw conclusions about other shapes in later chapters. In Section 4.1 we discuss reasoning and proof. In Sections 4.2 and 4.3, we study congruent triangles and relationships that can be derived from them. In Section 4.4 geometric constructions using a compass and straightedge are the focus. Note that we will use some of the results that we proved informally in Chapters 2 and 3 even though they are not proved formally until Chapter 5 or 6. This approach has the virtue of allowing us to explore a richer collection of applications, yet it doesn't introduce any circular reasoning.

4.1
REASONING AND PROOF IN GEOMETRY

Applied Problem

A surveyor lays out a traverse with four vertices. As measured, the interior angles of the quadrilateral would be as shown. How can deductive reasoning be used to determine whether the traverse "closes"?

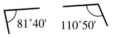

81°40'   110°50'

112°52'

54°38'

CONDITIONAL STATEMENTS AND DEDUCTIVE REASONING

In the first three chapters, many problems have been presented that require direct reasoning. **Direct reasoning**, or **deductive reasoning**, is used to draw a conclusion from a series of statements. **Conditional** statements, statements of the form "if p, then q," play a central role in direct reasoning. The conditional "if p, then q" is written symbolically as "$p \Rightarrow q$." The statement "p" is called the **hypothesis** of the conditional and "q" is called its **conclusion**. We also read "$p \Rightarrow q$" as "p implies q" and "p only if q."

EXAMPLE 4.1   What can be deduced from the following two statements?

1. If it is raining, then the street is wet.
2. It is raining.

SOLUTION   In conditional (1), p, the hypothesis, is the statement "it is raining," and q, the conclusion, is the statement "the street is wet." Because (2) states that it is raining—that is, that p holds—we can conclude from conditional (1) that q also must hold, or that the street is wet.   •

Example 4.1 illustrates one of the most common forms of direct reasoning. If we let p represent "it is raining" and q represent "the street is wet," then Example 4.1 and its solution may be represented as follows:

$$\begin{array}{ll} \text{If } p, \text{ then } q. & \qquad\qquad p \Rightarrow q \\ \underline{\quad p \qquad\qquad} & \text{or symbolically as} \qquad \underline{\quad p \qquad} \\ \text{Therefore } q. & \qquad\qquad \therefore q \end{array}$$

The **three-dot triangle** is a symbolic abbreviation for the word "therefore." The preceding form of argument is called the **Law of Detachment**. Notice how the q is "detached" from the conditional $p \Rightarrow q$ once we know p.

Besides displaying conditional statements in verbal and symbolic form, we also can view them diagrammatically using what are called **Venn Diagrams** or **Euler Circles** (Figure 4.1).

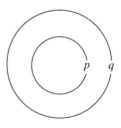

FIGURE 4.1

In Figure 4.1, because the interior of circle p is contained within the interior of circle q, we can say "if p, then q" is represented by this diagram.

A second common form of direct reasoning, called the Law of Syllogism, plays a central role in mathematics. The **Law of Syllogism** takes the following form:

$$\begin{array}{ll} \text{If } p, \text{ then } q. & \qquad\qquad p \Rightarrow q \\ \underline{\text{If } q, \text{ then } r.} & \text{or symbolically} \qquad \underline{q \Rightarrow r} \\ \text{Therefore if } p, \text{ then } r. & \qquad\qquad \therefore p \Rightarrow r \end{array}$$

EXAMPLE 4.2   What can be deduced from the following two statements?

1. If it is raining, then the street is wet.
2. If the street is wet, then the street is slippery.

SOLUTION   Here, p is the statement "it is raining," q is the statement "the street is wet," and r is the statement "the street is slippery." Both (1) and (2) are conditionals *and* the conclusion of (1) is the same as the hypothesis of (2). Therefore, by the Law of Syllogism we can conclude that "if it is raining, then the street is slippery." •

The Laws of Detachment and Syllogism are commonly used together. Consider the statements presented in the next example.

EXAMPLE 4.3   What can be deduced from the following statements?

1. If the circumference of a circle is 8π inches, then its diameter is 8 inches.
2. If the diameter of a circle is 8 inches, then its radius is 4 inches.
3. If the radius of a circle is 4 inches, then its area is 16π square inches.
4. The circumference of circle O is 8π inches.

SOLUTION   From statements 1 and 2 we know by the Law of Syllogism that if the circumference of a circle is 8π, then its radius is 4 inches. This result with statement 3 leads us to conclude that if the circumference of a circle is 8π, then its area must be 16π square inches. Finally, this statement with statement 4 and the Law of Detachment leads us to conclude that the area of circle O is 16π square inches. •

The verbal argument in Example 4.3 can be presented symbolically as follows.

$$
\begin{array}{l}
1.\ p \Rightarrow q \\
2.\ q \Rightarrow r \\
3.\ r \Rightarrow s \\
4.\ \underline{p\qquad\quad} \\
\quad\therefore\ s
\end{array}
$$

FIGURE 4.2

Using (1) and (2), we can deduce that $p \Rightarrow r$ using the Law of Syllogism. Using the Law of Syllogism with $p \Rightarrow r$ and $r \Rightarrow s$ from (3), we can conclude that $p \Rightarrow s$. Using the Law of Detachment with $p \Rightarrow s$ and p from (4), we can finally conclude s. This argument can be represented using Euler circles (Figure 4.2). Notice that the nested circles represent statements 1–3. Then, because circle p is within circle s, we can state "if p, then s."

When presented with a sequence of statements such as that given in Example 4.3, we do not always deliberate over each statement separately. Sometimes the conclusion will seem fairly automatic. However, it is important to recognize why such conclusions can be drawn and to know how to begin to analyze more complex arguments. In addition, some conclusions that may seem obvious may not really be valid. Symbolic presentations such as the previous one are often easier to analyze than their verbal counterparts.

The next example shows again how direct reasoning can be used to draw conclusions.

EXAMPLE 4.4   What can be deduced from the following statements?

1. If quadrilateral *ABCD* is a square, then the opposite sides of *ABCD* are parallel.
2. If the opposite sides of a quadrilateral are parallel, then the quadrilateral is a parallelogram.
3. *ABCD* is a square.

SOLUTION   We can apply the Law of Syllogism to (1) and (2) to deduce that "if *ABCD* is a square, then it is a parallelogram." Then, using the Law of Detachment with this latter statement and (3), we can conclude that *ABCD* is a parallelogram.

   Alternatively, using the Law of Detachment with (1) and (3), we can conclude that the opposite sides of *ABCD* are parallel. Combining this latter statement with (2) and using the Law of Detachment again, we can conclude that *ABCD* is a parallelogram. Thus, using either method, we can conclude that *ABCD* is a parallelogram.   •

OTHER FORMS OF THE CONDITIONAL

Many conditional statements do not appear explicitly in the recognizable "if . . . , then . . ." form, but they can be restated in that form and treated as conditionals.

EXAMPLE 4.5   Rewrite each of the following statements in "if . . . , then . . ." form.

(a) The sum of the squares of the lengths of the legs of a right triangle is equal to the square of the length of the hypotenuse.
(b) All squares are rectangles.

SOLUTION

(a) In the statement given, it is assumed that the triangle being discussed is a right triangle. This assumption is the hypothesis. Thus, we could say the following:
    If a triangle is a right triangle, then the sum of the squares of the lengths of its legs is equal to the square of the length of its hypotenuse.
(b) If a polygon is a square, then it is a rectangle.   •

   There are three commonly used variants of the conditional statement "If *p*, then *q*."

   Converse of $p \Rightarrow q$: $q \Rightarrow p$
   Inverse of $p \Rightarrow q$:   not $p \Rightarrow$ not q
   Contrapositive of $p \Rightarrow q$:   not $q \Rightarrow$ not p

EXAMPLE 4.6   Write the converse, inverse, and contrapositive of the statement "If *ABCD* is a square, then it has four right angles."

SOLUTION

Converse: If *ABCD* has four right angles, then it is a square.

Inverse: If *ABCD* is not a square, then it does not have four right angles.

Contrapositive: If *ABCD* does not have four right angles, then it is not a square.   •

The converse, inverse, and contrapositive of a conditional are covered in Topic 1 near the end of the book, but because the converse of a conditional will come up often, it is important to understand how a conditional relates to its converse. Notice that in Example 4.6, the original statement was true. Its converse was false because there are quadrilaterals with four right angles that are *not* squares.

On the other hand, there are instances when a conditional and its converse both hold. In the case when $p \Rightarrow q$ and $q \Rightarrow p$ both hold, we write $p \Leftrightarrow q$. This is read "*p* if and only if *q*." This form is called a **biconditional**. The following are examples of biconditionals.

1. The three sides of a triangle are congruent if and only if the three angles of the triangle are congruent.
2. The opposite sides of a quadrilateral are congruent if and only if the opposite angles of the quadrilateral are congruent.
3. Two triangles have the same shape if and only if their angles are congruent, respectively.

Remember that, for example, statement (1) means that *both* of the following statements are true:

If the three sides of a triangle are congruent, then the three angles of the triangle are congruent, and

If the three angles of a triangle are congruent, then the three sides of the triangle are congruent.

PROOF

At the heart of Euclid's axiomatic system of mathematics is the concept of proof. Simply put, a **proof** is a convincing mathematical argument. This means that any person who understands the terminology accepts the definitions and premises of the mathematics involved, and thinks in a logically correct fashion could not deny the validity of the conclusions drawn.

The Laws of Detachment and Syllogism will form the basis of our proofs or direct reasoning. When we write out proofs, we will use two different forms, paragraph and statement-reason forms. Our first example of a proof concerns angles formed by intersecting lines.

Two intersecting lines form many angles with the same vertex. Two angles that are opposite each other, such as $\angle 1$ and $\angle 3$ or $\angle 2$ and $\angle 4$ in Figure 4.3,

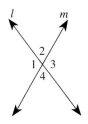

FIGURE 4.3

are called **vertical angles**. Each pair of vertical angles appears to be congruent. Next we will display a paragraph proof and a statement-reason proof of this fact. In each case, it is helpful to state what is given and what is to be proved. The "given" is the hypothesis of the conditional to be proved and the "prove" is the conclusion. We will prove the following result in two ways.

Theorem 4.1

If x and y are the measures of a pair of vertical angles, then $x = y$.

PARAGRAPH PROOF

Given Intersecting lines l and m with two vertical angles having measures x and y [Figure 4.4(a)]

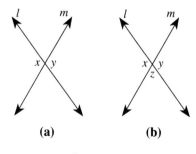

(a)   (b)

FIGURE 4.4

Prove $x = y$

Proof In Figure 4.4(a), lines l and m intersect to form a pair of vertical angles having measures x and y. Let one of the other two angles have measure z [Figure 4.4(b)]. Then $x + z = 180°$ and $y + z = 180°$ because each of these pairs of angles forms a straight angle. So by substitution $x + z = y + z$. Subtracting z from both sides of this equation yields $x = y$, which is what we wanted to prove.   •

STATEMENT-REASON PROOF

Given Intersecting lines l and m with two vertical angles having measures x and y [Figure 4.4(a)]

Prove $x = y$ [Figure 4.4(b)]

Proof

Statement	Reason
1. $x + z = 180°$	1. x and z are adjacent angles whose nonadjacent sides form a straight angle.
2. $y + z = 180°$	2. y and z are adjacent angles whose nonadjacent sides form a straight angle.
3. $x + z = y + z$	3. Substitution from steps 1 and 2.
4. $x = y$	4. z is subtracted from both sides of the equation in step 3. •

Both the paragraph and statement-reason forms of proof are acceptable. The advantage of paragraph proofs is that they are conversational. That is, you write out the proof in much the same way that you would make a convincing verbal argument. However, reasons are sometimes omitted in paragraph proofs, so when you read a proof, you must be able to justify for yourself parts of the proof. Statement-reason proofs have the advantage of being easier to read. In addition, precise reasons are provided for each step.

The next example illustrates another result proved using both forms of proof.

EXAMPLE 4.7   Prove the following conditional statement in two ways: If two angles are congruent and supplementary, then they are right angles.

SOLUTION

Given $\angle A$ and $\angle B$ where $\angle A \cong \angle B$ and $\angle A + \angle B = 180°$

Prove $\angle A = \angle B = 90°$.

PARAGRAPH PROOF

Let $\angle A = \angle B$ and $\angle A + \angle B = 180°$. By substitution, we have that $180° = \angle A + \angle B = \angle A + \angle A$. Thus, $2(\angle A) = 180°$, or $\angle A = 90°$. Because $\angle A = \angle B$, $\angle B = 90°$ also.

STATEMENT-REASON PROOF

Statement	Reason
1. $\angle A = \angle B$	1. Given
2. $\angle A + \angle A = \angle A + \angle B$	2. The same number may be added to both sides of an equation.
3. $\angle A + \angle B = 180°$	3. Given
4. $\angle A + \angle A = 180°$, or $2(\angle A) = 180°$	4. Substitution from steps 2 and 3, and simplification
5. $\angle A = 90°$	5. Both sides of an equation may be divided by the same number.
6. $\angle B = 90°$	6. Substitution from steps 1 and 5 •

Solution to Applied Problem

The surveyor knows that the traverse closes if and only if the sum of the measures of the angles is 360°. The sum of the angles is

$$81°40' + 110°50' + 54°38' + 112°52' = 360°$$

so the traverse does close.

GEOMETRY AROUND US

If you watch and listen carefully you will observe many examples of the use (and misuse) of conditionals and deductive reasoning in advertising. For example, a popular advertising jingle went something like "If you're out of Crunchy's, you're out of chips." Here the advertiser is attempting to get the viewer to think of Crunchy's and chips as equivalent.

A similar example is "Lo-Lo Yogurt is the breakfast that real women eat." This statement is equivalent to "If you are a real woman, then you eat Lo-Lo Yogurt." Here again, the advertiser wants the viewer to accept the converse statement "If you eat Lo-Lo Yogurt, then you are a real woman." However, the converse may or may not be true. Analyzing advertising for correct logic can help to make you a more discriminating consumer.

PROBLEM SET 4.1

EXERCISES/PROBLEMS

Draw a conclusion for each of Exercises 1–10. If impossible, write "no conclusion possible."

1. If three points are not collinear, then they lie in one and only one plane.

   The three vertices of a triangle are not collinear.

2. If a quadrilateral is an isosceles trapezoid, then it has two congruent sides.

   PQRS has two congruent sides.

3. If a cube has a volume of 27 in³, then it has an edge of length 3 in.

   If a cube has an edge of length 3 in., then it has a surface area of 54 in².

   ABCDEFGH is a cube with a volume of 27 in³.

4. If a quadrilateral has four right angles, then it is a rectangle.

   If a quadrilateral is a rectangle, then its opposite sides are congruent.

   ABCD is a rectangle.

5. If lines *l* and *m* are in the same plane and are not parallel, then *l* and *m* intersect.

   Lines *l* and *m* are in the same plane and are not parallel.

6. If a point is equidistant from the endpoints of a segment, then the point lies on the perpendicular bisector of the segment.

   Point *P* is equidistant from points *A* and *B*.

7. If two lines are perpendicular, then the lines form right angles.

   If two lines form right angles, then the lines are perpendicular.

8. If *XYZW* is a square, then the diagonals bisect each other.

   If the diagonals of a quadrilateral bisect each other, then the quadrilateral is a parallelogram.

   XYZW is a square.

9. If a triangle is scalene, then the triangle has no two sides congruent.

   △*ABC* is a scalene triangle.

10. If a quadrilateral has four congruent sides, then it is a rhombus.

    If a quadrilateral is a rhombus, then its diagonals are perpendicular.

    A quadrilateral has four congruent sides.

If possible, write a statement that can be deduced from the statements in Exercises 11–16. If impossible, write "no deduction possible."

11. If two lines are parallel, then the lines do not intersect.

    Two lines do not intersect.

12. If the sum of two angles is 180°, then the angles are supplementary.

    The sum of ∠*A* and ∠*B* is 90°.

13. If each angle in a triangle is less than 90°, then the triangle is acute.

    △*QRS* is equiangular.

14. If △*PQR* is equilateral, then all three angles of the triangle are less than 90°.

    A triangle has a 60° angle.

15. If *ABCD* is a square, then it is a quadrilateral.

    If *ABCD* is a polygon with four sides, then it is a quadrilateral.

    ABCD is a square.

16. If *XYZW* is a square, then it has four right angles.

    If a quadrilateral has four right angles, then it is a rectangle.

    If a quadrilateral is a rectangle, then its opposite sides are parallel.

    If a quadrilateral has opposite sides parallel, then it is a parallelogram.

    XYZW is a square.

For Exercises 17–32, decide if the statement is true or false, form the converse, and decide if the converse is true or false. If both the statement and its converse are true, write a biconditional equivalent to the two statements.

17. If a square has an area of 25 cm², then the length of its side is 5 cm.

18. If an angle measures 117°, then it is an obtuse angle.

19. If two angles are adjacent, then they share a common side.

20. If a triangle is equilateral, then it is isosceles.

21. If the side of a square measures 10 in., then the diagonal of the square is $10\sqrt{2}$ in.

22. If a polygon is a regular polygon, then all its sides are congruent.

23. If △*ABC* as shown has a right angle at *C*, then $a^2 + b^2 = c^2$.

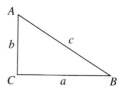

24. If $a = b$, then the kite shown is a rhombus.

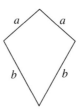

25. If a triangle has side lengths 3, 4, and 5, then it is a right triangle.

26. If two angles have a common vertex, then they are adjacent.

27. If a triangle is equilateral, then the triangle has rotation symmetry.

28. If two angles of a triangle are complementary, then the triangle is a right triangle.

29. If a triangle has three equal sides, then the triangle is equiangular.

30. If a triangle is equilateral, then the triangle has a line of symmetry.

31. If a pyramid has a base with n sides, then the pyramid has $n + 1$ vertices.

32. If a prism has bases with n sides each, then the prism has a total of $3n$ edges.

State a conclusion that can be drawn from each of Exercises 33–38.

33. $r \Rightarrow s$

$\quad \underline{r \qquad}$

$\quad \therefore$

34. not $p \Rightarrow t$

$\quad \underline{\text{not } p \qquad}$

$\quad \therefore$

35. $r \Rightarrow s$

$\quad s \Rightarrow p$

$\quad \underline{p \Rightarrow t}$

$\quad \therefore$

36. $p \Leftrightarrow q$

$\quad \underline{q \Leftrightarrow r}$

$\quad \therefore$

37. $p \Rightarrow \text{not } q$

$\quad \text{not } r \Rightarrow t$

$\quad \underline{\text{not } q \Rightarrow \text{not } r}$

$\quad \therefore$

38. $p \Rightarrow q$

$\quad t \Rightarrow q$

$\quad q \Rightarrow s$

$\quad \underline{p \qquad}$

$\quad \therefore$

For each statement given in problems 39–44, write an equivalent conditional statement in "if . . . , then . . ." form.

39. All parallelograms are quadrilaterals.

40. The height of an equilateral triangle with a side of length s is $\frac{s}{2}\sqrt{3}$.

41. A prism with a base having n sides has $2n$ vertices.

42. The only polyhedron with four vertices is a triangular pyramid.

43. Every regular polyhedron has faces that are regular polygons.

44. A cylinder is not a polyhedron.

PROOF

In problem 45, give reasons that justify each step.

45. Given $\triangle PQR$ with a right angle at R, prove that $\angle P$ and $\angle Q$ are complementary.

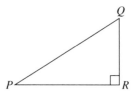

Proof

Statement	Reason
1. $\triangle PQR$ is a right triangle.	1. _____
2. $\angle P + \angle Q + \angle R = 180°$	2. _____
3. $\angle R = 90°$	3. _____
4. $\angle P + \angle Q + 90° = 180°$	4. _____
5. $\angle P + \angle Q = 90°$	5. _____
6. $\angle P$ and $\angle Q$ are complementary.	6. _____ •

4.2
TRIANGLE CONGRUENCE CONDITIONS

Applied Problem

A saw blade is made by cutting six right triangles out of a regular hexagon as shown. If a segment of length AB is cut at a right angle at each tooth, show that all the triangles that are cut out are congruent.

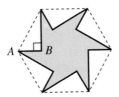

CONGRUENT TRIANGLES

Recall that two line segments $\overline{AB}$ and $\overline{CD}$ are congruent if they have the same length and two angles $\angle EFG$ and $\angle HIJ$ are congruent if they have the same angle measure. Also, remember that line segments of the same length and angles of the same measure are identified using special marks as illustrated in Figure 4.5.

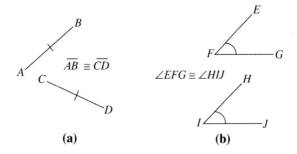

$\overline{AB} \cong \overline{CD}$

$\angle EFG \cong \angle HIJ$

(a)   **(b)**

FIGURE 4.5

Because the lengths of sides and measures of angles determine the size and shape of a triangle, congruent triangles can be defined using their various parts.

Definition

Congruent Triangles

$\triangle ABC$ is **congruent** to $\triangle DEF$ (written $\triangle ABC \cong \triangle DEF$) under the correspondence $A \leftrightarrow D$, $B \leftrightarrow E$, $C \leftrightarrow F$ if and only if
(i) all three pairs of corresponding angles are congruent, and
(ii) all three pairs of corresponding sides are congruent.

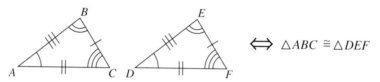

$\Longleftrightarrow$ $\triangle ABC \cong \triangle DEF$

$\triangle ABC \cong \triangle DEF$ if and only if $\angle A \cong \angle D$, $\angle B \cong \angle E$, and $\angle C \cong \angle F$, and $\overline{AB} \cong \overline{DE}$, $\overline{BC} \cong \overline{EF}$, and $\overline{AC} \cong \overline{DF}$.

The following relationships follow from the definition:

1. **Reflexive Property:** $\triangle ABC \cong \triangle ABC$ for all triangles $\triangle ABC$.
2. **Symmetric Property:** If $\triangle ABC \cong \triangle DEF$, then $\triangle DEF \cong \triangle ABC$.
3. **Transitive Property:** If $\triangle ABC \cong \triangle DEF$ and $\triangle DEF \cong \triangle GHI$, then $\triangle ABC \cong \triangle GHI$.

EXAMPLE 4.8   Decide if the following pairs of triangles are congruent under the implied correspondences in Figure 4.6.
(a) $\triangle ABC$ and $\triangle QRP$ (b) $\triangle GHI$ and $\triangle JKL$

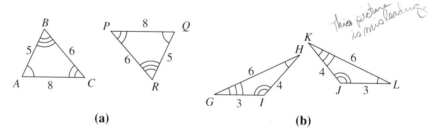

(a)   **(b)**

this picture is misleading.

FIGURE 4.6

SOLUTION
(a) $\triangle ABC \cong \triangle QRP$ because $\angle A \cong \angle Q$, $\angle B \cong \angle R$, $\angle C \cong \angle P$, $\overline{AB} \cong \overline{QR}$, $\overline{BC} \cong \overline{RP}$, and $\overline{AC} \cong \overline{QP}$.

(b) $\triangle GHI$ is not congruent to $\triangle JKL$ because, for example, $\overline{GH}$ and $\overline{JK}$ are not congruent. However, we do have $\triangle GHI \cong \triangle LKJ$ because all pairs of corresponding angles and sides are congruent under the correspondence $G \leftrightarrow L$, $H \leftrightarrow K$, $I \leftrightarrow J$.   •

Every triangle $\triangle ABC$ is congruent to itself under the correspondence $A \leftrightarrow A$, $B \leftrightarrow B$, $C \leftrightarrow C$, but it is interesting to observe that a triangle may be congruent to itself under different correspondences. For example, consider the equilateral triangle in Figure 4.7(a). Here $\triangle ABC \cong \triangle BCA$ because six pairs of the corresponding angles and sides are congruent under the correspondence $A \leftrightarrow B$, $B \leftrightarrow C$, $C \leftrightarrow A$. We could show in a similar fashion that $\triangle ABC \cong \triangle CBA$.

The different correspondences between a triangle and itself can be represented physically. When we say $\triangle ABC \cong \triangle ABC$, we can imagine a tracing of $\triangle ABC$ placed over $\triangle ABC$ with vertex A paired with A and so on. The two images will coincide. When we say $\triangle ABC \cong \triangle BCA$, we can imagine a tracing of $\triangle ABC$ rotated 120° clockwise [Figure 4.7(b)] and placed over $\triangle ABC$. In this way, vertex A is paired with vertex B of the original triangle and so on. Again, the two triangles will coincide.

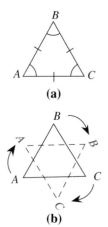

(a)

(b)

FIGURE 4.7

SAS CONGRUENCE

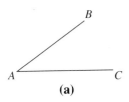

(a)

Now consider the partial triangle in Figure 4.8(a). To complete this triangle, we need only draw $\overline{BC}$ [Figure 4.8(b)]. That is, if two sides and the included angle between them are designated, the size and shape of the triangle are completely determined. Thus, to decide whether two triangles are congruent, it is sufficient to find only three appropriate pairs of parts (here, two sides and the included angle) that are congruent. This is the essence of the next postulate.

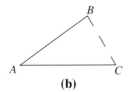

(b)

FIGURE 4.8

Postulate 4.1

SAS Congruence Postulate

If two sides and the included angle of one triangle are congruent respectively to two sides and the included angle of another triangle, then the two triangles are congruent.

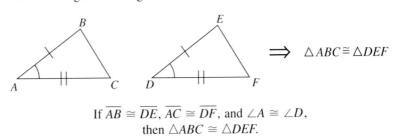

If $\overline{AB} \cong \overline{DE}$, $\overline{AC} \cong \overline{DF}$, and $\angle A \cong \angle D$,
then $\triangle ABC \cong \triangle DEF$.

The next theorem is an immediate consequence of Postulate 4.1 where the included angle is a right angle.

Theorem 4.2

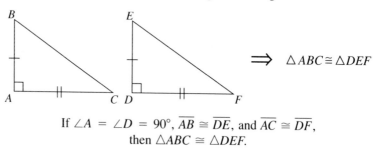

LL Congruence Theorem

If two legs of one right triangle are congruent respectively to two legs of another right triangle, then the two triangles are congruent.

If $\angle A = \angle D = 90°$, $\overline{AB} \cong \overline{DE}$, and $\overline{AC} \cong \overline{DF}$, then $\triangle ABC \cong \triangle DEF$.

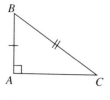

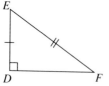

FIGURE 4.9

A similar congruence result holds if two right triangles have their hypotenuse and one of their legs congruent respectively (Figure 4.9). The Pythagorean Theorem can be used to show that the other two corresponding legs are congruent. Therefore, the two triangles are congruent by SAS. The proof of this theorem is left for the problem set.

Theorem 4.3

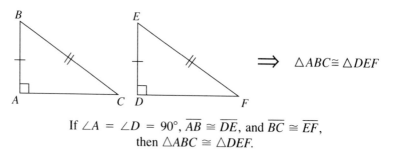

HL Congruence Theorem

If the hypotenuse and a leg of one right triangle are congruent respectively to the hypotenuse and a leg of another right triangle, then the two triangles are congruent.

If $\angle A = \angle D = 90°$, $\overline{AB} \cong \overline{DE}$, and $\overline{BC} \cong \overline{EF}$, then $\triangle ABC \cong \triangle DEF$.

The SAS Postulate can be used to verify many other relationships that appear in geometric figures. The advantage of using this postulate is that once two sides and the included angle of one triangle are found to be congruent respectively to two sides and the included angle of a second triangle, we can conclude that the other three parts of the triangles are also congruent. We will refer to this technique of proving that two angles or segments are congruent because they are corresponding parts of two congruent triangles by the phrase **corresponding parts of congruent triangles are congruent**. This method is abbreviated as **C.P.** (short for *c*orresponding *p*arts). The next example illustrates this technique.

EXAMPLE 4.9   Congruent parts are indicated in Figure 4.10. Prove that $\angle P \cong \angle S$.

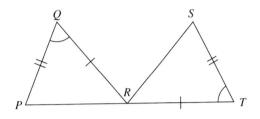

FIGURE 4.10

SOLUTION

Subgoal 1 Show that $\triangle PQR \cong \triangle STR$.

Subgoal 2 Show that $\angle P \cong \angle S$ by corresponding parts.
We will write this proof in paragraph form.

Proof of Subgoal 1 From Figure 4.10, $\angle Q \cong \angle T$, $\overline{PQ} \cong \overline{ST}$, and $\overline{QR} \cong \overline{TR}$. Thus, we have two sides and the included angle of one triangle congruent to two sides and the included angle of the other triangle. Therefore, $\triangle PQR \cong \triangle STR$ by the SAS Congruence Postulate.

Proof of Subgoal 2 Because the two triangles are congruent, all six parts of one triangle are congruent to the six corresponding parts of the other triangle. In particular, $\angle P \cong \angle S$ by C.P.   •

ASA CONGRUENCE

Now consider the partial triangle given in Figure 4.11(a), where two angles and their included side are specified. Notice how the third vertex B is established by extending the two sides [Figure 4.11(b)]. That is, only one triangle is possible for the given two angles and their included side. This figure suggests the following result.

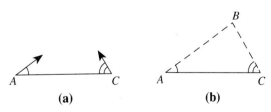

(a)   **(b)**

FIGURE 4.11

Postulate 4.2

ASA Congruence Postulate

If two angles and the included side of one triangle are congruent respectively to two angles and the included side of another triangle, then the two triangles are congruent.

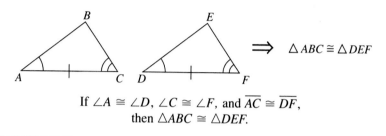

If $\angle A \cong \angle D$, $\angle C \cong \angle F$, and $\overline{AC} \cong \overline{DF}$,
then $\triangle ABC \cong \triangle DEF$.

Although we have assumed ASA Congruence as a postulate, it can be proved as a theorem using the SAS Congruence Postulate, but the proof is somewhat complex.

EXAMPLE 4.10   In Figure 4.12(a), $\angle A \cong \angle D$ and $\overline{AC} \cong \overline{DC}$. Prove that $\overline{AB} \cong \overline{DE}$.

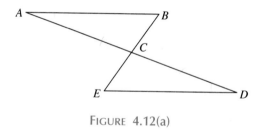

FIGURE 4.12(a)

SOLUTION   First we will label the figure using the information given [Fig. 4.12(b)], and then establish subgoals.

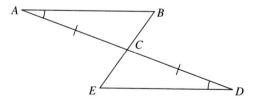

FIGURE 4.12(b)

Subgoal 1 Show that $\triangle ABC \cong \triangle DEC$.

Subgoal 2 Show that $\overline{AB} \cong \overline{DE}$ by corresponding parts.
We will write this proof in statement-reason form.

Proof

Statement	Reason
1. $\angle A \cong \angle D$	1. Given
2. $\overline{AC} \cong \overline{DC}$	2. Given
3. $\angle BCA \cong \angle ECD$	3. Vertical angles formed by intersecting lines are congruent.
4. $\triangle ABC \cong \triangle DEC$	4. ASA Congruence Postulate
5. $\overline{AB} \cong \overline{DE}$	5. C.P.

[NOTE: Statements 1–4 constitute a proof of Subgoal 1. Statement 5 proves Subgoal 2.]

SSS CONGRUENCE

Next consider the three line segments given in Figure 4.13(a). (Notice that the sum of the lengths of any two segments exceeds the length of the third segment.)

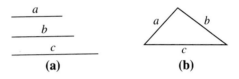

FIGURE 4.13

If we try to draw a triangle using these three segments for sides, the size and shape of the triangle are determined [Figure 4.13(b)]. In other words, only one triangle can be constructed. It is possible to prove SSS congruence. However, due to the complexity of the proof, we postulate it next.

Postulate 4.3

> **SSS Congruence Postulate**
>
> If three sides of one triangle are congruent respectively to three sides of another triangle, then the two triangles are congruent.
>
>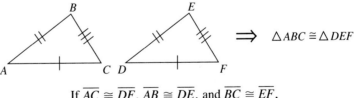
>
> If $\overline{AC} \cong \overline{DF}$, $\overline{AB} \cong \overline{DE}$, and $\overline{BC} \cong \overline{EF}$, then $\triangle ABC \cong \triangle DEF$.

The following example shows how Postulate 4.3 can be used to learn more about geometric figures.

EXAMPLE 4.11   For the kite in Figure 4.14(a), prove that $\angle A \cong \angle C$.

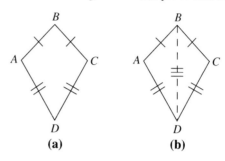

(a)   (b)

FIGURE 4.14

SOLUTION   First, draw in diagonal $\overline{BD}$ [Figure 4.14(b)]. Then $\triangle ABD \cong \triangle CBD$ by SSS because $\overline{AB} \cong \overline{CB}$, $\overline{AD} \cong \overline{CD}$, and $\overline{BD} \cong \overline{BD}$. Therefore, $\angle A \cong \angle C$ by C.P.   •

Now that we have the SSS Congruence Postulate to use, we can prove the converse of the Pythagorean Theorem, which was introduced informally in Chapter 3 as the Test for Right Triangles. It is fascinating to observe that we can use the Pythagorean Theorem to prove its converse.

Theorem 4.4

Converse of the Pythagorean Theorem

If the sum of the squares of the lengths of two legs of a triangle equals the square of the third side, then the triangle is a right triangle.

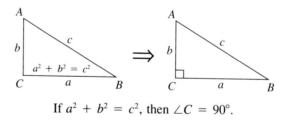

If $a^2 + b^2 = c^2$, then $\angle C = 90°$.

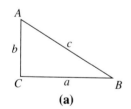

(a)

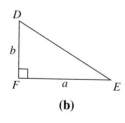

(b)

FIGURE 4.15

Given $\triangle ABC$ with $a^2 + b^2 = c^2$ [Figure 4.15(a)]

Prove $\angle C = 90°$.

Proof Construct a right triangle $\triangle DEF$ with legs of length a and b [Figure 4.15(b)]. By the Pythagorean Theorem, $a^2 + b^2 = DE^2$. Because $a^2 + b^2 = c^2$ is given, we have $c^2 = DE^2$, or $c = DE$. Therefore, $\triangle ABC \cong \triangle DEF$ by SSS. Consequently, $\angle C \cong \angle F$ by C.P., so $\angle C = 90°$.   •

Solution to Applied Problem

Because the saw blade is made out of a regular hexagon, the hypotenuses of the six triangles are all congruent. Also, the same length *AB* is cut at a right angle to form each tooth. Thus, all six triangles are congruent by the HL Congruence Theorem.

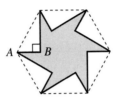

GEOMETRY AROUND US

The triangle is frequently seen in construction—in trusses, bridges, scaffolding, braces, and so on. The reason for its frequent use is its rigid shape. Any other polygon will shift and flex and change its shape. Only a triangle retains its fixed shape. Other polygons can be made rigid by adding diagonal sections to form triangles as in some railroad bridges such as the one illustrated.

PROBLEM SET 4.2

EXERCISES/PROBLEMS

1. Given that $\triangle PQR \cong \triangle XYZ$, list the six pairs of congruent parts for the two triangles.

2. Suppose $\triangle DEF \cong \triangle JKL$. List the six pairs of congruent parts for the two triangles.

3. For the pair of triangles shown, write an appropriate congruence statement. Be careful about the order in which you list the vertices.

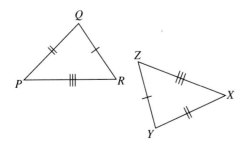

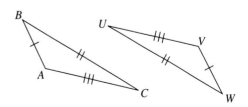

4. For the pair of triangles shown, write an appropriate congruence statement. Be careful about the order in which you list the vertices.

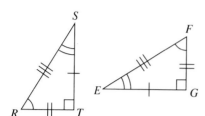

For each figure in Exercises 5–18, decide whether a pair of triangles is necessarily congruent. If so, write an appropriate congruence statement and specify which congruence principle applies. Be careful *not* to make any assumptions about triangles that just "appear" to be congruent.

5.

6.

7.

8.

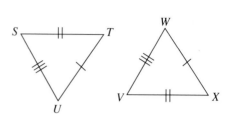

9.

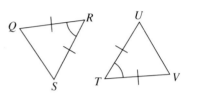

10.

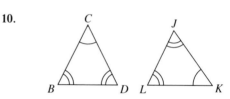

11.

12.

13.

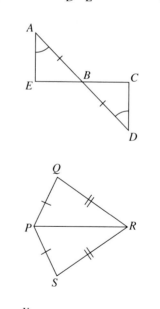

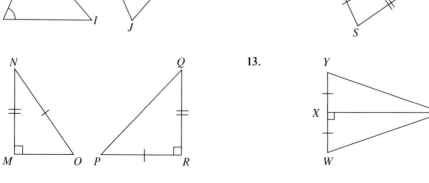

14.

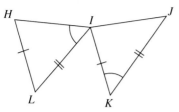

15.

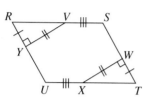

16.

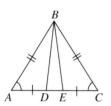

17.

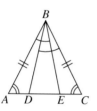

18.

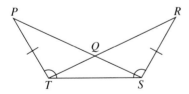

19. Two pairs of triangles are congruent in this figure. Give both of them, and state which congruence principles apply.

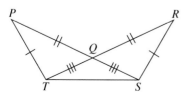

20. Two pairs of triangles are congruent in this figure. Give both of them, and state which congruence principles apply.

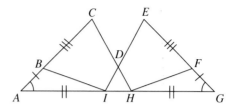

Refer to the following triangles for Exercises 21–26.

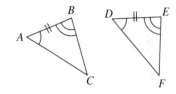

21. $\overline{AB} \cong$ _____

22. $\angle A \cong \angle$ _____

23. $\angle E \cong \angle$ _____

24. $\triangle FED \cong \triangle$ _____ . Which congruence principle guarantees this?

25. $\overline{CB} \cong$ _____ . Why?

26. $\angle F \cong \angle$ _____ . Why?

Refer to the following *regular* hexagon for Exercises 27–32.

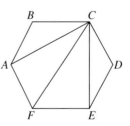

27. $\angle B \cong \angle$ _____

28. $\triangle BAC \cong \triangle$ _____ . Which congruence principle guarantees this?

29. $\overline{AC} \cong$ _____ . Why?

30. $\angle BAC \cong \angle$ _____

31. $\angle FEC \cong \angle$ _____

32. $\triangle FEC \cong \triangle$ _____. Which congruence principle guarantees this?

33. If possible, draw an example of two noncongruent triangles that satisfy the following conditions. If not possible, explain why.
   (a) Three pairs of corresponding parts are congruent.
   (b) Four pairs of parts are congruent.
   (c) Five pairs of parts are congruent.

PROOFS

34. In the figure shown $\angle BAC \cong \angle DAC$ and $\angle ACB \cong \angle ACD$. Prove that $\overline{AB} \cong \overline{AD}$. [HINT: First show that $\triangle ABC \cong \triangle ADC$.]

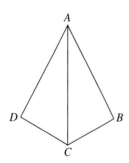

35. In the following figure $\overline{UW} \cong \overline{YW}$ and $\overline{VW} \cong \overline{XW}$. Prove that $\angle U \cong \angle Y$. [HINT: First show that $\triangle UVW \cong \triangle YXW$.]

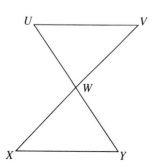

36. In the triangles shown, congruent parts are marked. Prove that $\angle Q \cong \angle T$.

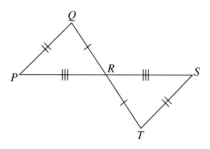

37. ABCD is a rhombus with diagonal $\overline{BD}$. Prove that the diagonal divides the rhombus into two congruent triangles.

38. In the following figure, $\angle EDH = \angle GHD = 90°$, and $\overline{DE} \cong \overline{HG}$. Prove that $\angle GDH \cong \angle EHD$.

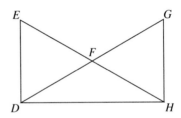

39. In ABCD, line segments $\overline{AC}$ and $\overline{BD}$ bisect each other at E. Prove that $\angle DBC \cong \angle BDA$.

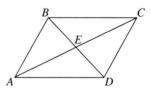

40. In ABCD, $\overline{BC} \cong \overline{AD}$ and $\overline{AB} \cong \overline{CD}$. Prove $\angle BCA \cong \angle DAC$.

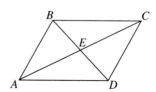

41. In $ABCD$, $\overline{AC}$ and $\overline{BD}$ bisect each other at E. Prove $\triangle DBC \cong \triangle BDA$.

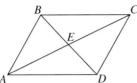

42. Given $\triangle ABC$, where $\overline{BD}$ is an altitude and $AD = CD$. Prove $\triangle ADB \cong \triangle CDB$.

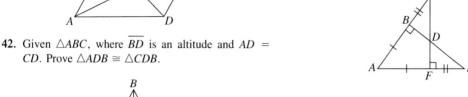

43. In the following figure, congruent segments are marked. Prove $\angle ACF \cong \angle AEB$.

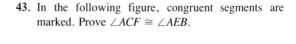

44. Prove Theorem 4.3: If the hypotenuse and a leg of one right triangle are congruent respectively to the hypotenuse and a leg of another right triangle, then the two triangles are congruent.

4.3
PROBLEM SOLVING USING TRIANGLE CONGRUENCE

Applied Problem

It is recommended that gardeners plant tomatoes with three feet between plants. A truck farmer planting many tomato plants might use a square arrangement of plants in rows, as shown next.

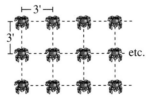

On the other hand, the farmer could make maximal use of the land by planting the tomatoes in a triangular grid, as shown next.

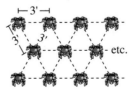

How much space between rows is saved by using the triangular planting arrangement?

In this section, we use triangle congruence to prove properties of geometric figures. In particular, we will derive some relationships that will be useful in later chapters.

ISOSCELES TRIANGLES

In Chapter 2 we used the reflection symmetry of an isosceles triangle to establish a relationship between its base angles. The next theorem is a formal verification of that relationship.

Theorem 4.5

In an isosceles triangle, the angles opposite the congruent sides are congruent.

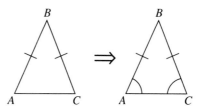

If $\overline{AB} \cong \overline{CB}$, then $\angle C \cong \angle A$.

Theorem 4.5 can be summarized as follows: Base angles of an isosceles triangle are congruent.

Given $\triangle ABC$ with $\overline{AB} \cong \overline{CB}$

Prove $\angle C \cong \angle A$

Subgoal 1 Prove that $\triangle ABC$ in Figure 4.16(a) is congruent to itself using the correspondence $\triangle ABC \cong \triangle CBA$.

Subgoal 2 Prove that the base angles are congruent by C.P.

(a)

(b)

FIGURE 4.16

Proof

Statement	**Reason**
1. $\overline{AB} \cong \overline{CB}$	1. Given
2. $\angle B \cong \angle B$	2. Identity
3. $\overline{CB} \cong \overline{AB}$	3. Given
4. $\triangle ABC \cong \triangle CBA$	4. SAS Postulate [Figure 4.16(b)]
5. $\angle A \cong \angle C$	5. C.P.

Because every equilateral triangle is also an isosceles triangle, Theorem 4.5 applies to equilateral triangles also. In fact, by viewing an equilateral triangle in two different ways, the following corollary shows that all the angles in an equilateral triangle are congruent. The proof of this corollary is left for the problem set.

Corollary 4.6

Every equilateral triangle is equiangular.

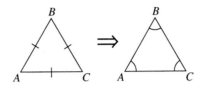

If $\overline{AB} \cong \overline{BC} \cong \overline{AC}$, then $\angle A \cong \angle B \cong \angle C$.

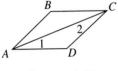

FIGURE 4.17

EXAMPLE 4.12   In the rhombus $ABCD$ in Figure 4.17 show that $\angle 1$ and $\angle 2$ formed by the diagonal $\overline{AC}$ are congruent.

SOLUTION   In a rhombus, all four sides are congruent. Therefore, $\triangle ADC$ is isosceles because $\overline{AD} \cong \overline{CD}$. By Theorem 4.5 the angles opposite these sides are congruent. Hence $\angle 1 \cong \angle 2$.

The next two results follow from Postulate 4.2, the ASA Congruence Postulate. Their proofs are similar to the ones for Theorem 4.5 and Corollary 4.6 and are left for the problem set.

Theorem 4.7

If two angles of a triangle are congruent, then the sides opposite those angles are congruent.

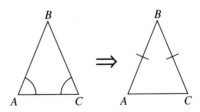

If $\angle C \cong \angle A$, then $\overline{AB} \cong \overline{CB}$.

This theorem can be summarized as follows: If two angles of a triangle are congruent, then the triangle is isosceles.

By extension, if all three angles of a triangle are congruent, then the triangle is equilateral. This result is stated in the next corollary.

Corollary 4.8

Every equiangular triangle is equilateral.

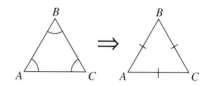

If $\angle A \cong \angle B \cong \angle C$, then $\overline{AB} \cong \overline{BC} \cong \overline{AC}$.

PERPENDICULAR BISECTOR OF A SEGMENT

The next two theorems display a relationship between isosceles triangles and perpendicular bisectors.

Theorem 4.9

In an isosceles triangle, the ray that bisects the vertex angle bisects the base and is perpendicular to it.

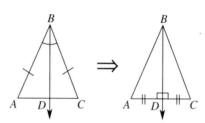

If $\overline{BA} \cong \overline{BC}$ and $\angle ABD \cong \angle CBD$,
then $\overline{BD}$ is the perpendicular bisector of $\overline{AC}$.

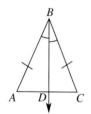

FIGURE 4.18

Given $\triangle ABC$ with $\overline{BA} \cong \overline{BC}$ and $\overline{BD}$ the angle bisector of $\angle B$ (Figure 4.18)

Prove $\overline{AD} \cong \overline{CD}$ and $\overline{BD} \perp \overline{AC}$

Subgoal Prove that $\triangle ABD \cong \triangle CBD$. Then use C.P.

Proof

Statement	Reason
1. $\overline{BA} \cong \overline{BC}$	1. Given
2. $\angle ABD \cong \angle CBD$	2. $\overline{BD}$ is the angle bisector.
3. $\overline{BD} \cong \overline{BD}$	3. Identity
4. $\triangle ABD \cong \triangle CBD$	4. SAS Congruence
5. $\overline{AD} \cong \overline{CD}$	5. C.P.
6. $\angle ADB \cong \angle CDB$	6. C.P.

7. ∠*ADB* and ∠*CDB* are right
angles, therefore $\overline{BD}$ is
perpendicular to $\overline{AC}$.

7. Two congruent angles that
are supplementary are right
angles.

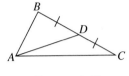

FIGURE 4.19

A **median** of a triangle is a line segment whose endpoints are a vertex of the triangle and the midpoint of the side opposite that vertex. In Figure 4.19, $\overline{AD}$ is a median of △*ABC*. Therefore, Theorem 4.9 can be summarized as follows: In an isosceles triangle, the bisector of the vertex angle is also the altitude to the base, the perpendicular bisector of the base, and the median to the base. Because every equilateral triangle is isosceles, this theorem also holds for any equilateral triangle. In the equilateral triangle any angle may be considered the "vertex angle" and any side the "base."

EXAMPLE 4.13

(a) In Figure 4.20(a), *B* is on the perpendicular bisector of $\overline{AC}$. Find *AB* and *BC*.

(b) In Figure 4.20(b), *H* is the midpoint of $\overline{EG}$. Find ∠*EHF*.

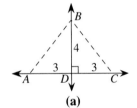

(a)

SOLUTION

(a) By the Pythagorean Theorem, $AB = \sqrt{3^2 + 4^2} = 5$. Similarly, $BC = 5$. Thus, $AB = BC$; that is, *B* is equidistant from *A* and *C*.

(b) First, △*EFH* ≅ △*GFH* by SSS. Thus, ∠*EHF* ≅ ∠*GHF* by corresponding parts. Because these angles are supplementary, ∠*EHF* = ∠*GHF* = 90°. This means that $\overline{FH}$ is the perpendicular bisector of $\overline{EG}$.

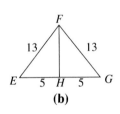

(b)

FIGURE 4.20

This example can be generalized to the following theorem.

Theorem 4.10

Perpendicular Bisector Theorem

A point is on the perpendicular bisector of a line segment if and only if it is equidistant from the endpoints of the segment.

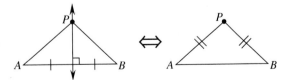

P is on the perpendicular bisector of $\overline{AB}$ if and only if $AP = BP$.

Because Theorem 4.10 is an "if and only if" theorem, we must prove two parts.

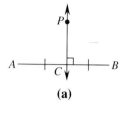

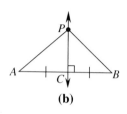

(b)

FIGURE 4.21

Subgoal 1 Prove that if a point is on the perpendicular bisector of a segment, then it is equidistant from the endpoints of the segment.

Subgoal 2 Prove that if a point is equidistant from the endpoints of a segment, then it is on the perpendicular bisector of the segment.

Proof of Subgoal 1 Let P be a point on the perpendicular bisector of $\overline{AB}$ [Figure 4.21(a)]. Then, in Figure 4.21(b), we can see that $\triangle ACP \cong \triangle BCP$ by LL Congruence. Therefore, $AP = BP$ by corresponding parts, which means P is equidistant from A and B.

Proof of Subgoal 2 Let P be equidistant from the endpoints of $\overline{AB}$ [Figure 4.22(a)]. Let C be the midpoint of $\overline{AB}$, and draw $\overline{PC}$ [Figure 4.22(b)]. Then, $\triangle ACP \cong \triangle BCP$ by SSS Congruence. By corresponding parts, $\angle PCA \cong \angle PCB$. Also, $\angle PCA$ and $\angle PCB$ are supplementary, so $\angle PCA = \angle PCB = 90°$. Thus, $\overline{PC}$ is the perpendicular bisector of $\overline{AB}$, so P lies on the perpendicular bisector of $\overline{AB}$.   •

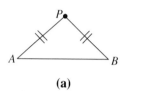

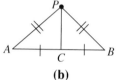

FIGURE 4.22

Theorem 4.10 will be applied in Section 7.4.

The proof of Theorem 4.10 was done in two parts due to the biconditional "if and only if." Because our goal was to prove this theorem, each of these two parts was considered as a subgoal to reach our goal. The next two theorems also will be organized into subgoals whose proofs will be combined to form a proof of the theorem.

EXTERIOR ANGLE OF A TRIANGLE

Although we showed informally in Chapter 2 that the angle sum in a triangle is 180°, we will not be able to prove this fact formally until Chapter 5. However, we can prove a theorem that shows a relationship between an exterior angle of a triangle and the two nonadjacent interior angles. Consider the exterior angle at C in Figure 4.23(a).

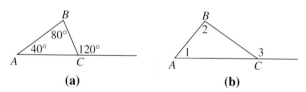

FIGURE 4.23

Notice that the exterior angle is larger than either of the two interior angles at A and B. This relationship seems true for any triangle. For example, in Figure 4.23(b) it appears that $\angle 3 > \angle 1$ and that $\angle 3 > \angle 2$. That is, the measure of an exterior angle at C is greater than the measure of either nonadjacent, interior angle. This relationship is stated in the next theorem.

Theorem 4.11

The measure of an exterior angle of a triangle is greater than the measure of either of the nonadjacent interior angles.

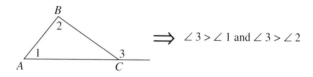

$\Rightarrow$ $\angle 3 > \angle 1$ and $\angle 3 > \angle 2$

Given $\triangle ABC$ with an exterior angle at C [Figure 4.24(a)]

Prove $\angle BCD > \angle A$ and $\angle BCD > \angle B$

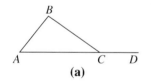

(a)

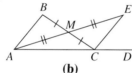

(b)

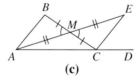

(c)

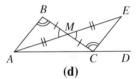

(d)

FIGURE 4.24

Plan Because this proof requires several steps, we will decompose it into subgoals that can be more easily proved. Then the proof of the theorem will follow from the proofs of the subgoals.

Subgoal 1 Draw $\overline{AE}$ through M, where M is the midpoint of $\overline{BC}$ and $AM = EM$ [Figure 4.24(b)]. Prove that $\triangle AMB \cong \triangle EMC$.

Subgoal 2 Using the result of Subgoal 1, prove that $\angle BCD > \angle B$.

Proof of Subgoal 1 Because $\angle AMB$ and $\angle EMC$ are vertical angles, they are congruent [Figure 4.24(c)]. Therefore, $\triangle AMB \cong \triangle EMC$ by SAS.

Proof of Subgoal 2 By corresponding parts, $\angle B \cong \angle BCE$ [Figure 4.24(d)]. Also, by Figure 4.24(d), $\angle BCD > \angle BCE$. Therefore, by substitution, $\angle BCD > \angle B$.

Similarly, it can be shown that $\angle BCD > \angle A$. This proof is left for the problem set.

Solution to Applied Problem

In the square arrangement there are 3′ between rows. Consider the situation where the plants are 3′ apart in triangular regions such as △ABC.

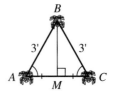

In equilateral △ABC, $AB = BC = AC = 3$ ft. If M is the midpoint of $\overline{AC}$, then $\overline{BM}$ is the perpendicular bisector of $\overline{AC}$. Therefore, $AM = MC = 3/2$, and BM, which represents the distance between rows, can be determined using the Pythagorean Theorem.

$$AB^2 = AM^2 + BM^2$$
$$3^2 = AM^2 + (3/2)^2$$
$$9 = AM^2 + 9/4$$
$$27/4 = AM^2$$

Therefore, $AM = \sqrt{27/4} \approx 2.6$ ft to the nearest tenth of a foot. So, by using the triangular arrangement, $3 - 2.6 = 0.4$ ft, or about 5 in., is saved between rows.

GEOMETRY AROUND US

The geodesic domes designed by Buckminster Fuller are special polyhedra composed of many congruent triangular faces of several types. The triangles provide rigidity to the structure. Also, the nearly spherical shape allows a minimum amount of material to be used in the construction to contain a given volume inside the structure.

PROBLEM SET 4.3

EXERCISES/PROBLEMS

1. In △XYZ, $\angle X = 70°$ and $\overline{YP}$ is the perpendicular bisector of $\overline{XZ}$. Find $\angle Z$. Justify your answer.

2. In the figure shown, △ADE and △ABE are isosceles with $\overline{AD} \cong \overline{DE}$ and $\overline{AB} \cong \overline{AE}$. If $\angle ADE = 50°$ and $\angle BAC = 45°$, find the measures of $\angle DEC$ and $\angle ACE$.

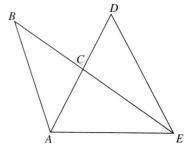

3. In the following figure, in addition to the congruent segments indicated, $\overline{BD} \cong \overline{DG}$. If A, G, and E are collinear and $\angle ADG = 28°$, find the measures of the following angles.
   (a) $\angle GBD$ (b) $\angle ACG$ (c) $\angle AGB$
   (d) $\angle FGE$ (e) $\angle CGD$ (f) $\angle BGC$

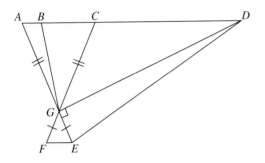

4. In the figure shown, $\overline{UX}$ is the perpendicular bisector of $\overline{SW}$. If $UX = 4$ in., $SX = 4$ in., and $TU = UV = 3$ in., complete the following.
   (a) $\triangle STU \cong \triangle$ _____ by _____
   (b) $SU =$ _____ in.
   (c) $VW =$ _____ in.

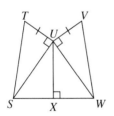

5. In isosceles $\triangle ABC$, $\angle ABD = \angle CBD$. Justify the following statements.

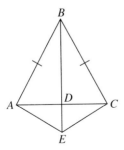

   (a) $\angle ADB = \angle BDC = 90°$ and $\overline{AD} \cong \overline{CD}$.
   (b) $\overline{AE} \cong \overline{CE}$.

PROOFS

6. Given that $\overline{XZ}$ bisects $\angle WZY$, $\overline{WX} \perp \overline{WZ}$, and $\overline{YX} \perp \overline{YZ}$. Prove that $\triangle XWZ \cong \triangle XYZ$.

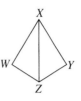

7. In the following figure, congruent angles are marked. Prove $\overline{WX} \cong \overline{YV}$.

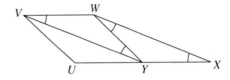

8. In the following figure $\overline{NO} \cong \overline{PO}$, $\overline{MS} \cong \overline{RQ}$, and $\angle OSR \cong \angle ORS$. Prove that $\triangle MNR \cong \triangle QPS$.

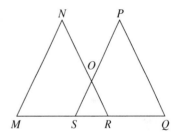

9. In isosceles $\triangle PQR$, $\overline{PY}$ and $\overline{RX}$ are altitudes, and $\overline{PX} \cong \overline{RY}$. Prove that $\triangle PXR \cong \triangle RYP$.

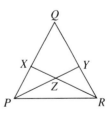

10. In $\triangle ABC$, $\overline{BD}$ is an altitude of $\triangle ABC$ and also the bisector of $\angle B$. Prove that $\triangle ABC$ is isosceles.

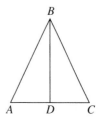

11. In $\triangle ABC$, $\overline{BD}$ is an altitude of $\triangle ABC$ and also the bisector of $\angle B$. Prove that $\overline{BD}$ is the perpendicular bisector of $\overline{AC}$.

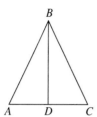

12. Prove or disprove: For any triangle $\triangle ABC$, if $\triangle ABC \cong \triangle CBA$, then $\triangle ABC$ is isosceles.

13. Prove that in any isosceles triangle, the median drawn from the vertex angle is the perpendicular bisector of the opposite side.

14. Prove that in any isosceles triangle, the median drawn from the vertex angle is the bisector of that angle.

15. Prove Corollary 4.6: Every equilateral triangle is equiangular.

16. Prove Theorem 4.7: If two angles of a triangle are congruent, then the sides opposite those angles are congruent.

17. Prove Corollary 4.8: Every equiangular triangle is equilateral.

18. Prove that $\angle BCD > \angle A$ in Theorem 4.11.

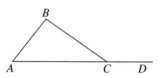

APPLICATION

19. A woman wants to align three pictures vertically along the middle of a wall as shown.

Once she has determined approximate locations for the pictures, explain how she can check the positions by using only a tape measure and a pencil.

4.4
THE BASIC GEOMETRIC CONSTRUCTIONS

Applied Problem

A drafting technician wishes to calculate the shortest distance from point P to the roadway using the drawing shown. If the scale is 1 cm:20 m, how can she calculate the distance?

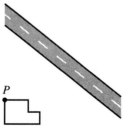

The ancient Greeks used a compass and straightedge to construct geometric figures. A **compass** is a device that is used to draw circles or arcs of circles (Figure 4.25). A **straightedge** is a device that is used to draw straight line segments. Neither of these two devices has marks for measuring. Some basic constructions made using a compass and straightedge are introduced in this section. We will use triangle congruence conditions to prove that those constructions are correct. Then the basic constructions will be combined to make more complex constructions.

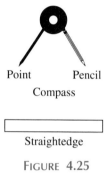

Point   Pencil
Compass

Straightedge

FIGURE 4.25

COPYING LINE SEGMENTS AND ANGLES

Construction 1.   *To copy a line segment*

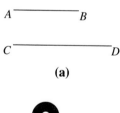

(a)

Procedure To copy $\overline{AB}$, first draw a line segment $\overline{CD}$, using a straightedge [Figure 4.26(a)]. To make a segment $\overline{CE}$ on $\overline{CD}$ that is congruent to $\overline{AB}$, spread your compass so that the point is on A and the pencil point is on B [Figure 4.26(b)]. Pick up your compass, put the point on C and swing an arc to intersect $\overline{CD}$ at some point E [Figure 4.26(c)]. Then $\overline{CE}$ will be congruent to $\overline{AB}$.

Justification Because the compass is rigid, the lengths AB and CE will be the same. Therefore, $\overline{AB} \cong \overline{CE}$.

Construction 2.   *To copy an angle*

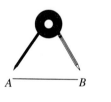

(b)

Procedure Consider $\angle A$ and $\overline{DE}$ [Figure 4.27(a)].

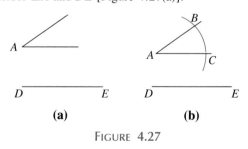

(a)   (b)

FIGURE 4.27

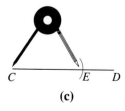

(c)

FIGURE 4.26

To copy $\angle A$ so that point A corresponds to D and one side of $\angle A$ corresponds to $\overline{DE}$, first swing an arc crossing the sides of $\angle A$ forming points B and C [Figure 4.27(b)].

Pick up your compass, put the point on D and, using the radius AC, swing an arc intersecting $\overline{DE}$ forming point F [Figure 4.27(c)].

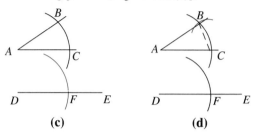

(c)   **(d)**

FIGURE 4.27

Use your compass to measure the distance BC [Figure 4.27(d)]. Set the opening of the compass as BC. Pick up your compass, put the point on F and swing an arc intersecting your arc through F forming G [Figure 4.27(e)].

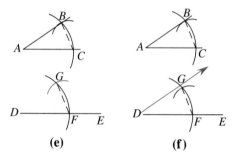

(e)   **(f)**

FIGURE 4.27

Draw $\overrightarrow{DG}$. Then $\angle D$ will be congruent to $\angle A$ [Figure 4.27(f)].

Justification   Using the larger arcs in Figure 4.27(f) we can conclude that $AB = AC = DG = DF$. Also, in Figure 4.27(f), because $\overline{GF}$ is copied from $\overline{BC}$, we can conclude that $GF = BC$. Thus, $\triangle ABC \cong \triangle DGF$ by SSS and $\angle A \cong \angle D$ by C.P.   •

BISECTING ANGLES

Construction 3.   *To bisect an angle*

Procedure   Consider $\angle A$ [Figure 4.28(a)]. To bisect $\angle A$, swing an arc crossing the sides of $\angle A$ forming B and C [Figure 4.28(b)].

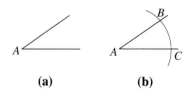

(a)   **(b)**

FIGURE 4.28

Pick up your compass, put the point on B and swing an arc [Figure 4.28(c)]. Then put the point on C and, using the same opening, swing an arc intersecting your first arc forming D [Figure 4.28(d)].

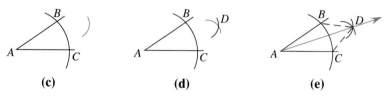

(c)   (d)   (e)

FIGURE 4.28

Draw $\overrightarrow{AD}$ [Figure 4.28(e)]. Then $\overrightarrow{AD}$ is the bisector of $\angle A$.

Justification By the construction in Figures 4.28(a)–(d), $AB = AC$ and $BD = CD$. Also, $AD = AD$ [Figure 4.28(e)]. Thus, $\triangle ABD \cong \triangle ACD$ by SSS. Therefore, $\angle BAD \cong \angle CAD$ by C.P., so $\overrightarrow{AD}$ bisects $\angle BAC$.   •

CONSTRUCTING PERPENDICULARS

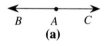

(a)

Construction 4. *To construct a perpendicular to a point on a line*

Procedure View $\angle BAC$ as a straight angle [Figure 4.29(a)]. Then we can construct a line perpendicular to $\overleftrightarrow{BC}$ at point A by bisecting $\angle BAC$. Follow the same procedure as in Construction 3 [Figure 4.29(b)].

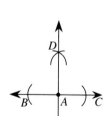

(b)

FIGURE 4.29

Justification This is a special case of Construction 3 where the angle is a straight angle. Thus, $\angle BAD \cong \angle CAD$ by Construction 3. Since these angles are supplementary, they must both be 90°.   •

Construction 5. *To construct a perpendicular to a line from a point not on the line*

Procedure Given point A and line l [Figure 4.30(a)], swing an arc with center A that intersects l forming B and C [Figure 4.30(b)].

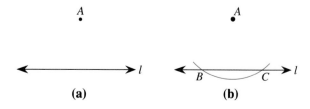

(a)   (b)

FIGURE 4.30

Now bisect $\angle BAC$ using Construction 3 [Figure 4.30(c)]. Then $\overrightarrow{AD}$ is perpendicular to l at E [Figure 4.30(d)].

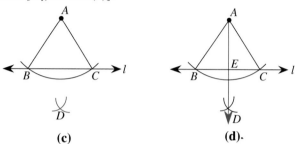

(c)   (d).

FIGURE 4.30

Justification According to Construction 3, $\overrightarrow{AD}$ bisects $\angle BAC$, so $\angle BAE \cong \angle CAE$ in Figure 4.30(d). Also, $AB = AC$ and $AE = AE$. So, by SAS, $\triangle BEA \cong \triangle CEA$, and $\angle BEA \cong \angle CEA$ by C.P. Because $\angle BEA$ and $\angle CEA$ are congruent and supplementary, they are right angles. Hence, $\overrightarrow{AD}$ is perpendicular to $\overline{BC}$.   •

Construction 6.   *To construct the perpendicular bisector of a segment*

Procedure To bisect $\overline{BC}$ [Figure 4.31(a)], first place the compass point at B and make arcs above and below $\overline{BC}$ [Figure 4.31(b)], then place the point of the compass at C and make arcs above and below using the same radius, forming points A and D [Figure 4.31(c)].

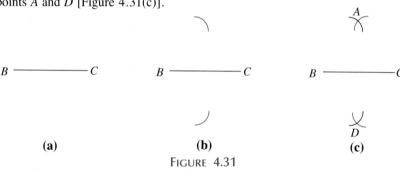

(a)   (b)   (c)

FIGURE 4.31

Draw $\overleftrightarrow{AD}$ [Figure 4.31(d)]. Then $\overrightarrow{AD}$ is perpendicular to $\overline{BC}$ at E.

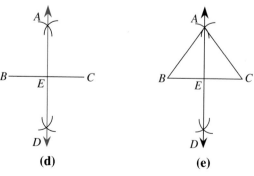

(d)   (e)

FIGURE 4.31

Justification Using an argument similar to that for Construction 5, we can conclude that $\triangle ABE \cong \triangle ACE$ [Figure 4.31(e)] and that $\overline{AD}$ is perpendicular to $\overline{BC}$. By C.P. we can conclude that $BE = CE$. Thus, $\overline{AD}$ is the perpendicular bisector of $\overline{BC}$.

We also can use Theorem 4.10 to justify this construction. Because the same opening on the compass was used for all arcs, we know that $AB = AC$. Thus, A is equidistant from B and C, so A lies on the perpendicular bisector of $\overline{BC}$. Likewise, D must lie on the perpendicular bisector of $\overline{BC}$ because $DB = DC$. Therefore, $\overleftrightarrow{AD}$ is the perpendicular bisector of $\overline{BC}$.   •

EXAMPLE 4.14   Using only a compass and straightedge, construct a 30-60 right triangle with hypotenuse of length c as given in Figure 4.32(a).

$$c$$

FIGURE 4.32(a)

SOLUTION   There are several ways to do this construction. We will first construct a right angle by using Construction 4 [Figure 4.32(b)].

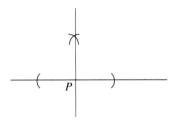

FIGURE 4.32(b)

Now, we can use the fact that in a 30-60 right triangle the length of the side opposite the 30° angle is half the length of the hypotenuse. So, one leg in this triangle must have length $c/2$. We bisect the segment of length c given using Construction 6 and copy half of that segment, making it leg $\overline{PQ}$ in the desired triangle [Figure 4.32(c)].

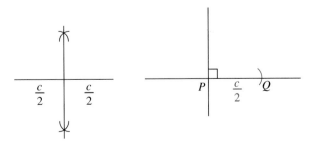

FIGURE 4.32(c)

Now the hypotenuse of the triangle must have length c. Placing the point of the compass at Q we make an arc with radius c. The point R where this arc intersects the perpendicular through P is the third vertex of the triangle [Figure 4.32(d)].

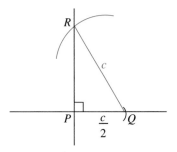

FIGURE 4.32(d)

EXAMPLE 4.15   Construct $\triangle ABC$ given two angles, $\angle A$ and $\angle B$, and their included side of length c as shown. Then construct the altitude from vertex C.

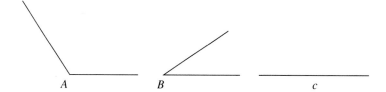

SOLUTION   We start by copying $\angle A$ [Figure 4.33(a)].

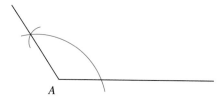

FIGURE 4.33(a)

Then mark off length c on one leg of A [Figure 4.33(b)]. The point of intersection will be vertex B.

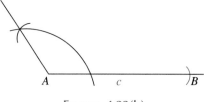

FIGURE 4.33(b)

Now copy $\angle B$. Extend the side of the angle at B so that it intersects the side from A forming point C [Figure 4.33(c)].

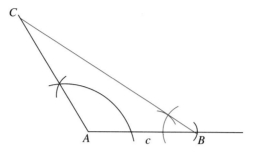

FIGURE 4.33(c)

Next we construct the altitude of $\triangle ABC$ from C [Figure 4.33(d)].

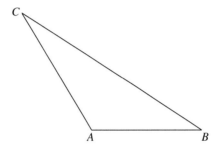

FIGURE 4.33(d)

The altitude from C is the line segment drawn from C that is perpendicular to $\overleftrightarrow{AB}$. We extend $\overline{AB}$ through A and construct a perpendicular from C to $\overleftrightarrow{AB}$ forming point D [Figure 4.33(e)]. Then $\overline{CD}$ is an altitude of $\triangle ABC$. •

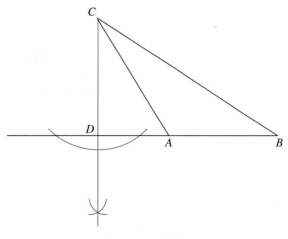

FIGURE 4.33(e)   •

Solution to Applied Problem

The shortest distance from P to the roadway is the length of the perpendicular segment from P to the roadway. She can construct a perpendicular from P to the line determined by the road and measure the length on the drawing.

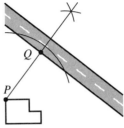

If $PQ = 2.3$ cm, for example, then the actual distance d is found as follows:

$$\frac{1 \text{ cm}}{20 \text{ m}} = \frac{2.3 \text{ cm}}{d}, \quad \text{or} \quad d = 20(2.3) = 46 \text{ m}$$

GEOMETRY AROUND US

The Penrose Triangle, a familiar example of an optical illusion, is composed of three triangles with congruent corresponding angles. The figure appears to be three-dimensional, but is actually impossible to construct.

PROBLEM SET 4.4

EXERCISES/PROBLEMS

1. Draw a line segment $\overline{AB}$. Use a compass and straightedge to find the midpoint of $\overline{AB}$.

2. Draw a line segment $\overline{PQ}$. Use a compass and straightedge to divide it into four congruent segments.

3. Draw a line $\overleftrightarrow{AB}$ and a point P that is not on $\overleftrightarrow{AB}$. Use your compass and straightedge to construct a line through P perpendicular to $\overleftrightarrow{AB}$.

4. Draw line $\overleftrightarrow{AB}$. Choose another point, C, on $\overleftrightarrow{AB}$. Use your compass and straightedge to construct a line that is perpendicular to $\overleftrightarrow{AB}$ and that passes through C.

5. Draw an angle, $\angle A$. Use your compass and straightedge to bisect $\angle A$.

6. Draw an acute angle, $\angle A$. Use your compass and straightedge to construct another angle whose measure is twice that of $\angle A$.

7. Draw a line segment $\overline{AB}$. Use your compass and straightedge to construct another line segment whose length is $4(AB)$.

8. Draw a line segment $\overline{AB}$. Use your compass and straightedge to construct another line segment whose length is $2.5(AB)$.

9. Construct an equilateral triangle given a side of length a.

a
_____

10. Construct a triangle given three sides of lengths a, b, and c.

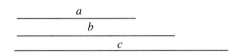

11. Construct a triangle given two sides of lengths a and b, and their included angle, $\angle C$.

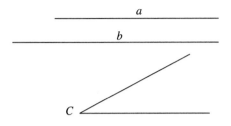

12. Construct a triangle given two angles, $\angle A$ and $\angle B$, and their included side of length c.

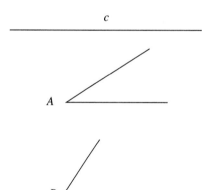

13. Use a compass and straightedge to construct a 60° angle.

14. Use a compass and straightedge to construct a 30° angle.

15. Use a compass and straightedge to construct a 45° angle.

16. Use a compass and straightedge to construct a 15° angle.

17. Use a compass and straightedge to construct a 120° angle.

18. Use a compass and straightedge to construct a 75° angle.

19. Construct a right triangle with hypotenuse of length c and a leg of length a.

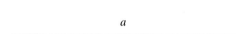

20. Construct a right isosceles triangle with legs of length a.

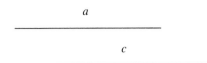

21. Trace $\triangle ABC$ and construct the altitude from B.

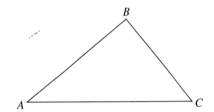

22. Trace $\triangle ABC$ and construct the median from B.

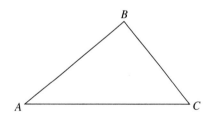

23. Trace $\triangle ABC$ and construct the perpendicular bisector to $\overline{AC}$ where $\triangle ABC$ is isosceles. Also construct the bisector of $\angle B$ in $\triangle ABC$. What do you observe?

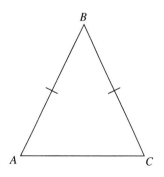

24. Trace square *ABCD* and construct the bisectors of ∠*B* and ∠*D*. What do you observe?

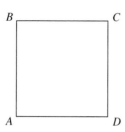

25. Trace △*ABC* and construct the perpendicular bisector of each side of △*ABC*. What do you observe?

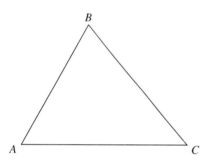

Repeat this construction with two other triangles of your choice. What generalization can you make?

26. Trace △*ABC* and construct the bisector of each angle of △*ABC*. What do you observe?

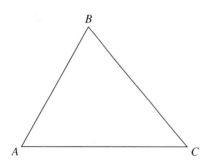

Repeat this construction with two other triangles of your choice. What generalization can you make?

27. Trace △*ABC* and construct the altitude to each side of △*ABC*. What do you observe?

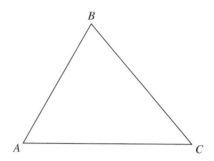

Repeat this construction with two other triangles of your choice. What generalization can you make?

28. Trace △*ABC* and construct the median to each side of △*ABC*. What do you observe?

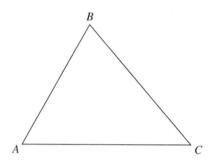

Repeat this construction with two other triangles of your choice. What generalization can you make?

PROOFS

29. Justify your construction in Exercise 9.

30. Justify your construction in Exercise 10.

31. Justify your construction in Exercise 11.

32. Justify your construction in Exercise 12.

33. Justify your construction in Exercise 21.

34. Justify your construction in Exercise 22.

APPLICATIONS

35. A method sometimes employed by drafters to bisect a line segment $\overline{AB}$ follows.

A ———————————————— B

(i) Angles of 45° are drawn at A and at B.

(ii) A perpendicular is drawn from the point of intersection C.

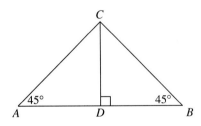

(iii) Point D divides $\overline{AB}$ into congruent segments. Explain why this method works.

36. Two neighbors at points A and B will share the cost of extending electric service from a road to their houses. A line will be laid that is equidistant from the two houses. How can they decide where the line should run?

• B

A •

37. A well will be dug to serve three residences located at points A, B, and C. So that the cost to each homeowner is about the same, the well should not be closer to one house than another.

• B

A •

• C

(a) Where can the well be dug if it must be equidistant from A and B?
(b) Where can the well be dug if it must be equidistant from A and C?
(c) Where should the well be dug if it must be equidistant from A, B, and C? How can this location be determined?

Solution to Initial Problem The following figure has a regular octagon adjacent to a regular pentagon. Find the measure of ∠ ABC.

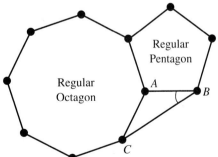

Because both the octagon and the pentagon are regular and they have a side in common, all their sides must be the same length. Therefore, △ABC is an isosceles triangle and ∠ACB = ∠ABC. If we can find the measure of ∠BAC, then we also can find the measure of ∠ABC; namely, ∠ABC = ½(180 − ∠BAC). But ∠ABC is 360° minus the sum of the measures of an angle of the octagon and an angle of the pentagon. Hence, if we solve the following two subgoals, we can solve the problem.

Subgoal 1 Find the measures of an angle of the octagon and an angle of the pentagon.

Subgoal 2 Find the measure of $\angle BAC$.

SOLUTION   The measure of an angle of a regular octagon is $\dfrac{(8 - 2)(180)}{8} =$

$135°$ and the measure of an angle of a regular pentagon is $\dfrac{(5 - 2)(180)}{5} =$

$108°$. Therefore, $\angle BAC = 360 - (135 + 108) = 117°$. So, $\angle ABC = \frac{1}{2}(180 - 117) = 31.5°$. •

Additional Problems Where the Strategy "Identify Subgoals" Is Useful

1. In the figure shown, $AB = BC = CD$, $AE = 4''$, and the area of $\triangle CDE = 6$ square inches. Find EC.

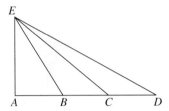

2. Asphalt roofing felt is sold in 36-inch rolls as shown. According to the label, one roll covers 324 square feet. Use the dimensions shown to determine the approximate thickness of a sheet of felt.

Writing for Understanding

1. Describe as many ways as you can to show that two triangles are congruent. By selecting appropriate proofs or problems, explain how you decide which congruence condition to apply in a given situation.

2. During the course of one day, pay close attention to conditionals and uses of deductive reasoning that you hear. Write at least five examples. Determine whether the arguments presented were valid.

3. Determine if the following are triangle congruence conditions. If they are, explain why; if they are not, explain why not. Examples and counterexamples may be used.
   (a) AAA   (b) SSA   (c) AAS

4. The "Geometry Around Us" section at the end of Section 4.2 discusses the construction of bridges. Do some research on the construction of at least three different types of bridges, focusing on the role geometry plays. Summarize your findings.

CHAPTER REVIEW

Following is a list of key vocabulary, notation, and ideas for this chapter. Mentally review these items and, where appropriate, write down the meaning of each term. Then restudy the material that you are unsure of before proceeding to take the chapter test.

Section 4.1—Reasoning and Proof in Geometry

Vocabulary/Notation

Direct (deductive) reasoning 165
Conditional statements 165
Hypothesis 165
Conclusion 165
Law of Detachment 166
Venn Diagram (Euler Circle) 166
Law of Syllogism 166

Converse 168
Inverse 168
Contrapositive 168
Biconditional 169
Vertical angles 169
Paragraph proof 170
Statement-reason proof 170–171

Main Ideas/Results

1. The Laws of Detachment and Syllogism form the basis for constructing proofs.

2. There are two common styles of proofs: paragraph and statement-reason.

3. If x and y are the measures of a pair of vertical angles, then $x = y$.

Section 4.2—Triangle Congruence Conditions

Vocabulary/Notation

Correspondence between vertices of triangles ($A \leftrightarrow D$) 175
Congruent triangles ($\triangle ABC \cong \triangle DEF$) 175

Reflexive Property 176
Symmetric Property 176
Transitive Property 176
Corresponding Parts (C.P.) 179
Converse of the Pythagorean Theorem 182

Main Ideas/Results

1. Two triangles are congruent if their corresponding sides and angles are congruent.
2. There are several triangle congruence conditions:

   SAS Congruence
   LL Congruence
   HL Congruence
   ASA Congruence
   SSS Congruence

3. In $\triangle ABC$, if $a^2 + b^2 = c^2$, then $\triangle ABC$ is a right triangle and $\angle C = 90°$.

Section 4.3—Problem Solving Using Triangle Congruence

Main Ideas/Results

1. A triangle is isosceles if and only if the angles opposite the congruent sides are congruent.
2. Every equilateral triangle is equiangular and conversely every equiangular triangle is equilateral.
3. In an isosceles triangle, the ray that bisects the vertex angle also bisects the base and is perpendicular to it.
4. A point is on the perpendicular bisector of a line segment if and only if it is equidistant from the endpoints of the segment.
5. An exterior angle of a triangle is greater than the measure of either of the nonadjacent interior angles.

Section 4.4—The Basic Geometric Constructions

Main Ideas/Results

1. Construction 1. To copy a line segment
2. Construction 2. To copy an angle
3. Construction 3. To bisect an angle
4. Construction 4. To construct a perpendicular to a point on the line
5. Construction 5. To construct a perpendicular to a line from a point not on the line
6. Construction 6. To construct the perpendicular bisector of a segment

PEOPLE IN GEOMETRY

Blaise Pascal was a precocious child, discovering for himself that the sum of the measures of the angles of a triangle is 180° by experimenting with folded paper triangles. At the age of 18 he invented the first mechanical calculating machine. It worked with a series of toothed wheels and was designed to assist his father, who was a tax official in France. Pascal went on to make important discoveries in the study of fluid pressure, probability, and conic sections (circles, ellipses, parabolas, and hyperbolas). One of his most remarkable results in geometry is what is called the Mystic Hexagram Theorem. It states that if one inscribes a hexagon in a conic section so that no two opposite sides of the hexagon are parallel and pairs of opposite sides are extended until they intersect, then the three points of intersection will be collinear.

CHAPTER 4 TEST

TRUE-FALSE

Mark as true any statement that is always true. Mark as false any statement that is never true or that is not necessarily true. Be able to justify your answers.

1. Every median of an equilateral triangle is also an altitude of the triangle.

2. If two angles of a triangle are congruent, then two sides of the triangle are congruent.

3. If a conditional statement and its inverse are both true, then they can be combined to form a true biconditional statement.

4. If four pairs of corresponding parts of two triangles are congruent, then the two triangles are congruent.

5. If one right triangle has a hypotenuse of 34 inches and one leg of 16 inches, and another right triangle has legs of 16 inches and 30 inches, then the triangles are congruent.

6. The "if" part of a conditional statement is called the conclusion.

7. If three angles of one triangle are congruent, respectively, to three angles of a second triangle, then the two triangles are congruent.

8. If P is a point on the perpendicular bisector of $\overline{MN}$ and $PM = 5$ cm, then $PN = 5$ cm.

9. Any two intersecting lines form two pairs of congruent vertical angles.

10. If two sides and one angle of one triangle are congruent, respectively, to two sides and one angle of a second triangle, then the two triangles are congruent.

EXERCISES/PROBLEMS

11. (a) Write the statement "Every integer is a real number" as a conditional in "if . . . , then . . ." form.

    (b) Write the converse of the conditional statement that you wrote in part (a).

    (c) Can the statements you wrote for parts (a) and (b) be combined to form a biconditional that is true?

12. In the figure shown next, BD is the perpendicular bisector of EC, and BE is the perpendicular bisector of AD. If $BC = 50$ cm, $EF = 13$ cm, and $AB = 43$ cm, find FD. Round your answer to the nearest tenth.

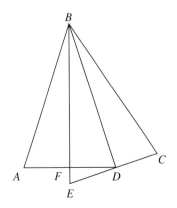

13. What conclusion, if any, can be deduced from the following statements?

 If Jan walks to work, then she carries her work shoes.

 If Jan carries her work shoes, then she doesn't carry a lunch.

 If Jan doesn't carry a lunch, then she eats out for lunch.

 If Jan eats out for lunch, then she overeats.

 If Jan overeats, then she gains weight.

 Jan walks to work.

 For problems 14–16, decide whether a pair of triangles in the figure is congruent. If so, write an appropriate congruence statement and specify which congruence principle applies. Be careful not to make any assumptions about triangles that just "appear" to be congruent.

14.

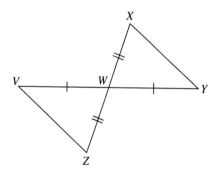

15.

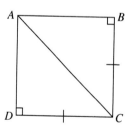

16.

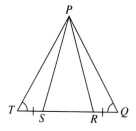

17. Use your compass and straightedge to construct an angle of 22.5°.

18. Use your compass and straightedge to construct the three medians of $\triangle PQR$.

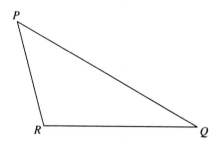

19. Use your compass and straightedge to construct a triangle with sides of length $2a$, $3a$, and $4a$, where length a is given.

——————— a ———————

20. By copying line segments and angles, construct a copy of quadrilateral $WXYZ$ as shown.

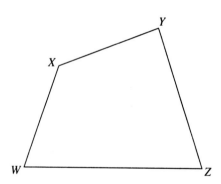

PROOFS

21. In the next figure, congruent segments are marked. Prove that $\triangle PST \cong \triangle RTS$.

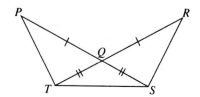

22. In isosceles trapezoid $ABCD$, $\overline{QA}$ bisects $\angle BAD$ and $\overline{PD}$ bisects $\angle CDA$. If $\angle BAD \cong \angle CDA$ and $\angle B \cong \angle C$, prove each of the following:

(a) $\triangle ARD$ is isosceles.

(b) $\triangle PCD \cong \triangle QBA$

(c) $\triangle PQR$ is isosceles.

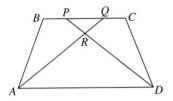

the arch. Round your answer to the nearest square foot.

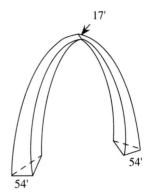

APPLICATIONS

23. The Gateway Arch in St. Louis, Missouri, has cross sections that are equilateral triangles pointing in at the base and pointing down at the top. The triangles at the base measure 54 feet on a side, and those at the top measure 17 feet on a side. Find the difference in the cross-sectional areas at the base and at the top of

24. A man starts jogging toward a friend's house. When he gets halfway there, he can't decide whether to continue or return home. What path from that point should he follow so that at any point in his run he will be the same distance from his home and his friend's house?

5

Parallel Lines and Quadrilaterals

EUCLID'S PARALLEL POSTULATE

The statement of one of Euclid's postulates was unusually long and complex. It stated the following.

If a straight line meets two straight lines, so as to make the two interior angles on the same side of it taken together less than two right angles, these straight lines, being continually produced, shall at length meet on the side on which are the angles which are less than two right angles.

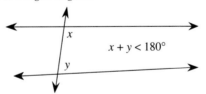

$$x + y < 180°$$

Many mathematicians who used *The Elements* were certain that this statement was provable from previously known theorems and postulates. However, it was shown that one could accept or reject this postulate.

Another way of thinking about parallel lines is to imagine three lines in a plane where two of the lines are perpendicular to the third line. It would seem that the two lines would be parallel. However, there are other geometries where this is not true. One of these geometries is spherical geometry. If you consider the Earth to be a sphere and great circles to be "lines," any two great circles from the North Pole form right angles with the equator. However, these two "lines" are not parallel. In fact, all "lines" perpendicular to the equator intersect at the North Pole. Other interesting types of geometries are studied in Topic 4 near the end of the book.

PROBLEM-SOLVING STRATEGIES

1. Draw a Picture
2. Guess and Test
3. Use a Variable
4. Look for a Pattern
5. Make a Table
6. Solve a Simpler Problem
7. Look for a Formula
8. Use a Model
9. Identify Subgoals
10. *Use Indirect Reasoning*

STRATEGY 10: USE INDIRECT REASONING

Consider the following situation. A room has two doors, say door X and door Y, as its only entrances, and we want to decide if a person enters the room through door X. To decide this directly, we could sit inside and watch door X until someone walked through—this would be a direct verification. Alternatively, we could watch door Y. If a person was observed in the room but didn't come through Y, we could conclude that door X was used—this is an indirect verification. Often in geometry there are proofs that are difficult to construct using direct reasoning. In those situations, indirect reasoning is especially useful.

INITIAL PROBLEM

The numbers 1 through 9 can be arranged in a 3 by 3 square array so that the sum of the numbers in each row, column, and diagonal is 15. Show that 1 cannot be in a corner. [HINT: Assume that 1 *can* be in a corner and show that this assumption leads to an impossible situation.]

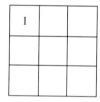

CLUES

The Use Indirect Reasoning strategy may be appropriate when

· Direct reasoning seems too complex or does not lead to a solution.
· Assuming the negation of what you are trying to prove narrows the scope of the problem.

A solution for the initial problem above is on page 263.

INTRODUCTION

Many important concepts, relationships, and results are presented in this chapter. First, indirect reasoning is discussed. This important technique can be used to construct proofs where direct reasoning is cumbersome, too difficult, or even impossible. Next, the Parallel Postulate is introduced. This postulate is the most famous and was the most controversial in Euclidean geometry. Then many results that follow from the Parallel Postulate and are related to parallelograms, rhombuses, rectangles, squares, and trapezoids are proved.

5.1
INDIRECT REASONING AND
THE PARALLEL POSTULATE

Applied Problem

In about 275 B.C., Eratosthenes developed a method for approximating the size of the earth. He observed that at noon on June 21 in Syene, Egypt, the sun was directly overhead. He determined that at the same time 500 miles north in Alexandria, the inclination of the sun's rays to the vertical was approximately 7.2°. Assuming that the sun's rays hit the earth in parallel lines, how did Eratosthenes calculate the circumference of the earth?

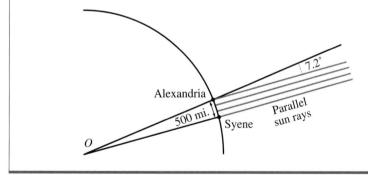

INDIRECT REASONING

In Chapter 4 direct reasoning was used to derive and prove many useful relationships in geometry. There are situations in mathematics where direct reasoning is difficult to apply. For this reason, we will begin to use another form of reasoning, indirect reasoning, in this chapter. The first example illustrates the use of both direct and indirect reasoning to prove the same result.

EXAMPLE 5.1   In $\triangle ABC$ and $\triangle DEF$, two angles of one triangle are congruent to two angles of the other triangle (Figure 5.1).

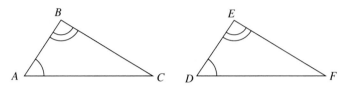

FIGURE 5.1

Show that the third angles are congruent. (Recall that in Chapter 2 we used an informal argument to show that the angle sum in a triangle is 180°.)

Given $\angle A = \angle D$ and $\angle B = \angle E$.

Prove $\angle C = \angle F$

SOLUTION

USING DIRECT REASONING

Proof First we have $\angle A + \angle B + \angle C = 180°$ and $\angle D + \angle E + \angle F = 180°$. So we have

(i) $\angle A + \angle B + \angle C = \angle D + \angle E + \angle F$

Because $\angle A = \angle D$ and $\angle B = \angle E$, we have

(ii) $\angle A + \angle B = \angle D + \angle E$

Subtracting (ii) from (i) we have $\angle C = \angle F$.

USING INDIRECT REASONING

Plan Assume that the given holds; that is, $\angle A = \angle D$ and $\angle B = \angle E$. We know that one, and only one, of the following can be true: (1) $\angle C < \angle F$, (2) $\angle C > \angle F$, or (3) $\angle C = \angle F$. We will assume that (1) is true and show that a contradiction results. Then we will do the same with (2). If neither of these statements is true, then (3) must be true, which is what we wanted to prove.

Proof *Assume* that the given holds, namely $\angle A = \angle D$ and $\angle B = \angle E$, and also assume that $\angle C < \angle F$. If $\angle C < \angle F$ and $\angle A = \angle D$, then $\angle A + \angle C < \angle D + \angle F$. Because $\angle B = \angle E$, we also have

(iii) $\angle A + \angle B + \angle C < \angle D + \angle E + \angle F$

But $\angle A + \angle B + \angle C = 180°$ and $\angle D + \angle E + \angle F = 180°$. Thus, by substitution in (iii), we have $180° < 180°$, a contradiction. If we assume that $\angle C > \angle F$, a similar contradiction arises. Thus, neither $\angle C < \angle F$ nor $\angle C > \angle F$ is possible. Hence we must conclude that $\angle C = \angle F$.   •

Notice that when using **indirect reasoning** (or making an **indirect proof**), we assume the hypothesis of the result that we are trying to prove and the *negation* (or opposite) of the conclusion. Then we show that this situation leads to a contradiction. Because this situation is not possible, we conclude that our assumption must be incorrect and the conclusion *as given in the original statement* must follow from the hypothesis.

The technique of indirect reasoning will be used often in this section. To use the technique to prove a theorem, we always assume that the hypothesis *and* the negation of the conclusion are true. Then, from these assumptions, we reach a statement that contradicts an earlier result or a known fact.

PARALLEL LINES

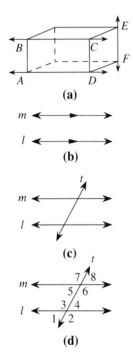

(a)

(b)

(c)

(d)

FIGURE 5.2

Recall that two lines in a plane are parallel if they do not intersect. It is important to note that the two parallel lines are said to lie in the same plane. Lines which do not lie in the same plane and do not intersect are called **skew lines**. For example, in Figure 5.2(a), *AD* and *BC* are parallel lines and *BC* and *EF* are skew lines. Throughout the rest of this book, we will be studying geometry of the *plane*. Therefore, we will not be considering skew lines. When two lines are parallel, we denote this by placing an arrowhead (or arrowheads) on each line [Figure 5.2(b)].

If two lines are intersected by a third line, the third line is called a **transversal**. In Figure 5.2(c), a pair of lines *l* and *m* and a transversal *t* are pictured. There are several pairs of angles formed that are of special interest. In Figure 5.2(d), the pair ∠3 and ∠6 and the pair ∠4 and ∠5 are called **alternate interior angles**. The pair ∠1 and ∠8 and the pair ∠2 and ∠7 are called **alternate exterior angles**. The pair ∠3 and ∠5 and the pair ∠4 and ∠6 are called **interior angles on the same side of the transversal**. The pair ∠1 and ∠7 and the pair ∠2 and ∠8 are called **exterior angles on the same side of the transversal**. In addition, the following pairs are called **corresponding angles**: ∠1 and ∠5, ∠3 and ∠7, ∠2 and ∠6, and ∠4 and ∠8.

In Figure 5.2(d), many pairs of angles appear to be congruent. In particular, it appears that ∠4 ≅ ∠5. In addition, lines *l* and *m* appear to be parallel. In fact, if a pair of lines and a transversal form congruent alternate interior angles, we can conclude that the lines are parallel. This result is proved next using indirect reasoning.

Theorem 5.1

If two lines are cut by a transversal to form a pair of congruent alternate interior angles, then the lines are parallel.

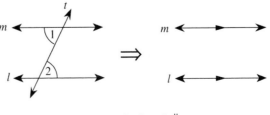

If ∠1 ≅ ∠2, then *l* ∥ *m*.

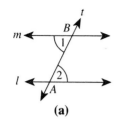

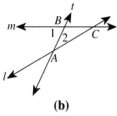

(a)

(b)

FIGURE 5.3

Given Lines l and m with transversal t, $\angle 1 = \angle 2$, and A and B points of intersection of t with l and m, respectively [Figure 5.3(a)].

Prove $l \parallel m$

Plan We will use indirect reasoning. We will assume that l and m are not parallel and try to arrive at a contradiction.

Proof If we assume l and m are not parallel, then they intersect in a point C [Figure 5.3(b)]. From the given we have $\angle 1 = \angle 2$ for $\triangle ABC$. But, by Theorem 4.11, $\angle 1 > \angle 2$ because $\angle 1$ is an exterior angle and $\angle 2$ is a nonadjacent interior angle. This is a contradiction because we cannot have $\angle 1 = \angle 2$ and $\angle 1 > \angle 2$. By indirect reasoning, we can conclude that l and m do *not* intersect. So they must be parallel.   •

The corollaries listed next follow from Theorem 5.1. Their proofs are left for the problem set.

Corollary 5.2

If two lines are both perpendicular to a transversal, then the lines are parallel.

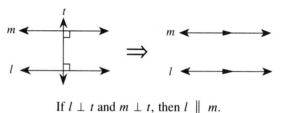

If $l \perp t$ and $m \perp t$, then $l \parallel m$.

Corollary 5.3

If two lines cut by a transversal form a pair of congruent corresponding angles with the transversal, then the lines are parallel.

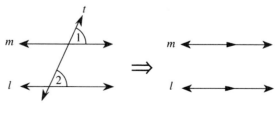

If $\angle 1 \cong \angle 2$, then $l \parallel m$.

Corollary 5.4

If two lines cut by a transversal form a pair of supplementary interior angles on the same side of the transversal, then the lines are parallel.

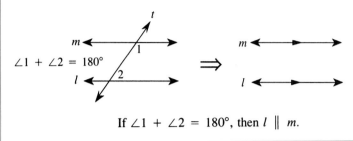

If $\angle 1 + \angle 2 = 180°$, then $l \parallel m$.

Any one of Theorem 5.1 or Corollaries 5.2–5.4 can be used to show that two lines are parallel. We will make frequent use of them.

EXAMPLE 5.2   Prove that *ABCD* is a parallelogram [Figure 5.4(a)].

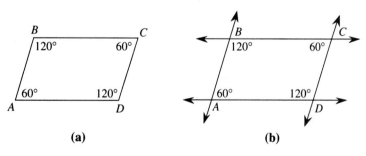

(a)   **(b)**

FIGURE 5.4

SOLUTION   Think of $\overline{AB}$ as a transversal for sides $\overline{AD}$ and $\overline{BC}$ [Figure 5.4(b)]. Because $60° + 120° = 180°$, *A* and *B* are supplementary interior angles on the same side of the transversal, and $\overline{AD} \parallel \overline{BC}$ by Corollary 5.4. By a similar argument, $\overline{AB} \parallel \overline{CD}$. Because both pairs of opposite sides are parallel, *ABCD* is a parallelogram.

THE PARALLEL POSTULATE

The preceding theorems and corollaries began with a pair of lines and a transversal. Suppose that we are given one line *l* and a point *P* not on that line [Figure 5.5(a)]. According to Theorem 5.1, a line *m* through *P* parallel to *l* can be constructed by forming congruent alternate interior angles with any transversal through *P* [Figure 5.5(b)].

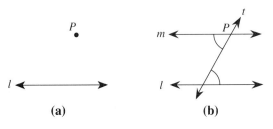

(a)   **(b)**

FIGURE 5.5

From this figure, it would appear that there is *only one* such line. For hundreds of years, mathematicians tried to prove this result from Euclid's postulates that preceded it. Finally it was shown that this result could not be proved in this way, so it was added as a postulate.

Postulate 5.1

The Parallel Postulate

Given a line *l* and a point *P* not on *l*, there is only one line *m* containing *P* such that *l* ∥ *m*.

The Parallel Postulate has several equivalent forms. However, the version stated here is popular because other geometries have been developed by replacing the phrase "only one line" with "no line" and "infinitely many lines." These geometries are called **elliptic** and **hyperbolic** geometries, respectively and are discussed in Topic 4 near the end of the book.

Now that we have the Parallel Postulate, we can prove the converse of Theorem 5.1.

Theorem 5.5

If a pair of parallel lines is cut by a transversal, then the alternate interior angles formed are congruent.

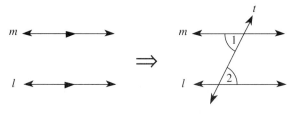

If *l* ∥ *m*, then ∠1 ≅ ∠2.

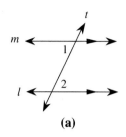

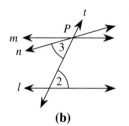

(a)

(b)

FIGURE 5.6

Given Parallel lines *l* and *m*, and transversal *t* [Figure 5.6(a)].

Prove ∠1 = ∠2

Plan Use indirect reasoning. We will assume that ∠1 ≠ ∠2 and try to arrive at a contradiction.

Proof Assume that ∠1 ≠ ∠2. Then ∠1 > ∠2 or ∠1 < ∠2. Suppose that ∠1 > ∠2. Then there is a line *n* through *P* that forms ∠3, where ∠3 = ∠2. [Figure 5.6(b)]. By Theorem 5.1, *n* ∥ *l*. However, because *m* ∥ *l* is given, we now have two lines (*m* and *n*) containing *P*, each parallel to *l*. This contradicts the Parallel Postulate, which states that there can be only one such line. Thus, by indirect reasoning, we have ∠1 = ∠2. [NOTE: In this proof, we assumed that ∠1 > ∠2 and then reached a contradiction. We would have reached the same contradiction if we had assumed that ∠1 < ∠2.]

EXAMPLE 5.3   In Figure 5.7, *l* ∥ *m* and ∠4 = 115°. Find the measures of the other numbered angles.

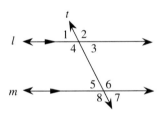

FIGURE 5.7

SOLUTION   Because ∠1 and ∠4 are supplementary,

$$\angle 1 = 180° - \angle 4 = 180° - 115° = 65°$$

In the same way, ∠3 = 65°. Notice that ∠2 and ∠4 are vertical angles, so ∠2 = ∠4 = 115°. By Theorem 5.5, ∠4 ≅ ∠6 because they are alternate interior angles and *l* ∥ *m*, so ∠6 = 115°. Then ∠5 and ∠7 are supplements of ∠6, so each of them measures 65°. Finally, ∠6 and ∠8 are vertical angles, so ∠8 = ∠6 = 115°. We can summarize these results as follows:

$$\angle 1 = \angle 3 = \angle 5 = \angle 7 = 65° \text{ and } \angle 2 = \angle 4 = \angle 6 = \angle 8 = 115°$$

Notice in Example 5.3 that each pair of corresponding angles was congruent. For instance, ∠1 ≅ ∠5 and ∠3 ≅ ∠7. Also, the pairs of interior angles on the same side of the transversal, such as ∠4 and ∠5, were supplementary. The corollaries stated next summarize results like these, which are the converses of Corollaries 5.2–5.4. Their proofs are left for the problem set.

Corollary 5.6

If two lines are parallel and a line is perpendicular to one of the two lines, then it is perpendicular to the other line.

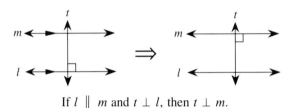

If $l \parallel m$ and $t \perp l$, then $t \perp m$.

Corollary 5.7

If two parallel lines are cut by a transversal, then each pair of corresponding angles formed is congruent.

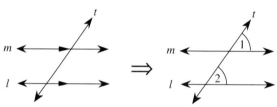

If $l \parallel m$, then $\angle 1 \cong \angle 2$.

Corollary 5.8

If two parallel lines are cut by a transversal, then both pairs of interior angles on the same side of the transversal are supplementary.

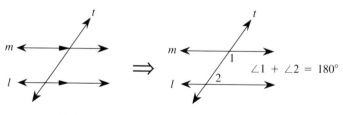

$\angle 1 + \angle 2 = 180°$

If $l \parallel m$, then $\angle 1 + \angle 2 = 180°$.

Solution to Applied Problem

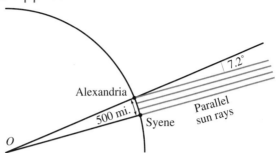

Because the sun's rays are assumed to be parallel, the angle formed at the center of the earth and the 7.2° angle are congruent corresponding angles.

Thus, $\angle O = 7.2°$, so the distance from Syene to Alexandria is $\dfrac{7.2}{360} = \dfrac{1}{50}$ of the circumference of the earth. This means that the circumference of the earth is about 50(500 mi), or about 25,000 miles. A more recent calculation showed the circumference to be about 24,820 miles.

GEOMETRY AROUND US

Flashlights and headlights use reflectors that are designed to take advantage of a unique property of parabolas. The reflector behind the bulb is made in the shape of a paraboloid. The bulb is located at a special point called the focus of the paraboloid so that the light rays from the bulb will be reflected in parallel lines, yielding one strong beam of light.

PROBLEM SET 5.1

EXERCISES/PROBLEMS

Use the following figure for Exercises 1–10.

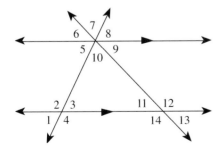

1. Identify two pairs of alternate interior angles.

2. Identify two pairs of interior angles on the same side of the transversal.

3. Identify two pairs of corresponding angles.

4. Identify two pairs of exterior angles on the same side of the transversal.

5. Find the measure of $\angle 6 + \angle 12$.

6. Find the measures of $\angle 2$ and $\angle 4$ if $\angle 5 = 52°$.

7. Find the measures of $\angle 3$ and $\angle 8$ if $\angle 5 = 65°$.

8. Find the measures of $\angle 13$, $\angle 12$, and $\angle 9$ if $\angle 6 = 40°$.

9. Find the measures of all the numbered angles given that $\angle 5 = 60°$ and $\angle 6 = 45°$.

10. Find the measures of all the numbered angles given that $\angle 14 = 140°$ and $\angle 1 = 50°$.

Exercises 11–18 refer to the following figure.

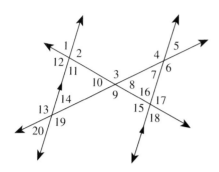

11. ∠16 and ∠ _____ are a pair of alternate interior angles.

12. ∠5 and ∠ _____ are a pair of alternate exterior angles.

13. ∠2 and ∠ _____ are a pair of corresponding angles.

14. ∠14 and ∠ _____ are a pair of interior angles on the same side of the transversal.

15. Find the measures of ∠13 and ∠5 if ∠6 = 140°.

16. Find the measures of ∠1 and ∠2 if ∠16 = 72°.

17. Find the measures of all the numbered angles if ∠1 = 80° and ∠4 = 125°.

18. Find the measures of all the numbered angles if ∠10 = 58° and ∠18 = 75°.

Use the next figure for Exercises 19–22, where congruent angles are marked. Decide if each statement is true or false. If true, explain why. Be careful not to make any assumptions about angles that just "appear" to be congruent.

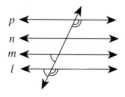

19. n ∥ p

20. n ∥ l

21. m ∥ l

22. p ∥ l

Use the following figure to find the measures of the angles specified in Exercises 23–28. Parallel segments are marked.

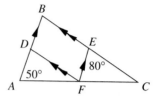

23. ∠B = _____ . Why?

24. ∠BDF = _____ . Why?

25. ∠ADF = _____ . Why?

26. ∠AFD = _____ . Why?

27. ∠DFE = _____ . Why?

28. ∠C = _____ . Why?

29. Find the measures of the numbered angles.

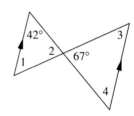

30. Find the measures of the numbered angles.

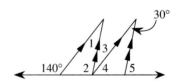

31. Find the measures of the numbered angles.

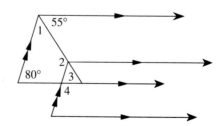

32. Find the measures of $\angle 1$, $\angle 2$, $\angle 3$, and $\angle 4$ if some angles formed are related as shown.

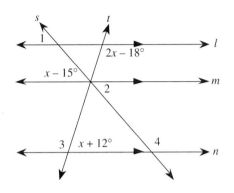

PROOFS

33. Prove Corollary 5.2: If two lines are both perpendicular to a transversal, then the lines are parallel.

34. Prove Corollary 5.3: If two lines cut by a transversal form a pair of congruent corresponding angles with the transversal, then the lines are parallel.

35. Prove Corollary 5.4: If two lines are cut by a transversal to form a pair of supplementary interior angles on the same side of the transversal, then the lines are parallel.

36. Prove: If two lines cut by a transversal form a pair of supplementary exterior angles on the same side of the transversal, then the lines are parallel.

37. Prove Corollary 5.6: If two lines are parallel and a line is perpendicular to one of the two lines, then it is perpendicular to the other line.

38. Prove Corollary 5.7: If two parallel lines are cut by a transversal, then each pair of corresponding angles formed is congruent.

39. Prove Corollary 5.8: If two parallel lines are cut by a transversal, then both pairs of interior angles on the same side of the transversal are supplementary.

40. Prove: If two parallel lines are cut by a transversal, then they form a pair of congruent alternate exterior angles with the transversal.

41. Prove: If two parallel lines are cut by a transversal, then they form a pair of supplementary exterior angles on the same side of the transversal.

5.2
APPLICATIONS OF PARALLEL LINES

Applied Problem

Two roads intersect as shown below. A rest stop will be constructed between them at a point that is equidistant from the two roads. Where can the rest stop be located, and how can the planners lay it out?

Rest
Stop

ANGLE SUM IN A TRIANGLE

In this section we apply the Parallel Postulate and the postulates and theorems related to parallel lines from Section 5.1. They will be used to prove some important theorems about triangles. For example, in Chapter 2 we determined informally that the sum of the angle measures in a triangle is 180°. Now that the Parallel Postulate has been assumed, we can give a deductive proof of that result.

Theorem 5.9

Angle Sum in a Triangle Theorem

The angle sum in a triangle is 180°.

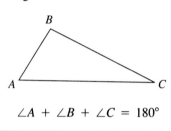

$$\angle A + \angle B + \angle C = 180°$$

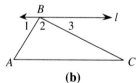
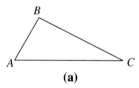

FIGURE 5.8

Given $\triangle ABC$ [Figure 5.8(a)]

Prove $\angle A + \angle B + \angle C = 180°$

Plan Draw a line l through B parallel to $\overline{AC}$ [Figure 5.8(b)] and use properties of parallel lines.

Proof

Statements	Reasons
1. $\angle 1 = \angle A$	1. Alternate interior angles (Theorem 5.5)
2. $\angle 3 = \angle C$	2. Alternate interior angles (Theorem 5.5)
3. $\angle 1 + \angle 2 + \angle 3 = 180°$	3. A straight angle has measure 180°.
4. $\angle A + \angle 2 + \angle C = 180°$	4. Substitution from Steps 1, 2, and 3
5. $\angle A + \angle B + \angle C = 180°$	5. Substitution from Step 4

EXTERIOR ANGLE OF A TRIANGLE

In Chapter 4, Theorem 4.11 states that an exterior angle of a triangle is greater than any nonadjacent interior angle. The next example will suggest a refinement of that theorem.

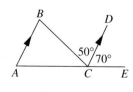

FIGURE 5.9

EXAMPLE 5.4   In Figure 5.9, $\overline{AB}$ is parallel to $\overline{CD}$. Show $\angle A + \angle B = 120°$.

SOLUTION $\overline{BC}$ is a transversal for segments $\overline{AB}$ and $\overline{CD}$. Thus $\angle B$ and $\angle BCD$ are alternate interior angles, hence $\angle B = 50°$. Consider $\overline{AE}$ as a transversal for

$\overline{AB}$ and $\overline{CD}$. Then $\angle A$ and $\angle DCE$ are corresponding angles. Hence $\angle A = 70°$. Thus, $\angle A + \angle B = 70° + 50° = 120°$.

Notice that we have just shown that the measure of $\angle BCE$ is the same as the sum of the measures of $\angle A$ and $\angle B$. Using the Angle Sum in a Triangle Theorem, we now prove the following refinement of Theorem 4.11.

Corollary 5.10

The Exterior Angle Theorem

An exterior angle of a triangle is equal to the sum of the two nonadjacent interior angles.

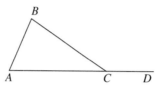

$$\angle BCD = \angle A + \angle B$$

Given $\triangle ABC$ (Figure 5.10)

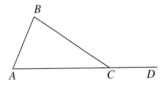

FIGURE 5.10

Prove $\angle BCD = \angle A + \angle B$

Proof

Statements	Reasons
1. $\angle BCA + \angle BCD = 180°$	1. A straight angle has measure $180°$.
2. $\angle A + \angle B + \angle BCA = 180°$	2. Angle Sum in a Triangle Theorem
3. $\angle BCA + \angle BCD =$ $\angle A + \angle B + \angle BCA$	3. Substitution from Steps 1 and 2
4. $\angle BCD = \angle A + \angle B$	4. Subtracting $\angle BCA$ from both sides in Step 3

AAS CONGRUENCE

The next two corollaries also follow from the Angle Sum in a Triangle Theorem. Their proofs will be left for the problem set.

Corollary 5.11

AAS Congruence Theorem

If two angles and a side of one triangle are congruent respectively to two angles and a side of another triangle, then the two triangles are congruent.

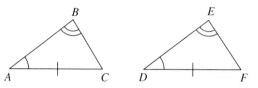

If $\angle A \cong \angle D$, $\angle B \cong \angle E$, and $\overline{AC} \cong \overline{DF}$,
then $\triangle ABC \cong \triangle DEF$.

Notice that Corollary 5.11 states that in order to show that two triangles are congruent we no longer have to insist on having two angles and their *included* side congruent respectively. That is, two angles and *any* side will suffice. Next is a special case of this corollary.

Corollary 5.12

HA Congruence Theorem

If the hypotenuse and an acute angle of one right triangle are congruent respectively to the hypotenuse and an acute angle of another right triangle, then the two triangles are congruent.

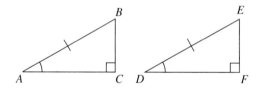

If $\angle C = \angle F = 90°$, $\angle A \cong \angle D$, and $\overline{AB} \cong \overline{DE}$,
then $\triangle ABC \cong \triangle DEF$.

BISECTOR OF AN ANGLE

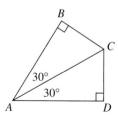

FIGURE 5.11

Earlier we showed that any point on the perpendicular bisector of a segment was equidistant from the endpoints of the segment. The next example and theorem show an analogous property of angle bisectors.

EXAMPLE 5.5   In Figure 5.11, $AC = 10$ inches. Find BC and DC.

SOLUTION $\triangle ABC$ is a 30-60 right triangle with hypotenuse of length 10″. Because $\overline{BC}$ is the leg opposite the 30° angle, $BC = \frac{1}{2}(10″) = 5″$. Similarly, since $\triangle ACD$ is a 30-60 right triangle, $DC = 5″$.   •

Notice in Example 5.5 that $\overline{AC}$ is the bisector of $\angle A$ and point C is equidistant from B and D. This result is stated more generally in the next theorem.

Theorem 5.13

Angle Bisector Theorem

A point is on the bisector of an angle if and only if it is equidistant from the sides of the angle.

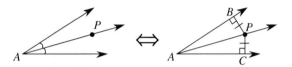

P is on the bisector of $\angle A$ if and only if $PB = PC$.

Because Theorem 5.13 is an "if and only if" theorem, we must prove two parts.

Subgoal 1 Prove: If a point is on the bisector of an angle, then it is equidistant from the sides of the angle.

Subgoal 2 Prove: If a point is equidistant from the sides of an angle, then it is on the bisector of the angle.

We will prove Subgoal 1 and leave the proof of Subgoal 2 for the problem set.

Given Point P on the bisector of $\angle A$ [Figure 5.12(a)]

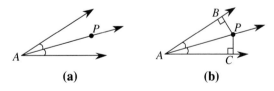

        (a)   **(b)**

FIGURE 5.12

Prove P is equidistant from the sides of $\angle A$.

Proof The distances from P to the sides of $\angle A$ are the lengths of the segments $\overline{PB}$ and $\overline{PC}$, which are perpendicular to the respective sides [Figure 5.12(b)]. Because $\overrightarrow{AP}$ bisects $\angle A$, we have $\angle BAP \cong \angle CAP$. Also, $\angle B = \angle C = 90°$ and $\overline{PA} \cong \overline{PA}$ [Figure 5.12(b)]. Therefore, we have $\triangle APB \cong \triangle APC$ by AAS Congruence (or by HA Congruence) and $PB = PC$ by corresponding parts. •

Solution to Applied Problem

Because the rest stop will be equidistant from the two roads, it must be placed at a point that is equidistant from the lines representing the two roads. Thus, by Theorem 5.13 it should be somewhere on the bisector of the angle formed by the two roads. The planners can construct the angle bisector on the drawing to examine the possible locations of the rest stop.

Rest Stop

GEOMETRY AROUND US

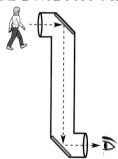

A simple periscope uses two parallel mirrors placed at each end of a tube. One mirror receives light from an object and reflects it to the other mirror. The second mirror then reflects the light to the viewer's eye. Recently, periscopes have been made of glass fibers that can be bent and twisted around corners. The glass fibers reflect the image down the length of the tube, so mirrors are unnecessary.

PROBLEM SET 5.2

EXERCISES/PROBLEMS

Use the following figure for Exercises 1–6.

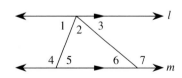

If $l \parallel m$, determine whether each statement is true or false. If true, explain why.

1. $\angle 4 = \angle 2 + \angle 6$

2. $\angle 3 = \angle 2 + \angle 5$

3. $\angle 1 + \angle 2 + \angle 6 = 180°$

4. $\angle 4 + \angle 1 = 180°$

5. $\angle 2 + \angle 6 = 180°$

6. $\angle 7 = \angle 1 + \angle 2$

Use the following figure for Exercises 7–12.

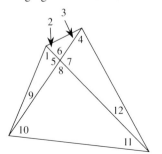

Determine whether each statement is true or false. If true, explain why.

7. $\angle 5 = \angle 2 + \angle 3$

8. $\angle 7 = \angle 1 + \angle 9$

9. $\angle 2 + \angle 3 = \angle 10 + \angle 11$

10. $\angle 1 + \angle 9 = \angle 4 + \angle 12$

11. $\angle 2 + \angle 3 + \angle 4 + \angle 12 = 180°$

12. $\angle 6 + \angle 10 + \angle 11 = 180°$

Use the following figure for Exercises 13–20. Congruent segments and parallel lines are indicated.

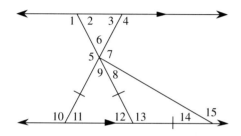

Determine whether each statement is true or false, explain why.

13. $\angle 12 = \angle 8$

14. $\angle 5 = \angle 11 + \angle 14$

15. $\angle 15 = \angle 8 + \angle 9 + \angle 11$

16. $\angle 7 + \angle 8 = \angle 11 + \angle 12$

17. $\angle 12 = 2(\angle 8)$

18. $\angle 2 = \angle 3$

19. $\angle 13 + \angle 11 = 180°$

20. $\angle 4 + \angle 7 = 180°$

21. Find the measure of each numbered angle in the following figure, where $l \parallel m$ and congruent segments are marked.

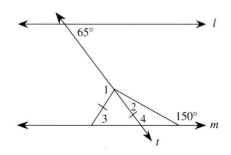

22. Find the measure of each numbered angle in the next figure, where $l \parallel m$.

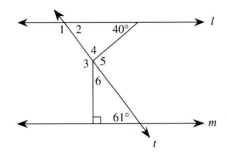

23. In the figure shown, $\angle ABF = \angle CBF$, $AB = 11''$, $AD = 3''$, $EC = 4''$, and $BF = 10''$. Find the area of $\triangle ABC$.

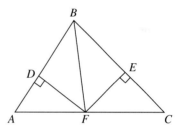

24. In the following figure, $ABCD$ is a trapezoid, $BC = 14$ cm, $AE = 11.9$ cm, $BG = 13$ cm, and $CD = 12$ cm. If $\angle EBG = \angle FBG$, find the area of trapezoid $ABCD$.

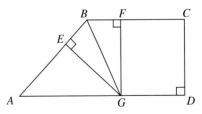

PROOFS

25. In the figure shown, $\angle ACD \cong \angle ACB$ and $\angle ADC \cong \angle ABC$. Prove that $CD = CB$.

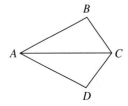

26. In isosceles $\triangle ABC$ with $\overline{AC} \cong \overline{AB}$, and $\angle BDC \cong \angle CEB$, prove that $\overline{BD} \cong \overline{CE}$.

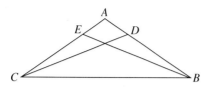

27. Prove Corollary 5.11: If two angles and a side of one triangle are congruent respectively to two angles and a side of another triangle, then the two triangles are congruent.

28. Prove Corollary 5.12: If the hypotenuse and an acute angle of one right triangle are congruent respectively to the hypotenuse and an acute angle of another right triangle, then the two triangles are congruent.

29. Prove Subgoal 2 of Theorem 5.13: If a point is equidistant from the sides of an angle, then it is on the bisector of the angle.

APPLICATION

30. A walking trail/bike path has a turn as shown. A marker is to be erected that will be the same distance from all three portions of the path. How can the position of the marker be determined?

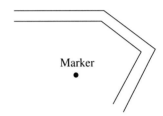

5.3
PARALLELOGRAMS AND RHOMBUSES

Applied Problem

A pool table measures $9'$ by $4\frac{1}{2}'$. A ball is hit from corner A at an angle of $60°$ as shown. How many times will the ball hit a side of the table before it goes into a pocket?

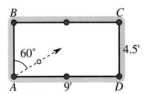

PARALLELOGRAMS

In Chapter 4 and Sections 5.1 and 5.2 many results involving triangles and parallel lines were derived. In this section, we focus on the study of parallelograms, that is, quadrilaterals with opposite sides parallel.

In Chapter 3 we observed that the diagonal of a parallelogram divides it into two congruent triangles. Here we will examine that idea more formally. Consider the parallelogram $ABCD$ in Figure 5.13(a) with diagonal $\overline{AC}$. Because $\overline{BC} \parallel \overline{AD}$, we have $\angle BCA \cong \angle DAC$ since they are alternate interior angles [Figure 5.13(b)]. Because $\overline{AB} \parallel \overline{CD}$, we also have $\angle BAC \cong \angle DCA$

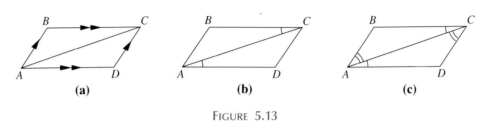

FIGURE 5.13

[Figure 5.13(c)]. Notice that $\overline{AC}$ is common to $\triangle ABC$ and $\triangle CDA$. Therefore, we can conclude that $\triangle ABC \cong \triangle CDA$ by ASA. This result is stated next.

Theorem 5.14

A diagonal of a parallelogram forms two congruent triangles.

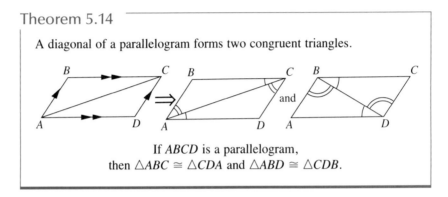

If $ABCD$ is a parallelogram,
then $\triangle ABC \cong \triangle CDA$ and $\triangle ABD \cong \triangle CDB$.

Notice that by corresponding parts of $\triangle ABC$ and $\triangle CDA$, $\angle B \cong \angle D$, $\overline{BC} \cong \overline{AD}$, and $\overline{AB} \cong \overline{CD}$. In a similar manner, using $\triangle ABD$ and $\triangle CDB$, we can conclude that $\angle A \cong \angle C$. This observation leads us to the next corollary.

Corollary 5.15

In a parallelogram, the opposite sides are congruent and the opposite angles are congruent.

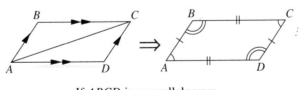

If $ABCD$ is a parallelogram,
then $\overline{AB} \cong \overline{CD}$, $\overline{AD} \cong \overline{BC}$, $\angle A \cong \angle C$, and $\angle B \cong \angle D$.

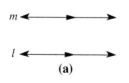

(a)

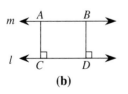

(b)

FIGURE 5.14

Another important result follows from Corollary 5.15. Let $l \parallel m$ [Figure 5.14(a)]. Let $\overline{AC}$ and $\overline{BD}$ be any two segments that are perpendicular to l (and hence m) [Figure 5.14(b)]. Therefore, $\overline{AC} \parallel \overline{BD}$ because corresponding angles are congruent. This means that $ABCD$ is a parallelogram. Consequently, $AC = BD$ and the two distances between l and m are equal. This conclusion is summarized next.

Corollary 5.16

Parallel lines are everywhere equidistant.

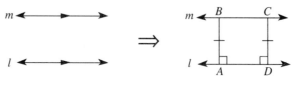

If $l \parallel m$, then $AB = CD$.

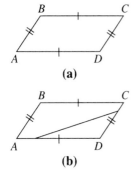

(a)

(b)

FIGURE 5.15

Now let's consider the converse of Corollary 5.15. That is, if the opposite sides of a quadrilateral are congruent, will it be a parallelogram? Consider quadrilateral $ABCD$ in which the opposite sides are congruent [Figure 5.15(a)]. If $\overline{AC}$ is drawn [Figure 5.15(b)], then two congruent triangles are formed, namely $\triangle ABC \cong \triangle CDA$, by SSS. By corresponding parts, $\angle BCA \cong \angle DAC$, so $\overline{BC} \parallel \overline{AD}$. Similarly, by drawing $\overline{BD}$ in Figure 5.15(a), we could show that $\overline{AB} \parallel \overline{CD}$. Thus, $ABCD$ is a parallelogram. This result is stated next.

Theorem 5.17

If the opposite sides of a quadrilateral are congruent, then the quadrilateral is a parallelogram.

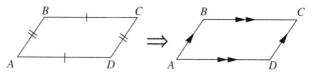

If $\overline{AB} \cong \overline{CD}$ and $\overline{AD} \cong \overline{BC}$, then $ABCD$ is a parallelogram.

It also turns out to be the case that if the opposite angles of a quadrilateral are congruent, then it is a parallelogram. This result is stated next and its proof is left for the problem set.

Theorem 5.18

If the opposite angles of a quadrilateral are congruent, then the quadrilateral is a parallelogram.

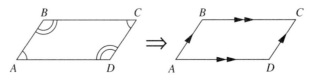

If $\angle A \cong \angle C$ and $\angle B \cong \angle D$, then $ABCD$ is a parallelogram.

What can be said about a quadrilateral in which two sides are both congruent and parallel? Consider the pair of parallel line segments $\overline{AD}$ and $\overline{BC}$ in Figure 5.16(a), where $\overline{AD} \cong \overline{BC}$.

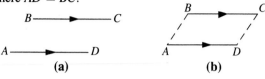

(a)   **(b)**

FIGURE 5.16

If $\overline{AB}$ and $\overline{CD}$ are drawn [Figure 5.16(b)], it appears that $ABCD$ is a parallelogram. This fact is proved next.

Theorem 5.19

If a quadrilateral has two sides that are parallel and congruent, then it is a parallelogram.

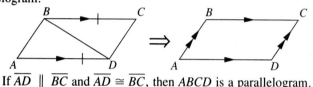

If $\overline{AD} \parallel \overline{BC}$ and $\overline{AD} \cong \overline{BC}$, then $ABCD$ is a parallelogram.

Given $\overline{AD} \parallel \overline{BC}$ and $\overline{AD} \cong \overline{BC}$ [Figure 5.17(a)]

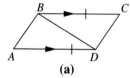
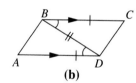

(a)   **(b)**

FIGURE 5.17

Prove $ABCD$ is a parallelogram.

Subgoal 1 Show that $\triangle ABD \cong \triangle CDB$.

Subgoal 2 Show that $\overline{AB} \parallel \overline{CD}$.

Proof

Statements	**Reasons**
1. $\overline{AD} \parallel \overline{BC}$ and $\overline{AD} \cong \overline{BC}$	1. Given
2. $\angle ADB \cong \angle CBD$	2. Alternate interior angles [Figure 5.17(b)]
3. $\overline{BD} \cong \overline{BD}$	3. Congruence is reflexive [Figure 5.17(b)]
4. $\triangle ABD \cong \triangle CDB$	4. SAS (Subgoal 1)
5. $\angle ABD \cong \angle CDB$	5. C.P.
6. $\overline{AB} \parallel \overline{CD}$	6. Congruent alternate interior angles (Subgoal 2)
7. $ABCD$ is a parallelogram.	7. Definition of a parallelogram (Opposite sides are parallel.)

EXAMPLE 5.6   Given parallelogram *ABCD* with midpoints *E*, *F*, *G*, and *H* on sides $\overline{AB}$, $\overline{BC}$, $\overline{CD}$, and $\overline{AD}$ respectively, prove that *EFGH* is a parallelogram [Figure 5.18(a)].

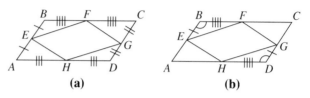

FIGURE 5.18

Plan Show that the opposite sides of *EFGH* are congruent using congruent triangles.

SOLUTION   First, $\angle B = \angle D$ because opposite angles of a parallelogram are congruent. Next, $AB = CD$ and $AD = BC$ because opposite sides of a parallelogram are congruent. Also, $EB = \frac{1}{2}AB$ and $DG = \frac{1}{2}CD$. Hence, $EB = DG$. Similarly, $BF = DH$ [Figure 5.18(b)]. Therefore, $\triangle BEF \cong \triangle DGH$ by SAS. By corresponding parts, $EF = GH$. In a similar manner, we can show that $FG = EH$. Thus, because the opposite sides of *EFGH* are congruent, *EFGH* is a parallelogram.   •

In Chapter 6 we will consider the figure formed by connecting the midpoints of the sides of any quadrilateral.

DIAGONALS OF A PARALLELOGRAM

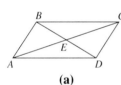

(a)

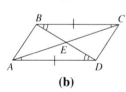

(b)

FIGURE 5.19

Next we will look at the diagonals of a parallelogram. Consider parallelogram *ABCD* with diagonals $\overline{AC}$ and $\overline{BD}$ [Figure 5.19(a)]. Because $\overline{AD} \parallel \overline{BC}$ and $\overline{CA}$ is a transversal, $\angle EAD = \angle ECB$. Similarly, $\angle EBC = \angle EDA$ [Figure 5.19(b)]. Because *ABCD* is a parallelogram, $\overline{BC} \cong \overline{AD}$. Thus, $\triangle ADE \cong \triangle CBE$ by ASA. By corresponding parts, $\overline{BE} \cong \overline{DE}$ and $\overline{AE} \cong \overline{CE}$, which means that *E* is the midpoint of both $\overline{BD}$ and $\overline{AC}$. This result is summarized next.

Theorem 5.20

The diagonals of a parallelogram bisect each other.

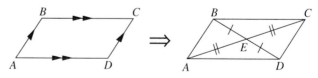

If *ABCD* is a parallelogram, then $AE = CE$ and $BE = DE$.

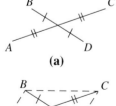

(a)

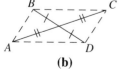

(b)

FIGURE 5.20

Now consider the two line segments that intersect at their midpoints in Figure 5.20(a). If segments $\overline{AB}$, $\overline{BC}$, $\overline{CD}$, and $\overline{AD}$ are drawn [Figure 5.20(b)], the resulting quadrilateral appears to be a parallelogram. This result, which is the converse of the previous theorem, is proved next.

Theorem 5.21

If the diagonals of a quadrilateral bisect each other, then it is a parallelogram.

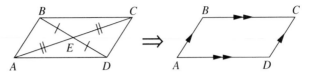

If $AE = CE$ and $BE = DE$, then $ABCD$ is a parallelogram.

Given $ABCD$ with $AE = CE$ and $BE = DE$ [Figure 5.21(a)]

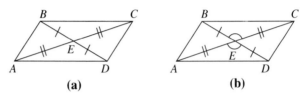

(a)   (b)

FIGURE 5.21

Prove $ABCD$ is a parallelogram.

Plan Use congruent triangles to show that two sides of $ABCD$ are parallel and congruent.

Proof

Statements	Reasons
1. $AE = CE$ and $BE = DE$	1. Given
2. $\angle AED = \angle CEB$	2. Vertical angles [Figure 5.21(b)]
3. $\triangle AED \cong \triangle CEB$	3. SAS
4. $AD = BC$	4. C.P.
5. $\angle ADE = \angle CBE$	5. C.P.
6. $\overline{AD} \parallel \overline{BC}$	6. Congruent alternate interior angles
7. $ABCD$ is a parallelogram.	7. Two opposite sides of the quadrilateral are parallel and congruent. •

RHOMBUSES

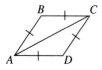

FIGURE 5.22

Recall that a rhombus was defined to be a quadrilateral with all sides congruent. Consider rhombus $ABCD$ (Figure 5.22), where the diagonal $\overline{AC}$ is shown. By SSS, $\triangle ABC \cong \triangle CDA$. Then the corresponding parts of these congruent triangles are congruent. In particular we have $\angle BAC \cong \angle DCA$. Taking these as alternate interior angles with $\overline{AC}$ as a transversal, we can conclude that $\overline{AB} \parallel \overline{CD}$. Because $\overline{AB}$ and $\overline{CD}$ are both congruent and parallel, $ABCD$ is a parallelogram. Thus we have the following theorem.

Theorem 5.22

Every rhombus is a parallelogram.

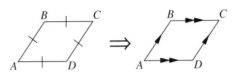

If $AB = BC = CD = AD$,
then $\overline{AB} \parallel \overline{CD}$ and $\overline{AD} \parallel \overline{BC}$.

Now that we have shown that every rhombus is a parallelogram, we can conclude that all the theorems about parallelograms also apply to rhombuses. That is, we can say that

1. The diagonals of a rhombus bisect each other.
2. The opposite angles of a rhombus are congruent.
3. A diagonal of a rhombus forms two congruent triangles.

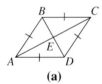

(a)

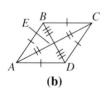

(b)

FIGURE 5.23

Consider rhombus $ABCD$ with diagonals $\overline{AC}$ and $\overline{BD}$ [Figure 5.23(a)]. Because $ABCD$ is a rhombus (hence a parallelogram), the diagonals bisect each other at E [Figure 5.23(b)]. By SSS, the two diagonals divide the rhombus into four congruent triangles. In particular, we have $\triangle AEB \cong \triangle AED$. Therefore, $\angle AEB = \angle AED$ by corresponding parts. Because they are also supplementary adjacent angles, they must be right angles. So $\overline{AC}$ is perpendicular to $\overline{BD}$. This result is summarized next.

Theorem 5.23

The diagonals of a rhombus are perpendicular to each other.

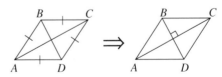

If $ABCD$ is a rhombus, then $\overline{AC} \perp \overline{BD}$.

The proof of the converse of this theorem, restricted to parallelograms, is given next.

Theorem 5.24

If the diagonals of a parallelogram are perpendicular to each other, then the parallelogram is a rhombus.

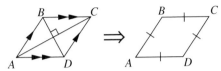

If *ABCD* is a parallelogram with $\overline{AC} \perp \overline{BD}$, then *ABCD* is a rhombus.

Given Parallelogram *ABCD* with $\overline{AC}$ perpendicular to $\overline{BD}$ [Figure 5.24(a)]

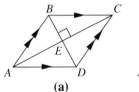 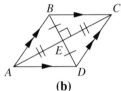

(a)   (b)

FIGURE 5.24

Prove *ABCD* is a rhombus.

Proof

Statements	Reasons
1. $\overline{AC}$ is perpendicular to $\overline{BD}$.	1. Given
2. $AE = CE$ and $BE = DE$	2. The diagonals of a parallelogram bisect each other [Figure 5.24(b)].
3. $\triangle AED \cong \triangle AEB \cong \triangle CED \cong \triangle CEB$	3. LL Congruence (or SAS)
4. $\overline{AB} \cong \overline{BC} \cong \overline{CD} \cong \overline{AD}$	4. C.P.
5. *ABCD* is a rhombus.	5. Definition of a rhombus (All four sides are congruent.)

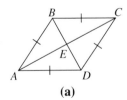

(a)

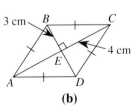

(b)

FIGURE 5.25

EXAMPLE 5.7   The diagonals in rhombus *ABCD* have lengths 6 cm and 8 cm. Find the length of the sides of the rhombus.

SOLUTION   Let *E* be the point where the diagonals intersect [Figure 5.25(a)]. Then, because the diagonals of a rhombus bisect each other, $CE = \frac{1}{2}(CA) = 4$ cm and $BE = 3$ cm [Figure 5.25(b)]. Because the diagonals of a rhombus are perpendicular, we can use the Pythagorean Theorem to find the length of a side. Thus, we have the following:

$$BC^2 = 3^2 + 4^2 = 9 + 16 = 25$$

so $BC = 5$ cm. Because the four sides of a rhombus are congruent, their lengths are 5 cm.

Using the proof given for Theorem 5.24, we can also conclude that the diagonals bisect the opposite angles of a rhombus. This is the final result of the section. Its proof is left for the problem set.

Theorem 5.25

A parallelogram is a rhombus if and only if its diagonals bisect the opposite angles.

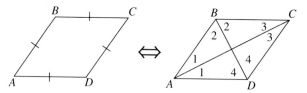

Parallelogram $ABCD$ is a rhombus if and only if $\overline{AC}$ bisects $\angle A$ and $\angle C$, and $\overline{BD}$ bisects $\angle B$ and $\angle D$.

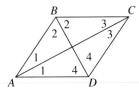

FIGURE 5.26

EXAMPLE 5.8   In rhombus $ABCD$, prove $\angle 1 = \angle 3$ and $\angle 2 = \angle 4$ (Figure 5.26).

SOLUTION   Because $ABCD$ is a rhombus, it is a parallelogram. Thus opposite angles are congruent, and $\angle A = \angle C$. So, by Theorem 5.25, $\frac{1}{2}\angle A = \frac{1}{2}\angle C$ or $\angle 1 = \angle 3$. Similarly, $\angle 2 = \angle 4$.   •

Solution to Applied Problem

The ball travels so that the angle of incidence always equals the angle of reflection as illustrated on the sides $\overline{BC}$ and $\overline{CD}$. That is, the angle at which the ball hits the side is the same as the angle at which the ball bounces off the side.

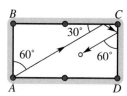
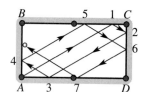

The ball will continue to hit a side and rebound at the same angle it hits, always travelling along parallel lines and forming parallelograms as shown. The ball will hit the sides 17 times before going into the pocket in the middle of side $\overline{AD}$. You can verify this result by making a scale drawing and sketching the path of the ball.

GEOMETRY AROUND US

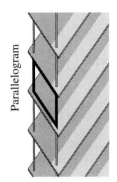

Parallelogram

Window blinds are threaded in such a way that the blinds always move in parallel. If the rod controlling the strings is twisted to adjust the blinds, they all move the same distance, remaining parallel to each other. At any given time the blinds and strings form parallelograms.

PROBLEM SET 5.3

EXERCISES/PROBLEMS

Use the following figure for Exercises 1–12.

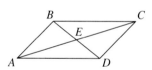

Determine whether the statement is true or false for parallelogram *ABCD*. If true, explain why.

1. $AC = BD$

2. $AB = CD$

3. $AD = BC$

4. $AE = CE$

5. $AE = BE$

6. $\angle A \cong \angle C$

7. $\angle A \cong \angle D$

8. $\angle A + \angle D = 180°$

9. $\triangle AED \cong \triangle BEC$

10. $\triangle ABE \cong \triangle CDE$

11. $\triangle ADC \cong \triangle CBA$

12. $\triangle BCA \cong \triangle CBD$

Use the following figure for Exercises 13–20.

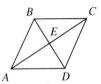

Determine whether the statement is true or false for rhombus *ABCD*. If true, explain why.

13. $AE = CE$

14. $AB = BC$

15. $\triangle ABC \cong \triangle ADC$

16. $\triangle BEA \cong \triangle DEA$

17. $\angle EBC + \angle ECB = 90°$

18. $\angle DAC \cong \angle BCA$

19. $\angle ABC \cong \angle CDA$

20. $AE = ED$

21. Find the measures of the angles in the following parallelogram.

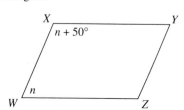

22. Find the measures of the angles ∠WXY and ∠XYZ in rhombus WXYZ.

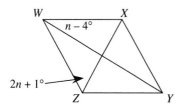

23. Find the measures of the numbered angles in the rhombus PQRS.

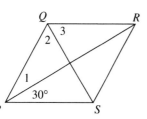

24. Find the measures of the numbered angles in rhombus ABCD if BE = 7″ and CD = 14″.

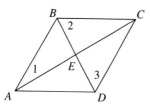

25. Find the measures of the numbered angles in parallelogram ABCD.

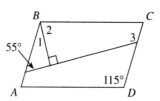

26. Find the measures of ∠1 and ∠2 in parallelogram PQRS.

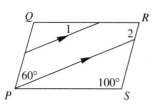

27. Find the length of a side of rhombus ABCD if AE = 6 in.

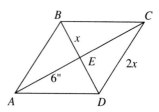

28. In rhombus ABCD shown next, BC = BD and AC = 16 cm. Find the measures of AB, DE, ∠ABC, and ∠DCA.

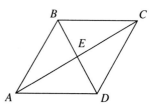

29. Find the measures of the numbered angles in the next figure where CD ∥ BE ∥ AF, BC ∥ DE, and AB ∥ EF. Congruent segments are marked.

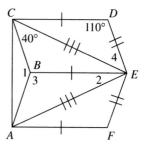

30. In the figure shown, ABDF and BCEF are parallelograms. If ∠CBD = 22°, ∠E = 56°, and ∠BAF = 60°, what are the measures of the numbered angles?

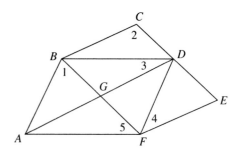

31. In parallelogram *PQRS*, find the height *h*. Then calculate the area of *PQRS*.

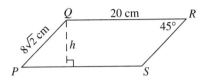

32. In rhombus *WXYZ*, find the height *h*. Then calculate the area of *WXYZ*.

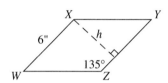

33. The side of a rhombus measures 10 cm and one diagonal measures 12 cm. Find the length of the other diagonal.

34. The diagonals of a rhombus measure 10 inches and 24 inches. Find the length of a side of the rhombus.

35. The measures of consecutive angles in a parallelogram are in the ratio 2:3. Find the measures of the angles of the parallelogram.

36. The measure of one vertex angle of a parallelogram is 18° more than twice the measure of the next vertex angle. Find the measures of the angles of the parallelogram.

PROOFS

37. Prove Theorem 5.18: If the opposite angles of a quadrilateral are congruent, then the quadrilateral is a parallelogram.

38. Prove that the diagonals of a rhombus bisect each other.

39. Prove: If *ABCD* is a rhombus, then ∠*CAD* and ∠*BDA* are complementary.

40. Prove: In parallelogram *ABCD*, if ∠*CAD* and ∠*BDA* are complementary, then *ABCD* is a rhombus.

41. Prove: The bisectors of any two consecutive angles of a parallelogram are perpendicular.

42. Prove Theorem 5.25: A parallelogram is a rhombus if and only if its diagonals bisect the opposite angles.

43. In parallelogram *ABCD*, $\overline{AE}$ bisects ∠*A* and $\overline{CF}$ bisects ∠*C*. Prove that *AECF* is a parallelogram.

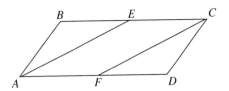

APPLICATIONS

44. A **linkage** is a system of rods and joints used to transmit motion. A four-bar linkage in which opposite links are of equal length is called a parallel-motion linkage. Using the following diagram, explain why the connecting link will always move in such a way that it is parallel to the fixed link.

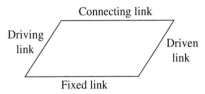

An example of this kind of linkage is the rod on the wheels of a locomotive. In this case, the two wheels take the place of the driving link and the driven link.

45. An ironing board and a scissor-bed truck utilize the same type of linkage. For example, by positioning the legs appropriately, an ironing board can be adjusted to different heights.

In the picture, suppose that the legs of the ironing board are connected at their midpoints and are able to pivot there. Why will the ironing board surface always be parallel to the floor, regardless of the height of the board from the floor? The same principle keeps the bed of a scissor-bed truck parallel to the ground.

46. Suppose that a ball is hit from point *A* at an angle of 45° on a pool table that measures 9′ by 4.5′ as shown.

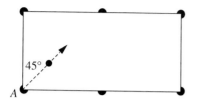

How many times will the ball hit a side of the table before going into a pocket? If the ball is hit at the same angle but the table measures 9′ by 6′, how many times will it hit a side? [HINT: Try making a scale drawing.]

47. Suppose that a ball is hit from *A* at an angle of 20° on a pool table measuring 9′ by 4.5′. How many times will the ball hit a side of the table before going into a pocket?

5.4
RECTANGLES, SQUARES, AND TRAPEZOIDS

Applied Problem

Harry wants to construct a frame for his favorite photograph. He cuts two pieces of oak 9″ long and two pieces 12″ long, miters the corners, and attaches the pieces. How can he verify that the resulting frame is "square," that is, has 90° corners?

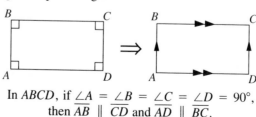

RECTANGLES

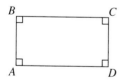

FIGURE 5.27

Recall that a rectangle was defined to be a quadrilateral with four right angles (Figure 5.27). By viewing ∠*A* and ∠*B* as supplementary angles formed by the transversal $\overline{AB}$, we can see that $\overline{AD} \parallel \overline{BC}$. Similarly, $\overline{AB} \parallel \overline{CD}$. Thus, we can say that *ABCD* is also a parallelogram.

Theorem 5.26

Every rectangle is a parallelogram.

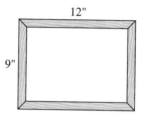

In *ABCD*, if ∠*A* = ∠*B* = ∠*C* = ∠*D* = 90°, then $\overline{AB} \parallel \overline{CD}$ and $\overline{AD} \parallel \overline{BC}$.

Because every rectangle is a parallelogram, we know the following results for parallelograms hold for rectangles, also:

1. The opposite sides are parallel.
2. The opposite sides are congruent.
3. The diagonals bisect each other.

A rectangle is a quadrilateral with four right angles. However, if we know that a quadrilateral is a parallelogram, then one right angle is sufficient to show that the parallelogram is also a rectangle. The proof of this fact is left for the problem set.

Theorem 5.27

A parallelogram with one right angle is a rectangle.

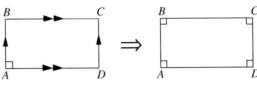

If $\overline{AB} \parallel \overline{CD}$, $\overline{AD} \parallel \overline{BC}$, and $\angle A = 90°$,
then $\angle A = \angle B = \angle C = \angle D = 90°$.

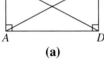

(a)

This theorem gives an alternate definition of a rectangle. That is, we could have defined a rectangle to be a parallelogram with one right angle.

Now let's consider rectangle $ABCD$ having diagonals $\overline{AC}$ and $\overline{BD}$ [Figure 5.28(a)]. Because opposite sides are congruent, we know that $\overline{AB} \cong \overline{CD}$. Because $\angle BAD = 90° = \angle CDA$, we can conclude that $\triangle BAD \cong \triangle CDA$ by SAS (or LL) [Figure 5.28(b)]. Thus, $\overline{AC} \cong \overline{BD}$ by corresponding parts. This discussion is summarized in the next theorem.

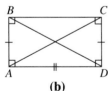

(b)

FIGURE 5.28

Theorem 5.28

The diagonals of a rectangle are congruent.

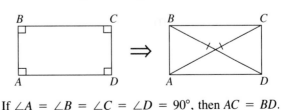

If $\angle A = \angle B = \angle C = \angle D = 90°$, then $AC = BD$.

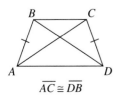

$\overline{AC} \cong \overline{DB}$

FIGURE 5.29

Does the converse of this theorem hold? That is, suppose that a quadrilateral has congruent diagonals. Is it necessarily a rectangle? Figure 5.29 shows an isosceles trapezoid that has congruent diagonals. Because a trapezoid is *not* a rectangle, a quadrilateral with congruent diagonals is not necessarily a rectangle. However, the following theorem *is* true.

Theorem 5.29

If a parallelogram has congruent diagonals, then it is a rectangle.

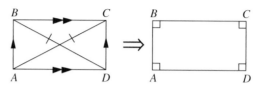

If $AC = BD$, then $ABCD$ is a rectangle.

Given Parallelogram $ABCD$ with $AC = BD$ [Figure 5.30(a)]

Prove $ABCD$ is a rectangle.

Proof

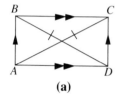

(a)

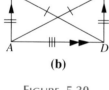

(b)

FIGURE 5.30

Statements	Reasons
1. $ABCD$ is a parallelogram with $AC = BD$.	1. Given
2. $\overline{AB} \cong \overline{DC}$	2. In a parallelogram, opposite sides are congruent.
3. $\overline{AD} \cong \overline{AD}$	3. Reflexive property
4. $\triangle BAD \cong \triangle CDA$	4. SSS [Figure 5.30(b)]
5. $\angle BAD \cong \angle CDA$	5. C.P.
6. $\angle BAD$ and $\angle CDA$ are supplementary.	6. Interior angles on the same side of the transversal
7. $\angle BAD$ and $\angle CDA$ are right angles.	7. Steps 5 and 6
8. $ABCD$ is a rectangle.	8. A parallelogram with one right angle is a rectangle. •

FIGURE 5.31

EXAMPLE 5.9   Given rectangle $ABCD$ with $AB = 3$ and $AD = 4$, find AC (Figure 5.31).

SOLUTION   Because $ABCD$ is a rectangle, $\angle A = 90°$. Thus, by the Pythagorean Theorem, $BD^2 = 3^2 + 4^2 = 9 + 16 = 25$, or $BD = 5$. Because the diagonals of a rectangle are congruent, $BD = AC$, or $AC = 5$. Notice that AC could also have been found directly by observing that $CD = 3$ and applying the Pythagorean Theorem to $\triangle ADC$. •

SQUARES

In Chapter 2, a square was defined to be a quadrilateral with all sides congruent and four right angles. Because all of its sides are congruent, a square is a rhombus, and because all four of its angles are right angles, a square is also a rectangle. Because both a rhombus and a rectangle are parallelograms, a square is also a parallelogram. Therefore, all the theorems that hold for rectangles, parallelograms, and rhombuses also apply to squares.

Definitions are chosen to best describe shapes or relationships. However, as in the case of the rectangle, different definitions could have been used for the square. Next we will state several theorems that could have served as definitions of a square. Their proofs will be left for the problem set.

A square is a rhombus with one right angle.

A square is a rectangle with two adjacent sides congruent.

A square is a rhombus with congruent diagonals.

A square is a rectangle with perpendicular diagonals.

ISOSCELES TRAPEZOIDS

(a)

(b)

FIGURE 5.32

Finally, let's turn our attention to trapezoids. Recall that a trapezoid is a quadrilateral with exactly one pair of parallel sides [Figure 5.32(a)]. An isosceles trapezoid has congruent legs [Figure 5.32(b)]. In Figure 5.32(b), $\angle A$ appears to be congruent to $\angle D$. The next theorem is suggested by that figure.

Theorem 5.30

The base angles of an isosceles trapezoid are congruent.

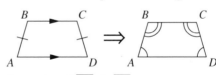

In $ABCD$, if $\overline{AD} \parallel \overline{BC}$ and $AB = CD$,
then $\angle A \cong \angle D$ and $\angle B \cong \angle C$.

Given Isosceles trapezoid $ABCD$ with $\overline{AD} \parallel \overline{BC}$ and with $AB = CD$ [Figure 5.33(a)]

 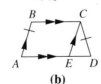

(a)   **(b)**   **(c)**   **(d)**

FIGURE 5.33

Prove $\angle A \cong \angle D$ and $\angle B \cong \angle C$

Plan Draw a segment through C parallel to $\overline{AB}$ to form a parallelogram and a triangle.

Proof Let $\overline{CE}$ be parallel to $\overline{AB}$ [Figure 5.33(b)]. Then $ABCE$ is a parallelogram, so $AB = CE$. Therefore, $\angle A \cong \angle CED$ because they are corresponding angles [Figure 5.33(c)]. But remember that $AB = CD$, so we must have $CE = CD$.

Thus, $\triangle CDE$ is an isosceles triangle, so $\angle CED \cong \angle D$ [Figure 5.33(d)]. Therefore, $\angle A \cong \angle D$. Because $\overline{AD} \parallel \overline{BC}$, it follows that $\angle B$ and $\angle C$ are supplementary to $\angle A$ and $\angle D$ respectively. Therefore, they are equal.   •

EXAMPLE 5.10   Trapezoid $ABCD$ is isosceles. Find the measure of angle x [Figure 5.34(a)].

SOLUTION   In $\triangle ACD$, the sum of the angles is $180°$. Hence, $\angle CAD + 60° + 70° = 180°$, and $\angle CAD = 50°$ [Figure 5.34(b)]. Because $ABCD$ is isosceles, $\angle A = \angle D$. This means that $x + \angle CAD = 70°$, or $x + 50° = 70°$. Thus, $x = 20°$.   •

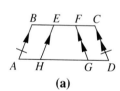

(a)

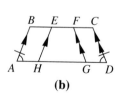

(b)

FIGURE 5.34

Next, let's consider the converse of Theorem 5.30. Suppose that $ABCD$ is a trapezoid with base angles congruent [Figure 5.35(a)]. Is it an isosceles trapezoid? Let $\overline{CE}$ be parallel to $\overline{AB}$ [Figure 5.35(b)]. Then $ABCE$ is a parallelogram,

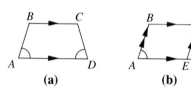

(a)   **(b)**   **(c)**

FIGURE 5.35

so $\overline{AB} \cong \overline{CE}$. Viewing $\overline{AD}$ as a transversal, we see that $\angle A$ and $\angle CED$ are corresponding angles, hence $\angle A \cong \angle CED$ [Figure 5.35(c)]. Because $\angle A \cong \angle D$ is given, we have $\angle CED \cong \angle D$. This means that $\overline{CE} \cong \overline{CD}$ in $\triangle CED$. From $\overline{AB} \cong \overline{CE}$ and $\overline{CE} \cong \overline{CD}$ we can conclude that $\overline{AB} \cong \overline{CD}$. This proves the following theorem.

(a)

Theorem 5.31

If the base angles of a trapezoid are congruent, then it is isosceles.

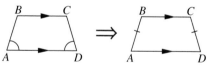

If $\overline{AD} \parallel \overline{BC}$ and $\angle A \cong \angle D$, then $\overline{AB} \cong \overline{CD}$.

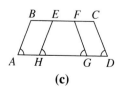

(b)

EXAMPLE 5.11   In isosceles trapezoid $ABCD$, $\overline{AB} \parallel \overline{EH}$ and $\overline{CD} \parallel \overline{FG}$ [Figure 5.36(a)]. Prove that $EFGH$ is an isosceles trapezoid.

SOLUTION $EFGH$ is a trapezoid because $\overline{EF} \parallel \overline{GH}$ and the other sides are not parallel. Because $ABCD$ is isosceles, $\angle A = \angle D$ [Figure 5.36(b)]. Because $\overline{AB} \parallel \overline{EH}$, we have $\angle A = \angle EHG$, and $\overline{CD} \parallel \overline{FG}$ implies that $\angle D = \angle FGH$ [Figure 5.36(c)]. Consequently, $\angle EHG = \angle FGH$, so by Theorem 5.31, $EFGH$ is isosceles.   •

(c)

FIGURE 5.36

ATTRIBUTES OF QUADRILATERALS

In Sections 5.3 and 5.4, many attributes of parallelograms, rhombuses, rectangles, squares, and trapezoids were discussed. The following table provides a summary of these attributes.

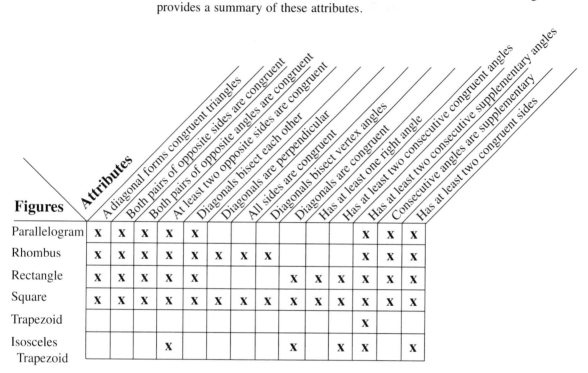

Figures / Attributes	A diagonal forms congruent triangles	Both pairs of opposite sides are congruent	Both pairs of opposite angles are congruent	At least two opposite sides are congruent	Diagonals bisect each other	Diagonals are perpendicular	All sides are congruent	Diagonals bisect vertex angles	Diagonals are congruent	Has at least one right angle	Has at least two consecutive congruent angles	Has at least two consecutive supplementary angles	Consecutive angles are supplementary	Has at least two congruent sides
Parallelogram	X	X	X	X	X							X	X	X
Rhombus	X	X	X	X	X	X	X	X				X	X	X
Rectangle	X	X	X	X	X				X	X	X	X	X	X
Square	X	X	X	X	X	X	X	X	X	X	X	X	X	X
Trapezoid												X		
Isosceles Trapezoid				X					X		X	X		X

Solution to Applied Problem

If the frame is to have four 90° angles, then it must be a rectangle. Because opposite sides are congruent, the frame is a parallelogram. To verify that *ABCD* is a rectangle, Harry can find *AC* and *BD*. If the lengths of these diagonals are the same, then the frame is a rectangle and thus has four "square" corners. He could also verify that each angle is 90° by using the Pythagorean Theorem to calculate what the length of a diagonal should be.

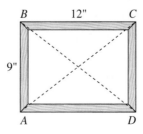

GEOMETRY AROUND US

The structures of most radio towers include isosceles trapezoids. To increase the strength and stability of the towers, the (congruent) diagonals of the trapezoids are included.

PROBLEM SET 5.4

EXERCISES/PROBLEMS

Use the following figure for Exercises 1–10.

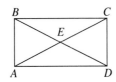

Determine whether each statement is true or false for rectangle *ABCD*. If true, explain why.

1. $\overline{AC} \perp \overline{BD}$

2. $\overline{AD} \parallel \overline{BC}$

3. $AE = CE$

4. $AE = BE$

5. $AC = BD$

6. $\angle BEC = 90°$

7. $\angle BAE = \angle DAE$

8. $\triangle BEC \cong \triangle AED$

9. $\triangle ABD \cong \triangle BAC$

10. $\triangle ABC \cong \triangle CDA$

Use the following figure for Exercises 11–20.

Determine whether each statement is true or false for square *ABCD*. If true, explain why.

11. $\overline{AC} \perp \overline{BD}$

12. $\overline{AB} \perp \overline{BC}$

13. $BE = ED$

14. $BE = AE$

15. $\angle ABD = \angle CDB$

16. $\angle BCA = \angle DCA$

17. $\triangle ABE \cong \triangle BCE$

18. $\triangle ABD \cong \triangle CDA$

19. $\angle BDA = 45°$

20. $\triangle AED$ is isosceles.

Use the following figure for Exercises 21–28.

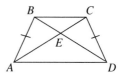

Determine whether the statement is true or false for trapezoid *ABCD*. If true, explain why.

21. $\angle BAD = \angle CDA$

22. $\angle BAC = \angle DAC$

23. $\triangle BAD \cong \triangle CDA$

24. $\triangle BAC \cong \triangle DCA$

25. $BD = CA$

26. $BE = DE$

27. $BE = CE$

28. $\angle ABD = \angle DCA$

29. A rectangle has a diagonal of 15 cm. The width of the rectangle is 3 cm less than its length. Find the area of the rectangle.

30. The length of a rectangle is one inch less than twice its width. The diagonal of the rectangle is two inches more than its length. Find the area of the rectangle.

31. The diagonal of a square is 16 mm. Find the length of a side of the square.

32. An isosceles trapezoid has legs of length $5\sqrt{2}$ cm and base angles that measure 45°. If the shorter base of the trapezoid measures 15 cm, find the length of the longer base.

33. Find the length of the legs in the following isosceles trapezoid.

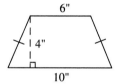

34. Find the area of the following isosceles trapezoid. Round to the nearest tenth of a square meter.

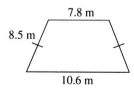

35. In isosceles trapezoid *EFGH*, *EF* = 20″, *FG* = 25″, and ∠*H* = 60°. Find *EH*.

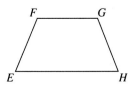

36. *PQRS* is an isosceles trapezoid. If ∠*TRS* = 100°, ∠*T* = 35°, and *PS* = *TS*, find the measures of the numbered angles.

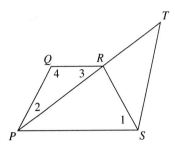

37. In isosceles trapezoid $\overline{WXYZ}$, *XP* = 5 in., *YZ* = 12 in., $\overline{XZ} \perp \overline{WX}$, and $\overline{WY} \perp \overline{YZ}$. Find *WZ*.

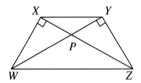

38. In the following figure, *BCDE* is a rectangle and *ABDE* is a parallelogram. Find the measures of angles 1 and 2.

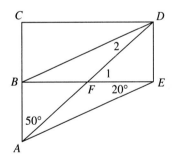

39. In the following figure, *BCDE* is a rectangle and *ABDE* is a parallelogram. If *AE* = 12 cm and ∠*CEB* = 30°, find *CF* and *CB*.

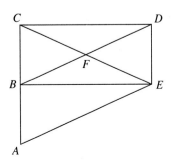

40. Quadrilateral *ABCD* is a kite. Draw several other examples of kites. Then refer to the attributes of quadrilaterals given in the table that precedes this problem set. Which of these attributes seem to hold true for kites?

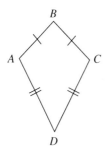

PROOFS

41. Prove: A square is a rhombus with one right angle.

42. Prove: A square is a rectangle with two adjacent sides congruent.

43. Prove: A square is a rhombus with congruent diagonals.

44. Prove: A square is a rectangle with perpendicular diagonals.

45. Prove: A parallelogram with one right angle is a rectangle. (Theorem 5.27)

46. Prove: The diagonals of an isosceles trapezoid are congruent.

47. Prove: If the diagonals of a trapezoid are congruent, then it is an isosceles trapezoid.

48. In parallelogram *ABCD*, $\overline{BP}$ and $\overline{CQ}$ are altitudes. Prove that *PBCQ* is a rectangle.

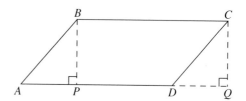

49. Refer to problem 40. For each attribute you found to be true for kites, prove your assertion.

APPLICATION

50. An antenna is erected at the center of a square, flat roof. Guy wires are attached at each corner of the roof as shown. Why are the lengths of the four guy wires the same?

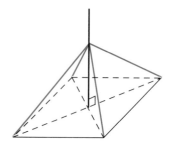

5.5
GEOMETRIC CONSTRUCTIONS INVOLVING PARALLEL LINES

Applied Problem

In drafting, a segment $\overline{AB}$ can be divided into 6 congruent segments using a ruler as follows:

A segment perpendicular to $\overline{AB}$ is constructed at B. Then a ruler is aligned as shown to the right. When perpendiculars are dropped to $\overline{AB}$ from marks at 1″, 2″, 3″, 4″, and 5″ on the ruler, $\overline{AB}$ is divided into 6 congruent segments. Why does this method work?

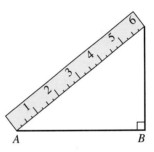

In Section 4.4, several basic geometric constructions were given together with their justifications. Those basic constructions together with the ones that follow will be used in this section to construct parallel lines and quadrilaterals satisfying various conditions. These constructions will rely heavily on theorems from the first four sections of this chapter.

CONSTRUCTING PARALLEL LINES

Construction 7. *To construct a line, through a point, parallel to a given line* [Figure 5.37(a)]

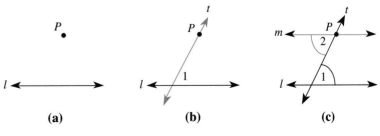

(a)   **(b)**   **(c)**

FIGURE 5.37

Construction Draw any line t through P that intersects l [Figure 5.37(b)]. Then copy $\angle 1$ to form $\angle 2$ with line m and line t as shown in Figure 5.37(c). Line m is parallel to line l.

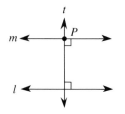

FIGURE 5.38

Justification $\angle 1 \cong \angle 2$ by construction. Because they form a pair of alternate interior angles with lines l and m and transversal t, lines l and m are parallel. [NOTE: We could have also constructed congruent corresponding angles.] •

Notice that rather than draw any line t through P, line t could have been constructed to be perpendicular to l. Then m could be constructed through P perpendicular to t (Figure 5.38).

CONSTRUCTING QUADRILATERALS

Now, our seven basic constructions can be applied to construct a variety of quadrilaterals.

a ————
      ———— b

FIGURE 5.39(a)

EXAMPLE 5.12   Construct a rectangle given the lengths a and b of its sides [Figure 5.39(a)].

SOLUTION
STEP 1. Construct two lines, m and n, perpendicular at A [Figure 5.39(b)].

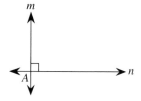

FIGURE 5.39(b)

STEP 2. Copy the lengths a and b, as illustrated in Figure 5.39(c).

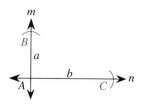

FIGURE 5.39(c)

STEP 3. From point B swing an arc of length b to the right of m, and from C swing an arc of length a above l to form point D [Figure 5.39(d)].

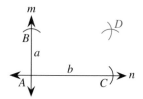

FIGURE 5.39(d)

Draw $\overline{BD}$ and $\overline{CD}$. Then *ABDC* is a rectangle [Figure 5.39(e)].

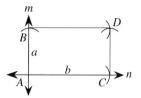

FIGURE 5.39(e)

Justification *ABDC* is a quadrilateral with opposite sides congruent and at least one right angle (at *A*). By Theorems 5.17 and 5.27, *ABDC* is a rectangle. •

EXAMPLE 5.13   Construct a rhombus given the lengths *a* and *b* of its two diagonals [Figure 5.40(a)].

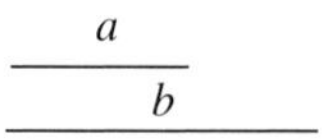

$$\overline{}^{\,a}$$
$$\underline{}_{\,b}$$

FIGURE 5.40(a)

SOLUTION
STEP 1. Construct a line segment $\overline{AB}$ of length *b* [Figure 5.40(b)].

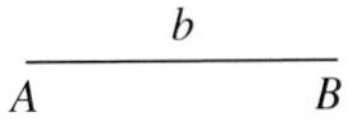

$$\underset{A}{}\overline{}^{\,b}\underset{B}{}$$

FIGURE 5.40(b)

STEP 2. Construct the perpendicular bisector *l* of $\overline{AB}$. Let *M* represent the midpoint of $\overline{AB}$ [Figure 5.40(c)].

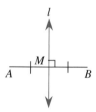

FIGURE 5.40(c)

STEP 3. Construct the perpendicular bisector of the original segment of length *a* [Figure 5.40(d)].

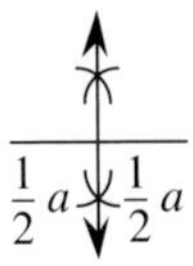

FIGURE 5.40(d)

Then mark off the points C and D on line l so that $CM = DM = \frac{1}{2}a$, as shown in Figure 5.40(e).

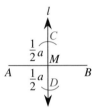

FIGURE 5.40(e)

STEP 4. Draw the segments $\overline{AC}$, $\overline{CB}$, $\overline{BD}$, and $\overline{AD}$. Then $ACBD$ is a rhombus [Figure 5.40(f)].

FIGURE 5.40(f)

Justification We constructed $\overline{AB}$ and $\overline{CD}$ to be perpendicular to each other where M is the midpoint of each segment. By Theorems 5.23 and 5.24, a quadrilateral is a rhombus if and only if its diagonals are perpendicular bisectors of each other.   •

SUBDIVIDING A LINE SEGMENT

In Chapter 4, a construction to bisect a segment was studied. By repeating that process, we could divide a line segment into four congruent segments, eight congruent segments, and so on. The following theorem shows how to divide a segment into any number of congruent subsegments.

Theorem 5.32

If three parallel lines form congruent segments on one transversal, then they form congruent segments on any transversal.

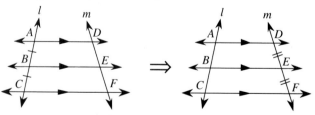

If $AB = BC$, then $DE = EF$.

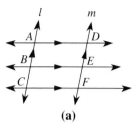

(a)

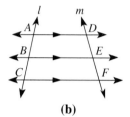

(b)

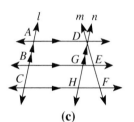

(c)

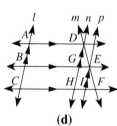

(d)

FIGURE 5.41

Given $\overleftrightarrow{AD} \parallel \overleftrightarrow{BE} \parallel \overleftrightarrow{CF}$ and $AB = BC$

Prove $DE = EF$

Proof There are two cases to consider. Case 1: l is parallel to m. Case 2: l is *not* parallel to m.

Case 1 l is parallel to m [Figure 5.41(a)].
   Because $l \parallel m$ and $\overleftrightarrow{AD} \parallel \overleftrightarrow{BE}$, we can conclude that $ADEB$ is a parallelogram. Consequently, $AB = DE$ because opposite sides of a parallelogram are congruent. Similarly, $BC = EF$. Because it is given that $AB = BC$, we can conclude from our three equations that $DE = EF$.

Case 2 l is *not* parallel to m [Figure 5.41(b)].
   An outline for the proof follows.

Subgoal 1 Let n be the line through D that is parallel to l, intersecting $\overleftrightarrow{BE}$ at G and intersecting $\overleftrightarrow{CF}$ at H [Figure 5.41(c)]. Use the result from Case 1 to show that $DG = GH$.

Subgoal 2 Let p be the line through E that is parallel to l and n, intersecting $\overleftrightarrow{CF}$ at I [Figure 5.41(d)]. Show that $BEIC$ is a parallelogram and hence $BC = EI$.

Subgoal 3 Prove that $\triangle DGE \cong \triangle EIF$ using ASA. By corresponding parts, we can conclude that $DE = EF$.   •

   The next construction uses this theorem to illustrate a method for dividing a segment into any number of congruent segments. It completes our collection of basic constructions.

Construction 8. *To divide a given segment into a specified number of congruent segments.*

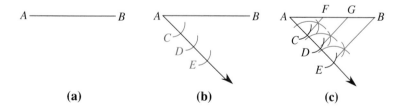

(a)   **(b)**   **(c)**

FIGURE 5.42

Construction Three congruent segments will be constructed on $\overline{AB}$ [Figure 5.42(a)].

STEP 1. Draw any ray with endpoint at A, but not containing $\overline{AB}$. Then mark off three congruent segments of any length on the ray as illustrated in Figure 5.42(b).

STEP 2. Draw $\overline{BE}$ and construct angles at C and D congruent to $\angle AEB$. Using these angles, draw lines through C and D that are parallel to $\overline{BE}$ [Figure 5.42(c)]. The points where the lines intersect $\overline{AB}$ will divide $\overline{AB}$ into the three congruent segments $\overline{AF}$, $\overline{FG}$, and $\overline{GB}$.

Justification Because we constructed $\angle ACF \cong \angle ADG \cong \angle AEB$, we have $\overline{CF} \parallel \overline{DG} \parallel \overline{EB}$ by congruent corresponding angles. We also know that $AC = CD = DE$, by construction. By Theorem 5.32, if parallel lines form congruent segments on one transversal, then they form congruent segments on any transversal. This means it must also be true that $AF = FG = GB$. •

Solution to Applied Problem

When the segments perpendicular to $\overline{AB}$ are drawn, we obtain the figure to the right. Because all of these segments are perpendicular to $\overline{AB}$, they are parallel. The ruler acts as a transversal. Because the segments on the ruler are congruent, the segments on another transversal, like $\overline{AB}$, will also be congruent.

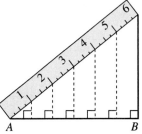

GEOMETRY AROUND US

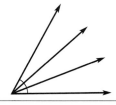

A line segment can be divided into three congruent line segments using only a compass and straightedge using Construction 8. Can an angle be similarly subdivided into three congruent angles using only a compass and straightedge? This is the famous "Trisection Problem" that mathematicians struggled to solve for 2000 years. Not until the 19th century was it proved by way of algebra that, in general, such a construction is impossible.

PROBLEM SET 5.5

EXERCISES/PROBLEMS

1. Construct a line through P that is parallel to line l.

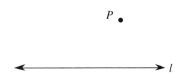

2. Trace $\triangle ABC$. Then construct a line through B parallel to $\overline{AC}$, a line through A parallel to $\overline{BC}$, and a line through C parallel to $\overline{AB}$. Extend each pair of lines so they meet at points D, E, and F. How does the area of $\triangle DEF$ compare to the area of $\triangle ABC$? Justify your answer.

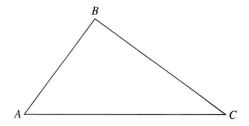

3. Construct a square given one of its sides *a*.

_____ *a* _____

4. Construct a rectangle given two adjacent sides *a* and *b*.

_____ *a* _____

_____ *b* _____

5. Construct a rhombus given one of its sides *a* and one of its angles *A*.

_____ *a* _____

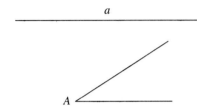

6. Construct a parallelogram given two adjacent sides, *a* and *b*, and the included angle *A*.

_____ *a* _____

_____ *b* _____

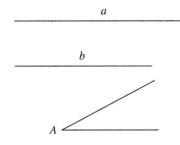

7. Construct a square given one of its diagonals *a*.

_____ *a* _____

8. Construct a rectangle given one of its sides *a* and one of its diagonals *b*.

_____ *a* _____

_____ *b* _____

9. Construct an isosceles trapezoid given one base *a* and an angle *A* at one end of the base. (There are many possibilities.)

_____ *a* _____

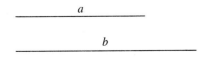

10. Construct an isosceles trapezoid with a base angle of 30°, one base of length *b*, and legs of length *a*.

_____ *b* _____

_____ *a* _____

11. Divide segment $\overline{AB}$ into five congruent segments using a compass and straightedge.

A ————————————————— *B*

12. Divide segment $\overline{PQ}$ into 6 congruent segments using a compass and straightedge.

P ————————————— *Q*

13. Using only a compass and straightedge, divide segment $\overline{MN}$ into two segments such that one segment is twice as long as the other.

M ————————————— *N*

14. Using only a compass and straightedge, divide segment $\overline{ST}$ into three segments whose lengths are in the ratio 1:2:3.

S ————————————— *T*

15. Construct a parallelogram with adjacent sides having the lengths as given and with a vertex angle of 60°.

_____ *a* _____

_____ *b* _____

16. Construct a rhombus with a vertex angle of 60° and a diagonal of length d.

17. Construct a rhombus given the length of a diagonal and the angle between that diagonal and a side of the rhombus.

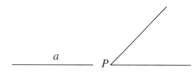

18. Construct a rhombus whose diagonals have the following lengths.

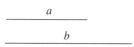

19. A parallelogram has one side of length a. Its diagonals have lengths b and c. Construct the parallelogram.

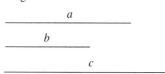

20. Trace △PQR. Using compass and straightedge, divide △PQR into five triangles, all with the same area.

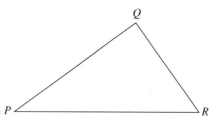

21. In the figure, T is the midpoint of $\overline{PS}$, $RU = 5.8$ cm, and $SQ = 7.4$ cm. Find the area of △PRS.

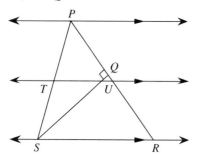

22. Find the value of x in the following figure.

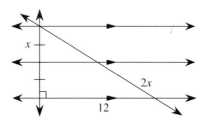

23. An interesting result called **Aubel's Theorem** can be observed in the following construction.

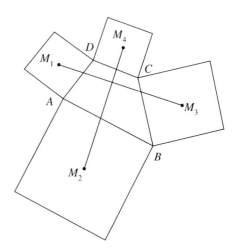

(i) Draw any quadrilateral $ABCD$.

(ii) Using compass and straightedge, construct a square on each side of quadrilateral $ABCD$.

(iii) Locate the midpoint of each square: M_1, M_2, M_3, M_4. Connect the midpoints of opposite squares with line segments.

   (a) Draw a quadrilateral like the one shown and perform the steps in (i), (ii), and (iii). Use your ruler to measure the lengths of segments M_1M_3 and M_2M_4. Use your protractor to measure the angles formed where M_1M_3 and M_2M_4 intersect.

   (b) Repeat the construction for two more quadrilaterals. Include one quadrilateral for which the squares overlap. Measure the segments and angles described in part (a).

   (c) What conclusion can you draw?

PROOFS

24. Justify your construction in Exercise 5.

25. Justify your construction in Exercise 6.

26. Justify your construction in Exercise 7.

27. Justify your construction in Exercise 8.

28. Justify your construction in Exercise 13.

29. Justify your construction in Exercise 16.

30. Justify your construction in Exercise 17.

31. Justify your construction in Exercise 19.

32. Justify your construction in Exercise 20.

APPLICATION

33. The following figure shows a type of bridge truss called a **Baltimore truss**. If members $\overline{SC}$, $\overline{RU}$, and $\overline{QD}$ must be parallel and if $SR = RQ$, what other members must have the same length by Theorem 5.32?

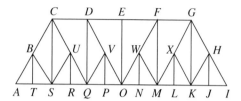

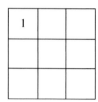

Solution to Initial Problem The numbers 1 through 9 can be arranged in a 3 by 3 square array so that the sum of the numbers in each row, column, and diagonal is 15. Show that 1 cannot be in one of the corners.

STRATEGY: USE INDIRECT REASONING

Suppose that 1 could be in a corner as shown. Then the row, column, and diagonal containing 1 must each have a sum of 15. Equivalently, there must be *three* pairs of numbers from 2 to 9 whose sum is 14. However, there are only *two* such pairs, $5 + 9 = 14$ and $6 + 8 = 14$. Therefore, by indirect reasoning, it is impossible to have 1 in a corner.

Additional Problems Where the Strategy "Use Indirect Reasoning" Is Useful

1. For right triangle $\triangle ABC$ with sides a, b, c as shown, if c^2 is a perfect square, then the lengths a and b cannot both be odd.

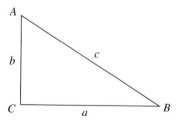

2. Given $\triangle ABC$ as shown, prove that there is only one altitude $\overline{BD}$ to $\overline{AC}$.

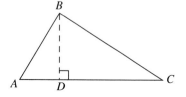

Writing for Understanding

1. Describe at least five examples in which properties of parallel lines are used in the real world. Explain how the parallel lines serve a useful function.

2. If asked to show that two lines are parallel, how might you do it? Describe as many ways as you can. Does one method have an advantage over other methods?

3. The "Geometry Around Us" at the end of Section 5.2 discusses reflectors. Do additional research on reflectors, such as the paraboloid. Summarize your findings, pointing out at least three important applications of the use of reflectors in daily life.

4. The "Geometry Around Us" at the end of Section 5.4 contains a discussion about towers. Do some research on the construction of at least three types of towers, focussing on the role geometry plays. Summarize your findings.

CHAPTER REVIEW

Following is a list of key vocabulary, notation, and ideas for this chapter. Mentally review these items and, where appropriate, write down the meaning of each term. Then restudy the material that you are unsure of before proceeding to take the chapter test.

Section 5.1—Indirect Reasoning and the Parallel Postulate

Vocabulary/Notation

Indirect reasoning (proof) 218
Skew lines 219
Transversal 219
Alternate interior angles 219
Alternate exterior angles 219

Interior angles on the same side
of the transversal 219
Exterior angles on the same side
of the transversal 219
Corresponding angles 219

Main Ideas/Results

1. Indirect reasoning
2. The Parallel Postulate
3. If a pair of parallel lines is cut by a transversal, several pairs of congruent angles are formed.

Section 5.2—Applications of Parallel Lines

Vocabulary/Notation

AAS Congruence 230

HA Congruence 230

Main Ideas/Results

1. Angle Sum in a Triangle Theorem
2. The Exterior Angle Theorem
3. AAS Congruence Theorem

4. Hypotenuse-Angle Congruence Theorem
5. The Angle Bisector Theorem

Section 5.3—Parallelograms and Rhombuses

Main Ideas/Results

1. There are congruence relationships among the sides and angles of a parallelogram.
2. The diagonals of a quadrilateral can be used to characterize it as a parallelogram.
3. Every rhombus is a parallelogram.
4. The diagonals of a parallelogram can be used to characterize a rhombus in two ways.

Section 5.4—Rectangles, Squares, and Trapezoids

Main Ideas/Results

1. Every rectangle is a parallelogram.
2. A parallelogram with one right angle is a rectangle.
3. Diagonals can be used to characterize a parallelogram as a rectangle.
4. A trapezoid can be characterized as isosceles using its base angles.

Section 5.5—Geometric Constructions Involving Parallel Lines

Main Ideas/Results

1. Construction 7. To construct a line through a point parallel to a given line
2. Construction 8. To divide a given segment into a specified number of congruent segments

PEOPLE IN GEOMETRY

George Riemann (1826–1866) developed a non-Euclidean geometry by modifying the Parallel Postulate. He supposed that through any given point on a plane, there are no lines parallel to a given line. The resulting system could be modeled as the geometry on the surface of a sphere, where every line is a great circle and line segments are portions of great circles. In this realm, it can be shown that in every triangle, the sum of the angle measures is greater than 180 degrees.

Riemann came from a poor family, and this poverty contributed to his early death at the age of forty. As a lecturer at the University of Göttingen, he received no salary and depended on contributions from his students. Yet it was during this time that he developed stunning new methods which cast all of geometry in a new light. His mathematical descriptions of curved space were later used in Einstein's theory of general relativity.

CHAPTER 5 TEST

TRUE-FALSE

Mark as true any statement that is always true. Mark as false any statement that is never true or that is not necessarily true. Be able to justify your answers.

1. If two parallel lines are cut by a transversal, the alternate exterior angles are congruent.

2. If two lines are cut by a transversal forming a pair of supplementary corresponding angles, then the lines are parallel.

3. The angle sum in a quadrilateral is 180°.

4. A diagonal of an isosceles trapezoid forms two congruent triangles.

5. A quadrilateral is a rhombus if and only if its opposite angles are congruent.

6. A quadrilateral is a parallelogram if and only if two sides are parallel and congruent.

7. If the diagonals of a quadrilateral bisect each other, then it is a parallelogram.

8. If the diagonals of a quadrilateral are congruent, then it is a rectangle.

9. If a rhombus has a side of length 10 cm and a diagonal of length 16 cm, then the other diagonal measures 12 cm.

10. In any △ABC, if P is a point on the bisector of ∠ABC, then PA = PC.

EXERCISES/PROBLEMS

11. In a parallelogram one angle has measure 20°. Find the measures of the other three angles.

12. In rectangle ABCD, AB = 1 and ∠BDA = 30°. Find the measures of ∠CBD, BD, and AC.

13. What type of quadrilateral is ABCD? Be as specific as you can. Use only the angle measures shown. That is, do not make your decision based on what the shape "appears" to be.

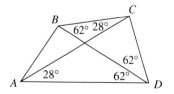

14. A rectangle has a length of 45 yards. The diagonal of the rectangle measures 3 yards more than twice the width. Find the width of the rectangle.

Refer to the following figure for questions 15–17.

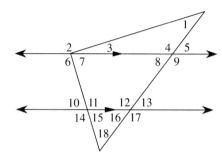

15. Identify four pairs of alternate interior angles in the figure.

16. Identify six pairs of corresponding angles in the figure.

17. Find the measures of ∠1, ∠5, and ∠10, given that ∠12 = 140°, ∠18 = 65°, and ∠2 = 170°.

18. Using only your compass and straightedge, divide $\overline{AB}$ into two parts such that the length of one segment is four times the length of the other segment.

A _____ B

19. Using only your compass and straightedge, construct a parallelogram with a diagonal of length d, one side of length a, and an angle between the side of length a and diagonal of length d that measures exactly 30°.

_____ a _____   _____ d _____

20. Using only your compass and straightedge, construct a rhombus whose diagonals have the following lengths.

_____ a _____   _____ b _____

PROOFS

21. In square ABCD, prove △ADE ≅ △ABE.

22. In parallelogram $ABCD$, E and F are the midpoints of $\overline{AB}$ and $\overline{CD}$, respectively. Prove that $AECF$ is a parallelogram.

APPLICATIONS

23. Plans for a patio in the shape of an isosceles trapezoid have dimensions as shown.

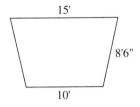

If a layer of concrete 4 inches thick will be poured, calculate the number of cubic feet of concrete that will be required for the project. Round your answer up to the nearest cubic foot.

24. A balance scale such as the one shown uses a parallel linkage. Explain why the pans of the scale remain horizontal, even when unequal weights are placed on the pans.

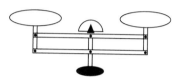

6

Similarity

Euclid's *Elements* contained the following construction: "to cut a line segment in extreme and mean ratio." In modern terms, this means that a point B divides $\overline{AC}$ so that $\dfrac{AB}{BC} = \dfrac{BC}{AC}$.

$$A \qquad\qquad B \qquad\qquad\qquad\qquad C$$

In this construction, the ratio $BC:AB = (1 + \sqrt{5})/2$, the golden ratio. The Greek historian Herodotus related that the Egyptians knew about this famous ratio before the Greeks. In the famous Pyramid of Gizeh, the dimensions of the square pyramid were chosen so that the square whose side is the height of the pyramid has the same area as any one of its triangular faces. A computation shows that the ratio of the altitude of a face to half the length of the base of a triangular face is the golden ratio.

PROBLEM-SOLVING STRATEGIES

1. Draw a Picture
2. Guess and Test
3. Use a Variable
4. Look for a Pattern
5. Make a Table
6. Solve a Simpler Problem
7. Look for a Formula
8. Use a Model
9. Identify Subgoals
10. Use Indirect Reasoning
11. *Solve an Equation*

STRATEGY 11: SOLVE AN EQUATION

Often, when the Use a Variable or Look for a Formula strategies are applied to solve a problem, the representation of the problem will result in an equation. After the equation has been solved using techniques from algebra, the answer should be substituted in the *original problem* as a check.

INITIAL PROBLEM

Aerial photographs are taken from a plane. A dam that is known to be 100 m long is 3 cm long in a photograph. A second picture taken from the same height shows another dam to be 5.7 cm long. How long is the second dam?

3 cm   5.7 cm

CLUES

The Solve an Equation strategy may be appropriate when

· A variable has been introduced.
· The phrase "is," "is equal to," or "is proportional to" appears in a problem.
· The stated conditions can easily be represented in an equation.

A solution for the initial problem above is on page 312.

INTRODUCTION

Geometric figures that have the same shape, but not necessarily the same size, are said to be **similar**. Similarity is used in making maps, scale drawings, enlargements of photos, and indirect measurements of distance. For example, to determine the height of a building, we could compare the shadow of the building with the shadow of a meter stick and then use the fact that similar triangles are formed to calculate the height of the building. The concepts of ratio and proportion are presented in Section 6.1 and are used in Sections 6.2 and 6.3, which contain the ideas central to the study of similar triangles. Section 6.4 introduces relationships within right triangles.

6.1
RATIO AND PROPORTION

Applied Problem

A computer-generated model of a car is to be sculpted into clay. The dimensions of the car are 400 cm long by 160 cm wide by 120 cm high. If the ratio of the dimensions of the model to the dimensions of the car is to be 1 to 4, what should be the dimensions of the model?

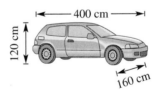

RATIOS

The solution to the applied problem requires an understanding of the concepts of ratio and proportion. A **ratio** is an ordered pair of numbers. The ratio of the number a to the number b is represented in four common ways:

$$a \text{ to } b \qquad a{:}b \qquad \frac{a}{b} \qquad a/b$$

[NOTE: The b in the ratio a to b can be zero. For example, the ratio of the number of stars to the number of circles on an American flag is 50 to 0. But we will assume that $b \neq 0$; that is, denominators are nonzero.]

The ratio 1 to 4 in the applied problem above could also be written as 1:4, $\frac{1}{4}$, or 1/4. When doing calculations with ratios, we express a ratio as a quotient of two numbers. For example, the ratio π:4.2 would be written $\pi/4.2$.

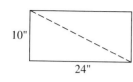

10"

24"

FIGURE 6.1

EXAMPLE 6.1   For the rectangle in Figure 6.1, calculate the following ratios.

(a) length to width   (b) width to length
(c) length of a diagonal to the perimeter   (d) perimeter to area

SOLUTION

(a) length:width $= 24":10" = \dfrac{24"}{10"} = \dfrac{12}{5}$

(b) width:length $= 10":24" = \dfrac{10"}{24"} = \dfrac{5}{12}$

(c) diagonal:perimeter $= \sqrt{10^2 + 24^2}$ in. : $(20 + 48)$ in.

$$= 26" : 68"$$

$$= \dfrac{26"}{68"}$$

$$= \dfrac{13}{34}$$

(d) perimeter:area $= 68$ in. : 240 in^2

$$= \dfrac{68 \text{ in.}}{240 \text{ in}^2}$$

$$= \dfrac{17 \text{ in.}}{60 \text{ in}^2}$$

Notice that each ratio in Example 6.1 was expressed in simplest form. Also notice that in parts (a)–(c), the units in the numerator and denominator were the same. In such cases, the units "cancel" and we express the ratio without units.

Ratios that include two different units of measures are called **rates**. Examples of rates are 12 miles per hour (12 miles:1 hour), 5% per year (5%:1 year), and $3.69 per pound ($3.69:1 pound). Applications often involve rates that must be converted into equivalent rates. As was shown in Section 2.5, dimensional analysis can be used to convert from one rate to another.

PROPORTIONS

When two ratios are involved, we often have a proportion. A **proportion** is an equation stating that two ratios are equal.

EXAMPLE 6.2   Show that the following are proportions.

(a) $\dfrac{1}{2} = \dfrac{2}{4}$ (b) $\dfrac{3.2}{4.8} = \dfrac{2}{3}$

(c) $\pi:3\pi = 1:3$ (d) $\dfrac{7x}{5x} = \dfrac{7}{5}$

SOLUTION

(a) $\dfrac{1}{2} = \dfrac{2}{4}$ because $\dfrac{1}{2} = \dfrac{2 \times 1}{2 \times 2} = \dfrac{2}{4}$

(b) $\dfrac{3.2}{4.8} = \dfrac{2}{3}$ because $\dfrac{3.2}{4.8} = \dfrac{2 \times 1.6}{3 \times 1.6} = \dfrac{2}{3}$

(c) $\pi{:}3\pi = 1{:}3$ because $\dfrac{\pi}{3\pi} = \dfrac{1 \times \pi}{3 \times \pi} = \dfrac{1}{3}$

(d) $\dfrac{7x}{5x} = \dfrac{7}{5}$ because $\dfrac{7x}{5x} = \dfrac{7 \cdot x}{5 \cdot x} = \dfrac{7}{5}$

When a proportion is written as $\dfrac{a}{b} = \dfrac{c}{d}$, the numbers a, b, c, and d are called **terms**. The two numbers a and d are called the **extremes**, and b and c are called the **means**. Next we show how the means and extremes of a proportion can be utilized in a convenient way to determine if two ratios form a proportion.

EXAMPLE 6.3   Determine if $\dfrac{10}{15} = \dfrac{8}{12}$ is a proportion.

SOLUTION   Rewrite $\dfrac{10}{15}$ and $\dfrac{8}{12}$ using the common denominator 15×12:

$\dfrac{10}{15} = \dfrac{10 \times 12}{15 \times 12}$ and $\dfrac{8}{12} = \dfrac{15 \times 8}{15 \times 12}$. Because the two fractions $\dfrac{10 \times 12}{15 \times 12}$ and $\dfrac{15 \times 8}{15 \times 12}$ have a common denominator, we need only determine if the numerators are equal. Because $10 \times 12 = 120$ and $15 \times 8 = 120$, we have $\dfrac{10}{15} = \dfrac{8}{12}$.

Notice that the numbers we compared were 10×12 (the product of the extremes) and 15×8 (the product of the means). This technique is summarized next.

Theorem 6.1

Cross-Multiplication Theorem

$\dfrac{a}{b} = \dfrac{c}{d}$ if and only if the product of the means equals the product of the extremes. That is,

$$\frac{a}{b} = \frac{c}{d} \qquad \text{if and only if} \qquad ad = bc$$

The proof of this theorem consists of two parts:

Subgoal 1 Prove that if $ad = bc$, then $\dfrac{a}{b} = \dfrac{c}{d}$.

Subgoal 2 Prove that if $\dfrac{a}{b} = \dfrac{c}{d}$, then $ad = bc$.

Subgoal 1 will be proved here and Subgoal 2 will be left for the problem set.

Prove If $ad = bc$, then $\dfrac{a}{b} = \dfrac{c}{d}$.

Proof If $ad = bc$, then we can divide both sides of the equation by the same nonzero number, bd, to obtain $\dfrac{ad}{bd} = \dfrac{bc}{bd}$. Simplifying, we have $\dfrac{a}{b} = \dfrac{c}{d}$. ▪

Using Theorem 6.1 we can see that $\dfrac{3}{5} = \dfrac{6}{10}$ because $3 \times 10 = 5 \times 6$. Notice that various terms of a proportion can be interchanged to form new proportions. So we can write

$\dfrac{10}{5} = \dfrac{6}{3}$ if the extremes of $\dfrac{3}{5} = \dfrac{6}{10}$ are interchanged

$\dfrac{3}{6} = \dfrac{5}{10}$ if the means of $\dfrac{3}{5} = \dfrac{6}{10}$ are interchanged

$\dfrac{5}{3} = \dfrac{10}{6}$ if both the ratios in $\dfrac{3}{5} = \dfrac{6}{10}$ are inverted

This discussion motivates the following theorem, which can be proved using Theorem 6.1.

Theorem 6.2

$\dfrac{a}{b} = \dfrac{c}{d}$ if and only if $\dfrac{d}{b} = \dfrac{c}{a}$ (Exchange the extremes.)

$\dfrac{a}{b} = \dfrac{c}{d}$ if and only if $\dfrac{a}{c} = \dfrac{b}{d}$ (Exchange the means.)

$\dfrac{a}{b} = \dfrac{c}{d}$ if and only if $\dfrac{b}{a} = \dfrac{d}{c}$ (Invert each ratio.)

EXAMPLE 6.4 Determine if the following are proportions.

(a) $\dfrac{21}{35} = \dfrac{27}{45}$ (b) $\dfrac{21}{27} = \dfrac{35}{45}$ (c) $\dfrac{45}{35} = \dfrac{27}{21}$

SOLUTION

(a) $21 \times 45 = 945$ and $35 \times 27 = 945$. Thus, $\dfrac{21}{35} = \dfrac{27}{45}$ is a proportion by the Cross-Multiplication Theorem. To work this problem in another way, we could simplify both ratios.

$$\dfrac{21}{35} = \dfrac{3 \times 7}{5 \times 7} = \dfrac{3}{5} \quad \text{and} \quad \dfrac{27}{45} = \dfrac{3 \times 9}{5 \times 9} = \dfrac{3}{5}$$

Because both ratios equal $\dfrac{3}{5}$, they form a proportion.

(b) From part (a) we know that $\dfrac{21}{35} = \dfrac{27}{45}$ is a proportion. By interchanging the means, we obtain $\dfrac{21}{27} = \dfrac{35}{45}$.

(c) From part (a) we know that $\dfrac{21}{35} = \dfrac{27}{45}$. By exchanging the extremes we have $\dfrac{45}{35} = \dfrac{27}{21}$. The proportion in (c) also could have been obtained from the one in (b) by inverting each ratio. We also could have used cross-multiplication in both parts (b) and (c).   •

SOLVING FOR A MISSING TERM IN A PROPORTION

The Cross-Multiplication Theorem can be used to solve for a missing term in a proportion. The following example shows how this can be done.

EXAMPLE 6.5   Solve for the variable in each proportion.

(a) $\dfrac{14}{17} = \dfrac{42}{y}$ (b) $\dfrac{x}{7} = \dfrac{31}{42}$

SOLUTION

(a) $\dfrac{14}{17} = \dfrac{42}{y}$ if and only if $14y = 17 \cdot 42$, or $y = \dfrac{17 \cdot 42}{14} = 51$.

(b) $\dfrac{x}{7} = \dfrac{31}{42}$ if and only if $42x = 7 \cdot 31$, or $x = \dfrac{7 \cdot 31}{42} = \dfrac{31}{6}$.   •

Many problems in mathematics and in geometry, in particular, can be solved using proportions.

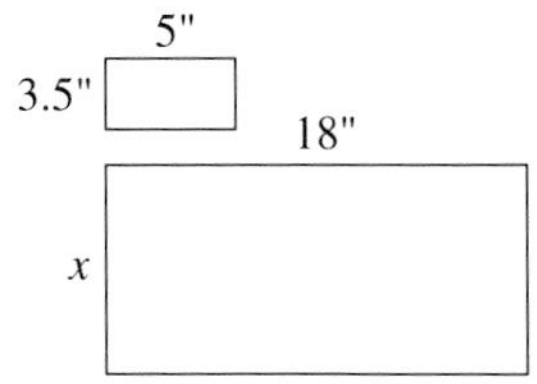

EXAMPLE 6.6   A rectangular snapshot measures 3.5 inches tall by 5 inches wide. An enlargement of the photo will be made, where the larger photo will have a width of 18 inches. If none of the original picture will be cropped (left out) and its dimensions are proportional to the dimensions of the original photo, what will be the height of the larger photograph (Figure 6.2)?

SOLUTION   We can use the proportion

$$\frac{\text{height of small photo}}{\text{width of small photo}} = \frac{\text{height of large photo}}{\text{width of large photo}}$$

So we have $\dfrac{3.5 \text{ inches}}{5 \text{ inches}} = \dfrac{x \text{ inches}}{18 \text{ inches}}$, or $\dfrac{3.5}{5} = \dfrac{x}{18}$.

Thus, $5x = (3.5)(18) = 63$ or $x = 63/5 = 12.6$ inches. Therefore, the height of the new photograph will be 12.6 inches.   •

FIGURE 6.2

THE GEOMETRIC MEAN

The **arithmetic average** or **mean** of two numbers a and c is $\dfrac{a + c}{2}$.

Analogously, the **geometric mean** or **mean proportional** of a and c is $\sqrt{ac}$ for positive a, c. Consider the proportion $a{:}b = b{:}c$. Here $ac = b^2$, or $b = \sqrt{ac}$. This result is stated as follows.

Theorem 6.3

If $\dfrac{a}{b} = \dfrac{b}{c}$, then b is the geometric mean of a and c.

The geometric mean appears often in geometry. In particular, there is an application of it in the next section.

EXAMPLE 6.7   Find the geometric mean of 3 and 27.

SOLUTION   Using Theorem 6.3, the geometric mean of 3 and 27 can be found by solving the equation $\dfrac{3}{x} = \dfrac{x}{27}$. Here $x^2 = 81$. Thus, $x = 9$ is the geometric mean.   •

Solution to Applied Problem

The dimensions of the car were 400 cm by 160 cm by 120 cm.

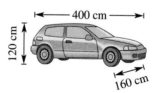

If the ratio of the clay model to the actual car was to be 1:4 and we let l be the length of the model, w its width, and h its height, then these are the appropriate ratios:

$$\frac{1}{4} = \frac{l}{400} = \frac{w}{160} = \frac{h}{120}$$

Solving $\dfrac{1}{4} = \dfrac{l}{400}$, we get $4l = 400$, so $l = 100$ cm. Similarly, $w = 40$ cm and $h = 30$ cm.

GEOMETRY AROUND US

The **golden ratio**, also called the **divine proportion**, was known to the Pythagoreans in 500 B.C. This ratio, the number $\dfrac{1 + \sqrt{5}}{2}$, occurs frequently in nature, art, and architecture. The Parthenon in Athens can be surrounded by a rectangle whose sides have length and width in the golden ratio. The Greeks believed that a rectangle with sides in this ratio had the most aesthetically pleasing shape.

PROBLEM SET 6.1

EXERCISES/PROBLEMS

Represent each ratio in Exercises 1–4 as a fraction in simplest form.

1. 16 to 64

2. 30 to 75

3. 82.5 to 16.5

4. $4\dfrac{1}{2}$ to $2\dfrac{1}{3}$

Solve for the unknown in Exercises 5–16.

5. $\dfrac{57}{95} = \dfrac{18}{n}$

6. $2{:}130 = x{:}5$

7. $n/70 = 6/21$

8. $\dfrac{r}{84} = \dfrac{3}{14}$

9. $\dfrac{7}{5} = \dfrac{42}{s}$

10. $\dfrac{12}{t} = \dfrac{18}{45}$

11. $\dfrac{3}{5}{:}6 = d{:}25$

12. $b/8 = 2.25/18$

13. $\dfrac{x}{100} = \dfrac{4.8}{1.5}$

14. $\dfrac{57.4}{39.6} = \dfrac{7}{4x}$ (Round to one decimal place.)

15. $\dfrac{5}{8} = \dfrac{x}{x + 9}$

16. $\dfrac{x}{30 - x} = \dfrac{2}{3}$

Find the geometric mean of each pair of numbers in Exercises 17–20.

17. 2, 18

18. 5, 7

19. π, 10

20. n, $4n$

21. An Egyptian papyrus listed π as $\left(2 \times \dfrac{8}{9}\right)^{2}$. Archimedes listed $\dfrac{22}{7}$ and $\dfrac{355}{113}$ as approximations to π. Determine if any of these ratios are equal.

PROOFS

22. If the length and width of a rectangle are both doubled, find the ratio of the area of the new rectangle to the area of the original rectangle. Prove your assertion.

23. If the radius of the base of a cylinder is doubled and the height of the cylinder is tripled, find the ratio of the volume of the original cylinder to the volume of the new cylinder. Prove your assertion.

24. Prove Subgoal 2 of Theorem 6.1: If $\dfrac{a}{b} = \dfrac{c}{d}$, then

$ad = bc$.

25. Prove Theorem 6.2:

$$\frac{a}{b} = \frac{c}{d} \quad \text{if and only if} \quad \frac{d}{b} = \frac{c}{a}$$

$$\frac{a}{b} = \frac{c}{d} \quad \text{if and only if} \quad \frac{a}{c} = \frac{b}{d}$$

$$\frac{a}{b} = \frac{c}{d} \quad \text{if and only if} \quad \frac{b}{a} = \frac{d}{c}$$

Prove each of the statements in problems 26–32.

26. If $\dfrac{a + b}{b} = \dfrac{c + d}{d}$, then $\dfrac{a}{b} = \dfrac{c}{d}$.

27. If $\dfrac{a}{b} = \dfrac{c}{d}$, then $\dfrac{a + b}{b} = \dfrac{c + d}{d}$.

28. If $\dfrac{a}{b} = \dfrac{c}{d}$, then $\dfrac{a - b}{b} = \dfrac{c - d}{d}$.

29. If $\dfrac{a}{b} = \dfrac{c}{b}$, then $a = c$.

30. If $\dfrac{a}{b} = \dfrac{a + c}{b + d}$, then $\dfrac{a}{b} = \dfrac{c}{d}$.

31. If $\dfrac{a}{b} = \dfrac{c}{d}$, then $\dfrac{a}{b} = \dfrac{a + c}{b + d}$.

32. If $\dfrac{a}{b} = \dfrac{c}{d} = \dfrac{e}{f}$, then $\dfrac{a + c + e}{b + d + f} = \dfrac{a}{b}$.

33. State and prove a generalization of problem 32.

APPLICATIONS

34. A map is drawn to scale such that 1/8 inch represents 65 feet. If the shortest route from your house to the grocery store measures $23\frac{7}{16}$ inches on the map, how many miles is it to the grocery store?

35. (a) If 1 inch on a map represents 35 miles, how many miles are represented by 3 inches? 10 inches? n inches?
(b) Los Angeles is about 1000 miles from Portland, Oregon. About how many inches apart would Portland and Los Angeles be on this map?

36. A man who weighs 175 pounds on earth would weigh 28 pounds on the moon. How much would his 30-lb dog weigh on the moon?

37. Suppose that you drive your car an average of 4460 miles every half-year. At the end of 2.75 years, how far will your car have gone?

38. When blood cholesterol levels are tested, sometimes a cardiac risk ratio is calculated.

$$\text{Cardiac Risk Ratio} = \frac{\text{total cholesterol level}}{\text{high density lipoprotein level (HDL)}}$$

For women, a ratio between 3.0 and 4.5 is desirable. A woman's blood test yields an HDL cholesterol level of 60 mg/dL and a total cholesterol level of 225 mg/dL. What is her cardiac risk ratio, expressed as a one-place decimal? Is her ratio in the normal range?

39. For the tax year 1989 the Internal Revenue Service audited 92 of every 10,000 individual returns.
(a) In a community in which 12,500 people filed 1989 returns, how many returns might be expected to be audited?
(b) In 1990 only 80 returns per 10,000 were audited. How many fewer of the 12,500 returns would be expected to be audited for 1990 than for 1989?

40. Sheila is climbing a hill that has a 17° slope. For every 5 feet she gains in altitude, she travels about 16.37 horizontal feet. If at the end of her uphill climb she has traveled 1 mile horizontally, how much altitude has she gained?

41. In an aquarium that is 18 inches tall, the pressure on the bottom of the tank is 93.6 lb/ft². In a large viewing aquarium, a tank is 8 ft deep. Find the pressure on the bottom of this larger tank.

42. (a) A baseball pitcher has pitched a total of 25 innings so far during the season and has allowed 18 runs. At this rate, how many runs, to the nearest hundredth, would he allow in 9 innings? This number is called the pitcher's earned run average or ERA.
(b) Orel Hershiser of the Los Angeles Dodgers had an ERA of 2.26 in 1988. At that rate, how many runs would he be expected to allow in 100 innings pitched? Round your answer to the nearest whole number.

43. The "Spruce Goose," a wooden flying boat built for Howard Hughes, had the world's largest wing span (319 ft 11 in.), according to the Guinness Book of World Records. It flew only once, in 1947 for a distance of about 1000 yards. Shelly wants to build a scale model of the Spruce Goose, which is 218-ft-8-in. long. If her model will be 20 inches long, what will its wing span be (to the nearest inch)?

44. The aspect ratio of a TV screen is the ratio of its width to its height. The usual aspect ratio is 4:3. The size of a TV screen is specified by the length of its diagonal. Find the width and height of a 19-inch TV screen with an aspect ratio of 4:3.

45. A fashion doll popular with young girls for the last few decades is $11\frac{1}{2}$ inches tall and has a waist measurement of about 3 inches. Some concern has been expressed that this doll does not reflect realistic body proportions. Can you explain why? Suppose a "life-size" model of the same proportions with height 5′6″ were constructed. What would its waist measurement be (to the nearest inch)?

46. Many tires come with 13/32 inch of tread on them. The first 2/32 inch wears off quickly (say during the first 1000 miles). From then on the tire wears uniformly and more slowly. A tire is considered "worn out" when only 2/32 inch of tread is left.
   (a) How many 32nds of an inch of useable tread does a tire have after 1000 miles?
   (b) A tire has traveled 20,000 miles and has 5/32 inch of tread remaining. At this rate, how many total miles should the tire last before it is considered worn out?

47. Two professional drag racers are speeding down a quarter-mile track. The lead driver is traveling 1.738 feet for every 1.67 feet that the trailing car travels, and the trailing car is going 198 mph. How fast in mph is the lead car traveling?

48. According to the "big-bang" hypothesis, the universe was formed approximately 10^{10} years ago. The analogy of a 24-hour day is often used to put the passage of this amount of time into perspective. Imagine that the universe was formed at midnight 24 hours ago. If the earth was formed approximately 10^5 years ago, to what time in the 24-hour day does this correspond?

49. A **golden rectangle** is a rectangle whose length and width are in the golden ratio, $\dfrac{1 + \sqrt{5}}{2}$. Designers and manufacturers sometimes utilize the aesthetic appeal of the golden rectangle in packaging and construction. Dimensions of some common rectangular objects are listed below. How does the ratio of the dimensions in each case compare to the golden ratio?
   (a) Box of macaroni: $5\frac{1}{8}$ in. by $8\frac{7}{8}$ in.
   (b) Box of cereal: $11\frac{1}{2}$ in. by $7\frac{1}{2}$ in.
   (c) Box of corn starch: 4 in. by $6\frac{1}{2}$ in.
   (d) Window: 2 ft by 3 ft
   (e) Credit card: $3\frac{3}{8}$ in. by $2\frac{1}{8}$ in.

50. The golden ratio can be related to the lengths of line segments as follows. If we start with a line segment AB and locate a point C on the segment such that $\dfrac{AB}{AC} = \dfrac{AC}{CB}$, then the value of each of these ratios is the golden ratio. Notice that AC is the geometric mean of AB and CB.

Let $AB = x$ and $AC = 1$. Use the proportion $\dfrac{AB}{AC} = \dfrac{AC}{CB}$ to find the numerical value of the golden ratio. [HINT: Express CB in terms of 1 and x.]

51. A golden rectangle can be constructed with compass and straightedge as follows. Use the procedure given to construct a golden rectangle. Then measure the length and width of the rectangle you constructed and compare their ratio to the golden ratio.
   (a) Construct a square $ABCD$ of any size.

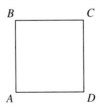

   (b) Construct the perpendicular bisector of $\overline{AD}$ and label the midpoint M.

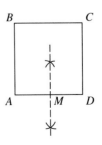

(c) Use your compass to draw an arc with radius MC and with center M. Extend $\overline{AD}$ so that it intersects the arc in P.

(d) Construct a segment perpendicular to $\overline{AP}$ at P. Extend $\overline{BC}$ to meet this segment at Q.

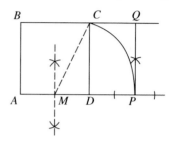

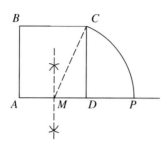

Now rectangle $ABQP$ is a golden rectangle. We could also say that D divides $\overline{AP}$ so that AD is the mean proportional of AP and DP.

6.2
SIMILAR TRIANGLES

Applied Problem

Campers walking along a river want to fell a tree tall enough so that they can walk on the tree to get across the river. How can they find the width of the river at its narrowest point without swimming across the river?

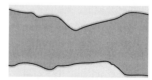

SIMILAR TRIANGLES

Recall that geometric shapes are similar if they have the same shape, though not necessarily the same size. For example, all squares are similar, as are all circles. However, a right triangle and an equilateral triangle do not have the same shape and so are not similar. In this section we focus our attention on similar triangles because they are often used for finding measurements indirectly, that is, without actually measuring the directed distance. Similar triangles will be used to solve the applied problem just stated.

Definition

Similar Triangles

$\triangle ABC$ is **similar** to $\triangle DEF$ (written $\triangle ABC \sim \triangle DEF$) under the correspondence $A \leftrightarrow D$, $B \leftrightarrow E$, $C \leftrightarrow F$ if and only if

1. All three pairs of corresponding angles are congruent.
2. All pairs of corresponding sides are proportional.

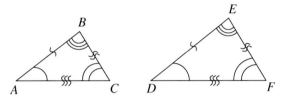

$\triangle ABC \sim \triangle DEF$ if and only if $\angle A \cong \angle D$, $\angle B \cong \angle E$, $\angle C \cong \angle F$, and

$$\frac{AB}{DE} = \frac{BC}{EF} = \frac{AC}{DF}.$$

Notice that curved marks are used to identify the pairs of corresponding sides that are proportional.

Congruent triangles are similar because their corresponding angles are congruent and the corresponding sides are congruent (hence are proportional in the ratio of 1:1). Also, as with congruence, the following relationships hold for similarity:

Reflexive Property: $\triangle ABC \sim \triangle ABC$ for all triangles $\triangle ABC$.

Symmetric Property: If $\triangle ABC \sim \triangle DEF$, then $\triangle DEF \sim \triangle ABC$.

Transitive Property: If $\triangle ABC \sim \triangle DEF$ and $\triangle DEF \sim \triangle GHI$, then $\triangle ABC \sim \triangle GHI$.

EXAMPLE 6.8   Find all pairs of triangles in Figure 6.3 that satisfy the definition of similar triangles. Also, identify other pairs that *appear* to be similar.

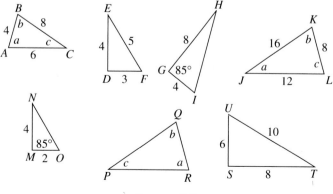

FIGURE 6.3

SOLUTION $\triangle ABC$ and $\triangle LKJ$ are similar because (a) corresponding angles are congruent ($\angle A \cong \angle L$, $\angle B \cong \angle K$, $\angle C \cong \angle J$) and (b) corresponding sides are proportional in the ratio of 1 to 2 ($AC{:}LJ = BC{:}KJ = AB{:}LK = 1{:}2$). Other pairs (such as $\triangle DEF$ and $\triangle STU$ or $\triangle GHI$ and $\triangle MNO$) appear to have the same shape. However, at this point we cannot conclude that they are similar from the definition because not all six parts of the triangles are known. We will be able to verify that other pairs are also similar as we complete this section.   •

AAA SIMILARITY

In Figure 6.3, all three of the corresponding angles of $\triangle LKJ$ and $\triangle PQR$ are congruent and the two triangles appear to have the same shape, or to be similar. The fact that two triangles having congruent corresponding angles have the same shape is stated next as a postulate.

Postulate 6.1

AAA Similarity Postulate

Two triangles are similar if and only if three angles of one triangle are congruent respectively to three angles of the other triangle.

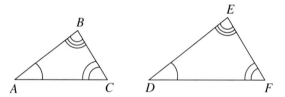

$\triangle ABC \sim \triangle DEF$ if and only if $\angle A \cong \angle D$, $\angle B \cong \angle E$, and $\angle C \cong \angle F$.

Now suppose that in $\triangle ABC$ and $\triangle DEF$ in Figure 6.4 we know $\angle A = \angle D = 30°$ and $\angle B = \angle E = 80°$. Then $\angle A + \angle B = \angle D + \angle E = 110°$.

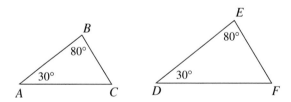

FIGURE 6.4

Moreover, because the sum of the measures of the angles in a triangle is 180°, the third angles must each equal $180° - 110° = 70°$. Consequently, all three pairs of corresponding angles are congruent. Therefore, $\triangle ABC \sim \triangle DEF$ by the AAA Postulate. The following theorem summarizes this result.

Theorem 6.4

AA Similarity Theorem

Two triangles are similar if two angles of one triangle are congruent respectively to two angles of the other triangle.

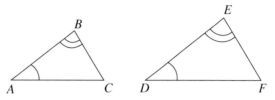

If $\angle A \cong \angle D$ and $\angle B \cong \angle E$, then $\triangle ABC \sim \triangle DEF$.

EXAMPLE 6.9   In $\triangle ABC$, $\overline{DE}$ is parallel to $\overline{AC}$ (Figure 6.5). If $AB = 6$, $CE = 1$, $AC = 8$, $BE = x$, and $BD = 4$, find (a) DE and (b) BC.

SOLUTION   First, $\overline{DE} \parallel \overline{AC}$ and $\angle C$ and $\angle BED$ are corresponding angles; hence they are congruent. Also, because $\angle B$ is common to $\triangle ABC$ and $\triangle DBE$, we can conclude that $\triangle ABC \sim \triangle DBE$ by AA Similarity.

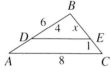

FIGURE 6.5

(a) Because $\triangle ABC \sim \triangle DBE$, we have $\dfrac{AB}{DB} = \dfrac{AC}{DE}$, or $\dfrac{6}{4} = \dfrac{8}{DE}$.

Therefore, $DE = \dfrac{32}{6} = 5\frac{1}{3}$.

(b) Similarly, $\dfrac{AB}{DB} = \dfrac{BC}{BE}$, or $\dfrac{6}{4} = \dfrac{1 + x}{x}$. It follows that $6x = 4 + 4x$, or $2x = 4$ and $x = 2$. Thus $BE = 2$ and $BC = x + 1 = 2 + 1 = 3$.   •

The following corollary is simply a special case of the AA Similarity Theorem where one of the angles is a right angle.

Corollary 6.5

Two right triangles are similar if an acute angle of one triangle is congruent to an acute angle of the other triangle.

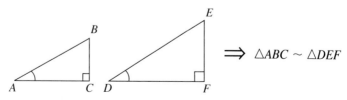

If $\angle C = \angle F = 90°$ and $\angle A \cong \angle D$, then $\triangle ABC \sim \triangle DEF$.

SAS SIMILARITY

$\triangle GHI$ and $\triangle MNO$ from Figure 6.3, reproduced here in Figure 6.6, appear to be similar. Note that we only know that two corresponding sides are proportional and the included angles are congruent and that the triangles appear to have the same shape. This suggests the following result whose proof will be omitted.

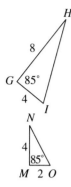

FIGURE 6.6

Theorem 6.6

SAS Similarity Theorem

Two triangles are similar if two sides of one triangle are proportional to two sides of another triangle and the angles included between the sides are congruent.

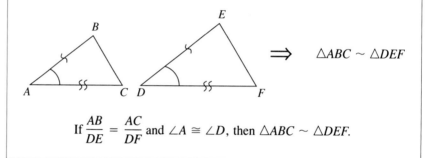

$\Rightarrow$ $\triangle ABC \sim \triangle DEF$

If $\dfrac{AB}{DE} = \dfrac{AC}{DF}$ and $\angle A \cong \angle D$, then $\triangle ABC \sim \triangle DEF$.

EXAMPLE 6.10   In Figure 6.7(a), $AB = 7$ cm, $BC = 10$ cm, and $AC = 14$ cm. If $DB = 5$ cm and $BE = 3.5$ cm,

(a) State which two triangles are similar.

(b) Find ED.

SOLUTION   Note that $\triangle ABC$ and $\triangle DBE$ share $\angle B$, so they have one angle congruent. When one triangle is drawn inside another as in this case, it is sometimes helpful to redraw the figure, separating the two triangles [Figure 6.7(b)].

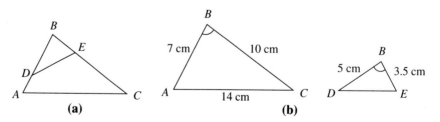

FIGURE 6.7

(a) Because $BC = 2(BD)$ and $AB = 2(EB)$, we have $\dfrac{BC}{BD} = \dfrac{AB}{EB} = \dfrac{2}{1}$. Also, $\angle B \cong \angle B$. So $\triangle ABC \sim \triangle EBD$ by SAS Similarity.

(b) By corresponding parts of similar triangles we have $\dfrac{AC}{ED} = \dfrac{AB}{EB}$, so $\dfrac{14}{ED} = \dfrac{2}{1}$. Therefore, $2(ED) = 14$ or $ED = 7$ cm.   •

The next corollary is a special case of the SAS Similarity Theorem.

Corollary 6.7

LL Similarity

Two right triangles are similar if the legs of one triangle are proportional respectively to the legs of the other triangle.

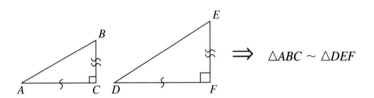

$\Rightarrow \quad \triangle ABC \sim \triangle DEF$

If $\angle C = \angle F = 90°$ and $\dfrac{AC}{DF} = \dfrac{BC}{EF}$, then $\triangle ABC \sim \triangle DEF$.

SSS SIMILARITY

$\triangle DEF$ and $\triangle STU$ from Figure 6.3 are reproduced in Figure 6.8.

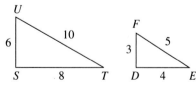

FIGURE 6.8

Although only the fact that their corresponding sides are proportional can be inferred from the figure, the two triangles appear to be similar. This suggests the following result. The proof of this theorem will be omitted.

Theorem 6.8

SSS Similarity Theorem

Two triangles are similar if three sides of one triangle are proportional to three sides of another triangle.

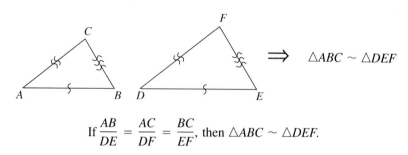

$$\Rightarrow \quad \triangle ABC \sim \triangle DEF$$

If $\dfrac{AB}{DE} = \dfrac{AC}{DF} = \dfrac{BC}{EF}$, then $\triangle ABC \sim \triangle DEF$.

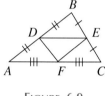

FIGURE 6.9

EXAMPLE 6.11   In Figure 6.9, *D*, *E*, and *F* are midpoints of the sides of $\triangle ABC$. Prove the following: $\triangle ABC \sim \triangle EFD$. (Note the correspondence of the vertices in this similarity statement.)

SOLUTION   Because *D*, *E*, and *F* are midpoints, we have midsegment $DE = \frac{1}{2}AC$, $EF = \frac{1}{2}AB$, and $DF = \frac{1}{2}BC$ by Theorem 4.12. Therefore, $\triangle ABC \sim \triangle EFD$ by SSS Similarity because the corresponding sides have the same ratio.

Solution to Applied Problem

Sight point *A* opposite *C*. Then measure from *C* to point *B*, where $\overline{BC}$ is perpendicular to $\overline{AC}$. Find point *E* collinear with *A* and *C* and a convenient distance (say 40 ft) from *C*. Then find *D*, where $\overline{ED}$ is perpendicular to $\overline{AE}$ and *A*, *B*, and *D* are collinear. Measure *ED*. Then $\triangle ABC \sim \triangle ADE$ by AA Similarity. Thus, if the width of the river is *x*, then $11/33 = x/(40 + x)$, or $440 + 11x = 33x$. Therefore, $440 = 22x$ and $x = 20$, so the river is 20 feet wide.

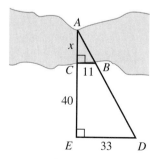

GEOMETRY AROUND US

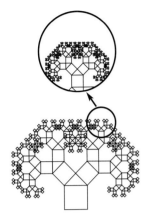

Many man-made structures have common geometric shapes such as circles, triangles, and rectangles, but many naturally occurring forms, such as clouds, mountains, lightning, trees, and coastlines, cannot be adequately represented by these shapes. A new branch of geometry uses the idea of fractional dimension, or fractal, to more accurately describe complex forms.

A fractal has an interesting property called self-similarity. That is, the shape of a portion of the object, if magnified, looks like the original object.

PROBLEM SET 6.2

EXERCISES/PROBLEMS

In Exercises 1–4, determine which pairs of triangles are similar, note the correspondence of the vertices, and write an appropriate similarity statement. Also, explain why the triangles are similar.

1.

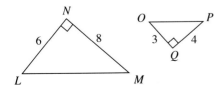

2.

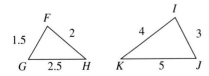

3.

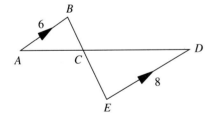

4.

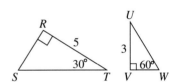

Pairs of similar triangles are shown in Exercises 5–12. Find the missing measures. If answers are not exact, round to two decimal places.

5.

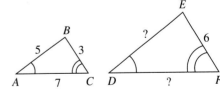

6.

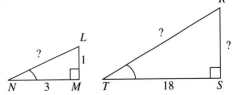

7.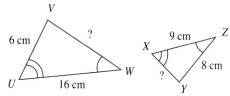

8.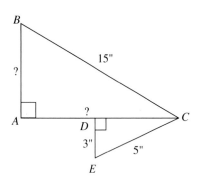

9.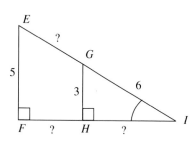

10.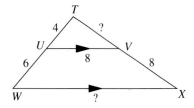

11.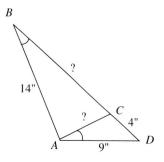

12.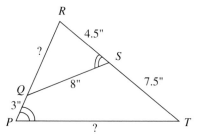

13. Compute the ratios requested in the following pairs of similar figures.

(a) Find the ratios of base:base, height:height, and area:area for this pair of similar triangles.

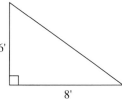

(b) Find the ratios of base:base, height:height, and area:area for this pair of similar triangles.

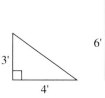

(c) Find the ratios of length:length, width:width, and area:area for this pair of similar rectangles.

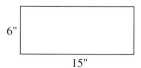

(d) Do you see a pattern in the relationship between the ratios of the dimensions and the ratios of the areas? Test your conjecture by drawing several other pairs of similar geometric figures.

14. The Pythagorean Theorem is often interpreted in terms of areas of squares as shown in the following figure.

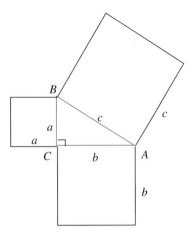

The equation $a^2 + b^2 = c^2$ means the sum of the areas of the squares drawn on the legs of the right triangle equals the area of the square drawn on the hypotenuse.

In fact, the theorem can be generalized to mean that if similar figures are drawn on each side of the right triangle, the sum of the areas of the two smaller figures equals the area of the larger figure. Verify that this is the case for each figure which follows.

(a)

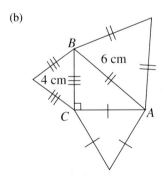

(b)

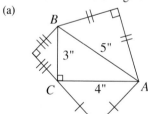

(c)

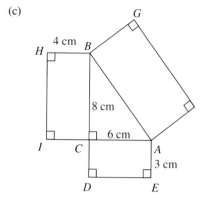

(d)

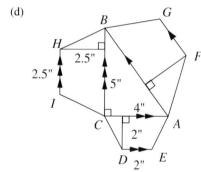

PROOFS

15. Prove that all equilateral triangles are similar.

16. Prove that all isosceles right triangles are similar to each other.

17. Prove that two isosceles triangles are similar if a base angle of one is congruent to a base angle of the other.

18. Prove that two isosceles triangles are similar if their vertex angles are congruent.

19. Let $ABCD$ be an isosceles trapezoid with diagonals intersecting in point E. Prove that $\triangle AED \sim \triangle CEB$.

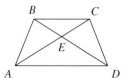

20. Prove that the diagonals of a trapezoid divide each other into proportional segments.

21. Prove that in similar triangles the ratio of the lengths of two corresponding medians is equal to the ratio of the lengths of any pair of corresponding sides.

22. Prove that in similar triangles, the ratio of the lengths of two corresponding altitudes is equal to the ratio of the lengths of their corresponding sides.

23. Prove that in similar triangles, the ratio of their areas is the square of the ratio of the lengths of their corresponding sides.

24. Prove that the perimeters of two similar triangles have the same ratio as the ratio of the lengths of any pair of corresponding sides.

25. The two right triangles shown are similar. Suppose that each triangle is rotated about a vertical axis, the dotted line in each case. Two right circular cones would be formed. Prove that the ratio of the volumes of these two cones is the cube of the ratio of the lengths of any two corresponding sides.

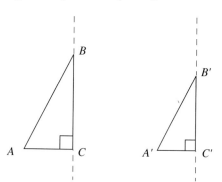

26. Given: $\angle A$ is supplementary to $\angle DEC$.
Prove: $\dfrac{BE}{AB} = \dfrac{BD}{BC}$.

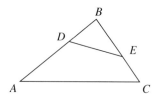

27. Prove that two triangles are similar if the lines containing their corresponding sides are perpendicular. That is, for the figure shown, prove that if $AB \perp A'B'$, $BC \perp B'C'$, and $AC \perp A'C'$ then $\triangle ABC \sim \triangle A'B'C'$.

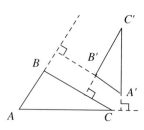

28. Show that $\triangle ABC$ is similar to $\triangle A'B'C'$ where x represents the measure of the angles.

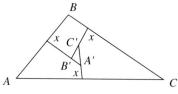

29. Prove that triangles are similar if their corresponding sides are parallel to each other.

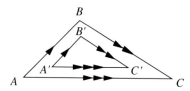

APPLICATIONS

30. Using the measurements given, find the distance, PQ, across the river.

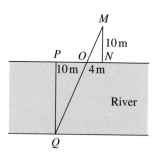

31. At a particular time, a tree casts a shadow 29 m long on horizontal ground. At the same time, a vertical pole 3 m high casts a shadow 4 m long. Calculate the height of the tree to the nearest meter.

32. Another method of determining the height of an object is illustrated next. A pan of water is located at point B. Name the two similar triangles in the diagram and explain why they are similar. Find the height of the tree if a person 1.5 m tall sees the

top of the tree when the pan of water is 18 m from the base of the tree and the person is 1 m from the pan of water.

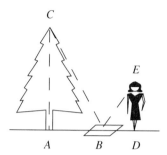

33. A folding workbench is constructed as shown and can be adjusted to various heights. If $AE = BE$ and $DE = CE$, explain why the workbench surface will always be parallel to the ground.

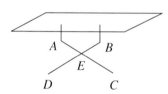

34. A boy and his friend wish to calculate the height of a flagpole. One boy holds a yardstick vertically at a point 40 feet from the base of the flagpole, as shown. The other boy backs away from the pole to a point where he sights the top of the pole over the top of the yardstick. If his position is 1 foot 9 inches from the yardstick and his eye level is 2 feet above the ground, find the height of the flagpole.

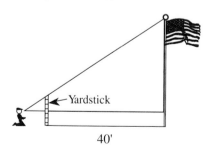

35. A new department store opens on a hot summer day. Kathy's job is to stand outside the main entrance and hand out a special supplement to incoming customers. The building is 30 feet tall and casts a shadow 8 feet wide in front of the store. If Kathy is 5'6" tall, how far away from the building can she stand and still be in the shade?

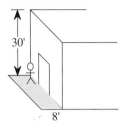

36. A water tank in the shape of an inverted cone has a radius of 5 m and a height of 15 m. What is the volume of water in the tank, to the nearest m³, when the water is 6 m deep?

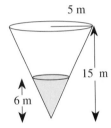

37. A paper cup is in the shape of a right circular cone with a diameter of 8 cm and a height of 12 cm.
   (a) Suppose the cup is filled with water to a depth of 8 cm. Calculate the volume of water in the cup. Round your answer to the nearest cubic centimeter. [HINT: Use similar triangles.]
   (b) To what depth must the cup be filled in order to be half full of water? Round your answer to the nearest tenth of a centimeter.

38. Sawdust is dropping from a conveyor belt onto a pile in the approximate shape of a right circular cone. At one time the height of the pile is 5 feet and the circumference of the base is 25 feet. Later the circumference of the base of the pile is measured as 62 feet. What is the height of the sawdust pile at the latter time and how many cubic feet of sawdust are in the pile? [HINT: Use similar triangles.]

39. Tom and Carol are playing a shadow game. Tom is 6' tall and Carol is 5' tall. If Carol stands at the "shadow top" of Tom's head their combined shadows total 15 feet. How long is each shadow?

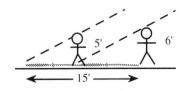

40. A farmer wishes to measure the distance, *EC*, across a lake. He stands on dry ground at *A* and sights point *B* which is in line with *C* on the other side of the lake. He then makes $\overline{DB}$ parallel to $\overline{EC}$. He finds that *EA* = 105 m, *AD* = 45 m, and *DB* = 30 m. How wide is the lake to the nearest meter?

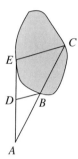

41. The following sequence of shapes which is composed of equilateral triangles produces a shape called the **Sierpinski Triangle** if continued indefinitely.

Each white triangle has as its vertices the midpoints of the sides of a larger black triangle. Notice how a portion of the curve at one stage is similar to a portion of the curve at another stage. This property is referred to as **self-similarity**. Suppose that the first equilateral triangle has sides of length 1.

(a) What are its perimeter and area?
(b) What are the total perimeter and total area of the black triangles in the second figure?
(c) What are the total perimeter and total area of the black triangles in the third figure?
(d) What are the total perimeter and total area of the black triangles in the fourth figure?
(e) What are the total perimeter and total area of the black triangles in the *n*th figure?

42. The following sequence of shapes composed of equilateral triangles produces a shape called the **Koch Curve** if continued indefinitely.

The sequence begins with an equilateral triangle. Next, equilateral triangles are constructed on the middle third of each side of the original triangle. The Koch Curve is self-similar. Suppose that the first equilateral triangle has sides of length 1.

(a) What are its perimeter and area?
(b) What are the perimeter and area of the second star?
(c) What are the perimeter and area of the third star?
(d) What are the perimeter and area of the fourth star?
(e) What are the perimeter and area of the *n*th star?

6.3
APPLICATIONS OF SIMILARITY

Applied Problem

A plastic kite available from a mail-order catalog must be packed and mailed in a rectangular envelope. How can the kite shown be easily folded to fit inside a standard 10″ by 13″ mailer?

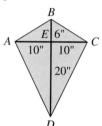

In this section we will use similar triangles to derive some properties of geometric figures. We will be applying all of the postulates and theorems about similarity that were developed in the last section.

MEAN PROPORTIONAL IN A RIGHT TRIANGLE

Recall that b is the mean proportional between a and c if $a{:}b = b{:}c$.

EXAMPLE 6.12   Name three pairs of similar triangles in Figure 6.10.

SOLUTION $\overline{QS}$ divides $\triangle PQR$ into two 30-60 right triangles. So by AA Similarity, we have $\triangle PQR \sim \triangle PSQ$, $\triangle PQR \sim \triangle QSR$, and $\triangle PSQ \sim \triangle QSR$.   •

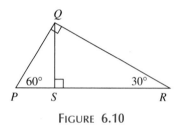

FIGURE 6.10

The fact that the two smaller triangles are similar to each other leads to the following result about the altitude to the hypotenuse in a right triangle.

Theorem 6.9

In a right triangle, the altitude to the hypotenuse is the mean proportional between the two segments formed by the altitude on the hypotenuse.

$$\Rightarrow \quad \frac{AD}{CD} = \frac{CD}{BD}$$

If $\overline{CD}$ is the altitude to $\overline{AB}$, then $\dfrac{AD}{CD} = \dfrac{CD}{BD}$.

Given $\triangle ABC$ with $\overline{CD}$ perpendicular to $\overline{AB}$ (Figure 6.11)

Prove $\dfrac{AD}{CD} = \dfrac{CD}{BD}$

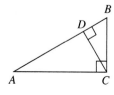

FIGURE 6.11

Proof Because $\angle A = \angle A$ and $\angle ADC = \angle ACB = 90°$, we can conclude that $\triangle ADC \sim \triangle ACB$ by AA Similarity. Also, $\triangle ACB \sim \triangle CDB$ by AA Similarity because $\angle ACB = \angle CDB$ and $\angle B = \angle B$. Thus we have $\triangle ADC \sim \triangle ACB$ and $\triangle ACB \sim \triangle CDB$, so, by transitivity, we have that $\triangle ADC \sim \triangle CDB$. Because corresponding sides of $\triangle ADC$ and $\triangle CDB$ are proportional, it follows that $\dfrac{AD}{CD} = \dfrac{CD}{BD}$.

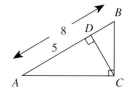

FIGURE 6.12

EXAMPLE 6.13   In $\triangle ABC$, $AD = 5$ and $AB = 8$ (Figure 6.12). Find CD.

SOLUTION   First, $BD = AB - AD = 8 - 5 = 3$. Next, by Theorem 6.9, $\dfrac{AD}{CD} = \dfrac{CD}{BD}$, so $\dfrac{5}{CD} = \dfrac{CD}{3}$. Thus, $CD^2 = 15$, or $CD = \sqrt{15}$.

SIDE SPLITTING THEOREM

The next theorem, which is a generalization of Example 6.9 in Section 6.2, can be proved using AA Similarity.

Theorem 6.10

Side Splitting Theorem

A line parallel to one side of a triangle forms a triangle similar to the original triangle and divides the other two sides of the triangle into proportional corresponding segments.

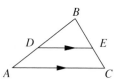

If $\overline{DE} \parallel \overline{AC}$, then $\triangle ABC \sim \triangle DBE$ and $\dfrac{BD}{DA} = \dfrac{BE}{EC}$.

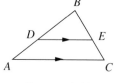

FIGURE 6.13

Given $\triangle ABC$ with $\overline{DE}$ parallel to $\overline{AC}$ (Figure 6.13)

Prove $\triangle ABC \sim \triangle DBE$ and $\dfrac{BD}{BA} = \dfrac{BE}{BC}$.

Proof $\angle BDE = \angle BAC$ and $\angle BED = \angle BCA$ because parallel lines form congruent corresponding angles. Therefore, $\triangle ABC \sim \triangle DBE$ by AA Similarity. Because corresponding sides are proportional, $\dfrac{BD}{BA} = \dfrac{BE}{BC}$. In the problem set you will show that $\dfrac{AD}{AB} = \dfrac{CE}{CB}$ and $\dfrac{AD}{DB} = \dfrac{CE}{EB}$.

The next example makes use of the Side Splitting Theorem.

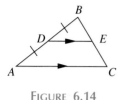

FIGURE 6.14

EXAMPLE 6.14   In $\triangle ABC$, $\overline{DE}$ is parallel to $\overline{AC}$ and D is the midpoint of $\overline{AB}$ (Figure 6.14). Show that E is the midpoint of $\overline{CB}$.

SOLUTION   By the Side Splitting Theorem, $\dfrac{BD}{AD} = \dfrac{BE}{CE}$. Because $\dfrac{BD}{AD} = 1$, it follows that $\dfrac{EB}{CB} = 1$, so $EB = CB$ and E is the midpoint of $\overline{CB}$.

THE MIDSEGMENT THEOREM

In a triangle, a line segment that joins the midpoints of two sides of the triangle is called a **midsegment** of the triangle. The next theorem shows two properties of midsegments.

Theorem 6.11

Midsegment Theorem

A midsegment of a triangle is parallel to the third side and is half as long.

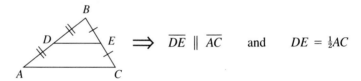

If $AD = DB$ and $BE = EC$, then $\overline{DE} \parallel \overline{AC}$ and $DE = \frac{1}{2}AC$.

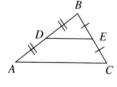

FIGURE 6.15

Given $\triangle ABC$ with midsegment $\overline{DE}$ (Figure 6.15)

Prove $\overline{DE} \parallel \overline{AC}$ and $DE = \frac{1}{2}AC$

Proof $\triangle ABC$ and $\triangle DBE$ are similar by SAS Similarity because they have $\angle B$ in common and the two pairs of corresponding sides have a common ratio of $\dfrac{2}{1}$; namely, $\dfrac{AB}{DB} = \dfrac{BC}{BE} = \dfrac{2}{1}$. Thus, by corresponding parts, $\angle BAC \cong \angle BDE$. Because they are corresponding angles with respect to $\overline{DE}$ and $\overline{AC}$, $\overline{DE} \parallel \overline{AC}$. Also, the ratio of DE to AC is 1:2, so $DE = \frac{1}{2}AC$.

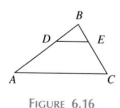

FIGURE 6.16

Although the Midsegment Theorem was proved using similarity, it can also be proved using triangle congruence results. However, the congruence proof is more complex. Not only is the similarity proof simpler, it also can be generalized to the situation where D and E are not necessarily midpoints of $\overline{AB}$ and $\overline{BC}$ respectively. In Figure 6.16, if $\dfrac{BD}{BA} = \dfrac{BE}{BC}$, then $DE \parallel AC$ and $\dfrac{DE}{AC} = \dfrac{BD}{BA}$. For example, if $\dfrac{BD}{BA} = \dfrac{BE}{BC} = \dfrac{1}{3}$, then $\overline{DE} \parallel \overline{AC}$ and $DE = \frac{1}{3}AC$. The proof of this result is analogous to the proof of the Midsegment Theorem.

Now suppose that *ABCD* is *any* quadrilateral and *E, F, G, H* are midpoints [Figure 6.17(a)]. When $\overline{BD}$, $\overline{EH}$, and $\overline{FG}$ are drawn, $\triangle ABD$ and $\triangle CBD$ have midsegments $\overline{EH}$ and $\overline{FG}$ respectively [Figure 6.17(b)].

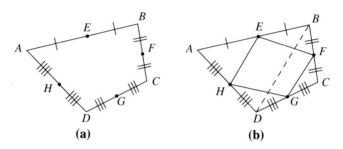

FIGURE 6.17

Using the Midsegment Theorem, $\overline{EH} \parallel \overline{BD}$ and $\overline{FG} \parallel \overline{BD}$, so $\overline{EH} \parallel \overline{FG}$. Also, $EH = FG$ because they are each equal to $\frac{1}{2}BD$. Therefore, *EFGH* is a parallelogram because it is a quadrilateral with one pair of opposite sides parallel and congruent. Because *EFGH* is a quadrilateral joining the midpoints of the sides of *ABCD*, we call it the **midquad** of *ABCD*. Thus, we have the following surprising result.

Corollary 6.12

Midquad Theorem

If *ABCD* is *any* quadrilateral and *E, F, G, H* are midpoints as shown, then *EFGH* is a parallelogram.

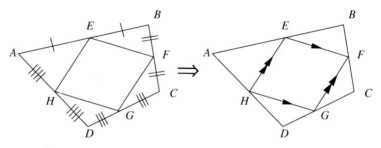

In *ABCD*, if $AE = BE$, $BF = CF$, $CG = DG$, and $AH = DH$, then *EFGH* is a parallelogram.

Solution to Applied Problem

If the midpoints of the sides are joined as shown, a rectangle of 10″ by 13″ is formed (by Theorem 6.11, the Midsegment Theorem). Furthermore, if the kite is folded on the dashed lines, each triangle outside the rectangle folds exactly onto the rectangle where points A, B, C, and D meet at point E.

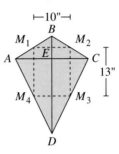

GEOMETRY AROUND US

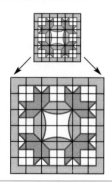

Patterns for quilting, crafts, woodworking, and so on must frequently be enlarged from the size printed in a book or instruction sheet to actual size. The larger figure must be similar to the original figure. Such an enlargement is often made using square grids. The pattern is printed on a grid of small squares, and the larger figure is made by duplicating the figure, square by square, on a grid of larger squares.

PROBLEM SET 6.3

EXERCISES/PROBLEMS

For Exercises 1–6, find the given length if $\overline{BC} \parallel \overline{DE}$.

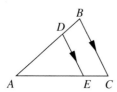

1. If $AE = 3$, $AB = 5$, and $AD = 4$, find AC.

2. If $AD = 5$, $AB = 8$, and $AC = 7$, find AE.

3. If $BD = 3$, $EC = 4$, and $AC = 9$, find AD.

4. If $AD = EC$, $DB = 4$, and $AE = 9$, find AD.

5. If $DE = 5$, $BC = 7$, and $AD = 4$, find AB.

6. If $ED = 7$, $BC = 10$, and $AE = 2 + EC$, find AE.

For Exercises 7–12, use the figure shown to find the given length.

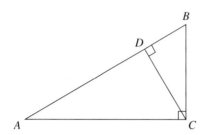

7. If $AD = 24$ and $BD = 9$, find CD.

8. If $AD = 8$ and $BD = 4$, find CD.

9. If $AD = 6$ and $CD = 4$, find BD.

10. If $BD = 10$ and $CD = 15$, find AD.

11. If $AB = 20$ and $CD = 6$, find BD.

12. If $CD = 9$ and $AB = 30$, find AD.

For Exercises 13–16, use the figure shown where parallel lines are marked.

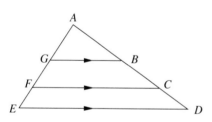

13. If $AB = 4$, $CD = 3$, $AG = 3$, and $GF = 2$, find EF and BC.

14. If $BG = 5$, $DE = 15$, and $AD = 12$, find BD.

15. If $AB = x$, $AG = x - 2$, $EG = 7$, and $BD = 10.5$, find AB and AG.

16. If $BG = x$, $AB = x + 2$, $BC = x + 3$, $DE = 18$, and $CD = x - 1$, find BG and AD.

17. In $\triangle PQR$, M and N are midpoints of $\overline{PQ}$ and $\overline{QR}$, respectively. In addition, $PQ = 20$ cm and $PR = 12$ cm. Find MN in two different ways.

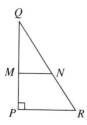

18. In rectangle $ABCD$, F and I are the midpoints of $\overline{AB}$ and $\overline{CD}$, respectively. Segments $\overline{BE}$, $\overline{FI}$, and $\overline{CE}$ are drawn. Congruent segments are indicated. If $BG = 10$, $CH = 15$, and $ID = 8$, find GH.

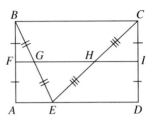

19. In trapezoid $ABCD$, M and N are midpoints of $\overline{AB}$ and $\overline{CD}$, respectively. If $MN = 15$ cm and $BE = 7$ cm, find the area of $ABCD$.

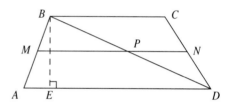

20. In parallelogram $PQRS$, points A, B, C, and D are midpoints of the sides. If the area of $\triangle AQB$ is 5 in.^{2}, find the area of hexagon $PABRCD$.

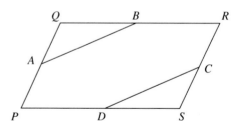

21. In the figure shown, $\angle A \cong \angle C$, M and N are midpoints of sides $\overline{AB}$ and $\overline{BC}$, and $MNPQ$ is a rectangle. Show that $\triangle MQA \cong \triangle NPC$.

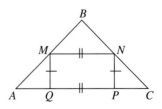

Find the area of $MNPQ$, where $MN = 14$ m and $BC = 22$ m.

22. In equilateral $\triangle STU$, $ST = 10$ and V and W are midpoints of $\overline{ST}$ and $\overline{TU}$, respectively. What is the area of $\triangle TVW$ if $\overline{TX} \perp \overline{SU}$?

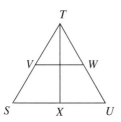

23. Corollary 6.12 in this section states that the midquad of any quadrilateral $PQRS$ is a parallelogram.
(a) Use the fact that each side of the midquad is parallel to and half the length of a diagonal of the quadrilateral to determine under what conditions the midquad $M_1M_2M_3M_4$ will be a rectangle.

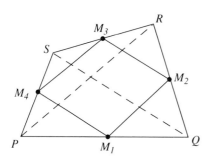

(b) Under what conditions will the midquad $M_1M_2M_3M_4$ be a rhombus?
(c) Under what conditions will the midquad $M_1M_2M_3M_4$ be a square?

PROOFS

24. In $\triangle ABC$, $\overline{DE} \parallel \overline{AC}$. Prove that $\dfrac{AD}{DB} = \dfrac{CE}{EB}$ and $\dfrac{AD}{AB} = \dfrac{CE}{CB}$.

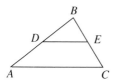

25. Explain how the figure shown can be used to find $\sqrt{a}$ for a given length a. [HINT: Find x.]

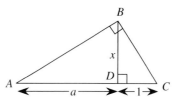

26. Complete the proof of Theorem 6.10: A line parallel to one side of a triangle forms a triangle similar to the original triangle and divides the other two sides of the triangle into proportional corresponding segments.

27. Prove that the midpoint of the hypotenuse of a right triangle is equidistant from its vertices.

28. Given: $(AC)^2 = AB \cdot AD$

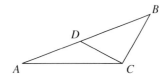

Prove: $\triangle ABC$ is similar to $\triangle ACD$.

29. $\overline{BE}$ bisects $\angle ABC$ and $\dfrac{AB}{BE} = \dfrac{BD}{BC}$. Prove that $\angle A = \angle D$.

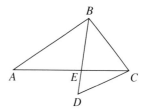

30. Prove that if three parallel lines are cut by intersecting transversals, then the corresponding segments of the transversals are proportional. That is, show $AD{:}DE{:}EG = BC{:}CE{:}EF$.

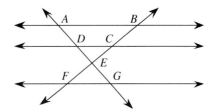

31. Using Theorem 6.9 and the following figure, prove the Pythagorean Theorem.

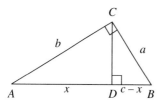

32. In $\triangle ABC$, $\dfrac{DB}{AB} = \dfrac{EB}{CB}$. Prove that $\overline{DE}$ is parallel to $\overline{AC}$.

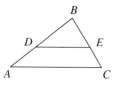

33. Prove that the midsegment of an equilateral triangle forms another equilateral triangle.

34. Prove that the quadrilateral formed by joining the midpoints of adjacent sides of a rectangle is a rhombus.

35. Prove: If the midpoints of the sides of a rhombus are connected consecutively, the figure formed is a rectangle.

36. In the next figure, M is the midpoint of $\overline{AB}$ and N is the midpoint of $\overline{BC}$. If $MN = NP$, prove that $AMPC$ is a parallelogram.

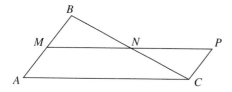

37. In $\triangle ABC$, points D, E, and F are midpoints. Prove that the four smaller triangles are congruent.

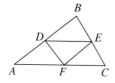

38. In $\triangle ABC$, prove that $\dfrac{AD}{AC} = \dfrac{AC}{AB}$.

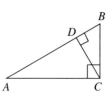

39. One method of constructing the geometric mean of a and b, for $a \geq b$, is described next.

(i) Construct $\overline{AC}$ of length a.

(ii) Locate B such that $\overline{BC}$ has length b and B is between A and C.

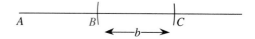

(iii) Construct $\overline{BD}$ of length a, where D is to the right of C.

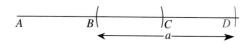

(iv) Draw two large arcs with centers A and D, and with radius a. Label their point of intersection E.

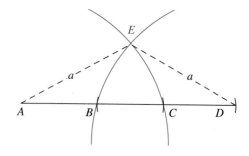

(v) The geometric mean of *a* and *b* is *x*, where *x* = *EB*.

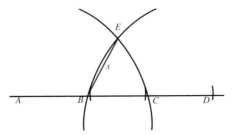

(a) Construct the geometric mean of *a* and *b* where the following segments are lengths *a* and *b*.

_____   _____

       a   *b*

(b) If the segment shown below has length 1, construct the geometric mean between 1 and 2. What is the length you have constructed?

_____

        1

(c) Prove that the length *x* in the construction described is in fact the geometric mean between *a* and *b*. [HINT: Consider △*AEC* and △*ECB*.]

APPLICATION

40. A quilt piece looks like the figure shown, where △*ABC* is an equilateral triangle and points *M, N, O, P, Q,* and *R* are midpoints.

(a) Which triangles are congruent and why?

(b) How does the area of the shaded triangles compare to the area of the unshaded triangles?

6.4
RIGHT TRIANGLE TRIGONOMETRY

Applied Problem

An approach ramp is being planned to lead to a suspension bridge. Given the dimensions in the figure, at what angle from the horizontal should the ramp be sloped?

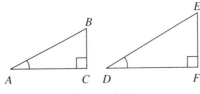

THE TANGENT RATIO

This section's applied problem can be solved using methods from right triangle trigonometry. **Trigonometry**, as its name suggests, is the study of the measures (metry) of triangles (*triang*). Consider the right triangles △*ABC* and △*DEF* in Figure 6.18 where ∠*A* = ∠*D*.

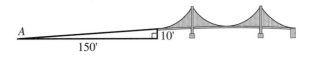

FIGURE 6.18

By AA Similarity, $\triangle ABC \sim \triangle DEF$. Hence the following ratios are equal:

$$\frac{BC}{AC} = \frac{EF}{DF} \qquad \frac{AC}{AB} = \frac{DF}{DE} \qquad \frac{BC}{AB} = \frac{EF}{DE}$$

This suggests that associated with any acute angle in a right triangle (such as $\angle A$ in Figure 6.18), there are three unique ratios that depend only on the measure of the angle, *not* on the size of the triangle. As shown in Figure 6.19, these ratios are $\dfrac{a}{b}$, $\dfrac{a}{c}$, and $\dfrac{b}{c}$.

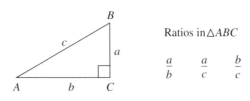

Ratios in $\triangle ABC$

$$\frac{a}{b} \qquad \frac{a}{c} \qquad \frac{b}{c}$$

FIGURE 6.19

These ratios form the basis of right triangle trigonometry and are given special names. The first of these ratios, the tangent ratio, is studied next. In right triangle $\triangle ABC$, the **tangent** of acute $\angle A$, written **tan A**, is the ratio a/b. That is, the tangent of $\angle A$, is the ratio of the length of the leg *opposite* $\angle A$ to the length of the leg *adjacent* to $\angle A$.

Definition

Tangent of an Acute Angle

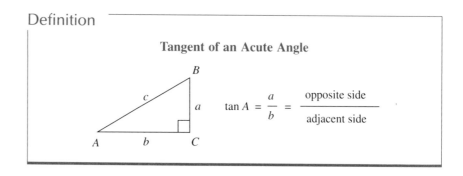

$$\tan A = \frac{a}{b} = \frac{\text{opposite side}}{\text{adjacent side}}$$

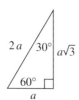

(b)

FIGURE 6.20

EXAMPLE 6.15   Calculate the following:

(a) tan 60°   (b) tan 30°   (c) tan 45°

SOLUTION

(a) Recall that in a 30-60 right triangle, if the length of the shorter leg is a, then the length of the hypotenuse is $2a$. The longer leg has length $a\sqrt{3}$ [Figure 6.20(a)]. We can find tan 60° as follows.

$$\tan 60° = \frac{\text{opp}}{\text{adj}} = \frac{a\sqrt{3}}{a} = \sqrt{3} \approx 1.732$$

(b) We also can find tan 30° using the triangle in Figure 6.20(a), as follows:

$$\tan 30° = \frac{\text{opp}}{\text{adj}} = \frac{a}{a\sqrt{3}} = \frac{1}{\sqrt{3}} \approx 0.577$$

(c) In a 45-45 right triangle, both legs have the same length a and the hypotenuse has length $a\sqrt{2}$ [Figure 6.20(b)]. Thus, we have

$$\tan 45° = \frac{a}{a} = 1$$

Figure 6.21 shows a right triangle with one leg of length 1. The other leg has length a, where $a > 0$ and a can be arbitrarily large. Therefore, $\tan A = \frac{a}{1} = a$. This shows that the tangent of an acute angle must be greater than zero and may be arbitrarily large.

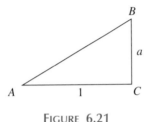

FIGURE 6.21

A scientific calculator can be used to find the tangent of any angle. Because there are various ways to measure angles other than using degrees, be sure that your calculator is set for degree measure.

EXAMPLE 6.16   Find tan 37° using a calculator.

SOLUTION

Press: 37 [TAN]

Result: [0.75355405]

The calculator is accurate to as many places as the calculator can display but we often round answers to four decimal places. Thus, tan 37° ≈ 0.7536. This means that in any right triangle with a 37° angle, the ratio of the shorter leg to the longer leg is about 0.7536.

[NOTE: The sequence of keystrokes on your calculator may differ from the one given here. Consult the manual for your calculator to be sure of the correct sequence.]

The tangent ratio can be used to solve problems involving right triangles.

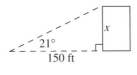

FIGURE 6.22

EXAMPLE 6.17   At a horizontal distance of 150 feet from the base of a building, the line of sight to the top of the building makes an angle of 21° with level ground (Figure 6.22). That is, the angle of elevation to the top of the building is 21°. About how tall is the building? Assume that the building is perpendicular to the ground.

SOLUTION   Let x represent the height of the building. Then $\tan 21° = \dfrac{\text{opp}}{\text{adj}} = \dfrac{x}{150}$, or $x = 150 \tan 21°$. On a calculator, $150 \tan 21°$ can be found as follows:

$$x = 150 \;\boxed{\times}\; 21 \;\boxed{\text{TAN}}\; = \;\boxed{57.57960526}$$

So the building is about 58 feet tall.   •

THE SINE AND COSINE RATIOS

The sine and cosine ratios compare the lengths of the legs of a right triangle to the length of its hypotenuse. In the right triangle $\triangle ABC$, the **sine** of $\angle A$, written **sin A**, is the ratio $\dfrac{a}{c}$. That is, the sine of $\angle A$ is the ratio of the length of the leg *opposite A* to the length of the *hypotenuse*.

Definition

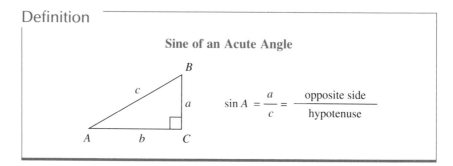

Sine of an Acute Angle

$$\sin A = \frac{a}{c} = \frac{\text{opposite side}}{\text{hypotenuse}}$$

In the right triangle $\triangle ABC$, the **cosine** of $\angle A$, written **cos A**, is the ratio $\dfrac{b}{c}$. That is, the cosine of $\angle A$ is the ratio of the length of the leg *adjacent to A* to the length of the *hypotenuse*.

Definition

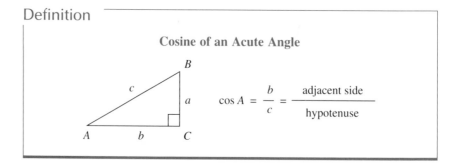

Cosine of an Acute Angle

$$\cos A = \frac{b}{c} = \frac{\text{adjacent side}}{\text{hypotenuse}}$$

[NOTE: Although it is common to speak of the tangent, sine, or cosine of an *angle*, it is also customary to find the tangent, sine, or cosine of the *number* (of degrees in this book) that is the measure of an angle. Ratios such as the tangent, sine, and cosine are referred to as **trig ratios**.]

EXAMPLE 6.18   Calculate the following.

(a) sin 60°   (b) sin 30°   (c) cos 45°

SOLUTION

(a) In Figure 6.23(a),

$$\sin 60° = \frac{\text{opp}}{\text{hyp}} = \frac{a\sqrt{3}}{2a} = \frac{\sqrt{3}}{2} \approx 0.866$$

(b) In Figure 6.23(a),

$$\sin 30° = \frac{\text{opp}}{\text{hyp}} = \frac{a}{2a} = 0.5$$

(c) In Figure 6.23(b),

$$\cos 45° = \frac{\text{adj}}{\text{hyp}} = \frac{a}{a\sqrt{2}} = \frac{1}{\sqrt{2}} \approx 0.707$$

[NOTE: Because the hypotenuse of a right triangle is longer than either of the legs, sin A and cos A will always be between 0 and 1 for any acute angle.]

EXAMPLE 6.19   Find sin 37° and cos 37° using a calculator.

SOLUTION

Press: 37 SIN
Result: 0.601815023
Press: 37 COS
Result: 0.79863551

So, to four decimal places, sin 37° ≈ 0.6019 and cos 37° ≈ 0.7986.

EXAMPLE 6.20   To determine the feasibility of constructing a tunnel through a mountain, surveyors want to measure the north-south distance BC through the mountain. They found that $AB = 4284$ ft, and $\angle A = 57°$, and $\angle C$ was a right angle based on compass measurement (Figure 6.24). How long will the tunnel be?

SOLUTION

$$\sin A = \frac{\text{opposite}}{\text{hypotenuse}} = \frac{BC}{AB}. \text{ Thus, } \sin 57° = \frac{BC}{4284}.$$

So, $BC = 4284 \sin 57°$. On a calculator,

$$BC = 4284 \;\boxed{\times}\; 57 \;\boxed{\text{SIN}}\; \boxed{=}\; \boxed{3592.864913}$$

or about 3593 ft.

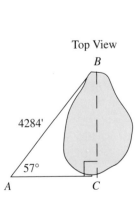

FIGURE 6.23

Top View

FIGURE 6.24

Examples 6.16 and 6.19 show how to find the tangent, sine, and cosine of given angles (numbers). On the other hand, a scientific calculator can also be used to find the approximate measure of an angle *given* one of its trig ratios.

INVERSE TRIGONOMETRIC FUNCTIONS

Next we investigate inverse trigonometric functions that are useful for solving applications involving angles of triangles.

EXAMPLE 6.21   Find an approximate value for $\angle A$ in each of the following.

(a) $\tan A = 0.8391$ (b) $\sin A = 0.3581$

SOLUTION   We must find the angle that has a tangent of 0.8391 or a sine of 0.3581.

(a) Press: .8391 $\boxed{TAN^{-1}}$
    Result: $\boxed{40.0000124}$
    Therefore, $A \approx 40°$.
This means that an angle which has a tangent of 0.8391 is about 40°.

(b) Press: .3581 $\boxed{SIN^{-1}}$
    Result: $\boxed{20.9835563}$
    Therefore, $A \approx 21°$.   •

[NOTE: On most calculators, the $\boxed{TAN^{-1}}$ key is the second function on the $\boxed{TAN}$ key. Thus, the second function key must be pressed *before* pressing the $\boxed{TAN}$ key. Again, the sequence of keystrokes for your calculator may differ. Consult your manual to be sure.]

Example 6.21 used the $\boxed{TAN^{-1}}$ and $\boxed{SIN^{-1}}$ keys to find an angle given its tangent and sine respectively. These keys, together with the $\boxed{COS^{-1}}$ key, are called **inverse trig function keys** because they can be used to find the measure of an angle given its corresponding trig ratio. Some calculators have an $\boxed{ARCTAN}$ or use the two keys $\boxed{INV}$ $\boxed{TAN}$ in place of $\boxed{TAN^{-1}}$.

EXAMPLE 6.22   A familiar right triangle is the 3-4-5 triangle as shown in Figure 6.25. Find the measures of $\angle A$ and $\angle B$.

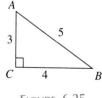

FIGURE 6.25

SOLUTION   Using the tangent ratio and the given side lengths we have tan $B = \dfrac{3}{4} = 0.75$. To find $\angle B$ using the calculator we evaluate .75 $\boxed{TAN^{-1}}$, which gives a result of about 36.8699. Therefore, $\angle B \approx 36.9°$ and $\angle A \approx 90° - 36.9° = 53.1°$.   •

RELATING THE SINE, COSINE, AND TANGENT RATIOS

The trig ratios are interrelated in two interesting ways. First, consider all three ratios $\frac{a}{b}$, $\frac{a}{c}$, and $\frac{b}{c}$ for the triangle in Figure 6.26.

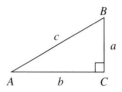

FIGURE 6.26

Notice that $\frac{a}{c} \div \frac{b}{c} = \frac{a}{c} \times \frac{c}{b} = \frac{a}{b}$

Because $\frac{a}{c} = \sin A$, $\frac{b}{c} = \cos A$, and $\frac{a}{b} = \tan A$, we can conclude that $\sin A \div \cos A = \tan A$.

Theorem 6.13

In a right triangle with acute angle A, sin A divided by cos A is tan A.

$$\frac{\sin A}{\cos A} = \tan A$$

Because the trig ratios pertain to right triangles, the Pythagorean Theorem can be applied to derive another relationship among these ratios. For $\triangle ABC$ in Figure 6.26, we have $a^2 + b^2 = c^2$. Dividing both sides of this equation by c^2 yields

$$\frac{a^2}{c^2} + \frac{b^2}{c^2} = \frac{c^2}{c^2} \quad \text{or} \quad \left(\frac{a}{c}\right)^2 + \left(\frac{b}{c}\right)^2 = 1$$

Because $\frac{a}{c} = \sin A$ and $\frac{b}{c} = \cos A$, we have $(\sin A)^2 + (\cos A)^2 = 1$. It is customary to write $(\sin A)^2$, $(\cos A)^2$, and $(\tan A)^2$ as $\sin^2 A$, $\cos^2 A$, and $\tan^2 A$ respectively. Thus, we have the following result.

Theorem 6.14

In a right triangle with acute angle A, the sum of $\sin^2 A$ and $\cos^2 A$ is 1.

$$\sin^2 A + \cos^2 A = 1$$

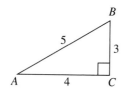

FIGURE 6.27

EXAMPLE 6.23   Verify Theorems 6.13 and 6.14 in the case of a 3-4-5 right triangle (Figure 6.27).

SOLUTION   First, $\sin A = \dfrac{3}{5}$, $\cos A = \dfrac{4}{5}$, and $\tan A = \dfrac{3}{4}$. So,

$$\frac{\sin A}{\cos A} = \frac{3/5}{4/5} = \frac{3}{4} = \tan A$$

which verifies Theorem 6.13. We also have

$$\sin^2 A + \cos^2 A = \left(\frac{3}{5}\right)^2 + \left(\frac{4}{5}\right)^2 = \frac{9}{25} + \frac{16}{25} = \frac{25}{25} = 1$$

which verifies Theorem 6.14.   •

EXAMPLE 6.24   Given that $\sin A = 0.5736$, find $\cos A$.

SOLUTION   Because $\sin^2 A + \cos^2 A = 1$, we have $\cos^2 A = 1 - \sin^2 A$, or $\cos A = \sqrt{1 - \sin^2 A}$. Thus, $\cos A = \sqrt{1 - (0.5736)^2}$, or $\cos A \approx 0.8191$.   •

SOLVING RIGHT TRIANGLES

If the lengths of any two of the sides of a right triangle are known, the length of the third side can be found using the Pythagorean Theorem. Moreover, the measures of the two acute angles can also be found using the sine or cosine ratios. Similarly, given the measure of one angle and the length of one side of a right triangle, the lengths of the other sides may be found using trigonometry. The next two examples illustrate the technique known as **solving right triangles** which is used to find the measures of all of the angles and sides of a right triangle.

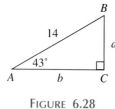

FIGURE 6.28

EXAMPLE 6.25   Solve the right triangle $\triangle ABC$ in Figure 6.28.

SOLUTION   First, $\angle B = 90° - 43° = 47°$. Then since $\sin 43° = a/14$, we have $a = 14 \sin 43°$, or $a = 43 \boxed{\text{SIN}} \boxed{\times} 14 \boxed{=} \boxed{9.547977041}$. Also, because $\sin 47° = b/14$, we have $b = 14 \sin 47°$, or $b = 47 \boxed{\text{SIN}} \boxed{\times} 14 \boxed{=} \boxed{10.23895182}$.

In summary,

$$a \approx 9.55 \qquad b \approx 10.2 \qquad c = 14$$
$$\angle A = 43° \qquad \angle B = 47° \qquad \angle C = 90°$$

•

Notice that both a and b could also have been found using the cosine as follows:

$\cos 43° = b/14$, so $b = 43 \boxed{\text{COS}} \boxed{\times} 14 \boxed{=} \boxed{10.23895182}$.

$\cos 47° = a/14$, so $a = 47 \boxed{\text{COS}} \boxed{\times} 14 \boxed{=} \boxed{9.547977041}$.

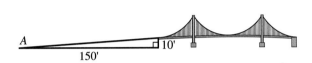

B

c

7

A   9   C

FIGURE 6.29

EXAMPLE 6.26   Solve the right triangle $\triangle ABC$ in Figure 6.29.

SOLUTION   By the Pythagorean Theorem, $c^2 = 7^2 + 9^2$. Therefore $c = \sqrt{49 + 81} \approx 11.4$. Also, $\tan A = 7/9 \approx 0.7778$. The value of A can be found as follows: 0.7778 [TAN^{-1}] [37.87577696]. So $A \approx 38°$, hence $B \approx 90° - 38° = 52°$. In summary, we have the following:

$$a = 7 \qquad b = 9 \qquad c \approx 11.4$$
$$\angle A \approx 38° \qquad \angle B \approx 52° \qquad \angle C = 90°$$

Solution to Applied Problem

A

150'

10'

In the figure, $\tan A = \dfrac{\text{opp}}{\text{adj}} = \dfrac{10}{150} = \dfrac{1}{15}$, so $A = \tan^{-1}\left(\dfrac{1}{15}\right)$, or

$$A = 1 \boxed{\div} 15 \boxed{=} \boxed{\text{TAN}^{-1}} \boxed{3.814074834} \approx 3.8°$$

So the ramp has an angle of inclination of approximately $3.8°$.

GEOMETRY AROUND US

Modern house construction involves the use of a framework called a roof truss system. Trusses are constructed at a factory and then delivered to the building site and lifted onto the tops of the walls. As illustrated in the figure, right triangles play a critical role in the construction of trusses.

PROBLEM SET 6.4

EXERCISES/PROBLEMS

Use the right triangle shown for Exercises 1–6.

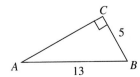

C

5

A   13   B

Find the following.

1. $\sin A$

2. $\cos A$

3. $\tan A$

4. $\sin B$

5. $\cos B$

6. $\tan B$

Use this figure for Exercises 7–24.

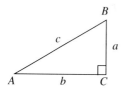

Find the following.

7. tan A if $a = 7$ and $b = 5$

8. sin B if $a = 5$ and $c = 12$

9. cos A if $b = 7$ and $c = 11$

10. tan B if $a = 5$ and $c = 13$

11. cos B if $a = b$

12. sin A if $a = 5$ and $b = 12$

13. a if sin $A = 0.503$ and $c = 10$

14. b if cos $A = 0.7$ and $c = 12$

15. c if tan $A = 0.6$ and $a = 5$

16. b if sin $B = 0.5$ and $a = 12$

17. $\angle A$ if $a = 5$ and $b = 7$

18. $\angle B$ if $a = 7$ and $c = 9$

19. $\angle A$ if $b = \pi$ and $c = \sqrt{21}$

20. $\angle B$ if $a = \sqrt{2}$ and $b = \sqrt{3}$

21. cos B if $b = a\sqrt{3}$

22. tan A if $c = 3a$

23. c if $\angle B = 70°$ and $b = 0.8$

24. b if $\angle A = 27.3°$ and $c = 5$

Solve each of the right triangles in Exercises 25–28.

25.

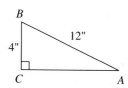

26.

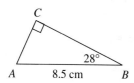

27.

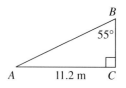

28.

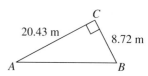

Verify Theorems 6.13 and 6.14 for Exercises 29–32.

29. $\angle A = 45°$

30. $\angle A = 13°$

31. $\angle A = 30°$

32. $\angle A = 49°$

In Exercises 33–40, solve for the missing parts of the following right triangle using the information given.

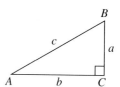

33. $c = 10$ and $\angle A = 30°$

34. $b = 8$ and $\angle B = 45°$

35. $a = 24$ and $b = 10$

36. $c = 7$ and $b = 5$

37. $\angle A = 23°$ and $b = 4$

38. $a = 9$ and $\angle B = 41°$

39. $\angle B = 12°$ and $c = 10.3$

40. $c = 21.7$ and $\angle A = 49°$

Find the area of each triangle in Exercises 41–44.

41.

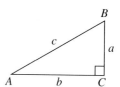

42.

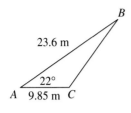

23.6 m
22°
A   9.85 m   C
B

43.

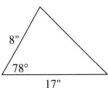

8″
78°
17″

44.

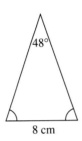

48°
8 cm

Find the area of each quadrilateral in Exercises 45–46.

45.

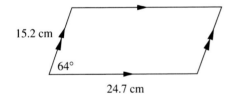

15.2 cm
64°
24.7 cm

46.

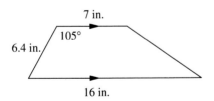

7 in.
105°
6.4 in.
16 in.

PROOFS

47. Prove: $\sin A = (\cos A)(\tan A)$

48. Prove: If A and B are the acute angles in a right triangle, then $\sin A = \cos B$.

49. Prove: If A is an acute angle in an isosceles right triangle, then $\sin A = \cos A$.

50. Show that $\sin A = \cos (90° - A)$.

51. Show that $\cos A = \sin (90° - A)$.

Three other trig ratios are defined as follows:

$$\cot A = \frac{\text{adjacent}}{\text{opposite}} \quad \text{(the \textbf{cotangent} ratio)}$$

$$\sec A = \frac{\text{hypotenuse}}{\text{adjacent}} \quad \text{(the \textbf{secant} ratio)}$$

$$\csc A = \frac{\text{hypotenuse}}{\text{opposite}} \quad \text{(the \textbf{cosecant} ratio)}$$

Using these definitions, prove the following:

52. $\cot A = \dfrac{1}{\tan A}$

53. $\sec A = \dfrac{1}{\cos A}$

54. $\csc A = \dfrac{1}{\sin A}$

55. $\cot A = \dfrac{\cos A}{\sin A}$

56. $\cot A = \dfrac{\csc A}{\sec A}$

57. $\csc^2 A - \cot^2 A = 1$

58. $\sec^2 A - \tan^2 A = 1$

APPLICATIONS

59. A hill makes an angle of 38° with the horizontal. A 150-foot-long fence is built from the top of the hill straight down to the base of the hill. How high is the hill, to the nearest foot?

60. A ramp leading to a freeway overpass is 300 feet long and rises 20 feet. What is the angle of inclination of the ramp (the acute angle between the ramp and the horizontal), to the nearest tenth of a degree?

61. A guy wire is attached to the top of a radio antenna and anchored to the ground at a point 20′ from the base of the antenna. If the wire makes an angle of 65° with the level ground, how high is the antenna?

62. A surveyor at point *P* wishes to measure the distance *PQ* in the drawing that follows. The surveyor sights point *Q*, makes a right angle at *P*, steps off 100 ft to *R*, and again sights point *Q*. ∠*PRQ* is determined to be 64°. Find the distance *PQ* to the nearest foot.

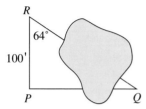

63. A girl flying a kite lets out all 120 feet of string. If the string forms an angle of 42° with level ground, how high is the kite at this moment?

64. A forester standing 50 feet from the base of a Douglas fir measures the angle of elevation to the top of the tree as 71.5°. Find the height of the tree to the nearest foot.

65. A conveyor used to put bales of hay into storage is 100 feet long. The recommended angle of elevation for the conveyor is 30°.

(a) To what height can the hay be moved?

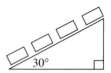

(b) If the conveyor is positioned at an angle of 28°, to what height can the hay be moved? Round to the nearest tenth of a foot.

66. A truck passes a sign that reads 8% GRADE NEXT 10 MILES. What is the angle of elevation of the road to the nearest tenth of a degree? [NOTE: Percentage grade is the slope of the road expressed as a percentage.]

67. A surveyor with a transit stands 100 ft away from a flagpole. The angle of elevation to the top of the flagpole is measured as 21°45′, and the angle of depression to the base of the flagpole is measured as 1°22′. Find the height of the flagpole to the nearest tenth of a foot.

68. A detective discovered that a bullet passed through a window at a point 4′2″ from the floor. It lodged in a wall 14′8″ from the hole in the window and 7′ up on the wall. At what angle from the horizontal, to the nearest tenth of a degree, was the bullet fired?

69. To determine the height of a building, a surveyor stands at a point *P* and measures the angle of elevation to the top of the building as 19.42°. From a point *Q*, 60 m further away from the building, the angle of elevation to the top of the building is 16.8°. Find the height of the building to the nearest meter.

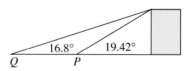

Solution to Initial Problem

Aerial photographs are taken from a plan. A dam that is known to be 100 m wide appears to be 3 cm long in a photograph. A second picture taken from the same height shows another dam to be 5.7 cm long. How wide is the second dam?

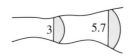

The ratio $\dfrac{3 \text{ cm}}{100 \text{ m}} = \dfrac{5.7 \text{ cm}}{x \text{ m}}$ leads to the equation $3x = 100(5.7)$. Solving for x, we have $x = \dfrac{570}{3} = 190$. Thus, the second dam is 190 m wide.

Additional Problems Where the Strategy "Solve an Equation" Is Useful

1. One leg of a right triangle is 2 cm less than twice as long as the other leg. The hypotenuse of the triangle is 2 cm more than twice as long as the shorter leg. Find the lengths of the sides of the triangle.

2. Find the length of diagonal AC in the rhombus $ABCD$ given that $AD = 5$ in. and $\angle B = 100°$. Round your answer to two decimal places.

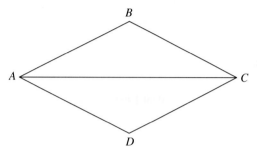

Writing for Understanding

1. Describe at least five ways in which ratios and proportions are used in everyday life. Write one problem illustrating each application that you mentioned.

2. Find at least five examples of the use of similar figures in the real world. Explain how you know that the figures are similar and how the concept of similarity is useful in each situation.

3. Research and write a brief article on the golden ratio (also called the divine proportion) as discussed in the "Geometry Around Us" at the end of Section 6.1.

4. Research and write a brief article on the geometry of fractals as discussed in the "Geometry Around Us" at the end of Section 6.2.

CHAPTER REVIEW

Following is a list of key vocabulary, notation, and ideas for this chapter. Mentally review these items and, where appropriate, write down the meaning of each term. Then restudy the material that you are unsure of before proceeding to take the chapter test.

Section 6.1—Ratio and Proportion

Vocabulary/Notation

Similar 271
Ratio 271

Rates 272
Proportion 272

Main Ideas/Results

1. Ratios and proportions are useful in solving problems.
2. The Cross-Multiplication Theorem can be used to solve a proportion.

Section 6.2—Similar Triangles

Vocabulary/Notation

Main Ideas/Results

1. There are three main ways to show that triangles are similar: AA Similarity, SAS Similarity, SSS Similarity
2. LL Similarity can be used to show that two right triangles are similar.

Section 6.3—Applications of Similarity

Vocabulary/Notation

Main Ideas/Results

1. The altitude to the hypotenuse of a right triangle is the mean proportional to the segments formed on the hypotenuse.
2. A line segment parallel to one side of a triangle divides the other two sides proportionally.
3. The midsegment of a triangle is parallel to the third side and half its length.
4. The line segments joining consecutive midpoints of the sides of any quadrilateral form a parallelogram.

Section 6.4—Right Triangle Trigonometry

Vocabulary/Notation

Main Ideas/Results

1. Tangent, sine, and cosine ratios are useful in solving problems.
2. The tangent, sine, and cosine ratios are related by the equation

$$\frac{\sin A}{\cos A} = \tan A$$

3. The sine and cosine ratios are related by the equation

$$\sin^2 A + \cos^2 A = 1$$

4. Right triangles can be solved using trig ratios.

PEOPLE IN GEOMETRY

Benoit Mandelbrot (b. 1924) once said "Clouds are not spheres, mountains are not cones, coastlines are not circles, and bark is not smooth, nor does lightning travel in a straight line." This is how he describes the inspiration for fractal geometry, a new field of mathematics that finds order in chaotic, irregular shapes and processes. Mandelbrot was largely responsible for developing the mathematics of fractal geometry while at IBM's Watson Research Center. Before Mandelbrot, other mathematicians had considered the requisite mathematics, but they described it as "pathological," giving rise to "mathematical monstrosities." Mandelbrot once said "The question I raised in 1967 is, 'How long is the coast of Britain,' and the correct answer is 'it all depends.' It depends on the size of the instrument used to measure length. As the measurement becomes increasingly refined, the measured length will increase. Thus, the coastline is of infinite length in some sense."

CHAPTER 6 TEST

TRUE-FALSE

Mark as true any statement that is always true. Mark as false any statement that is never true or that is not necessarily true. Be able to justify your answers.

1. If $\dfrac{a}{b} = \dfrac{c}{d}$, then $\dfrac{a}{c} = \dfrac{d}{b}$.

2. In a proportion, the product of the means equals the product of the extremes.

3. Two triangles are similar if two sides of one triangle are proportional to two sides of the other triangle.

4. A midsegment of a triangle is perpendicular to one of the sides of the triangle.

5. If $\angle A$ is an acute angle that measures less than 45°, then $\sin A < \cos A$.

6. Any two right isosceles triangles are similar.

7. If four angles of one quadrilateral are congruent, respectively, to four angles of another quadrilateral, then the quadrilaterals are similar.

8. In a right triangle, the altitude to the hypotenuse creates two right triangles, each of which is similar to the original triangle.

9. If the ratio of the lengths of the sides of two equilateral triangles is 3:5, then the ratio of their areas is 9:25.

10. The cosecant ratio is the reciprocal of the secant ratio.

EXERCISES/PROBLEMS

11. Find x in the proportion $\dfrac{x}{24} = \dfrac{5}{16}$.

12. Find the geometric mean of 7 and 13.

13. Suppose $\triangle ABC \sim \triangle DEF$, $AB = 5$, $BC = 9$, and $DE = 35$. Find EF.

14. (a) Draw a right triangle with sides of lengths 5 units, 12 units, and 13 units. Verify that this triangle is a right triangle.

    (b) Determine the sine and cosine of each acute angle in the triangle.

    (c) Verify that the relationship $\sin^2 A + \cos^2 A = 1$ holds for each acute angle in the triangle.

15. Square *ABCD* has *AC* = 20 cm. Find the perimeter of the quadrilateral *EFGH* formed by joining the midpoints of *ABCD*.

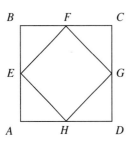

16. In a right triangle, one leg has length 10 and its opposite angle is 48°. Find the length of the hypotenuse to the nearest tenth.

17. Find *x* in the figure where $\overline{BC} \parallel \overline{DE}$.

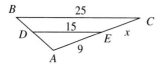

18. Find two different pairs of similar triangles in the figure where $\overline{DE} \parallel \overline{AC}$. Justify your answers.

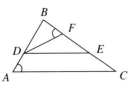

19. Find the area of the parallelogram shown. Round your answer to the nearest square meter.

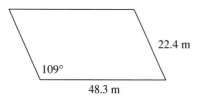

20. Solve for the missing parts of the triangle shown given that *b* = 55 and *c* = 73. Round angle measures to the nearest tenth of a degree.

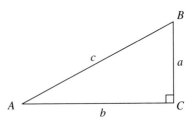

PROOFS

21. (a) If the height of a right circular cone is doubled and the radius of the cone is cut in half, what is the ratio of the volume of the new cone to the volume of the original cone?

(b) Prove your assertion in part (a)

22. Two parallel lines are crossed by two different transversals that are not parallel. The two transversals intersect in a point that is between the two parallel lines. Prove that the two triangles formed are similar.

APPLICATIONS

23. An observer in a lighthouse is 70 feet above sea level. She sees a whale spout and measures the angle of depression to the whale as 16.2°. To the nearest foot, what was the horizontal distance from the whale to the lighthouse when the woman saw it?

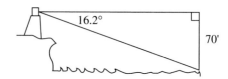

24. A wire is strung between two posts as shown. Find the length of the wire from *A* to *B* to *E* to *F* to the nearest tenth of a foot.

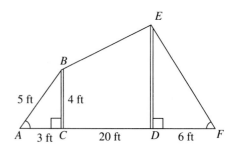

7

Circles

A **cycloid** is the path traced out by a point P on the circumference of a circle as the circle rolls (without slipping) along a horizontal line l.

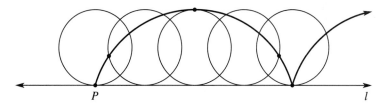

During the 17th century, a number of arguments among prominent mathematicians of the time centered on this curve. Galileo, Pascal, Descartes, Johann and Jakob Bernoulli, Leibniz, Newton, and others studied the mechanical properties of the cycloid. Among their discoveries were the following:

1. An inverted cycloid is the curve of equal descent. That is, if two marbles at different points on an inverted cycloid are released at the same time, they will reach the bottom at exactly the same time.

2. An inverted cycloid is also the curve of *fastest* descent. That is, if a marble at one point is to roll down to another point in the shortest amount of time, the path it should follow is, surprisingly, not a straight line, but an inverted cycloid.

PROBLEM-SOLVING STRATEGIES

1. Draw a Picture
2. Guess and Test
3. Use a Variable
4. Look for a Pattern
5. Make a Table
6. Solve a Simpler Problem
7. Look for a Formula
8. Use a Model
9. Identify Subgoals
10. Use Indirect Reasoning
11. Solve an Equation
12. *Use Cases*

STRATEGY 12: USE CASES

Many problems can be solved more easily by separating the problem into various cases. For example, to determine if a certain property is true for all triangles, it may be easier to first examine all right triangles for this property, then all acute triangles, and, finally, all obtuse triangles.

INITIAL PROBLEM

The circumscribed circle of a triangle is the unique circle that contains all three vertices of the triangle. For a given triangle, develop a simple test to determine if the center of the circumscribed circle is inside, on, or outside the triangle. (The center is the point of intersection of the perpendicular bisectors of the sides of the triangle.)

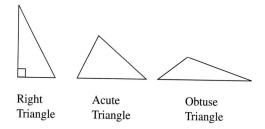

Right   Acute   Obtuse
Triangle   Triangle   Triangle

CLUES

The Use Cases strategy may be appropriate when

· A problem can be separated into several distinct categories.
· Investigations in specific cases can be generalized.

A solution for the initial problem above is on page 363.

INTRODUCTION

This chapter contains a study of circles and portions of circles as well as special lines and angles associated with circles. This material is particularly rich in that its organization ties together many ideas and consequently helps one see many interesting mathematical connections.

7.1
CENTRAL ANGLES AND INSCRIBED ANGLES

Applied Problem

A lawn sprinkler sprays water over a radius of 30 feet. If the sprinkler is set to turn through an angle of 110°, calculate the area that will be watered by the sprinkler.

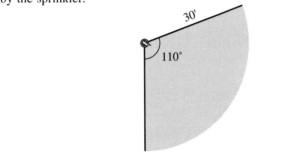

The definitions of circle, radius, and diameter were given in Chapter 2. In this section, we introduce some new terms associated with circles.

ARCS OF A CIRCLE

An **arc $\overset{\frown}{AB}$** of a circle consists of the points A and B of a circle together with a portion of the circle contained between the two points. Points A and B are called the **endpoints** of the arc [Figure 7.1(a)]. A **semicircle** is an arc whose endpoints are the endpoints of a diameter [$\overset{\frown}{AB}$ in Figure 7.1(b)]. An arc of a circle that is shorter than a semicircle is called a **minor arc** of the circle [$\overset{\frown}{AB}$ in Figure 7.1(c)].

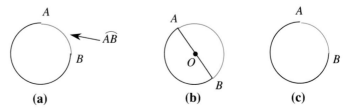

FIGURE 7.1

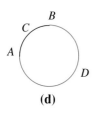

FIGURE 7.1

An arc that is longer than a semicircle is called a **major arc** [$\widearc{ADB}$ in Figure 7.1(d)].

Notice that due to possible ambiguity, arcs may also be named using a third letter on the arc. In Figure 7.1(d), the major arc is named $\widearc{ADB}$ or $\widearc{BDA}$ and the minor arc is named $\widearc{ACB}$. If there are only two letters, we agree that a minor arc is being named, so $\widearc{AB}$ describes the same arc as $\widearc{ACB}$. Arcs can be added or subtracted much like line segments. For example, in Figure 7.1(d), $\widearc{AB} = \widearc{AC} + \widearc{CB}$ and $\widearc{ADB} - \widearc{DB} = \widearc{AD}$.

CENTRAL ANGLES IN A CIRCLE

FIGURE 7.2

An angle whose vertex is the center of a circle and whose sides are radii is called a **central angle**. If O is the center of a circle containing points A and B, then $\angle AOB$ is a central angle (Figure 7.2). When the sides of an angle such as $\angle AOB$ intersect a circle at the two points A and B as in Figure 7.2, we say that $\angle AOB$ **intercepts** the arc $\widearc{AB}$. The measure of an arc of a circle is related to the central angle that intercepts it.

Definition

Measure of an Arc

The measure of an arc is the measure of the central angle that intercepts the arc.

$$\widearc{AB} = x° \text{ if and only if } \angle AOB = x°.$$

The measure of a semicircle is 180°. This means the measure of a minor arc will be less than 180° and the measure of a major arc is greater than 180° but less than 360°. Just as in the case of angle measure, the measure of $\widearc{AB}$ is written as m($\widearc{AB}$) or simply $\widearc{AB}$ when the context is clear.

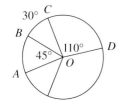

FIGURE 7.3

EXAMPLE 7.1   Use circle O in Figure 7.3 to find the measures of the following.

(a) $\widearc{AB}$ (b) $\angle BOC$ (c) $\widearc{ACD}$ (d) $\widearc{AD}$

SOLUTION
(a) $\widearc{AB} = 45°$ because $\angle AOB = 45°$.
(b) $\angle BOC = 30°$ because $\widearc{BC} = 30°$.
(c) $\widearc{ACD} = \widearc{AB} + \widearc{BC} + \widearc{CD} = 45° + 30° + 110° = 185°$
(d) $\widearc{ACD} + \widearc{AD} = 360°$. Therefore, $\widearc{AD} = 360° - \widearc{ACD} = 360° - 185° = 175°$.

LENGTH OF AN ARC OF A CIRCLE

We have seen that the degree measure of an arc of a circle is the same as the degree measure of the central angle that intercepts it. The degree measure of an arc can be used to describe the fraction of a circle an arc comprises. For example, an arc of 90° would be equivalent to one quarter of a circle. But two arcs of 90° will not necessarily be the same *length*. The length of an arc of a circle depends on both the central angle *and* the radius of the circle.

EXAMPLE 7.2   A clock has a pendulum 32″ long. If the pendulum swings through an arc of 12°, how far does the pendulum travel?

SOLUTION   The pendulum swings through an arc of a circle as shown in Figure 7.4. The central angle of 12° means that $\overset{\frown}{AB}$ is $\dfrac{12°}{360°} = \dfrac{1}{30}$ of a circle with radius 32″. So the length of arc $\overset{\frown}{AB}$ is $\dfrac{1}{30}$ of the circumference of a circle with radius 32″. Therefore,

$$\overset{\frown}{AB} = \frac{1}{30}(2\pi r) = \frac{1}{30}(2\pi)(32 \text{ in.}) \approx 6.7 \text{ in.} \qquad \bullet$$

This example can be generalized to yield the next postulate.

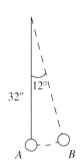

FIGURE 7.4

Postulate 7.1

Length of an Arc

The ratio of the length of an arc of a circle to the circumference of the circle equals the ratio of the central angle of the arc to 360°.

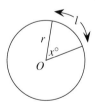

$$\frac{l}{C} = \frac{x}{360} \qquad \text{or} \qquad l = \frac{x}{360}(2\pi r)$$

AREA OF A SECTOR OF A CIRCLE

The measure of a central angle also can be used to calculate the area of a portion of a circular region called a sector. A **sector** of a circle is a region bounded by two radii of the circle and their intercepted arc.

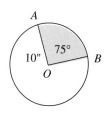

FIGURE 7.5

EXAMPLE 7.3   Find the area of the sector shaded in Figure 7.5.

SOLUTION   The region bounded by $\overline{OA}$, $\overline{OB}$, and $\overparen{AB}$ is a sector of circle O. Its area is some fraction of the area of circle O. Because $\angle AOB = 75°$, the area of the sector should be $\dfrac{75°}{360°} = \dfrac{5}{24}$ of the area of circle O. Therefore, the area of the shaded region will be $\left(\dfrac{5}{24}\right) \cdot \pi(10^2) \approx 65.4 \text{ in}^2.$

This example can be generalized to yield the following postulate.

Postulate 7.2

Area of a Sector

The ratio of the area of a sector of a circle to the area of the circle is equal to the ratio of the central angle of the sector to 360°.

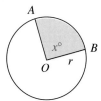

$$\frac{A_{\text{sector}}}{A_{\text{circle}}} = \frac{x}{360} \quad \text{or} \quad A_{\text{sector}} = \frac{x}{360}(\pi r^2)$$

INSCRIBED ANGLES IN A CIRCLE

FIGURE 7.6

A central angle has the center of the circle as its vertex. Next we will study an angle whose vertex is a point of the circle. An angle is an **inscribed angle** in a circle if its vertex is on the circle and its sides each intersect the circle in another point. In Figure 7.6, $\angle ABC$ is an inscribed angle in the circle shown. As was the case with a central angle, the measure of an inscribed angle is related to the measure of the arc it intercepts. The next example shows how.

EXAMPLE 7.4   Find the measure of $\angle BAC$ in circle O (Figure 7.7).

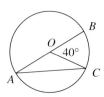

FIGURE 7.7

SOLUTION   We know that $\widehat{BC} = 40°$ because $\angle BOC = 40°$. $\triangle AOC$ is isosceles because $OA = OC$, so $\angle OAC = \angle OCA$. By the Exterior Angle Theorem, $\angle OAC + \angle OCA = 40°$. Therefore $\angle OAC = 20°$. Notice that this means that $\angle BAC = 20° = \frac{1}{2}\widehat{BC}$.

Example 7.4 suggests the following theorem.

Theorem 7.1

Inscribed Angle Theorem

The measure of an inscribed angle in a circle is equal to half the measure of its intercepted arc.

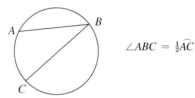

$$\angle ABC = \frac{1}{2}\widehat{AC}$$

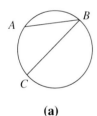

(a)

FIGURE 7.8

Given $\angle ABC$ inscribed in a circle [Figure 7.8(a)]

Prove $\angle ABC = \frac{1}{2}\widehat{AC}$

Proof There are three cases to consider:

(a) Where the center O is on $\overline{CB}$ [Figure 7.8(b)]
(b) Where O is outside $\angle ABC$ [Figure 7.8(c)]
(c) Where O is in the interior of $\angle ABC$ [Figure 7.8(d)]

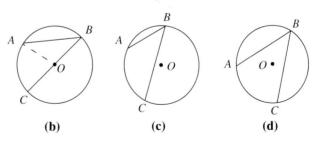

(b)   **(c)**   **(d)**

FIGURE 7.8

The proof of (a) is analogous to the solution of Example 7.4 if radius $\overline{OA}$ is drawn. We will prove the case where O is in the interior of $\angle ABC$ and leave the other two cases for the problem set. Draw the diameter $\overline{BD}$ [Figure 7.8(e)]. Then $\angle ABD = \frac{1}{2}\widehat{AD}$ and $\angle CBD = \frac{1}{2}\widehat{DC}$. Therefore,

$$\angle ABC = \angle ABD + \angle CBD = \frac{1}{2}\widehat{AD} + \frac{1}{2}\widehat{DC} = \frac{1}{2}(\widehat{AD} + \widehat{DC}) = \frac{1}{2}\widehat{AC}$$

(e)

FIGURE 7.8

The next example shows how to use the Inscribed Angle Theorem to find missing angle measures.

EXAMPLE 7.5

(a) Find x in Figure 7.9(a).

(b) Find y in Figure 7.9(b) where O is the center of the circle.

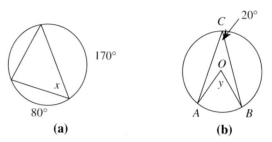

(a)   (b)

FIGURE 7.9

SOLUTION

(a) In Figure 7.9(a), $80° + 170° = 250°$, so the unlabeled arc has measure $360° - 250° = 110°$. By Theorem 7.1, $x = \frac{1}{2}(110°) = 55°$.

(b) In Figure 7.9(b), the measure of $\widehat{AB}$ is $40°$ by Theorem 7.1 because $\angle ACB$ is an inscribed angle. Therefore, $y = 40°$ because it is the measure of the central angle which intercepts $\widehat{AB}$.   •

The next two corollaries about inscribed angles follow from Theorem 7.1.

Corollary 7.2

Inscribed angles that intercept the same arc (or congruent arcs) are congruent.

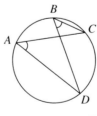

$$\angle A = \angle B = \tfrac{1}{2}\widehat{CD}$$

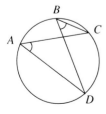

FIGURE 7.10

Given Inscribed angles $\angle CAD$ and $\angle CBD$ (Figure 7.10)

Prove $\angle CAD = \angle CBD$

Proof By the Inscribed Angle Theorem, we have $\angle CAD = \frac{1}{2}\widehat{CD}$ and $\angle CBD = \frac{1}{2}\widehat{CD}$. Therefore, $\angle CAD = \angle CBD$.   •

Corollary 7.3

An angle is inscribed in a semicircle if and only if it is a right angle.

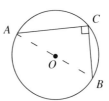

∠C is inscribed in semicircle $\overset{\frown}{ACB}$ if and only if ∠C is a right angle.

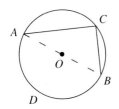

FIGURE 7.11

Given ∠ACB inscribed in a circle (Figure 7.11)

Prove

    (a) If $\overset{\frown}{ACB}$ is a semicircle, then ∠C is a right angle.

    (b) If ∠C is a right angle, then $\overset{\frown}{ACB}$ is a semicircle.

Proof

(a) If $\overset{\frown}{ACB}$ is a semicircle, then $\overset{\frown}{ADB}$ is a semicircle. Because an arc that is a semicircle has measure 180°, ∠$C = \frac{1}{2}(180°) = 90°$.

(b) If ∠C is a right angle, then it intercepts an arc whose measure is $2(90°) = 180°$. Thus, $\overset{\frown}{ADB}$ is a semicircle and $\overset{\frown}{ACB}$ must also be a semicircle. •

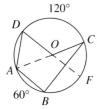

FIGURE 7.12

EXAMPLE 7.6   Find the measures of (a) ∠COD, (b) ∠B, (c) ∠C, (d) ∠DAB, and (e) ∠D in Figure 7.12, where O is the center of the circle and $\overline{AC}$ and $\overline{DF}$ are diameters.

SOLUTION

(a) ∠$COD = 120°$ because it is a central angle and $\overset{\frown}{DC} = 120°$.

(b) Because $\overline{AC}$ is a diameter, ∠$B = 90°$ by Corollary 7.3.

(c) ∠$C = \frac{1}{2}\overset{\frown}{AB} = \frac{1}{2}(60°) = 30°$ by the Inscribed Angle Theorem.

(d) Because $\overset{\frown}{CD} = 120°$ and $\overset{\frown}{BFC} = 180° - 60° = 120°$, we have $\overset{\frown}{DCB} = 120° + 120° = 240°$. Thus, ∠$DAB = \frac{1}{2}(240°) = 120°$ by the Inscribed Angle Theorem.

(e) Because ∠$COD = 120°$ and ∠$DOA = 180° - 120° = 60°$, we have $\overset{\frown}{AD} = 60°$. It follows that $\overset{\frown}{BF} = 180° - 60° - 60° = 60°$. Because $\overset{\frown}{ABF} = 120°$, we have ∠$D = \frac{1}{2}(120°) = 60°$ by the Inscribed Angle Theorem. •

Solution to Applied Problem

The region watered by the sprinkler is a sector of a circle. Because the central angle of the sector is 110° and the radius of the circle is 30 feet, the irrigated area is $A = \dfrac{110}{360} (\pi \cdot 30^2) \approx 864 \text{ ft}^2$

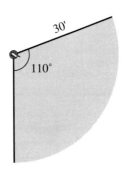

GEOMETRY AROUND US

Circle graphs such as the one shown appear frequently in newspapers to depict relative amounts of a whole. In a circle graph, the area of each sector is proportional to the fraction or percentage that it represents. In the graph shown, for example, revenue from federal sources represents less than one quarter of the state's revenue because the area of that sector is less than one quarter of the circle.

To construct a circle graph, the percentages are calculated first and are used to determine the central angle for each sector.

PROBLEM SET 7.1

EXERCISES/PROBLEMS

Refer to the following figure for Exercises 1–10.

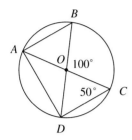

Segments $\overline{AC}$ and $\overline{BD}$ are diameters.

1. What kind of angle is $\angle BOC$ with respect to circle O?

2. What kind of angle is $\angle ABD$ with respect to circle O?

3. What is the measure of $\angle BAC$?

4. What is the measure of $\angle COD$?

5. What is the measure of $\overset{\frown}{DC}$?

6. What is the measure of $\angle DAC$?

7. What is the measure of $\overset{\frown}{AD}$?

8. What is the measure of $\angle ABD$?

9. What is the measure of $\angle DAB$?

10. What is the measure of $\angle AOB$?

Refer to the following figure for Exercises 11–18.

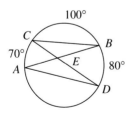

11. What is the measure of $\angle C$?

12. What is the measure of $\angle A$?

13. What is the measure of $\angle B$?

14. What is the measure of $\angle D$?

15. What is the measure of $\angle CEB$?

16. What is the measure of $\angle CEA$?

17. What is the measure of $\overset{\frown}{AD}$?

18. What is the measure of $\overset{\frown}{BDA}$?

Refer to the following figure for Exercises 19–24 where $\angle D = 30°$ and $\overline{AB} \parallel \overline{CD}$.

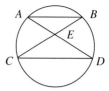

19. What are the measures of $\angle A$, $\angle B$, and $\angle C$?

20. Why is $\triangle ABE$ isosceles?

21. What is the measure of $\angle AEB$?

22. What is the measure of $\overset{\frown}{AC}$?

23. Why can we not necessarily conclude that $\overset{\frown}{AB} = $ m$(\angle AEB)$?

24. If $\overset{\frown}{CD} = 2\overset{\frown}{AB}$, what is the measure of $\overset{\frown}{CD}$?

25. In circle O, if $\angle BAC = 40°$, what is the measure of $\angle BOC$?

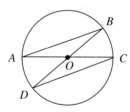

26. In circle O, if $\overset{\frown}{BC} = 38°$, what is the measure of $\overset{\frown}{AB}$?

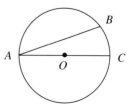

27. If $AB = 36''$ and $BC = 15''$, what is the radius of circle O?

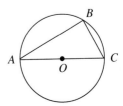

28. In the figure shown, the measure of $\overset{\frown}{AB}$ is twice the measure of $\overset{\frown}{BC}$ and the measure of $\overset{\frown}{BC}$ is 40° less than the measure of $\overset{\frown}{AC}$. Find $\angle A$.

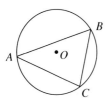

29. If $\overset{\frown}{AB} = 110°$, $\overset{\frown}{BC} = 80°$, and $\overset{\frown}{CD} = 50°$, what is the measure of $\angle B$?

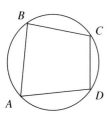

30. In circle O, $\angle PRS = 50°$ and $\overset{\frown}{SR} = \frac{2}{3}\overset{\frown}{PQ}$. Find the measure of $\angle QPR$.

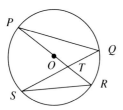

31. For quadrilateral $PQRS$, find PS.

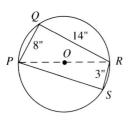

32. In circle O, $\overset{\frown}{AB} = 130°$. Find the measure of $\angle OBC$.

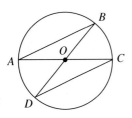

33. In circle O, $\overset{\frown}{AE} = 120°$ and $\overset{\frown}{CD} = 42°$. Find the measures of each of the five angles in pentagon $ABCDE$.

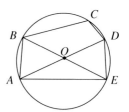

34. Find the measure of each angle of the star shown in the figure.

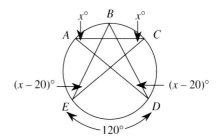

PROOFS

35. In the figure given, $ABCD$ is a rectangle. Prove that $\overline{AC}$ and $\overline{BD}$ are diameters of the circle.

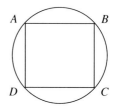

36. In the figure shown $\overline{AB} \parallel \overline{CD}$. Prove that $\triangle CED$ and $\triangle AEB$ are isosceles triangles.

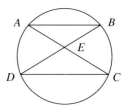

37. In the figure shown $\overline{BD}$ is a diameter. Prove that $\angle ADC$ and $\angle ABC$ are supplementary angles.

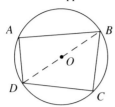

38. In the figure shown $\overline{AB} \parallel \overline{CD}$. Prove that $ABCD$ is an isosceles trapezoid.

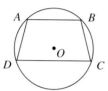

39. Prove Theorem 7.1, case (a): Show $\angle ABC = \frac{1}{2}\overparen{AC}$.

40. Prove Theorem 7.1, case (b): Show $\angle ABC = \frac{1}{2}\overparen{AC}$.

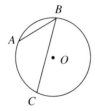

APPLICATIONS

41. The pendulum of a clock is 40 inches long and swings through an arc of 8°. Find the length of arc that the pendulum traces out.

42. A windshield wiper is 18 inches long and has a blade 12 inches long. If the wiper sweeps through an angle of 130°, how large an area does the wiper blade clean? Round your answer to the nearest square inch.

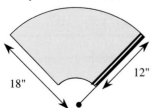

43. A wheel of cheese is 22 inches in diameter and 4 inches thick. If a wedge of cheese with a central angle of 14° is cut from the wheel, find the volume of the cheese wedge.

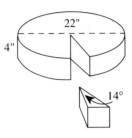

44. A paper cup has the shape shown (called a frustum of a right circular cone).

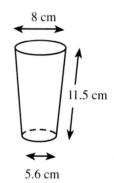

If the cup is sliced open and flattened, the sides of the cup have the following shape.

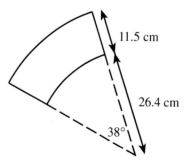

Use the dimensions given to calculate the number of square meters of paper used in the construction of the sides of 10,000 of these cups. Round your answer to the nearest square meter.

Parallels are circles in an east-west direction, parallel to the equator, that are used to measure how far north or south of the equator a particular place is. That distance is specified in degrees of **latitude** N or S. For example, Anchorage, Alaska has a latitude of approximately 61°N.

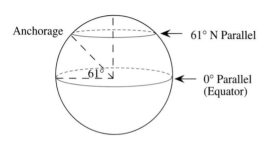

The distance between two points with the same **longitude**, or lying on the same meridian, can be calculated using their latitudes. **Meridians** are north-south circles through the poles that measure distance east or west of the Prime Meridian. For problems 45–49, use the fact that the radius of the earth is approximately 3960 miles to find distances between the pairs of cities. In each case, the longitudes of the cities are approximately the same, but their latitudes differ. That is, the cities lie on the same meridian, but on different parallels.

45. Quito, Ecuador, latitude 0°, and Lockport, New York, latitude 43°N

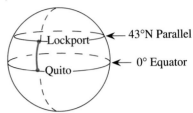

46. St. Louis, Missouri, latitude 38°N, and Ruth, Mississippi, latitude 31°N

47. Seattle, Washington, latitude 48°N, and San Francisco, California, latitude 38°N

48. Valparaiso, Chile, latitude 33°S, and Quebec, Canada, latitude 47°N

49. Brazzaville, Congo, latitude 4°S, and Örebro, Sweden, latitude 59°N

50. A drafter might use a right triangle like the one shown to find the center of a given circle. How can that be done? Explain why the technique works.

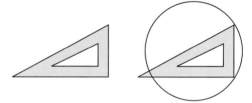

7.2
CHORDS OF A CIRCLE

Applied Problem

Some large circles turned up mysteriously in farmers' fields in Iowa and England. Conjectures concerning their origin include UFOs, freak wind storms, and magnetic fields. If the researchers who examined the fields needed to find the centers of the circles, how could they do it?

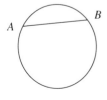

FIGURE 7.13

Radii and diameters are line segments that contain the centers of their circles. A **chord** is any line segment whose endpoints are on a circle. In Figure 7.13, $\overline{AB}$ is a chord of the circle. Notice that, according to the definition, diameters are chords that contain the center of the circle.

PERPENDICULAR BISECTOR OF A CHORD

In Chapter 4, we showed that if a point is equidistant from the endpoints of a line segment, then it is on the perpendicular bisector of the segment. The next theorem and its corollary show how this result can be used to locate the center of a circle.

Theorem 7.4

The perpendicular bisector of a chord contains the center of the circle.

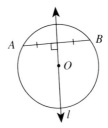

If l is the perpendicular bisector of $\overline{AB}$, then the center O is on l.

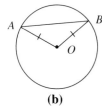

(a)

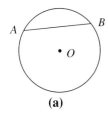

(b)

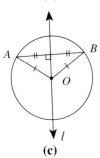

(c)

FIGURE 7.14

Given Chord $\overline{AB}$ of circle O [Figure 7.14(a)]

Prove O lies on the perpendicular bisector of $\overline{AB}$.

Proof Consider $\overline{AO}$ and $\overline{BO}$. Because $\overline{AO}$ and $\overline{BO}$ are radii, we have $AO = BO$ [Figure 7.14(b)]. This means that O is equidistant from A and B. Therefore, O is on l, the perpendicular bisector of $\overline{AB}$ [Figure 7.14(c)].   •

The next corollary extends Theorem 7.4 and states a relationship between the perpendicular bisectors of two different chords.

Corollary 7.5

The intersection of the perpendicular bisectors of any two nonparallel chords of a circle is the center of the circle.

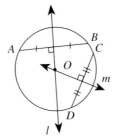

If l and m are perpendicular bisectors of $\overline{AB}$ and $\overline{CD}$, then l and m intersect at the center O.

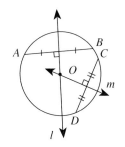

FIGURE 7.15

Given Two nonparallel chords $\overline{AB}$ and $\overline{CD}$ and their perpendicular bisectors *l* and *m* (Figure 7.15)

Prove The lines *l* and *m* intersect in the center.

Proof Because the chords are not parallel, neither are their perpendicular bisectors. Thus, the perpendicular bisectors intersect in one point. By Theorem 7.4, this must be the center because both lines contain the center.   •

This theorem provides a practical method for locating the center of a given circle. If any two nonparallel chords are drawn and their perpendicular bisectors constructed, the center of the circle will be the point of intersection of the bisectors. We will utilize this method in Section 7.4. Theorem 7.4 also can be used to prove the following corollary.

Corollary 7.6

If two circles, *O* and *O'*, intersect in two points *A* and *B*, then the line containing *O* and *O'* is the perpendicular bisector of $\overline{AB}$.

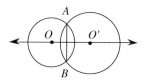

If *O* and *O'* intersect in points *A* and *B*, then $\overleftrightarrow{OO'}$ is the perpendicular bisector of $\overline{AB}$.

[NOTE: The line connecting the centers of two circles is called the **line of centers**.]

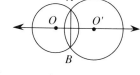

FIGURE 7.16

Given Circles *O* and *O'* intersecting in points *A* and *B* (Figure 7.16)

Prove $\overleftrightarrow{OO'}$ is the perpendicular bisector of $\overline{AB}$.

Proof By Theorem 7.4, the perpendicular bisector of $\overline{AB}$ contains the centers of both circle *O* and *O'*. Thus, the line of centers $\overline{OO'}$ is the perpendicular bisector of $\overline{AB}$.   •

EXAMPLE 7.7   Show three different ways to construct the perpendicular bisector to the chord $\overline{AB}$ in the circle with center *O* [Figure 7.17(a)].

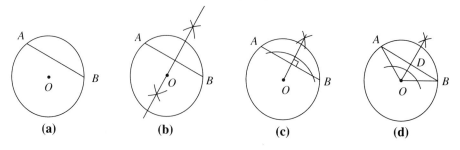

FIGURE 7.17

SOLUTION
(1) Construct the perpendicular bisector of $\overline{AB}$ using Construction 6 of Section 4.4 [Figure 7.17(b)].
(2) Drop a perpendicular from O to $\overline{AB}$ using Construction 5 of Section 4.4 [Figure 7.17(c)] . There is a unique perpendicular from a point to a line, and by Theorem 7.4, O lies on the perpendicular bisector of $\overline{AB}$. So this segment is the perpendicular bisector of $\overline{AB}$.
(3) Bisect $\angle AOB$ using Construction 3 of Section 4.4 [Figure 7.17(d)]. Because $\triangle AOB$ is isosceles, $\overline{OD}$ is the perpendicular bisector of $\overline{AB}$.   •

MEASURES OF ANGLES FORMED BY CHORDS

Next consider the following example that examines the measures of angles formed by chords.

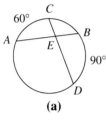

EXAMPLE 7.8   In Figure 7.18(a), $\widehat{AC} = 60°$ and $\widehat{BD} = 90°$. Find the measure of $\angle AEC$.

SOLUTION   Draw $\overline{BC}$ [Figure 7.18(b)]. Then $\angle B = \frac{1}{2}(60°) = 30°$, and $\angle C = \frac{1}{2}(90°) = 45°$. Because $\angle AEC$ is an exterior angle of $\triangle BCE$, $\angle AEC = \angle B + \angle C = 30° + 45° = 75°$.   •

The technique used to find $\angle AEC$ in Example 7.8 is generalized to any pair of intersecting chords in the next theorem.

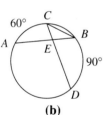

(b)

FIGURE 7.18

Theorem 7.7

If two chords intersect, then the measure of any one of the vertical angles formed is equal to half the sum of the two arcs intercepted by the angle and its vertical angle.

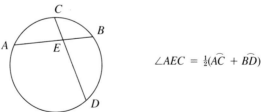

$$\angle AEC = \frac{1}{2}(\widehat{AC} + \widehat{BD})$$

Given Chords $\overline{AB}$ and $\overline{CD}$ intersecting at E (Figure 7.19)

Prove $\angle AEC = \frac{1}{2}(\widehat{AC} + \widehat{BD})$

Proof In Figure 7.19, $\angle C = \frac{1}{2}\widehat{BD}$ and $\angle B = \frac{1}{2}\widehat{AC}$ by the Inscribed Angle Theorem. Because $\angle AEC$ is an exterior angle to $\triangle BCE$, $\angle AEC = \angle C + \angle B$ by the Exterior Angle Theorem. Therefore,

$$\angle AEC = \frac{1}{2}\widehat{AC} + \frac{1}{2}\widehat{BD} = \frac{1}{2}(\widehat{AC} + \widehat{BD})$$ •

Notice that $\angle AEC = \angle BED$ because they are vertical angles. But this does *not* mean that $\widehat{AC}$ must be congruent to $\widehat{BD}$.

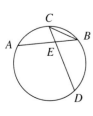

FIGURE 7.19

MEASURES OF SEGMENTS OF CHORDS

The next theorem develops a relationship between the four segments formed by intersecting chords.

Theorem 7.8

If two chords of a circle intersect, the product of the lengths of the two segments formed on one chord is equal to the product of the lengths of the two segments formed on the other chord.

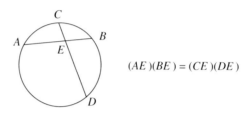

$(AE)(BE) = (CE)(DE)$

Given Chords $\overline{AB}$ and $\overline{CD}$ intersecting at E (Figure 7.20)

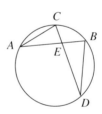

FIGURE 7.20

Prove $(AE)(BE) = (CE)(DE)$

Plan Show $\triangle AEC \sim \triangle DEB$. Then use C.P.

Proof

Statements	Reasons
1. $\angle AEC \cong \angle DEB$	1. Vertical angles
2. $\angle C = \angle B = \frac{1}{2}\overarc{AD}$	2. Inscribed Angle Theorem
3. $\triangle AEC \sim \triangle DEB$	3. AA Similarity
4. $AE:CE = DE:BE$	4. Corresponding parts
5. $(AE)(BE) = (CE)(DE)$	5. Cross multiplication

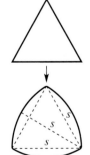

FIGURE 7.21

EXAMPLE 7.9 In Figure 7.21, find (a) $\angle AEC$ and (b) *EB*.

SOLUTION

(a) $\angle AEC = \frac{1}{2}(40° + 60°) = 50°$ by Theorem 7.7.

(b) By Theorem 7.8, $8(EB) = 6(16)$, so $EB = 12$.

Solution to Applied Problem

By Corollary 7.5, if any two nonparallel chords are drawn, their perpendicular bisectors will intersect in the center of the circle. Thus, we would draw two nonparallel chords of the circle for which we wish to find the center and construct their perpendicular bisectors. The point of intersection of the perpendicular bisectors is the center of the circle.

GEOMETRY AROUND US

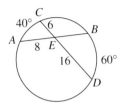

A circle is a **curve of constant width**. That is, if a circle rolls in a channel of parallel lines, the circle will always touch the top and bottom of the channel as shown.

Interestingly, there are many other curves of constant width that are not circles. One such curve is the shape of the rotor found in the Wankel (or rotary) engine. This shape is formed by connecting the vertices of an equilateral triangle with an arc of a circle whose center is at a vertex and whose radius is the length of a side of the triangle. These shapes are called **Reuleaux triangles** after the man who observed their constant width property. One amazing application of these triangles was devised by an English engineer, who used a Reuleaux triangle to design a drill that makes *square* holes.

PROBLEM SET 7.2

EXERCISES/PROBLEMS

For problems 1–8, refer to the following figure.

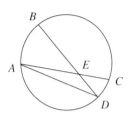

1. If $\overset{\frown}{AD} = 125°$ and $\overset{\frown}{BC} = 173°$, find $\angle BEC$.

2. If $\angle AEB = 35°$ and $\overset{\frown}{AB} = 24°$, find $\overset{\frown}{CD}$.

3. If $\angle CAD = 30°$ and $\angle BDA = 40°$, find $\angle CED$.

4. If $\angle CED = 41°$ and $\angle DAC = 25°$, find $\overset{\frown}{AB}$.

5. If $BE = 5$ cm, $AE = 4$ cm, and $CE = 2$ cm, find DE.

6. If $AE = 9.8$ m, $CE = 5.7$ m, and $DE = 3.8$ m, find BE.

7. If $BE = 2(EC)$, $AC = 12''$, and $DE = 3.5''$, find CE.

8. If $AC = 10$ cm, $BE = 8$ cm, and $DE = 3$ cm, find AE.

For problems 9–12, refer to the following figure.

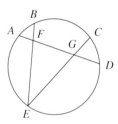

9. If $BF = 7''$, $EF = 18''$, and $FD = 15''$, find AF.

10. If $AF = 2$ cm, $GD = 3$ cm, $GC = 4$ cm, and $EG = 9$ cm, find FG.

11. If $\angle BEC = 42°$, $\overset{\frown}{CD} = 30°$, and $\overset{\frown}{AE} = 95°$, find $\angle AFE$.

12. If $\angle AFB = 70°$, $\overset{\frown}{AE} = 80°$, $\overset{\frown}{ED} = 120°$, and $\overset{\frown}{CD} = 15°$, find $\overset{\frown}{BC}$.

For problems 13–16, refer to circle O below.

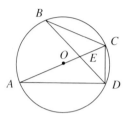

13. If $AD = 15$ cm, $CD = 8$ cm, $BE = 10$ cm, and $DE = 6$ cm, find AE.

14. If $\angle ADB = 65°$ and $\angle BEC = 80°$, find $\overset{\frown}{AD}$.

15. If $AE = 14'$, $BE = 12'$, $DE = 7'$, and $CD = 12'$, find AD.

16. If $\angle CBD = 30°$, $CD = 6''$, $BE = 5''$, and $ED = 3''$, find CE.

17. There is one, and only one, chord joining two points on a circle. How many different chords can be drawn for a set of 3 points on a circle? 4 points? 5 points? n points?

PROOFS

18. Prove that there is only one circle determined by three noncollinear points.

19. Prove that in the same or congruent circles, congruent chords intercept congruent arcs.

20. Prove that in the same or congruent circles, congruent arcs have congruent chords.

21. Prove that in the same or congruent circles, congruent central angles have congruent chords.

22. Prove that in the same or congruent circles, congruent chords determine congruent central angles.

23. Prove that two arcs are congruent if they are contained in congruent circles and have the same central angle.

24. Prove that in a circle, parallel lines intercept congruent arcs.

25. Prove that in the same or congruent circles, congruent chords are equidistant from the center.

26. Given circle O with diameter $\overline{AB}$ and C a point in the exterior of the circle, determine if $\angle ACB$ is acute, right, or obtuse and prove your assertion.

27. Prove that if a kite is inscribed in a circle, then its two congruent angles must be right angles.

28. Prove that a quadrilateral can be inscribed in a circle if and only if its opposite angles are supplementary.

29. Prove that a line drawn from the center of a circle through the midpoint of a chord is the perpendicular bisector of the chord.

30. Prove that a line containing the center of a circle and perpendicular to a chord contains the midpoint of the chord.

31. Prove that a line perpendicular to a chord bisects the arcs determined by the chord.

APPLICATIONS

32. The bridge shown has dimensions as given where the arch measures $18'$ at its highest point. The arch of the bridge is an arc of a circle, though *not* a semicircle. Find the radius of the circle that contains the arc.

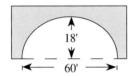

33. Wally, Jane, and Steve decide to get up a game of wastepaper basketball. They are seated at their desks in positions W, J, and S. Where should the waste-paper basket be placed so that each person has to throw the same distance to the basket?

J .   . S

$\overset{.}{W}$

34. A piece of a broken wheel is shown. It was taken to a machine shop to be replaced with a new "whole" wheel. Find the radius of the wheel if $\underline{AC} = 10$ cm, $BD = 3$ cm, and D is the midpoint of $\overline{AC}$.

7.3
SECANTS AND TANGENTS

Applied Problem

Three planets are aligned as shown. The diameter of the smallest planet is 3000 miles and the diameter of the planet in the middle is 8000 miles. Given the other dimensions in the figure, what is the diameter of the largest planet?

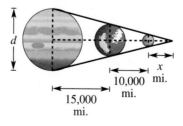

In the previous section we discussed chords of circles and associated results. This section presents similar results about secants and tangents.

SECANT LINES

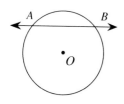

FIGURE 7.22

A **secant** is a line that intersects a circle in two points. In Figure 7.22, $\overleftrightarrow{AB}$ is a secant line for circle O. The next example shows how the measure of the angle between two secants can be found.

EXAMPLE 7.10   Find the measure of $\angle AEC$ in Figure 7.23 where $\widehat{AC} = 80°$ and $\widehat{BD} = 20°$.

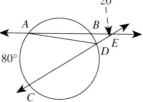

FIGURE 7.23

SOLUTION   Because $\widehat{AC} = 80°$, $\angle ADC = \frac{1}{2}(80)° = 40°$. Also, $\widehat{BD} = 20°$ implies that $\angle DAB = \frac{1}{2}(20)° = 10°$. So by the Exterior Angle Theorem, $\angle ADC = \angle DAB + \angle AEC$ and $\angle AEC = \angle ADC - \angle DAB = 40° - 10° = 30°$. •

This example is generalized in the next theorem.

Theorem 7.9

If two secants intersect outside a circle, the measure of the angle of intersection is half the difference of the measures of the intercepted arcs.

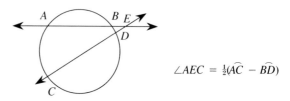

$$\angle AEC = \tfrac{1}{2}(\widehat{AC} - \widehat{BD})$$

Given   Secants $\overleftrightarrow{AB}$ and $\overleftrightarrow{CD}$ intersecting outside a circle at point E [Figure 7.24(a)]

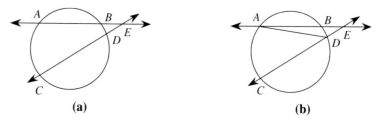

(a)   (b)

FIGURE 7.24

Prove $\angle AEC = \frac{1}{2}(\overset{\frown}{AC} - \overset{\frown}{BD})$

Proof Draw $\overline{AD}$ [Figure 7.24(b)].

Statements	**Reasons**
1. $\angle ADC = \frac{1}{2}\overset{\frown}{AC}$	1. Inscribed Angle Theorem
2. $\angle DAB = \frac{1}{2}\overset{\frown}{BD}$	2. Inscribed Angle Theorem
3. $\angle ADC = \angle DAB + \angle AEC$	3. Exterior Angle Theorem
4. $\angle AEC = \angle ADC - \angle DAB$	4. Subtraction
5. $\angle AEC = \frac{1}{2}\overset{\frown}{AC} - \frac{1}{2}\overset{\frown}{BD} =$ $\frac{1}{2}(\overset{\frown}{AC} - \overset{\frown}{BD})$	5. Substitution/Simplification from steps 1, 2, and 4

In the last section, we found a relationship between the lengths of the segments formed by intersecting chords (Theorem 7.8). A similar relationship exists between the lengths of segments formed by intersecting secants.

Theorem 7.10

If two secants intersect outside a circle, then the product of the lengths of two of the segments formed on one secant is equal to the product of the lengths of the corresponding segments on the other secant.

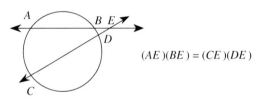

$(AE)(BE) = (CE)(DE)$

Given Secants $\overleftrightarrow{AB}$ and $\overleftrightarrow{CD}$ intersecting outside a circle at point E

Prove $(AE)(BE) = (CE)(DE)$

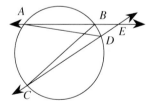

FIGURE 7.25

Proof Draw $\overline{AD}$ and $\overline{BC}$. Consider $\triangle AED$ and $\triangle CEB$ in Figure 7.25. Because E is common to both triangles and $\angle ABC$ and $\angle CDA$ intercept the same arc, we have $\angle ABC = \angle CDA$. Now, $\angle ADE = \angle CBE$ because they are supplements of the congruent angles $\angle CDA$ and $\angle ABC$ respectively. Thus, we can conclude that $\triangle AED \sim \triangle CEB$ by AA Similarity. Therefore, $AE{:}DE = CE{:}BE$, or $(AE)(BE) = (CE)(DE)$.

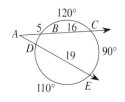

120°
5 B 16 C
A
D
19   90°
110°   E

FIGURE 7.26

EXAMPLE 7.11   In Figure 7.26, find the measures of the following.

(a) $\angle CAE$ (b) $\overline{AD}$

SOLUTION

(a) $\overset{\frown}{BD} = 360° - (120° + 110° + 90°) = 40°$. Thus, by Theorem 7.9,

$$\angle CAE = \frac{1}{2}(90° - 40°) = 25°$$

(b) By Theorem 7.10, $(AE)(AD) = (AC)(AB)$. Let x represent AD.

$$(19 + x)(x) = (5 + 16)(5)$$

$$19x + x^2 = 105$$

$$x^2 + 19x - 105 = 0$$

$$x = \frac{-19 \pm \sqrt{19^2 - 4(-105)}}{2}$$

$$x = \frac{-19 + \sqrt{781}}{2} \approx 4.47 \text{ in.}$$

TANGENT LINES

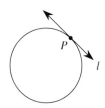

FIGURE 7.27

Chords and secants each intersect a circle in exactly two points. A **tangent** is a line that intersects a circle in exactly one point. This point is called the **point of tangency**. In Figure 7.27, l is a tangent line and P is the point of tangency. Notice that all points on a tangent line except the point of tangency lie outside the circle. Rays and line segments are said to be tangent to a circle if they lie on a line tangent to the circle and if they intersect the circle in one point .

Suppose a radius of a circle is drawn to the point of tangency (Figure 7.28). Notice that $\overline{OA}$ appears to be perpendicular to the tangent line $\overleftrightarrow{AB}$. The next theorem verifies this fact. This theorem will be useful when we want to construct a tangent line to a point on a circle.

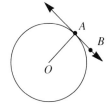

FIGURE 7.28

Theorem 7.11

A radius or diameter of a circle is perpendicular to a tangent line at its point of tangency.

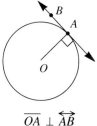

$$\overline{OA} \perp \overleftrightarrow{AB}$$

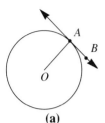

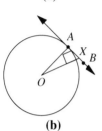

(a)

(b)

FIGURE 7.29

Given Circle O with $\overline{OA}$ a radius to tangent line $\overleftrightarrow{AB}$ [Figure 7.29(a)]

Prove $\overline{OA} \perp \overleftrightarrow{AB}$

Plan We will prove this result indirectly.

Proof Suppose that $\angle OAB$ is *not* a right angle. Then construct the perpendicular $\overline{OX}$ to the tangent line [Figure 7.29(b)] where X is on the tangent line $\overleftrightarrow{AB}$. Then, because $\triangle OAX$ is a right triangle with hypotenuse $\overline{OA}$, we have $OA > OX$. However, because X is outside the circle, it must be true that $OX > OA$. Thus, we have a contradiction, so $\angle OAB$ must be a right angle. •

ANGLES FORMED WITH TANGENT LINES

The next three theorems describe results involving tangents, secants, and angle measure based on the measure of intercepted arcs. As you proceed, observe how each succeeding theorem uses the previous result.

Theorem 7.12

The measure of an angle formed when a chord intersects a tangent line at the point of tangency is half the measure of the arc intercepted by the chord and the tangent line.

$$\angle ABC = \tfrac{1}{2}\widehat{AB}$$

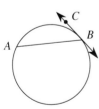

(a)

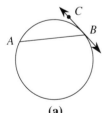

(b)

FIGURE 7.30

Given Chord $\overline{AB}$ and tangent line $\overleftrightarrow{CB}$ [Figure 7.30(a)]

Prove $\angle ABC = \tfrac{1}{2}\widehat{AB}$

Proof Consider $\overline{AB}$, tangent line $\overleftrightarrow{CB}$, and diameter BD [Figure 7.30(b)]. Because, by Theorem 7.11, a diameter is perpendicular to a tangent line at its point of tangency, $\angle CBD = 90° = \tfrac{1}{2}\widehat{DAB}$. Also, $\angle ABD = \tfrac{1}{2}\widehat{AD}$ because it is an inscribed angle. Therefore, we have

$$\angle ABC = \angle CBD - \angle ABD = \tfrac{1}{2}\widehat{DAB} - \tfrac{1}{2}\widehat{AD} = \tfrac{1}{2}(\widehat{DAB} - \widehat{AD}) = \tfrac{1}{2}\widehat{AB} \quad •$$

Theorem 7.13

If a secant and a tangent line intersect outside a circle, the measure of the angle formed is half the difference of the measures of the intercepted arcs.

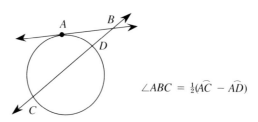

$\angle ABC = \frac{1}{2}(\overset{\frown}{AC} - \overset{\frown}{AD})$

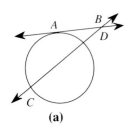

(a)

Given Secant line $\overset{\leftrightarrow}{CD}$ and tangent line $\overset{\leftrightarrow}{AB}$ where A is the point of tangency [Figure 7.31(a)]

Prove $\angle ABC = \frac{1}{2}(\overset{\frown}{AC} - \overset{\frown}{AD})$

Proof Draw $\overline{AD}$ [Figure 7.31(b)].

Statements	**Reasons**
1. $\angle ADC = \frac{1}{2}\overset{\frown}{AC}$	1. Inscribed Angle Theorem
2. $\angle BAD = \frac{1}{2}\overset{\frown}{AD}$	2. Theorem 7.12
3. $\angle ADC = \angle BAD + \angle ABC$	3. Exterior Angle Theorem
4. $\angle ABC = \angle ADC - \angle BAD$	4. Subtraction
5. $\angle ABC = \frac{1}{2}\overset{\frown}{AC} - \frac{1}{2}\overset{\frown}{AD} = \frac{1}{2}(\overset{\frown}{AC} - \overset{\frown}{AD})$	5. Substitution/Simplification

(b)

FIGURE 7.31

Theorem 7.13 can be used to prove a similar result about two tangent lines.

Theorem 7.14

If two tangent lines intersect outside a circle, the measure of the angle formed is half the difference of the measures of the intercepted arcs.

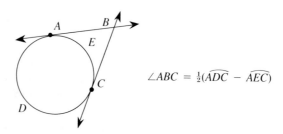

$\angle ABC = \frac{1}{2}(\overset{\frown}{ADC} - \overset{\frown}{AEC})$

Given Tangent lines $\overset{\leftrightarrow}{AB}$ and $\overset{\leftrightarrow}{BC}$ intersecting outside a circle where A and C are the points of tangency [Figure 7.32(a)]

Prove $\angle ABC = \frac{1}{2}(\widehat{ADC} - \widehat{AC})$

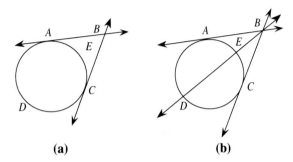

(a)   **(b)**

FIGURE 7.32

Proof Draw $\overset{\leftrightarrow}{BD}$ where $\overset{\leftrightarrow}{BD}$ is any secant line through B [Figure 7.32(b)]. Using Theorem 7.13, we have $\angle ABD = \frac{1}{2}\widehat{AD} - \frac{1}{2}\widehat{AE}$ and $\angle DBC = \frac{1}{2}\widehat{DC} - \frac{1}{2}\widehat{EC}$. Because $\angle ABC = \angle ABD + \angle DBC$, we have the following:

$$\begin{aligned}
\angle ABC &= \tfrac{1}{2}\widehat{AD} - \tfrac{1}{2}\widehat{AE} + \tfrac{1}{2}\widehat{DC} - \tfrac{1}{2}\widehat{EC} \\
&= \tfrac{1}{2}\widehat{AD} + \tfrac{1}{2}\widehat{DC} - \tfrac{1}{2}\widehat{AE} - \tfrac{1}{2}\widehat{EC} \\
&= \tfrac{1}{2}(\widehat{AD} + \widehat{DC}) - \tfrac{1}{2}(\widehat{AE} + \widehat{EC}) \\
&= \tfrac{1}{2}\widehat{ADC} - \tfrac{1}{2}\widehat{AEC}
\end{aligned}$$

EXAMPLE 7.12   In Figure 7.33(a), if $\widehat{CD} = 120°$, find $\angle BAO$.

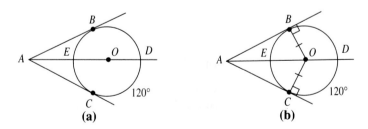

(a)   **(b)**

FIGURE 7.33

SOLUTION   Draw $\overline{OB}$ and $\overline{OC}$ [Figure 7.33(b)]. Both are perpendicular to tangent lines. Because $\overline{OB}$ and $\overline{OC}$ are radii of the same circle, $\overline{OB} \cong \overline{OC}$. Also, $\overline{AO} \cong \overline{AO}$. Therefore, we have $\triangle ABO \cong \triangle ACO$ by HL (Theorem 4.3). So $\angle BAO \cong \angle CAO$ by corresponding parts. But we know that

$$\angle CAO = \tfrac{1}{2}(\widehat{CD} - \widehat{CE}) = \tfrac{1}{2}[120° - (180° - 120°)] = \tfrac{1}{2}(120° - 60°) = 30°$$

So $\angle BAO$ is also 30°.

SEGMENTS ON TANGENT LINES

Notice that because $\triangle ABO \cong \triangle ACO$ in Example 7.12, $\overline{AB} \cong \overline{AC}$. This result is summarized next. The proof of the corollary is left for the problem set.

Corollary 7.15

If two tangents are drawn to a circle from the same point in the exterior of the circle, the distances from the common point to the points of tangency are equal.

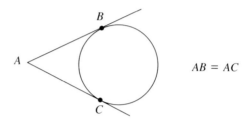

$AB = AC$

The next example shows an application of Corollary 7.15 to finding lengths of segments in a triangle.

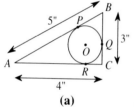

(a)

EXAMPLE 7.13   In Figure 7.34(a), the sides of $\triangle ABC$ are each tangent to circle O. We say that circle O is inscribed in $\triangle ABC$. If P, Q, and R are points of tangency, find AR and BP.

SOLUTION   Let $AR = x$ and $BP = y$. Because $\overline{AP}$ and $\overline{AR}$ are tangents to circle O from a common point, $AR = AP = x$, by Corollary 7.15. Likewise $BQ = y$ [Figure 7.34(b)]. Because $AC = 4''$, we have $RC = 4 - x$. Because $BC = 3''$, we have $QC = 3 - y$ [Figure 7.34(c)]. Because $\overline{QC}$ and $\overline{RC}$ are tangents to circle O from point C, by Corollary 7.15 we have $QC = RC$. So $4 - x = 3 - y$, or

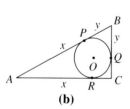

(b)

$$x - y = 1 \tag{1}$$

Because $AB = 5''$, we have

$$x + y = 5 \tag{2}$$

Solving equations (1) and (2) simultaneously, we get

$$
\begin{array}{r}
x - y = 1 \\
\underline{x + y = 5} \\
2x = 6
\end{array}
$$

(c)

or $x = 3$ in. and $y = 2$ in. So, $AR = 3''$ and $BP = 2''$.

The next theorem is an analogue to Theorem 7.14.

Theorem 7.16

If tangent and secant lines are drawn to a circle from the same point in the exterior of the circle, the length of the tangent segment is the mean proportional between the length of the external secant segment and the length of the secant.

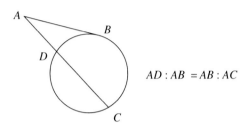

$$AD : AB = AB : AC$$

Given Tangent $\overline{AB}$ and secant $\overline{AC}$ [Figure 7.35(a)]

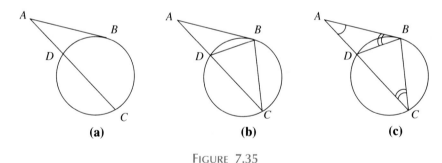

(a)   **(b)**   **(c)**

FIGURE 7.35

Prove $AD{:}AB = AB{:}AC$

Subgoal Show that $\triangle ABC \sim \triangle ADB$. [Figure 7.35 (b)]

Proof

Statements	**Reasons**
1. $\angle A \cong \angle A$	1. Reflexive property
2. $\angle C = \frac{1}{2}\widehat{BD}$	2. Inscribed Angle Theorem
3. $\angle ABD = \frac{1}{2}\widehat{BD}$	3. Theorem 7.12 [Figure 7.28(c)]
4. $\angle ABD = \angle C$	4. Substitution
5. $\triangle ADB \sim \triangle ABC$	5. AA Similarity
6. $AD{:}AB = AB{:}AC$	6. Corresponding sides in similar triangles are proportional.

Solution to Applied Problem

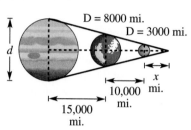

D = 8000 mi.
D = 3000 mi.

d

10,000 mi.

15,000 mi.

x

All three triangles formed are similar. Therefore, we have that $\dfrac{x}{3000} = \dfrac{(x + 10{,}000)}{8000}$, which yields $8000x = 3000(x + 10{,}000)$, or $x = 6000$ mi. Comparing the smallest triangle with the largest one, we have $\dfrac{x}{3000} = \dfrac{25{,}000 + x}{d}$, or $\dfrac{6000}{3000} = \dfrac{31{,}000}{d}$. So $d = 15{,}500$ miles.

GEOMETRY AROUND US

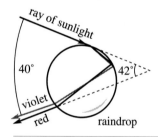

ray of sunlight

40°

42°

violet

red   raindrop

A rainbow occurs when rays of sunlight are reflected and refracted (bent) by drops of water in rain or mist. As a ray enters a raindrop, it is refracted and separated into colors. Then it is reflected and bent again as it leaves the raindrop. Each color of the spectrum is bent through a different angle, for example, 42° for red and 40° for violet.

PROBLEM SET 7.3

EXERCISES/PROBLEMS

Refer to the following figure for Exercises 1–6.

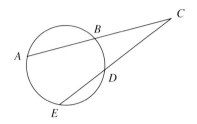

1. If $\widehat{AE} = 90°$ and $\widehat{BD} = 38°$, find the measure of $\angle C$.

2. If $\angle C = 25°$ and $\widehat{BD} = 34°$, find the measure of $\widehat{AE}$.

3. If $AC = 10$ cm, $AB = 4$ cm, and $DE = 3$ cm, find EC.

4. If $AC = 10$ cm, $AB = 4$ cm, and $DC = 3$ cm, find EC.

5. If $DE = 6$ in., $DC = 9$ in., and $\overline{AB}$ is 8 in. shorter than $\overline{BC}$, find AB.

6. If the length of $\overline{DC}$ is 2 cm more than the length of $\overline{DE}$, $AB = 3$ cm, and $BC = 7$ cm, find DC.

Refer to the following figure for Exercises 7–9 where $\overleftrightarrow{ST}$ is tangent to the circle at P.

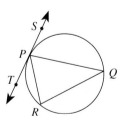

7. If $\widehat{PRQ} = 230°$, find the measure of $\angle SPQ$.

8. If $\angle PQR = 24°$, find the measure of $\angle TPR$.

9. If $\angle TPR = 28°$ and $\angle RPQ = 44°$, find $\widehat{PQ}$.

10. If $\angle PRQ = 70°$ and $\widehat{RQ} = 60°$, find the measure of $\angle TPR$.

Refer to the following figure for Exercises 11 and 12, where $\overline{UP}$ is tangent to the circle at P.

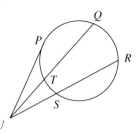

11. If $\angle SUP = 24°$ and $\widehat{PTS} = 40°$, find the measure of $\widehat{PQR}$.

12. If $\widehat{PQ} = 55°$ and $\widehat{PT} = 31°$, find the measure of $\angle TUP$.

Refer to the following figure for Exercises 13–18. Point O is the center of the circle.

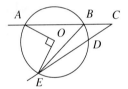

13. What is the measure of $\widehat{AE}$?

14. What is the measure of $\angle ABE$?

15. If $\angle OED = \angle ABE$ and BE bisects $\angle OED$, what is the measure of $\widehat{BD}$?

16. Using your results in Exercises 14 and 15, what is the measure of $\angle C$?

17. If $AB = 10$, $BC = 6$, and $CD = 7$, what is DE?

18. If $AB = 12$, $AC = 20$, and $CE = 18$, what are DE and CD?

Refer to the following figure for Exercises 19–24, where l and m are tangent to circle O and A and B are points of tangency.

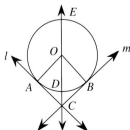

19. What is the measure of $\angle CBO$?

20. What is the measure of $\angle CAO$?

21. Why is $\triangle ACO \cong \triangle BCO$?

22. If $CD = 2$ and $BC = 6$, what is DE?

23. If $\angle AOD = 30°$, what is the measure of $\widehat{AE}$?

24. Find the measure of $\widehat{AED}$ in two different ways.

25. In the following figure, $\overline{AB}$ is tangent to the circle with center O. If $OA = 4$ cm and $OB = 12$ cm, find AB.

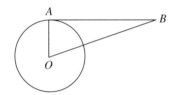

26. $\overline{XY}$ is tangent to the circle O. If $OZ = ZX$ and $XY = 6$ in., find the radius of the circle.

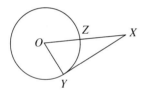

27. $\triangle ABC$ is isosceles with the inscribed circle as shown. $AB = BC = 12$ ft and $AC = 7$ ft. Find AP and BQ.

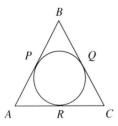

28. Given the following $\triangle ABC$ and its inscribed circle, find BX and AZ.

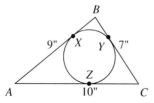

29. If $\angle A = 48°$, $\angle B = 72°$, $\angle C = 60°$, and the measures of the arcs are as indicated in the figure below, find the measures of $\widehat{EF}$, $\widehat{GH}$, and $\widehat{DI}$.

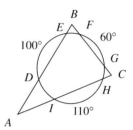

30. In each of the following two figures, the two circles have the same center and the sides of the square are tangent to the smaller circle.
   (a) If the radius of the larger circle is 1, find the area of the shaded region.

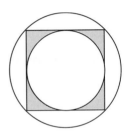

   (b) If the radius of the smaller circle is 1, find the area of the shaded region.

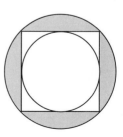

PROOFS

31. Prove Corollary 7.15: If two tangents are drawn to a circle from the same point in the exterior of the circle, the distances from the common point to the points of tangency are equal.

32. What figure would be formed by connecting the midpoints of all chords of a circle of a given length? Justify your answer.

33. Prove that a line perpendicular to a radius at the point where the radius intersects the circle is a tangent line.

34. Prove that if a tangent line and a secant are parallel, then the arcs they intercept are congruent.

35. Prove that if two secants of a circle are parallel, then the arcs between them are congruent.

36. Given are two circles with centers O and O' and two common tangents from P.

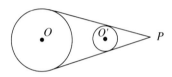

Prove that O, O', and P are collinear.

37. Two circles are **internally tangent** if they intersect in one point and one circle is in the interior of the other as shown next.

Prove that if two circles are internally tangent, then their centers and their point of intersection are collinear.

38. Two circles are **externally tangent** if they intersect in one point and each circle is in the exterior of the other as shown next.

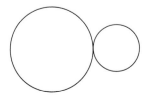

Prove that if two circles are externally tangent, then their centers and their point of intersection are collinear.

APPLICATIONS

39. In the corner of a fenced yard is a small triangular storage area closed off from the rest of the yard by a gate as shown. If the dimensions of the storage area are as shown, what is the diameter of the largest circular trash can that could be stored inside?

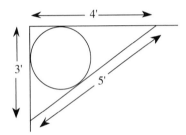

40. Suppose that a person is standing at an altitude h above sea level, as shown in the figure. The distance d the person can see is the distance to the horizon and can be calculated using the radius r of the earth and a property of circles.

(a) Express the distance, d, in terms of h and r.

(b) Use 3960 miles as the radius of the earth, and determine the distance to the horizon for an observer at an altitude of 590 feet. Round your answer to the nearest mile.

(c) Use 3960 miles as the radius of the earth, and determine the distance to the horizon for a pilot flying at 12,000 feet. Round your answer to the nearest mile.

7.4
CONSTRUCTIONS INVOLVING CIRCLES

Applied Problem

A woodworker needs to make a circular cut at a corner as shown. The arc should be tangent to each edge of the board. How can she draw the arc?

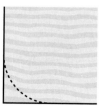

The basic constructions presented in Sections 4.4 and 5.5 can also be used to perform constructions involving circles. In this section we construct particular kinds of circles and tangents.

CONSTRUCTING THE CIRCUMSCRIBED CIRCLE OF A TRIANGLE

The **circumscribed circle** of a triangle is the unique circle that contains the vertices of the triangle. Next we will show how the circumscribed circle of a triangle can be constructed with a compass and straightedge.

Construction 9.   *To construct the circumscribed circle of a triangle.*

Procedure   Consider $\triangle ABC$ [Figure 7.36(a)]. Construct the perpendicular bisectors of any two sides, say $\overline{AC}$ and $\overline{AB}$ [Figure 7.36(b)]. Using P as center and AP as radius, construct a circle. The point P, the point of intersection of the three perpendicular bisectors of the sides of the triangle, is called the **circumcenter** of the triangle.

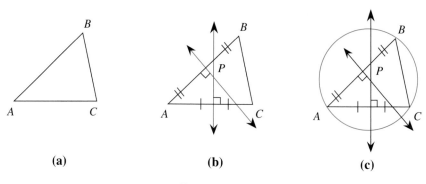

(a)   (b)   (c)

FIGURE 7.36

Justification Because P is on the perpendicular bisector of $\overline{AC}$, by Theorem 4.10, it is equidistant from A and C, so $AP = CP$. Similarly, P is equidistant from A and B, hence $AP = BP$. Because $AP = CP$ and $AP = BP$, we can conclude that $CP = BP$. This says that P is equidistant from C and B. Thus, we can conclude that P is on the perpendicular bisector of $\overline{BC}$. Because $AP = BP = CP$, the circumscribed circle with center P and radius AP can be constructed around $\triangle ABC$ [Figure 7.36(c)].

This justification leads to the following theorem.

Theorem 7.17

Circumcenter of a Triangle

The perpendicular bisectors of a triangle intersect in a single point, the circumcenter.

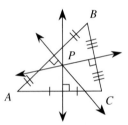

P is the circumcenter of $\triangle ABC$.

CONSTRUCTING THE ORTHOCENTER OF A TRIANGLE

The altitudes of a triangle also intersect in a common point called the **orthocenter** of the triangle. Although this result may not seem surprising, the proof of this fact follows from the previous discussion!

Construction 10. *To construct the orthocenter of a triangle.*

Procedure Consider $\triangle ABC$ [Figure 7.37(a)]. Construct any two altitudes in $\triangle ABC$. They intersect in P, the orthocenter of the triangle [Figure 7.37(b)].

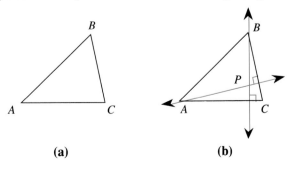

(a)   **(b)**

FIGURE 7.37

Justification Using the triangle in Figure 7.37(b), draw the lines through the vertices, making each one parallel to the opposite side [Figure 7.37(c)].

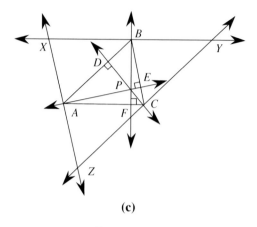

(c)

FIGURE 7.37

Now we'll prove that $\overline{BF}$, $\overline{AE}$, and $\overline{CD}$ are perpendicular bisectors of $\overline{XY}$, $\overline{XZ}$, and $\overline{YZ}$. Because $\overline{AC} \parallel \overline{XY}$ and $\overline{BF} \perp \overline{AC}$, we can conclude that (1) $\overline{BF} \perp \overline{XY}$. Also, because the opposite sides are parallel, $XBCA$ and $BYCA$ are both parallelograms. Because opposite sides of parallelograms are congruent, we have $XB = CA$ and $BY = CA$, or that (2) $XB = BY$. From (1) and (2) it follows that $\overline{BF}$ is the perpendicular bisector of $\overline{XY}$ (keep in mind that $\overline{BF}$ is also the altitude from vertex B). Similarly, we can show that $\overline{AE}$ is the perpendicular bisector of $\overline{XZ}$ and $\overline{CD}$ is the perpendicular bisector of $\overline{YZ}$. Thus, by Construction 9, $\overline{BF}$, $\overline{AE}$, and $\overline{CD}$ intersect in the point P. Because these are also the altitudes of $\triangle ABC$, the altitudes are concurrent. •

This justification provides a proof of the next theorem.

Theorem 7.18

Orthocenter of a Triangle

The altitudes of a triangle intersect in a single point, the orthocenter.

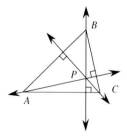

P is the orthocenter of $\triangle ABC$.

The orthocenter does not serve as a center of a circle related to the original triangle. However, if P is the orthocenter of $\triangle ABC$, the points P, A, B, and C form an **orthocentric set**, which has the property that the triangle formed by three of the four points in the set has the fourth point as its orthocenter.

CONSTRUCTING THE INSCRIBED CIRCLE OF A TRIANGLE

Every triangle contains a unique circle, called the **inscribed circle**, that is tangent to all three sides of the triangle. The center of the inscribed circle, the **incenter**, is the intersection of the angle bisectors of the triangle.

Construction 11.   *To construct the inscribed circle of a triangle.*

Procedure Consider $\triangle ABC$ [Figure 7.38(a)]. Construct any two angle bisectors of $\triangle ABC$. They meet in the incenter, that is, the point that is equidistant from the sides of the triangle [Figure 7.38(b)].

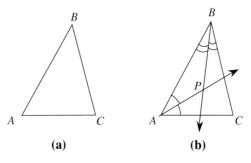

(a)   **(b)**

FIGURE 7.38

Next, using Construction 3 in Section 4.4, construct a perpendicular from P to $\overline{AC}$; call the point of intersection D [Figure 7.38(c)]. Using PD as a radius, construct the inscribed circle [Figure 7.38(d)].

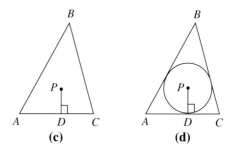

(c)   **(d)**

FIGURE 7.38

Justification This justification, which is analogous to the one for constructing the circumcenter, is left for the problem set. First show that the angle bisectors intersect in a common point.   •

The next result follows from this justification.

Theorem 7.19

Incenter of a Triangle

The angle bisectors of a triangle intersect in a single point, the incenter.

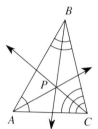

P is the incenter of $\triangle ABC$.

CONSTRUCTING THE CENTROID OF A TRIANGLE

Interestingly, the medians of any triangle are also concurrent; their point of intersection is called the **centroid** of the triangle. The centroid is the **center of gravity**, or balance point, of a triangle.

Construction 12.   *To construct the centroid of a triangle.*

Procedure Consider triangle $\triangle ABC$ [Figure 7.39(a)]. Construct the midpoints of any two sides of $\triangle ABC$ and draw the medians [Figure 7.39(b)]. Point P is the centroid of $\triangle ABC$.

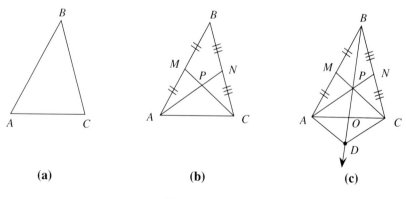

FIGURE 7.39

Justification We will show that the medians are concurrent. Consider $\triangle ABC$ with medians from A and C [Figure 7.39(b)]. Draw $\overrightarrow{BP}$ where O is the point where $\overrightarrow{BP}$ crosses $\overline{AC}$ [Figure 7.39(c)]. Let D be the point on $\overrightarrow{BP}$ where P is the midpoint of $\overline{BD}$. Next we'll show that $APCD$ is a parallelogram. Because the diagonals of a parallelogram bisect each other, this will show that 0 is the midpoint of $\overline{AC}$, thus $\overline{BO}$ is the third median through P. Because $\overline{MP}$ joins the midpoints of two sides of $\triangle ABD$, $\overline{MP}$ is parallel to $\overline{AD}$. Also, $\overline{PC}$ is parallel to $\overline{AD}$ since $\overline{PC}$ is an extension of $\overline{MP}$. Similarly, $\overline{NP}$ is parallel to $\overline{DC}$ and hence $\overline{AP}$ is parallel to $\overline{DC}$. Thus, $APCD$ is a parallelogram since opposite sides are parallel.   ●

The preceding justification can also be used to show that the point P divides each median into segments whose lengths are in a ratio of 2:1. The proof of that fact is left for the problem set. The preceding discussion is summarized in the next theorem.

Theorem 7.20

Centroid of a Triangle

The medians of a triangle intersect in a single point, the centroid, which is two thirds of the way from any vertex to the other endpoint of the median from that vertex.

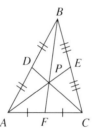

P is the centroid of $\triangle ABC$.
$BP = \frac{2}{3}(BF)$, $AP = \frac{2}{3}(AE)$, and $CP = \frac{2}{3}(CD)$.

The following table summarizes the four centers we have constructed for a triangle.

Name	Point of Intersection	Significance
Circumcenter	Perpendicular bisectors	Center of circumscribed circle
Orthocenter	Altitudes	Forms orthocentric set (together with vertices)
Incenter	Angle bisectors	Center of inscribed circle
Centroid	Medians	Center of gravity

CONSTRUCTING TANGENTS TO CIRCLES

Next we will look at constructions involving tangents to a circle. Theorem 7.11 provides a method for constructing tangents to a given circle.

Construction 13. *To construct a tangent to a circle at a point on the circle.*

Procedure To construct a tangent line at point P of circle O, first draw $\overrightarrow{OP}$ [Figure 7.40(a)]. Then, at P, construct a line, l, perpendicular to $\overrightarrow{OP}$ using Construction 4 in Section 3.4 [Figure 7.40(b)].

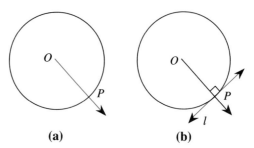

(a)   **(b)**

FIGURE 7.40

Justification By Theorem 7.11, l is tangent to the circle at P.   •

The construction of a tangent line to a circle from a point outside the circle is demonstrated next.

Construction 14. *To construct a tangent to a circle from a point outside the circle.*

Procedure Consider a circle with center O having P outside the circle. Draw $\overline{OP}$ [Figure 7.41(a)].

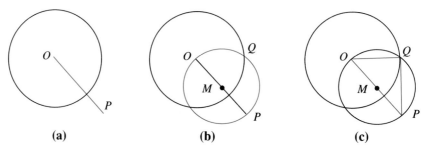

(a)   **(b)**   **(c)**

FIGURE 7.41

Construct the midpoint M of $\overline{OP}$, and construct the circle with center M and radius $MP = MO$ [Figure 7.41(b)]. Let Q be one of the points of intersection of the two circles [Figure 7.41(b)]. Draw $\overline{PQ}$ [Figure 7.41(c)]. $\overline{PQ}$ will be tangent to circle O.

Justification Because $\triangle QOP$ is inscribed in a semicircle, $\angle OQP$ is a right angle, so $\overleftrightarrow{QP}$ is tangent to circle O at Q.   •

If two circles do not intersect and have no parts of their interiors in common, there are two types of tangents that can be constructed. When the circles are on opposite sides of a common tangent line, the line is called a **common internal tangent** [Figure 7.42(a)]. When the circles are on the same side of the tangent line, the line is called a **common external tangent** [Figure 7.42(b)].

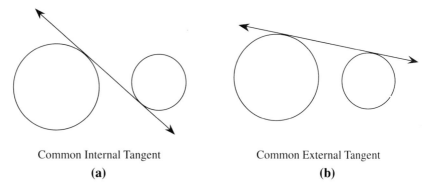

Common Internal Tangent
(a)

Common External Tangent
(b)

FIGURE 7.42

The following discussion shows how to construct these two types of tangent lines.

Construction 15. *To construct a common internal tangent to two circles.*

Procedure Consider circles O and P [Figure 7.43(a)].

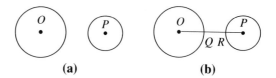

(a)   **(b)**

FIGURE 7.43

STEP 1. Draw $\overline{OP}$, giving points of intersection Q and R [Figure 7.43(b)].

STEP 2. Using O as center, construct a circle whose radius is $OQ + PR$ [Figure 7.43(c)].

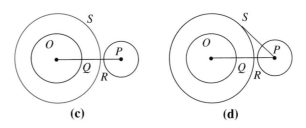

(c)   **(d)**

FIGURE 7.43

STEP 3. Construct a tangent line from P to the new circle at S using Construction 14 [Figure 7.43(d)].

STEP 4. Draw $\overline{OS}$, calling T the point where $\overline{OS}$ intersects the original circle with center O [Figure 7.43(e)].

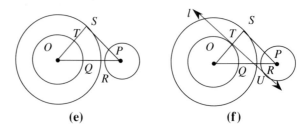

(e)   (f)

FIGURE 7.43

STEP 5. Construct a line, l, through T, parallel to $\overline{SP}$ [Figure 7.43(f)]. Line l is the common internal tangent to the original two circles.

Justification Because $l \parallel \overline{SP}$ and $\overline{SP} \perp \overline{OS}$, we have $l \perp \overline{OT}$ at T. Hence, l is tangent to the original circle at T. Let $\overline{PU} \perp l$. Then $PSTU$ is a rectangle because it has four right angles. Therefore, $PU = ST = PR$ is a radius perpendicular to l at U. Therefore, l is tangent to circle P at U. •

Construction 16.   *To construct a common external tangent to two circles.*

Procedure Consider circles O and P [Figure 7.44(a)]

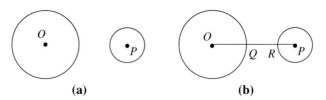

(a)   (b)

FIGURE 7.44

STEP 1. Draw $\overline{OP}$, yielding points Q and R [Figure 7.44(b)].

STEP 2. Using O as center, construct a circle of radius $OQ - PR$ [Figure 7.44(c)].

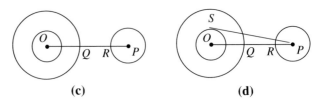

(c)   (d)

FIGURE 7.44

STEP 3. Construct a tangent line from P to the new circle at S using Construction 14 [Figure 7.44(d)].

STEP 4. Construct perpendiculars to $\overline{SP}$ at S and P forming points T and U [Figure 7.44(e)].

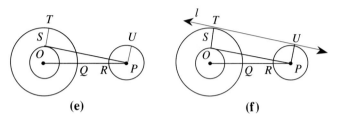

(e)   (f)

FIGURE 7.44

STEP 5. Draw line l through T and U [Figure 7.44(f)]. Then line l is the common external tangent to the original two circles.

Justification By construction, $TS = UP$ and $\overline{TS} \parallel \overline{UP}$ because they are both perpendicular to $\overline{SP}$. Therefore, $PSTU$ is a parallelogram. In addition, $PSTU$ contains a right angle, hence it is a rectangle. Thus, l is a common tangent since it is perpendicular to radii $\overline{OT}$ and $\overline{PU}$.   •

Solution to Applied Problem

First, a desired radius for the circular arc must be specified. Using that radius, an arc can be made with the vertex at the corner C giving points of intersection A and B. A square can be constructed with A, B, and C as vertices. The fourth vertex, D, will be the center of the circle containing the desired arc. The arc can now be drawn using the same radius.

Because $ADBC$ is a square, $AD \perp AC$ and $BD \perp CB$. Thus, the arc is tangent to the edges of the board, as required.

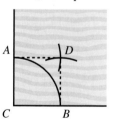

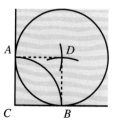

GEOMETRY AROUND US

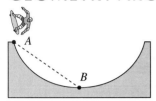

The cross section of an optimal skateboard ramp is the shape of an inverted cycloid called a **brachistochrone**. It is the curve of fastest descent. Although it might seem that a straight line ramp would be the fastest path from *A* to *B*, the inverted cycloid allows a skater to get from *A* to *B* in the shortest time of any curve.

PROBLEM SET 7.4

EXERCISES/PROBLEMS

For Exercises 1–3, trace each triangle and construct its circumscribed circle.

1.

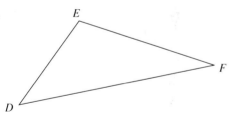

2.

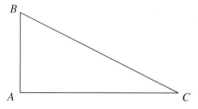

3.

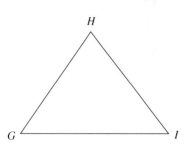

4. What did you notice about the circumcenter in Exercises 1–3? Was it always inside the triangle? Try a few more examples to test your conjecture.

For Exercises 5–7, trace each triangle and construct its inscribed circle.

5.

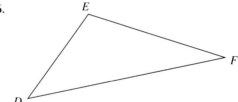

6.

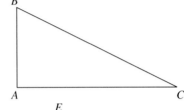

7.

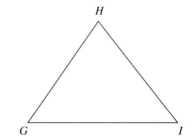

8. Why will the incenter of a triangle always be inside the triangle? Explain.

9. Find the center of this circle.

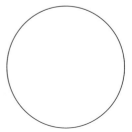

10. Find the center of the circle to which this arc belongs.

11. Trace △*ABC* and find its centroid using your compass and straightedge. By measuring, verify that the centroid is two thirds of the way from each vertex to the midpoint of the opposite side.

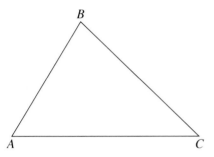

12. Trace △*ABC* and find its centroid using your compass and straightedge. By measuring, verify that the centroid is two thirds of the way from each vertex to the midpoint of the opposite side.

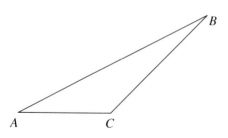

13. Trace △*PQR* and locate its orthocenter using your compass and straightedge.

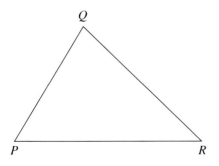

14. Trace △*PQR* and locate its orthocenter using your compass and straightedge.

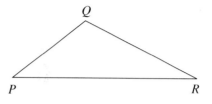

15. Construct a tangent line to circle *O* at *P*.

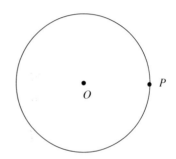

16. Construct a tangent line from *P* to the circle *O*.

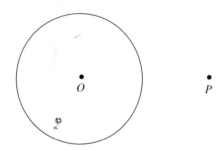

17. Using a theorem (or theorems) from this chapter, construct an angle at *D* such that ∠*D* = ∠*A*.

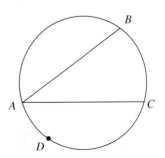

18. Using a theorem (or theorems) from this chapter, construct an angle at D such that $\angle D = 2\angle A$.

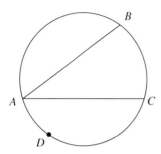

19. Construct a common internal tangent to circles O and P.

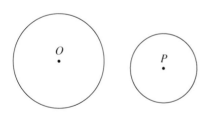

20. Construct a common external tangent to circles O and P.

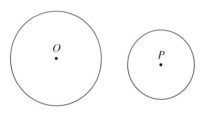

21. Draw a circle and construct a square inscribed in your circle. [HINT: Recall that an angle inscribed in a semicircle is a right angle.]

22. Draw a circle and construct a square circumscribed around your circle. [HINT: Recall that a radius is perpendicular to the tangent line at the point of tangency.]

23. Draw a circle and construct an octagon circumscribed about your circle.

24. Construct a central angle of 90° whose legs are two radii of a circle. Construct a tangent line at each of the endpoints of the radii where they intersect the circle. Give the most complete description possible of the quadrilateral formed by the two radii and the portions of the tangent lines contained between the circle and their point of intersection.

25. Construct a central angle of 60° whose legs are two radii of a circle. Construct the tangents to the circle where the legs intersect the circle. Find the measure of the angle formed by the two tangents in two different ways.

26. The **Euler line** of a triangle is the line that contains the circumcenter, orthocenter, and centroid of the triangle. Draw a triangle. Show, by construction, that these three points are collinear, and draw its Euler line.

27. A theorem proved by Karl Feuerbach states that for any triangle, one circle contains the following nine points: The three midpoints of the sides, the three feet of the altitudes, and the midpoints of the three segments joining the vertices of the triangle to its orthocenter. Draw a triangle, construct the nine points described here, and find a circle that fits the nine points.

28. An interesting result attributed to Napolean Bonaparte can be observed in the following construction.
  (1) Draw any triangle PQR.
  (2) Using compass and straightedge, construct an equilateral triangle on each side of $\triangle PQR$.
  (3) Use your compass and straightedge to locate the centroid of each equilateral triangle: C_1, C_2, C_3.
  (4) Connect C_1, C_2, and C_3 to form a triangle.

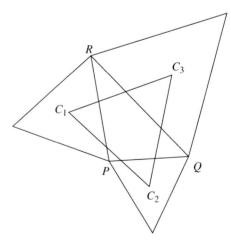

  (a) Draw a triangle and perform the steps described above. What kind of triangle is $\triangle C_1 C_2 C_3$? Use your ruler and protractor to check.

(b) Repeat the construction for two different types of triangles *PQR*. What kind of triangle is $\triangle C_1C_2C_3$?

(c) Summarize your observations.

PROOFS

29. Prove Theorem 7.19: The angle bisectors of a triangle intersect in a single point, the incenter.

30. Prove Theorem 7.17: The perpendicular bisectors of a triangle intersect in a single point, the circumcenter.

31. Justify Construction 11: To construct the inscribed circle of a triangle.

32. Prove Theorem 7.20: The medians of a triangle intersect in a single point, the centroid, which is two thirds of the way from any vertex to the other endpoint of the median from that vertex.

33. To inscribe a regular hexagon in a circle, set the opening of the compass to equal the radius. Starting at any point on the circle, mark arcs around the circle by placing the point of the compass on the arc just made.

The points of intersection will be vertices of an inscribed hexagon. Justify this construction.

34. (a) Construct two circles with the same center but with different radii. Choose any point *P* on the inner circle. Construct a tangent to the inner circle at point *P*. Label the points of intersection with the outer circle *Q* and *R*. Construct a third circle with center *P* and diameter $\overline{QR}$.

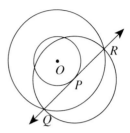

(b) Using a ruler, measure appropriate dimensions and calculate the area of circle *P*. Calculate the area of the region between the two circles centered at *O*.

(c) Repeat your construction and calculations from parts (a) and (b) for another pair of concentric circles.

(d) What conjecture can you make about the area of circle *P* and the area between the two circles centered at *O*?

(e) Prove your conjecture.

35. Construct the tangent lines at the endpoints of a diameter of a circle. Prove that the two tangent lines are parallel.

36. Prove that the orthocenter of a triangle forms an orthocentric set with the vertices of the triangle.

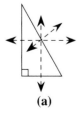

(a)

FIGURE 7.45

Solution to Initial Problem

The circumscribed circle of a triangle is the unique circle that contains all three of the vertices of the triangle. For a given triangle, develop a simple test to determine if the center of a circumscribed circle is inside, on, or outside the triangle. (The center is the intersection of the perpendicular bisectors of the sides of the triangle.)

Case 1: Right triangle [Figure 7.45(a)]. In this case, the midpoint of the hypotenuse is the center of the circumscribed circle; hence, it is on the triangle.

(b)

(c)

FIGURE 7.45

Case 2: Acute triangle [Figure 7.45(b)] In this case, the center is inside the triangle.

Case 3: Obtuse triangle [Figure 7.45(c)] Here the center is outside the triangle, opposite the obtuse angle.

Hence, the center is on, inside, or outside a triangle, depending on whether the triangle is right, acute, or obtuse respectively.

Additional Problems Where the Strategy "Use Cases" Is Useful

1. Determine whether the centroid of a triangle will always lie inside the triangle.

2. The measures of the angles of a triangle are consecutive integers. Show that the measures of two of the angles must be odd.

Writing for Understanding

1. The many angles involving circles and lines that were discussed in this chapter can be confusing to a beginning geometry student. Describe how the types of angles might be classified and their measures determined.

2. Choose one construction involving a circle from this chapter. Carefully describe the steps involved in that construction and discuss where such a construction might be useful.

3. The Wankel rotary engine, which is found in the Mazda RX-7, was mentioned in "Geometry Around Us" at the end of Section 7.2. Do some research on how circles and other curves of constant width, such as Reuleaux triangles, are important in the design of car engines. Summarize your findings.

4. One use of a brachistochrone is described in the "Geometry Around Us" at the end of Section 7.4. Do some research on this curve, focussing on its applications and the history of the "brachistochrone problem." Summarize your findings.

CHAPTER REVIEW

Following is a list of key vocabulary, notation, and ideas for this chapter. Mentally review these items and, where appropriate, write down the meaning of each term. Then restudy the material that you are unsure of before proceeding to take the chapter test.

Section 7.1—Central Angles and Inscribed Angles

Vocabulary/Notation

Arc 319
Endpoints of an arc 319
Semicircle 319

Minor arc 319
Major arc 320
Central angle 320

Main Ideas/Results

1. The measure of an arc of a circle is equal to the measure of its central angle.
2. The length of an arc is determined by its central angle and the radius of its circle.
3. The area of a sector is determined by its central angle and the radius of its circle.
4. The measure of an angle inscribed in a circle is equal to half the measure of its intercepted arc.

Section 7.2—Chords of a Circle

Vocabulary/Notation

Main Ideas/Results

1. The perpendicular bisector of a chord contains the center of its circle.
2. The measures of the angles formed by two intersecting chords are determined by their intercepted arcs.
3. If two chords of a circle intersect, the product of the lengths of the segments of one chord equals the product of the corresponding lengths of the segments of the other chord.

Section 7.3—Secants and Tangents

Vocabulary/Notation

Main Ideas/Results

1. The measures of angles formed when two secants intersect is determined by their intercepted arcs.
2. If two secants intersect outside a circle, then the product of the lengths of segments formed on one secant equals the product of the lengths of the corresponding segments of the other secant.
3. A radius (diameter) of a circle is perpendicular to a tangent line at its point of tangency.
4. The measure of an angle formed when a chord intercepts a tangent line is determined by the arc intercepted.
5. If a secant and a tangent line intersect outside a circle, the measure of the angle formed is determined by the arcs intercepted.

6. If two tangent lines intersect outside a circle, the measure of the angle formed is determined by the arcs intercepted.

7. If a tangent and a secant are drawn to a circle from the same point in the exterior of the circle, the length of the tangent segment is the mean proportional between the lengths of the two secant segments.

Section 7.4—Constructions Involving Circles

Vocabulary/Notation

Main Ideas/Results

1. The following centers can be constructed:
   a. Circumcenter
   b. Orthocenter
   c. Incenter
   d. Centroid
2. Tangents can be constructed to circles from points outside the circle.

PEOPLE IN GEOMETRY

Carl Friedrich Gauss (1777–1855), according to the historian E. T. Bell, "lives everywhere in mathematics." His contributions to geometry, number theory, and analysis were deep and wide-ranging. Yet he also made crucial contributions in applied mathematics. When the tiny planet Ceres was discovered in 1800, Gauss developed a technique for calculating its orbit, based on meager observations of its direction from earth at several known times. Gauss contributed to the modern theory of electricity and magnetism, and with the physicist W. E. Weber constructed one of the first practical electric telegraphs. In 1807 he became director of the astronomical observatory at Göttingen, where he served until his death. At age 18, Carl devised a method for constructing a 17-sided regular polygon, using only compass and straightedge. Remarkably, he then derived a general rule which predicted which regular polygons are likewise constructible.

CHAPTER 7 TEST

TRUE-FALSE

Mark as true any statement that is always true. Mark as false any statement that is never true or that is not necessarily true. Be able to justify your answers.

1. If a right triangle is inscribed in a circle, then the hypotenuse of the triangle must be a diameter of the circle.

2. The incenter of an obtuse triangle lies outside the triangle.

3. The circumcenter of an obtuse triangle lies outside the triangle.

4. If the radius of a circle is 2 in., then the longest possible chord has length 6 in.

5. The measure of an inscribed angle in a circle equals the measure of its intercepted arc.

6. If two central angles of a circle are congruent, then the arcs they intercept are congruent.

7. If the area of a sector of a circle is 2π square inches and the radius of the circle is 4 inches, then the central angle of the sector is 45°.

8. If $\overleftrightarrow{AB}$ is tangent to circle O at point P, then $\angle APO = 90°$.

9. If circle O_1 and circle O_2 intersect in points A and B, then O_1O_2 is the perpendicular bisector of $\overline{AB}$.

10. If chord $\overline{AB}$ of circle O measures 8 cm and chord $\overline{PQ}$ of circle O measures 5 cm, then the measure of $\angle AOB$ is less than the measure of $\angle POQ$.

EXERCISES/PROBLEMS

11. Using the figure shown, give one example of each of the following. [NOTE: O is the center of the circle.]

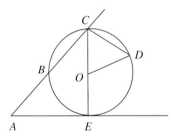

   (a) A secant line
   (b) An inscribed angle
   (c) A major arc
   (d) A right angle
   (e) A chord

For problems 12–14, refer to the following figure, where $\overset{\frown}{AC} = 126°$, $\angle BAC = 35°$, and $\overset{\frown}{AD} = 140°$.

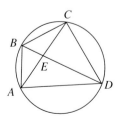

12. Find the measure of $\overset{\frown}{AB}$.

13. Find the measure of $\angle CAD$.

14. Find the measure of $\angle BEC$.

15. (a) Imagine the largest square plug that fits into a circular hole. How well does the plug fit? That is, what percentage of the circular hole does the square plug occupy?

   (b) Now imagine the largest circular plug that fits into a square hole. How well does this plug fit? That is, what percentage of the square hole does the circular plug occupy?

   (c) Which of the two plugs described in (a) and (b) is a better fit?

16. For the figure shown, determine the measures of each of the following where $\overset{\frown}{CE} = 14°$ and $\angle ADB = 8°$.
   (a) $\overset{\frown}{AB}$ (b) $\angle EDF$

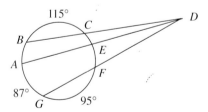

17. In the following figure, $PT = 12$ ft and $TR = 5$ ft. If $QS = 19$ ft, find TS and QT.

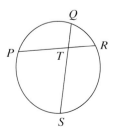

18. In the figure shown $\overleftrightarrow{AB}$ and $\overleftrightarrow{BC}$ are tangent to circle O, and A and C are the points of tangency. For $AB = 7$ and $EB = 3$, find DE.

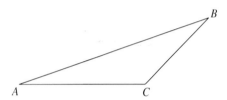

19. Trace $\triangle ABC$ and use your compass and straightedge to construct its inscribed circle.

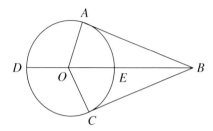

20. Use your compass and straightedge to construct a tangent line to circle O from point P.

• P

PROOFS

21. A line drawn through the midpoint of an arc and the center of its circle is perpendicular to the chord determined by the arc.

22. In the following figure, O is the center of the circle. The measure of $\angle OBC$ is twice the measure of $\angle ODC$.

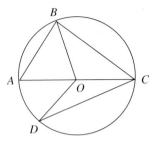

(a) Prove that the measure of $\angle AOB$ is four times the measure of $\angle OCD$.
(b) Prove that $\angle AOD$ and $\angle ABO$ are complementary.

APPLICATIONS

23. The maximum safe speed on a curve is related to the curvature of the road. Suppose that a road has a curve as shown, where the curve is an arc of a circle. The **radius of curvature** is defined as the radius of a circle that would contain the arc. Use the dimensions shown to find the radius of curvature r. Round your answers to the nearest foot.

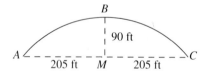

24. A baseball diamond is located on a field which is a sector of a circle as shown. To calculate the amount of fertilizer to be used on the grass, the area must be determined. Use the dimensions shown to find the area, to the nearest hundred square feet.

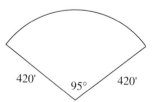

8

Coordinate Geometry

The subjects of algebra and geometry evolved on parallel tracks until the French philosopher and mathematician René Descartes (1596–1650) developed a method of linking them. This integration of algebra and geometry made possible the development of the calculus. Because of his important contribution to the evolution of mathematics, Descartes has been called the father of modern mathematics. The coordinate system used in analytic geometry is called the Cartesian coordinate system in his honor. The following timeline identifies some of the significant geometric contributions up to the time of Descartes.

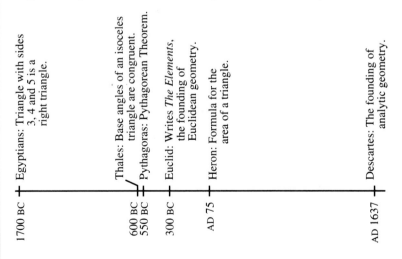

STRATEGY 13: USE COORDINATES

In many two-dimensional geometry problems, a problem may be simplified by replacing a figure with an equation. To accomplish this, a coordinate system of some type is introduced. Coordinate systems are also useful in three dimensions and on curved surfaces, such as a sphere.

INITIAL PROBLEM

A treasure has been hidden in a desert location. A map shows $\triangle ABC$, where B is 50 miles NE of A and C is 80 miles east of A. You are told that the treasure is at the midpoint of the median from A (Figure 8.1). Locate the treasure.

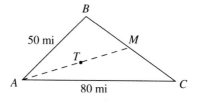

FIGURE 8.1

CLUES

The Use Coordinates strategy may be appropriate when

· A problem can be represented using two variables.
· A geometry problem cannot easily be solved by traditional Euclidean methods.
· Lines, circles, or other curves must be represented algebraically.
· A problem involves slope, parallel lines, perpendicular lines, and so on.
· The location of a geometric shape with respect to other shapes is important.
· A problem involves maps.

A solution for the initial problem above is on page 413.

INTRODUCTION

In this chapter we study geometry using the coordinate plane. By imposing a coordinate system on the plane, we are able to prove many useful results about lines, polygons, circles, and so on. In Section 8.1 we initiate the study of coordinates and the concept of distance. Then slope is used in Section 8.2 to derive results about parallel and perpendicular lines. In Section 8.3 we apply the study of coordinates to lines and circles. Finally, in Section 8.4 we prove many interesting geometric results from a coordinate point of view.

8.1
COORDINATES AND DISTANCE IN THE PLANE

Applied Problem

A lot has corners with coordinates as shown, where measurements are given in feet. Find the perimeter of the lot.

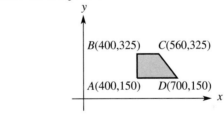

COORDINATES

Suppose that we choose two perpendicular real number lines x and y in the plane and use their intersection point as a reference point, O, called the **origin** [Figure 8.2(a)]. The horizontal line is called the **x-axis** and the vertical line is called the

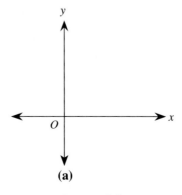

(a)

FIGURE 8.2

y-axis (plural of axis is **axes**). The plane with an (*x*, *y*)-coordinate system on it is called a **coordinate plane**. Each point *P* on the coordinate plane is identified by an **ordered pair** (*a*, *b*). The ordered pair (*a*, *b*) is also denoted *P*(*a*, *b*) [Figure 8.2(b)].

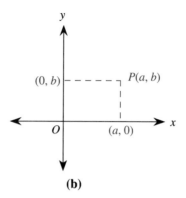

(b)

FIGURE 8.2

In the ordered pair (*a*, *b*) pictured in Figure 8.2(b), the first number, *a*, is called the **x-coordinate** and the second number, *b*, is called the **y-coordinate**. The *x*-coordinate locates the position of a point in relation to the *y*-axis. That is, it specifies the distance of the point from the *y*-axis. Similarly, the *y*-coordinate specifies the distance of the point from the *x*-axis. The *a* is also called the **projection** of the point *P* on the *x*-axis, and *b* is the projection of *P* on the *y*-axis. In general, the projection of a point on a line is the point obtained when a perpendicular line from *P* intersects the given line.

In Figure 8.3(a), several points are graphed and their coordinates are shown.

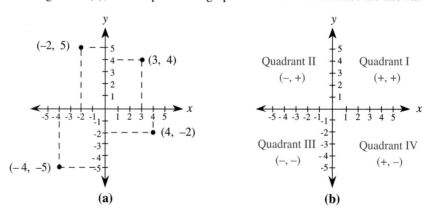

(a)   **(b)**

FIGURE 8.3

Notice how signs on the coordinates in the ordered pairs in Figure 8.3(a) are determined by the location of the points relative to the *x*- and *y*-axes. The two axes divide the plane into four quadrants. These quadrants are named in Figure 8.3(b).

EXAMPLE 8.1   Plot the points with the following coordinates: $A(-2, 4)$, $B(4, 3)$, $C(5, -2)$, and $D(-4, -3)$. Then draw the following segments: $\overline{AB}$, $\overline{BC}$, $\overline{CD}$, and $\overline{AD}$.

SOLUTION   See Figure 8.4.

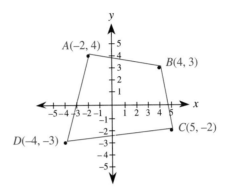

FIGURE 8.4

DISTANCE

Distances between points in the coordinate plane can be determined using the coordinates of the points and the Pythagorean Theorem.

EXAMPLE 8.2   Find the distance between $P(-2, 1)$ and $Q(4, 3)$.

SOLUTION   Consider the points as graphed in Figure 8.5(a). Draw in the line segments as shown in Figure 8.5(b) to form a right triangle.

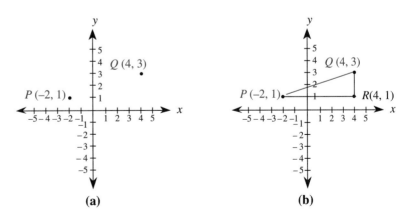

FIGURE 8.5

By the Pythagorean Theorem, we have $(PR)^2 + (QR)^2 = (PQ)^2$. Since $PR = 4 - (-2) = 6$ and $QR = 3 - 1 = 2$, we have $6^2 + 2^2 = 40 = (PQ)^2$. Therefore, $PQ = \sqrt{40}$.

The method in Example 8.2 is generalized in the following theorem whose proof is left for the problem set.

Theorem 8.1

Distance Formula

If $P(a, b)$ and $Q(c, d)$ are two points in the coordinate plane, then

$$PQ = \sqrt{(c - a)^2 + (d - b)^2}$$

Since $(c - a)^2 = (a - c)^2$ and $(d - b)^2 = (b - d)^2$, PQ also can be found using $(a - c)^2$ and $(b - d)^2$.

We can use the coordinates of the vertices of a geometric figure together with the distance formula to classify the polygon by type. The following example illustrates this process.

EXAMPLE 8.3   The vertices of a quadrilateral $ABCD$ are $A(-7, -3)$, $B(0, 3)$, $C(5, 0)$ and $D(2, -5)$. Describe this quadrilateral as specifically as possible.

SOLUTION $ABCD$ is graphed in Figure 8.6.

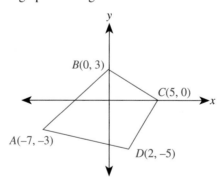

FIGURE 8.6

Using the distance formula, we can calculate the lengths of the sides of the quadrilateral.

$$AB = \sqrt{(-7 - 0)^2 + (-3 - 3)^2} = \sqrt{49 + 36} = \sqrt{85}$$
$$BC = \sqrt{(0 - 5)^2 + (3 - 0)^2} = \sqrt{25 + 9} = \sqrt{34}$$
$$CD = \sqrt{(5 - 2)^2 + [0 - (-5)]^2} = \sqrt{9 + 25} = \sqrt{34}$$
$$AD = \sqrt{(-7 - 2)^2 + [-3 - (-5)]^2} = \sqrt{81 + 4} = \sqrt{85}$$

Because $AB = AD$ and $BC = CD$, $ABCD$ is a kite.   •

TEST FOR COLLINEARITY

Recall that three points are collinear if there is a line that contains all three points. Notice that points P, Q, and R in Figure 8.7 are collinear and $PQ + QR = PR$. This suggests the following result whose proof is left for the problem set.

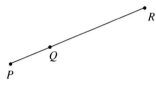

FIGURE 8.7

Theorem 8.2

Test for Collinearity

The points P, Q, and R are collinear, with Q between P and R, if and only if $PQ + QR = PR$.

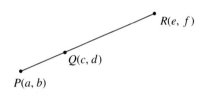

EXAMPLE 8.4   Determine whether the points $A(-2, 6)$, $B(1, 2)$, and $C(7, -6)$ are collinear (Figure 8.8).

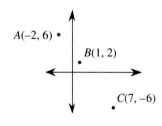

FIGURE 8.8

SOLUTION   Although the points in the figure appear to be collinear, we cannot say for certain without further investigation.

$$AB = \sqrt{[1 - (-2)]^2 + (2 - 6)^2} = \sqrt{9 + 16} = 5$$
$$AC = \sqrt{[7 - (-2)]^2 + (-6 - 6)^2} = \sqrt{81 + 144} = 15$$
$$BC = \sqrt{(7 - 1)^2 + (-6 - 2)^2} = \sqrt{36 + 64} = 10$$

Because $5 + 10 = 15$, we have $AB + BC = AC$. This shows that A, B, and C are collinear and that B is between A and C.

MIDPOINT OF A SEGMENT

Using the distance formula we can also find the midpoint of any line segment as the next example shows.

EXAMPLE 8.5   Find the midpoint of the segment whose endpoints are $A(3, 1)$ and $B(1, 5)$ [Figure 8.9(a)].

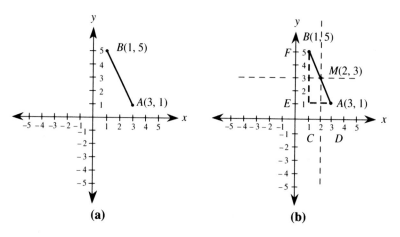

(a)   **(b)**

FIGURE 8.9

SOLUTION   Notice that in Figure 8.9(b) the vertical lines through $(1, 0)$, and $(2, 0)$ are parallel. By the Side Splitting Theorem, the line through $(2, 0)$ bisects $\overline{AB}$. Similarly, the horizontal line through the midpoint of $\overline{EF}$ bisects $\overline{AB}$. Thus, because the x-projection of the midpoint of $\overline{AB}$ is 2 and the y-projection is 3, $M(2, 3)$ is the midpoint of $\overline{AB}$.   •

This technique of finding midpoints is generalized as follows. Its proof is left for the problem set.

Theorem 8.3

Midpoint Formula

If $P(a, b)$ and $Q(c, d)$ are endpoints of a line segment, then the midpoint of $\overline{PQ}$ is $M\left(\dfrac{a + c}{2}, \dfrac{b + d}{2}\right)$.

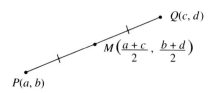

EXAMPLE 8.6   Find the coordinates of center O for the circle shown in Figure 8.10.

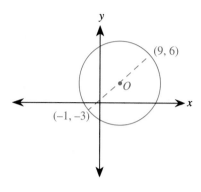

FIGURE 8.10

SOLUTION   Since the segment joining $(9, 6)$ and $(-1, -3)$ is a diameter of the circle, O is the midpoint of the segment. The coordinates of O are
$$\left(\frac{-1 + 9}{2}, \frac{-3 + 6}{2}\right) = \left(4, \frac{3}{2}\right).$$

Solution to Applied Problem

$AB = 325 - 150 = 175$ ft, $BC = 160$ ft, and $AD = 300$ ft. Using the distance formula, we find
$$CD = \sqrt{(700 - 560)^2 + (150 - 325)^2} = \sqrt{140^2 + (-175)^2},$$
which is approximately 224 feet. So the perimeter of the lot is approximately $175 + 160 + 130 + 224 = 859$ feet.

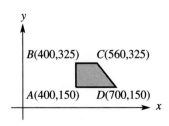

GEOMETRY AROUND US

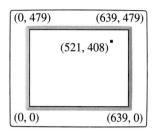

Coordinates are essential to graphics programming on any computer. Positions of pixels on the screen are identified by horizontal and vertical components, or x- and y-coordinates. One possible graphics screen is shown. It measures 640 pixels horizontally by 480 pixels vertically. Notice that the screen is oriented so that the origin is at the lower left-hand corner. The highlighted pixel is at the position described by (521, 408).

PROBLEM SET 8.1

EXERCISES/PROBLEMS

1. Plot the following sets of points on a coordinate system.
   (a) (3, 2), (−3, 2), (−3, −2), (3, −2)
   (b) (0, 4), (4, 0), (3, −5), (−2, 4)

2. Plot the following sets of points on a coordinate system.
   (a) (−3, −4), (−1, 5), (2, −4), (3, 6)
   (b) (−1, −1), (−4, −4), (3, 3), (5, 5)

3. Find the distance between the given pairs of points.
   (a) (0, 0), (3, 4)   (b) (−1, 2), (3, 5)
   (c) (−1, −3), (−5, −3)   (d) (3, −4), (3, 5)

4. Find the distance between the given pairs of points.
   (a) (0, 0), (5, 12)   (b) (3, −2), (−3, 2)
   (c) (−1, −3), (−1, 3)   (d) (−4, −4), (3, 3)

5. Use the distance formula to determine if points P, Q, and R are collinear.
   (a) $P(−1, 4)$, $Q(−2, 3)$, and $R(−4, 1)$
   (b) $P(−2, 1)$, $Q(3, 4)$, and $R(12, 10)$
   (c) $P(−2, −3)$, $Q(2, −1)$, and $R(10, 3)$

6. The length of $\overline{AB}$ is $2\sqrt{10}$, where A has coordinates $(−3, 4)$ and B has coordinates $(−1, k)$. Find the value of k.

7. The distance from $S(5, 2)$ to $T(2n, n)$ is $\sqrt{2}$ units. Find the value of n.

8. Which of the given points are collinear with $A(−2, −3)$ and $B(4, −1)$?
   (a) (1, −2) (b) (7, 1) (c) (12, 2) (d) (−8, −5)

9. The endpoints of a segment are given. Find the coordinates of the midpoint of the segment.
   (a) (0, 2) and (−3, 2)   (b) (−5, −1) and (3, 5)
   (c) (−2, 3) and (−3, 6) (d) (3, −5) and (3, 7)
   (e) (1, 5) and (3, 9)   (f) (6, −2) and (−3, 5)

10. The coordinates of two points are given. If M is the midpoint of $\overline{AB}$, find the coordinates of the third point.
    (a) $A(−2, 4)$, $B(−1, 10)$
    (b) $A(−1, −3)$, $B(5, 12)$
    (c) $A(3, −5)$, $B(3, 7)$
    (d) $A(2, 6)$, $M(4, −3)$
    (e) $A(1, −3)$, $M(5, 2)$
    (f) $M(−2, −5)$, $B(3, −4)$

11. Draw triangles that have vertices with the given coordinates. Describe each triangle as scalene, isosceles, equilateral, acute, right, and/or obtuse. Justify your answer.
    (a) (0, 0), (0, −5), (5, −5)
    (b) (−3, 1), (1, 3), (5, −5)
    (c) (−2, −1), (2, 2), (6, −1)
    (d) (−4, −2), (−1, 3), (4, −2)

12. Given are the vertices of triangles. Use the distance formula to determine if each triangle is a right triangle.
    (a) (−2, 5), (0, −1), (12, 3)
    (b) (2, 3), (−2, −3), (−6, 1)
    (c) (−3, −2), (5, −2), (1, 2)

13. A quadrilateral $ABCD$ has vertices $A(−1, 1)$, $B(2, 4)$, $C(6, 1)$, and $D(3, −2)$. Use the distance formula to verify that $ABCD$ is a parallelogram.

14. A quadrilateral $PQRS$ has vertices $P(0, −3)$, $Q(−1, 2)$, $R(4, 1)$, and $S(5, −4)$. Use the distance formula to verify that $PQRS$ is a rhombus.

15. Use the distance formula to determine whether *PQRS* is a rectangle for points $P(1, 1)$, $Q(4, 2)$, $R(2, 5)$, and $S(-1, 4)$. [HINT: Use the diagonals of *PQRS*.]

16. What general type of quadrilateral is *ABCD*, for $A(0, 0)$, $B(-4, 3)$, $C(-1, 7)$, and $D(10, 5)$. Describe it as completely as possible.

17. Two vertices of an equilateral triangle $\triangle ABC$ are $A(0, 0)$ and $B(0, 10)$. Find the coordinates of vertex C.

18. Three vertices of a parallelogram are $(0, 0)$, $(2, 5)$, and $(8, 0)$. Find the coordinates of the fourth vertex.

19. Draw quadrilateral *ABCD* whose vertices are $A(3, 0)$, $B(6, 6)$, $C(6, 9)$, and $D(0, 6)$. Divide each of the coordinates by 3 and graph the new quadrilateral $A'B'C'D'$. For example, A' has coordinates $(1, 0)$. How do the lengths of corresponding sides compare?

20. Draw $\triangle ABC$ for points $A(2, 0)$, $B(-1, 2)$, and $C(0, 0)$. Multiply each of the coordinates by 2 and graph $\triangle A'B'C'$. For example, A' has coordinates $(4, 0)$.
    (a) How do the lengths of corresponding sides compare?
    (b) How do the areas of the two triangles compare?

21. Draw quadrilateral *ABCD* for $A(4, -2)$, $B(4, 2)$, $C(-2, 2)$, and $D(-2, -2)$. Multiply each coordinate by 3 and graph the resulting quadrilateral $A'B'C'D'$. For example, A' has coordinates $(12, -6)$.
    (a) How do the perimeters of *ABCD* and $A'B'C'D'$ compare?
    (b) How do the areas of *ABCD* and $A'B'C'D'$ compare?

22. Draw quadrilateral *ABCD* for $A(-8, 0)$, $B(-6, 4)$, $C(6, 4)$, and $D(8, 0)$.
    (a) Describe *ABCD* as completely as possible.
    (b) Calculate the area of *ABCD*.
    (c) Multiply each coordinate of *ABCD* $\frac{1}{2}$ and graph the resulting quadrilateral $A'B'C'D'$. How does the area of $A'B'C'D'$ compare to the area of *ABCD*?

23. Cartesian coordinates may be generalized to three-dimensional space. A point in space is determined by giving its location relative to three axes as shown. Point *P* with coordinates (a, b, c) is plotted by going a units along the *x*-axis, b units along the *y*-axis, and c units along the *z*-axis as shown. Plot the following points in three-dimensional space.

(a) $(2, 1, 3)$ (b) $(-2, 1, 0)$ (c) $(3, -1, -2)$

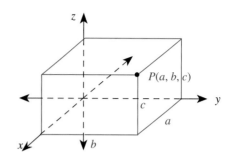

24. The three coordinate axes taken in pairs determine three coordinate planes:
    The (horizontal) *xy*-plane, where $z = 0$.
    The (vertical) *yz*-plane, where $x = 0$.
    The (vertical) *xz*-plane, where $y = 0$.

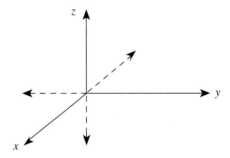

These three planes divide space into eight regions, called **octants**. Looking above the *xy*-plane, the front right octant is octant 1. Octants 2, 3, and 4 are found by going counterclockwise through the other upper-level octants. The octant directly below octant 1 is octant 5, 6 is below 2, 7 is below 3, and 8 is below 4. Given the following points, indicate in which octant each is found.
    (a) $(3, 2, 1)$
    (b) $(-3, -3, 1)$
    (c) $(-1, 2, -3)$
    (d) $(-5, -3, -2)$
    (e) $(6, -3, 5)$
    (f) $(8, 4, -2)$

25. In octant 1, the *x*-, *y*-, and *z*-coordinates are all positive. Characterize the coordinates in the remaining seven octants.

26. The distance formula can be generalized to three-dimensional space. Let P have coordinates (a, b, c) and Q have coordinates (r, s, t). The faces of the prism shown are parallel to the coordinate planes.
    (a) What are the coordinates of point R?
    (b) What is the distance PR? [HINT: Use the Pythagorean theorem.]
    (c) What is the distance QR?
    (d) What is the distance PQ?

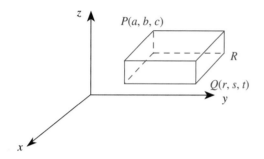

27. Use the distance formula in three dimensions to find the distance between the following pairs of points.
    (a) $(1, 2, 3)$ and $(2, 3, 1)$
    (b) $(-1, 0, 5)$ and $(6, 2, -1)$

28. The midpoint formula can be generalized to points in three-dimensional space. The midpoint of PQ for $P(a, b, c)$ and $Q(d, e, f)$ is M

$$\left(\frac{a + d}{2}, \frac{b + e}{2}, \frac{c + f}{2} \right).$$

(a) Find the midpoint of the segment having endpoints $(2, 0, -3)$ and $(6, 4, -7)$.
(b) If the midpoint of PQ is $M(1, -1, 3)$ and Q has coordinates $(4, -3, 0)$, find the coordinates of P.
(c) The endpoints of a diameter of a sphere are $(1, 1, 4)$ and $(2, -1, 5)$. Find the coordinates of the center of the sphere.

PROOFS

29. In the development of the midpoint formula, P and Q have coordinates (a, b) and (c, d) respectively, and M is the point $\left(\dfrac{a + c}{2}, \dfrac{b + d}{2} \right)$. Use the distance formula to verify that P, M, and Q are collinear and that $PM = MQ$.

30. Verify the Test for Collinearity by proving parts (a) and (b).
    (a) If Q is on $\overline{PR}$, then $PQ + QR = PR$. [HINT: Draw a picture and add corresponding lengths.]
    (b) If $PQ + QR = PR$, then Q is on $\overline{PR}$. [HINT: Use indirect reasoning: Assume that Q is not on $\overline{PR}$. Draw a picture showing the various possibilities and compare lengths.]

31. Suppose that l is a line with an equation of the form $y = mx + b$. Show that if a point (x, y) in the plane satisfies the equation $y = mx + b$, then (x, y) is on line l. [HINT: Choose two points on l by giving x two values. Then show that the point $(x, mx + b)$ is collinear with the points you have chosen.]

8.2
SLOPE

Applied Problem

A conveyor belt is used to move bundles of shingles to the roof of a house. If the conveyor belt is 60 feet long and makes an angle of 35° with the horizontal, what is the slope of the conveyor? That is, how many feet is a bundle of shingles elevated for every one foot it travels horizontally?

SLOPE

When graphing lines or line segments on a coordinate system, we often want to compare the steepness or slant of two line. For that purpose, we talk about the slope of two lines. The slope of a line is a measure of the inclination of the line with respect to the horizontal.

Definition

Slope of a Line

Given the coordinates of any two points that determine a nonvertical line, the slope of the line is the ratio of the change in y to the change in x.

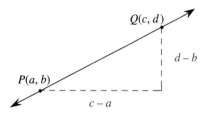

$$\text{Slope of } \overline{PQ} = \frac{d - b}{c - a}$$

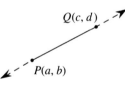

FIGURE 8.11

The difference $d - b$ is called the **rise** and $c - a$ is called the **run** of $\overleftrightarrow{PQ}$. Thus, the slope of a nonvertical line could also be defined to be the ratio of its rise to its run. Notice that the slope of $\overleftrightarrow{PQ}$ can also be written as $\frac{b - d}{a - c}$ because $\frac{d - b}{c - a} = \frac{b - d}{a - c}$. The **slope of a line segment** $\overline{PQ}$ is the slope of $\overleftrightarrow{PQ}$ (Figure 8.11). If $\overleftrightarrow{PQ}$ is vertical, then $a = c$, and the denominator in the slope definition, namely $c - a$, is zero. Therefore, we say the slope of a vertical line is undefined

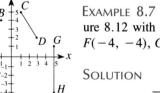

FIGURE 8.12

EXAMPLE 8.7   Find the slopes, if defined, of each of the line segments in Figure 8.12 with endpoints $A(-5, 4)$, $B(-1, 4)$, $C(1, 5)$, $D(3, 2)$, $E(-2, -2)$, $F(-4, -4)$, $G(5, 1)$, and $H(5, -4)$.

SOLUTION

(a) Slope of $\overline{AB} = \dfrac{4 - 4}{-5 - (-1)} = 0.$

(b) Slope of $\overline{CD} = \dfrac{5 - 2}{1 - 3} = \dfrac{3}{-2} = \dfrac{-3}{2}.$

(c) Slope of $\overline{EF} = \dfrac{-2 - (-4)}{-2 - (-4)} = \dfrac{2}{2} = 1.$

(d) Slope of $\overline{GH}$ is undefined because $\overline{GH}$ is vertical.

Several generalizations can be made about slopes of lines by reviewing Example 8.7. First, notice that the slope of the horizontal line segment $\overline{AB}$ is zero. Because the y-coordinates of any two points on a horizontal line are equal, the slope of *any* horizontal line is zero. The slope of $\overline{CD}$ is negative. This is true of *any* line that falls from left to right. The slope of $\overline{EF}$ is positive, which is true for *any* line that rises from left to right. Figure 8.13 gives a visual summary of slopes. More generalizations regarding slopes will be discussed in the problem set.

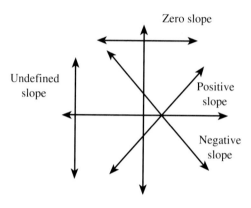

FIGURE 8.13

COLLINEARITY AND SLOPE

We have seen how to calculate the slope of a line or line segment given the coordinates of two points contained in it. If we know the coordinates of three different points on a line, we can determine the slope of that line by choosing any two of the three points and using the slope formula. In Figure 8.14, the three points P, Q, and R are collinear. The slopes of $\overline{PQ}$, $\overline{QR}$, and $\overline{PR}$ will be the same. That is, the choice of points does not affect the value of the slope. This idea will be discussed further in the problem set.

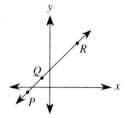

FIGURE 8.14

In Section 8.1, we developed a test for collinearity using the distance formula. An alternate form of the test, and one that is often easier to use, involves slope. It will be shown in the problem set that if the slope of $\overline{PQ}$ equals the slope of $\overline{QR}$, then P, Q, and R are collinear (Figure 8.14).

EXAMPLE 8.8   An experiment yielded the data points graphed in Figure 8.15. Determine whether the three points are collinear.

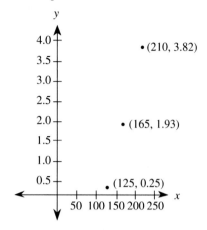

FIGURE 8.15

SOLUTION   If we label the three points $A(125, 0.25)$, $B(165, 1.93)$, and $C(210, 3.82)$, then we have

$$\text{Slope of } \overline{AB} = \frac{1.93 - 0.25}{165 - 125} = \frac{1.68}{40} = 0.042$$

$$\text{Slope of } \overline{BC} = \frac{3.82 - 1.93}{210 - 165} = \frac{1.89}{45} = 0.042$$

Because the slopes of $\overline{AB}$ and $\overline{BC}$ are the same, the three points are collinear. Notice that the two segments must share a common point, B in this case, for this test to work.

SLOPES OF PARALLEL AND PERPENDICULAR LINES

Because slope is a measure of the inclination of a line with respect to the horizontal, two (nonvertical) parallel lines must have the same slope. Also, vertical lines are parallel (Figure 8.16). The next theorem summarizes these relationships.

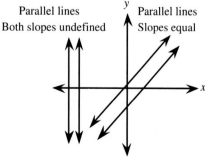

FIGURE 8.16

Theorem 8.4

Slopes of Parallel Lines

Two lines in a coordinate plane are parallel if and only if
(a) their slopes are equal, or
(b) their slopes are undefined.

We can use slope to classify and to verify properties of particular geometric figures as the next example demonstrates.

EXAMPLE 8.9   For quadrilateral $ABCD$ with $A(-2, 5)$, $B(6, 3)$, $C(8, -2)$, and $D(0, 0)$,

(a) Use slope to verify that $ABCD$ is a parallelogram.

(b) Verify that the diagonals of $ABCD$ bisect each other.

SOLUTION $ABCD$ is graphed in Figure 8.17.

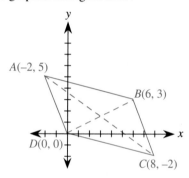

FIGURE 8.17

(a) While the figure *appears* to be a parallelogram, we will verify that it is one by using slope.

$$\text{Slope of } \overline{AB} = \frac{3-5}{6+2} = \frac{-2}{8} = -\frac{1}{4}$$

$$\text{Slope of } \overline{BC} = \frac{-2-3}{8-6} = \frac{-5}{2}$$

$$\text{Slope of } \overline{CD} = \frac{0+2}{0-8} = \frac{-2}{8} = -\frac{1}{4}$$

$$\text{Slope of } \overline{DA} = \frac{5-0}{-2-0} = \frac{5}{-2} = \frac{-5}{2}$$

Because both pairs of opposite sides are parallel, $ABCD$ is a parallelogram. [NOTE: We could also use the distance formula to verify that $ABCD$ is a parallelogram, as was done in Section 8.1. However, the slope formula is often easier to use.]

(b) To show that the diagonals bisect each other, we must show that the midpoint of diagonal $\overline{AC}$ is the same point as the midpoint of diagonal $\overline{BD}$.

$$\text{Midpoint of } \overline{AC} = \left(\frac{-2 + 8}{2}, \frac{5 + (-2)}{2}\right) = \left(3, \frac{3}{2}\right)$$

$$\text{Midpoint of } \overline{BD} = \left(\frac{0 + 6}{2}, \frac{0 + 3}{2}\right) = \left(3, \frac{3}{2}\right)$$

Because their midpoints are the same, diagonals $\overline{AC}$ and $\overline{BD}$ must intersect at their midpoints. Therefore, the diagonals of $ABCD$ bisect each other.   •

Just as we are able to use slope to determine whether two lines are parallel, we can also use slope to determine whether two lines are perpendicular as the next example shows.

EXAMPLE 8.10   In Figure 8.18, prove that $\triangle ABC$ has a right angle at C. Then find the product of the slopes of $\overline{AC}$ and $\overline{BC}$.

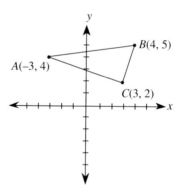

FIGURE 8.18

SOLUTION   To prove that $\triangle ABC$ is a right triangle, we'll use the converse of the Pythagorean Theorem (Test for Right Angles):

$$AB = \sqrt{(4 - (-3))^2 + (5 - 4)^2} = \sqrt{49 + 1} = \sqrt{50}. \text{ Thus } (AB)^2 = 50.$$
$$AC = \sqrt{[(-3) - 3]^2 + (4 - 2)^2} = \sqrt{36 + 4} = \sqrt{40}. \text{ Thus } (AC)^2 = 40.$$
$$BC = \sqrt{(4 - 3)^2 + (5 - 2)^2} = \sqrt{1 + 9} = \sqrt{10}. \text{ Thus } (BC)^2 = 10.$$

Because $(AC)^2 + (BC)^2 = 40 + 10 = 50 = (AB)^2$, we have that $\triangle ABC$ is a right triangle by the converse of the Pythagorean Theorem. The slope of $\overline{AC}$ equals $\dfrac{4 - 2}{(-3) - 3} = -\dfrac{1}{3}$ and the slope of $\overline{BC}$ is $\dfrac{5 - 2}{4 - 3} = 3$. Thus, the product of their slopes is -1.   •

In general, the product of the slopes of two perpendicular lines is -1, if neither line is vertical. This result is summarized in the next theorem.

Theorem 8.5

Slopes of Perpendicular Lines

Two lines in a coordinate plane are perpendicular if and only if
(a) one line is horizontal and the other is vertical, or
(b) the product of their slopes equals -1.

The "only if" part of this theorem will be proved next and the "if" part will be left for the problem set.

Given Two perpendicular lines l and m, neither of which is vertical [Figure 8.19(a)]

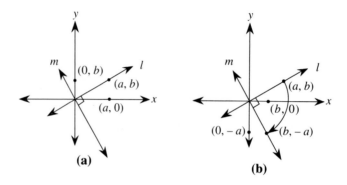

(a)   **(b)**

FIGURE 8.19

Prove The product of the slopes of l and m is -1.

Proof Imagine line l being rotated 90° clockwise. The point (a, b) on l would correspond to the point $(b, -a)$ on m [Figure 8.19(b)]. Using points $(0, 0)$ and (a, b), the slope of l is $\dfrac{b}{a}$ and the slope of m is $\dfrac{-a}{b}$. Therefore, the product of their slopes is $\left(\dfrac{b}{a}\right)\left(\dfrac{-a}{b}\right) = -1$, which is what we wanted to show. Although we only proved this result for the case when the two lines intersect at the origin, this result holds for any pair of perpendicular lines (neither of which is horizontal).

We can use slope to prove that segments are perpendicular or that geometric figures contain right angles, as the next example shows.

EXAMPLE 8.11   Prove that the diagonals of rhombus *PQRS* in Figure 8.20 are perpendicular using a coordinate proof.

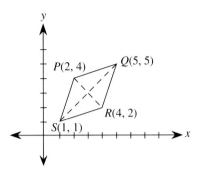

FIGURE 8.20

SOLUTION   The slope of diagonal $\overline{PR}$ is $\dfrac{4 - 2}{2 - 4} = -1$ and the slope of diagonal $\overline{QS}$ is $\dfrac{5 - 1}{5 - 1} = 1$. Since the product of these two slopes is -1, the diagonals are perpendicular. [NOTE: The result that the diagonals of *any* rhombus are perpendicular will be proved in the Section 8.4.]   •

Solution to Applied Problem

Using the given information, we can calculate the "rise" and the "run" in the right triangle shown.

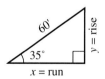

We know that $\sin 35° = \dfrac{y}{60}$, so $y = 60 \sin 35° \approx 34.4$ ft. Similarly, $x = 60 \cos 35° \approx 49.1$ ft. The slope of the conveyor is

$$m = \frac{\text{rise}}{\text{run}} = \frac{34.4}{49.1} \approx 0.70$$

So, for every one foot a bundle of shingles travels horizontally, it rises 0.7 ft. [NOTE: We can also find the slope *m* by using $\tan 35° = 0.70$, because $\tan 35° = \dfrac{y}{x}$.]

GEOMETRY AROUND US

The steepness of a sloping roof is described by its pitch, which is similar to the slope of a line. A "1 in 2" pitch, for example, means that the roof rises one foot vertically for every two feet of horizontal distance that it runs. In other words, the roof has slope $\frac{1}{2}$. Other common roof pitches are 5 in 12 and 5 in 8.

PROBLEM SET 8.2

EXERCISES/PROBLEMS

1. Find the slopes of the lines containing the following pairs of points.
   (a) (3, 2) and (5, 3)
   (b) (−2, 1) and (−5, −3)
   (c) (3, −5) and (−6, −5)
   (d) (4, −1) and (4, 2)

2. Use the ratios $\dfrac{d-b}{c-a}$ and $\dfrac{b-d}{a-c}$ to compute the slopes of the lines containing the following points. Do both ratios give the same result?
   (a) (−4, 5) and (6, −3)
   (b) (−2, −3) and (3, 2)

3. Classify the slope of each line shown as positive, negative, zero, or undefined.
   (a)

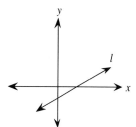

   (b)

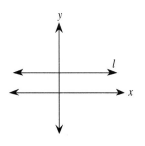

(c)

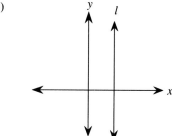

(d)

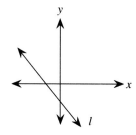

4. Given are the slopes of several lines. Indicate whether each line is horizontal or vertical, rises from left to right, or falls from left to right.
   (a) $\dfrac{3}{4}$ (b) undefined   (c) 0   (d) $-\dfrac{5}{6}$

5. Use slopes to determine if $\overline{AB} \parallel \overline{PQ}$.
   (a) A(−1, 0), B(4, 5), P(3, 9), Q(−2, 4)
   (b) A(0, 4), B(6, 8), P(−4, −6), Q(2, −2)

6. Give the slope of a line parallel to $\overline{PQ}$.
   (a) P(3, 2), Q(5, −7)
   (b) P(−3, 1), Q(−1, 9)
   (c) P(1, −5), Q(1, 4)
   (d) P(−2, −3), Q(−5, −2)

7. Determine if the following pairs of line segments are parallel.
   (a) The segment from $(0, 2)$ to $(1, 3)$ and the segment from $(2, 0)$ to $(3, 2)$
   (b) The segment from $(-5, 3)$ to $(-1, 2)$ and the segment from $(-3, 1)$ to $(0, 0)$
   (c) The segment from $(-1, 1)$ to $(-1, 4)$ and the segment from $(1, 0)$ to $(1, 4)$
   (d) The segment from $(-5, -3)$ to $(-3, -1)$ and the segment from $(-5, -4)$ to $(-2, -1)$

8. Use slopes to determine if the quadrilaterals with given vertices are parallelograms.
   (a) $(1, 4)$, $(4, 4)$, $(5, 1)$, and $(2, 1)$
   (b) $(1, -1)$, $(6, -1)$, $(6, 4)$, and $(1, -4)$
   (c) $(1, -2)$, $(4, 2)$, $(6, 2)$, and $(3, -2)$
   (d) $(-10, 5)$, $(-5, 10)$, $(10, -5)$, and $(5, -10)$

9. Three vertices of a parallelogram are given. Find the coordinates of a fourth vertex.
   (a) $(-2, 1)$, $(-4, 6)$, $(8, 1)$
   (b) $(-2, 1)$, $(0, 3)$, $(7, 2)$

10. Three vertices of a rectangle are given. Find the coordinates of the fourth vertex.
    (a) $(-1, 4)$, $(3, 5)$, $(0, 0)$
    (b) $(6,7)$, $(9, -2)$, $(-3, 4)$

11. A quadrilateral $ABCD$ has vertices $A(-1, 1)$, $B(2, 4)$, $C(6, 1)$, and $D(3, -2)$. Use slopes to verify that $ABCD$ is a parallelogram.

12. A quadrilateral $ABCD$ has vertices $A(-3, 2)$, $B(3, 4)$, $C(6, 0)$, and $D(-3, -3)$. Use slopes and the distance formula to verify that $ABCD$ is an isosceles trapezoid.

13. Use slopes to determine if the three points given are collinear.
    (a) $A(3, -2)$, $B(1, 2)$, and $C(-3, 10)$
    (b) $A(0, 7)$, $B(2, 11)$, and $C(-2, 1)$

14. Three of the four points given are collinear. Use slopes to determine which three points are collinear.
    $A(2, 3)$ $B(5, -2)$ $C(-2, -1)$ $D(6, 7)$

15. Give the slope of a line perpendicular to $\overline{AB}$.
    (a) $A(1, 6)$, $B(2, 5)$ (b) $A(0, 4)$, $B(-6, -5)$
    (c) $A(4, 2)$, $B(-5, 2)$ (d) $A(-1, 5)$, $B(-1, 3)$

16. In each part, determine if the line segments joining the given points are perpendicular.
    (a) $(3, 3)$, $(3, 9)$, and $(12, 9)$
    (b) $(2, 8)$, $(7, 1)$, and $(14, 6)$
    (c) $(-8, -3)$, $(8, 2)$, and $(6, 9)$
    (d) $(-7, -2)$, $(-4, 4)$, and $(10, -3)$

17. Find a so that the slope of the line segment having endpoints $P(-11, 4)$ and $Q(a, -3)$ is $-\dfrac{1}{3}$.

18. Find b so that the slope of the line segment having endpoints $P(5, 1)$ and $Q(-6, b)$ is perpendicular to the line segment having endpoints $(1, 4)$ and $(-1, 0)$.

19. One diagonal of a rhombus $ABCD$ contains the vertices $A(-4, -1)$ and $C(7, -2)$. Find the slope of the other diagonal.

20. One side of a rectangle $ABCD$ contains the vertices $A(9, -3)$ and $C(6, 1)$. Find the slope of an adjacent side.

21. Use slopes to determine which, if any, of the triangles with the given vertices is a right triangle.
    (a) $(2, 2)$, $(8, 6)$, and $(4, 8)$
    (b) $(1, -1)$, $(9, -4)$, and $(10, 11)$
    (c) $(8, 1)$, $(6, 3)$ and $(1, -6)$

22. Use slopes to determine which, if any, of the quadrilaterals with the given vertices is a rectangle.
    (a) $(0, 0)$, $(4, -2)$, $(8, 4)$, and $(6, 6)$
    (b) $(-3, 8)$, $(0, 12)$, $(12, 3)$, and $(9, -1)$
    (c) $(-10, -5)$, $(-6, 15)$, $(14, 11)$, and $(10, -9)$

23. Which of the following properties are true of the two kites having the given vertices?
    (i) $(0, 0)$, $(2, 5)$, $(0, 8)$, $(-2, 5)$
    (ii) $(0, 0)$, $(2, 3)$, $(3, -2)$, $(10, 2)$
    (a) The diagonals are congruent.
    (b) The diagonals are perpendicular to each other.
    (c) The diagonals bisect each other.
    (d) The kite has two right angles.

24. Which of the following properties are true of the two trapezoids having the given vertices?
    (i) $(0, 0)$, $(16, 0)$, $(6, 4)$, $(10, 4)$
    (ii) $(-4, 0)$, $(-4, 5)$, $(7, 5)$, $(9, 0)$
    (a) The diagonals have the same length.
    (b) The diagonals are perpendicular to each other.
    (c) The diagonals bisect each other.
    (d) The trapezoid has two right angles.

PROOFS

25. Recall that a midsegment of a triangle is a segment joining the midpoints of two sides of the triangle. For $\triangle ABC$, with $A(-3, 0)$, $B(0, 4)$, and $C(6, 2)$, verify that the midsegment connecting midpoints of $\overline{AB}$ and $\overline{BC}$ is parallel to and half the length of $\overline{AC}$.

26. Recall that the midquad of a quadrilateral is obtained by connecting consecutive midpoints of the sides of the quadrilateral. For quadrilateral $ABCD$, with $A(1, 1)$, $B(-2, 2)$, $C(5, 5)$, and $D(7, 1)$, verify that the midquad is a parallelogram.

27. A quadrilateral $ABCD$ has vertices $A(-2, 1)$, $B(3, 3)$, $C(7, 2)$, and $D(3, 1)$. Use slopes to verify that $ABCD$ is a rhombus.

28. Let P, Q, R, and S be any points on the line as shown. By drawing horizontal and vertical segments, draw right triangles $\triangle PQO$ and $\triangle RST$. Follow the given steps to prove that the slope of the line is independent of the pairs of points selected.

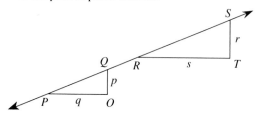

(a) Show that $\triangle PQO \sim \triangle RST$.

(b) Show that $\dfrac{p}{q} = \dfrac{r}{s}$.

(c) Is the slope of $\overline{PQ}$ equal to the slope of $\overline{RS}$? Explain.

29. If the slope of $\overline{PQ}$ is equal to the slope of $\overline{PR}$, we can show that P, Q, and R are collinear. The following justification uses similar triangles.

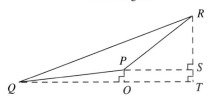

(a) Horizontal lines are drawn through points P and Q and vertical lines through P and R, intersecting at points O, S, and T as illustrated. If the slope of $\overline{PQ}$ and the slope of $\overline{PR}$ are equal, what two ratios of sides are equal?

(b) Show that $\triangle QOP \sim \triangle PSR$.

(c) Which sides are proportional?

(d) Verify that $\dfrac{QO + PS}{PS} = \dfrac{PO + RS}{RS}$.

(e) Show that $\triangle PSR \sim \triangle QTR$.

(f) Which sides are proportional?

(g) Verify that $QR = PR \cdot QT/PS$.

(h) Verify that $QP + PR = PR \cdot QT/PS$.

(i) Verify that points P, Q, and R are collinear.

30. Complete the proof of Theorem 8.5 by showing the following: If the product of the slopes of l and m is -1, then l and m are perpendicular. [HINT: Show that $\triangle OPQ$ is a right triangle.]

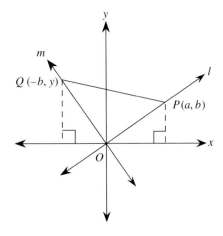

APPLICATIONS

31. A road rises 11.2 feet in a horizontal distance of 140 feet. What is the slope of the road? What is its angle of inclination (the angle formed with the horizontal)?

32. What is the slope of the roof shown? What is its pitch?

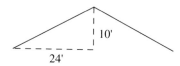

33. The **percent grade** of a highway is the amount that the highway rises (or falls) in a given horizontal distance. For example, a highway with a 4% grade rises 0.04 mile for every 1 mile of horizontal distance.

(a) How many feet does a highway with a 6% grade rise in 2.5 miles of horizontal distance?

(b) How many feet would a highway with a 6% grade rise in 90 miles of horizontal distance if the grade remained constant?

(c) How is percent grade related to slope?

34. A freeway ramp connects a highway to an overpass 10 meters above the ground. The ramp, which starts 150 meters from the overpass, is 150.33 meters long and 9 meters wide. What is the percent grade of the ramp to three decimal places?

8.3
EQUATIONS OF LINES AND CIRCLES

Applied Problem

A triangular metal plate of uniform thickness and density can be placed on a pair of coordinate axes so that it has vertices $A(0, 0)$, $B(2, 4)$, and $(8, 0)$. Find the center of mass of the plate.

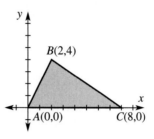

EQUATIONS OF LINES

The slope of a line can be used to derive its equation as illustrated in the next example.

EXAMPLE 8.12   Given points $(-4, 2)$ and $(2, 5)$ on line l (Figure 8.21), find an equation that describes all of the points on l.

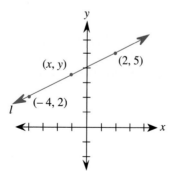

FIGURE 8.21

SOLUTION   Because $(-4, 2)$ and $(2, 5)$ are points on l, the slope of l is $\dfrac{5 - 2}{2 - (-4)} = \dfrac{3}{6} = \dfrac{1}{2}$. Let (x, y) represent any point on l (Figure 8.21). Because points $(-4, 2)$, $(2, 5)$, and (x, y) are collinear, the slope determined by *any* pair of them must be $\frac{1}{2}$. Thus, for points $(-4, 2)$ and (x, y), we have the slope of l is $\frac{y - 2}{x - (-4)}$ which, in turn, must be equal to $\frac{1}{2}$. Thus, we have $\frac{y - 2}{x - (-4)} = \frac{1}{2}$, or

(i) $(y - 2) = \frac{1}{2}[x - (-4)]$

This equation is usually simplified and written in one of two other equivalent forms:

(ii) $y = \frac{1}{2}x + 4$
(iii) $x - 2y + 8 = 0$ •

Equation (i), $(y - 2) = \frac{1}{2}[x - (-4)]$, displays the slope, $1/2$, and the coordinates of one of the points, $(-4, 2)$, that we used to derive this equation. For this reason, equation (i) is called the **point-slope form** of line l. Equation (ii), $y = \frac{1}{2}x + 4$, displays the slope as the coefficient of x. If 0 is substituted for x in equation (ii), we obtain the equation $y = 4$. The point $(0, 4)$ is where line l intersects the y-axis. For this reason, $(0, 4)$, or simply 4, is called the y-intercept of this equation. In general, the **y-intercept** of a line is the point where the line intersects the y-axis. Because it shows the slope and the y-intercept of the line, equation (ii) is called the **slope-intercept form** of line l. Equation (iii) is called the **standard form** of line l.

Theorem 8.6

Forms of Equations of Lines

Every nonvertical line in the plane can be expressed in either of the following two forms where m represents the slope of the line, b is the y-intercept of the line, and (r, s) is a point on the line.

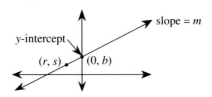

Point-Slope Form: $y - s = m(x - r)$
Slope-Intercept Form: $y = mx + b$

EXAMPLE 8.13   Find the equation of the perpendicular bisector of $\overline{AB}$, for $A(-5, -2)$ and $B(1, 7)$ [Figure 8.22(a)].

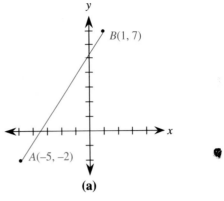

(a)

FIGURE 8.22(a)

SOLUTION   We will demonstrate two different approaches to solving this problem.

METHOD 1   First we find the midpoint of $\overline{AB}$ and the slope of a line perpendicular to $\overline{AB}$. The midpoint, M, of $\overline{AB}$ is $\left(\dfrac{-5 + 1}{2}, \dfrac{-2 + 7}{2}\right) = \left(-2, \dfrac{5}{2}\right)$ [Figure 8.22(b)]. The slope of $\overline{AB}$ is $\dfrac{7 - (-2)}{1 - (-5)} = \dfrac{9}{6} = \dfrac{3}{2}$, so the slope of any line perpendicular to $\overline{AB}$ is $-\dfrac{2}{3}$. We want the equation of the line through the point $\left(-2, \dfrac{5}{2}\right)$ having slope $-\dfrac{2}{3}$. Using the point-slope form, we obtain $y - \dfrac{5}{2} = -\dfrac{2}{3}(x + 2)$, which simplifies to $y - \dfrac{5}{2} = -\dfrac{2}{3}x - \dfrac{4}{3}$, or $y = -\dfrac{2}{3}x + \dfrac{7}{6}$.

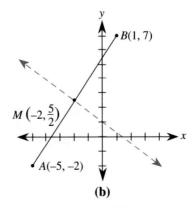

(b)

FIGURE 8.22(b)

METHOD 2   Next we use the distance formula and a property of points on the perpendicular bisector of a segment. We know that any point on the perpendicular bisector of a segment is equidistant from the endpoints of the segment. Let $P(x, y)$ be any point on the perpendicular bisector of $\overline{AB}$ [Figure 8.22(c)].

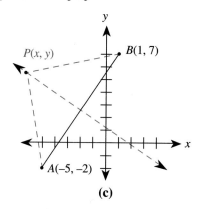

(c)

FIGURE 8.22(c)

Thus, $PA = PB$. Using the distance formula, this means that $\sqrt{(x + 5)^2 + (y + 2)^2} = \sqrt{(x - 1)^2 + (y - 7)^2}$. Squaring both sides of this equation and simplifying, we have

$$x^2 + 10x + 25 + y^2 + 4y + 4 = x^2 - 2x + 1 + y^2 - 14y + 49$$
$$10x + 4y + 29 = -2x - 14y + 50$$
$$12x + 18y = 21$$
$$4x + 6y = 7$$

This equation is equivalent to the one obtained using Method 1. Here the equation is in standard form whereas the one obtained by Method 1 is written in slope-intercept form.   •

SYSTEMS OF EQUATIONS

Two or more equations, called a **system of equations**, are said to have a **simultaneous solution** if there is an ordered pair that is a solution of all of both the equations. For example, since the ordered pair $(2, 3)$ is a solution of $3x + 4y = 18$ *and* $2x - y = 1$, it is a *simultaneous* solution of the equations. Using coordinate geometry, we can determine how many simultaneous solutions equations have because these solutions will be points common to each of their graphs. The following example shows how to find the simultaneous solutions of systems of linear equations, that is, equations of lines.

EXAMPLE 8.14   Determine the number of simultaneous solutions of the following systems of linear equations.

(a) $y = 3x - 2$ and $y = 8 - x$

(b) $y = 2x + 7$ and $y = 2x - 3$

(c) $y = x - 5$ and $2x = 2y + 10$

SOLUTION

(a) The graph of $y = 3x - 2$ has slope 3 and the graph of $y = 8 - x$ has slope -1. Because their slopes are different, their graphs must intersect in a single point. Therefore, they have one simultaneous solution [Figure 8.23(a)].

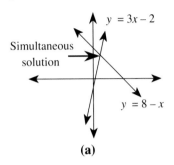

(a)

FIGURE 8.23

(b) The graphs of $y = 2x + 7$ and $y = 2x - 3$ have the same slopes, namely 2, but *different* y-intercepts. Thus, they are parallel lines with no common solutions [Figure 8.23(b)].

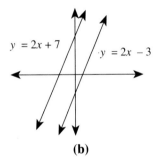

(b)

FIGURE 8.23

(c) The equations $y = x - 5$ and $2x = 2y + 10$ are two forms of the same equation. Therefore, their graphs are the same. Because every ordered pair that is a solution of one is also a solution of the other, every ordered pair on the graph is a simultaneous solution of both equations. So the equations have infinitely many solutions in common.   •

Example 8.14 is summarized as follows:

Theorem 8.7

Solutions of a System of Two Linear Equations

Two linear equations have 0, 1, or infinitely many simultaneous solutions if and only if the lines they represent are parallel, intersect in one point, or coincide, respectively.

EXAMPLE 8.15   Find the point M where the diagonals of $ABCD$ intersect (Figure 8.24).

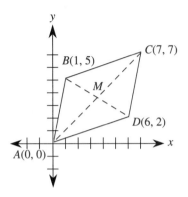

FIGURE 8.24

SOLUTION   We must find equations for the lines containing $\overline{AC}$ and $\overline{BD}$ and then solve a system of equations. The slope of $\overline{BD}$ is $\dfrac{2 - 5}{6 - 1} = -\dfrac{3}{5}$. Using this slope and the point $B(1, 5)$ in the point-slope form, we have

$$y - 5 = -\frac{3}{5}(x - 1)$$

$$y - 5 = -\frac{3}{5}x + \frac{3}{5}$$

$$y = -\frac{3}{5}x + \frac{28}{5}$$

The line containing $\overline{AC}$ has equation $y = x$. So, to find the point of intersection, we must solve the system of linear equations

$$y = -\frac{3}{5}x + \frac{28}{5} \qquad \text{and} \qquad y = x$$

The point of intersection satisfies both of these equations. This means that at the point of intersection y equals x and y also equals $\dfrac{-3}{5}x + \dfrac{28}{5}$. Because $y = x$, we can substitute y for x in the other equation to obtain the following:

$$x = -\frac{3}{5}x + \frac{28}{5}$$

$$5x = -3x + 28$$

$$8x = 28, \text{ or } x = \frac{7}{2}.$$

Thus, the x-coordinate of the point of intersection is $\dfrac{7}{2}$. Because $y = x$ at the point of intersection, we know that $y = \dfrac{7}{2}$. Therefore, the point of intersection is $\left(\dfrac{7}{2}, \dfrac{7}{2}\right)$, which is consistent with the graph in Figure 8.24.

The method used to solve the system of equations in Example 8.15 is called the **substitution method** because we solve for a variable in one equation and then substitute the value of that variable in the other equation. Two other methods for solving a system of equations will be discussed in the problem set.

EQUATION OF A CIRCLE

We have seen in Example 8.13 how the distance formula can be applied to derive an equation of the perpendicular bisector of a line segment. The equation of a circle can also be derived using the distance formula.

EXAMPLE 8.16   Find the equation that describes all points of the plane at a distance of 5 from the point $(1, 2)$.

SOLUTION   The figure described is the circle with center $(1, 2)$ and radius 5 (Figure 8.25). If (x, y) is any point on the circle, then by the distance formula we have $5 = \sqrt{(x - 1)^2 + (y - 2)^2}$. Squaring both sides of this equation, we obtain the equation $(x - 1)^2 + (y - 2)^2 = 25$, which is the equation of the circle.

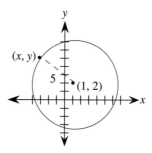

FIGURE 8.25

The form of the equation for a circle shown in Example 8.16 is called the **standard form** of the equation of a circle. This example can be modified to derive the equation of any arbitrary circle whose center is (a, b) and whose radius is r. This result is stated next and its proof is left for the problem set.

Theorem 8.8

Equation of a Circle

The circle with center (a, b) and radius r has the equation

$$(x - a)^2 + (y - b)^2 = r^2$$

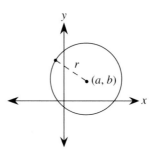

The next example shows how the equation of a circle can be used in conjunction with the equation of a line to find their points of intersection.

EXAMPLE 8.17   Determine where the circle with center $(-5, 3)$ and radius 6 intersects the line whose slope is 1 and whose y-intercept is 2.

SOLUTION   First we find the equations of the circle and the line. Their graphs are sketched in Figure 8.26. We see that there are two points of intersection.

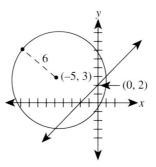

FIGURE 8.26

Equation of the circle: $[x - (-5)]^2 + (y - 3)^2 = 6^2$
Equation of the line: $y = x + 2$

To find the points of intersection, we must find the simultaneous solution of the system of equations $(x + 5)^2 + (y - 3)^2 = 36$ and $y = x + 2$.

We can substitute $x + 2$ for y in the equation of the circle and simplify the resulting equation.

$$(x + 5)^2 + [(x + 2) - 3]^2 = 36$$
$$(x + 5)^2 + (x - 1)^2 = 36$$
$$x^2 + 10x + 25 + x^2 - 2x + 1 = 36$$
$$2x^2 + 8x + 26 = 36$$
$$2x^2 + 8x - 10 = 0$$
$$x^2 + 4x - 5 = 0$$
$$(x + 5)(x - 1) = 0.$$

Therefore, $x = -5$ or $x = 1$. These are the x-coordinates of the two points of intersection of the line and the circle.

Substituting these values for x in the equation of the line, we obtain the corresponding values of y: $y = (-5) + 2 = -3$ and $y = 1 + 2 = 3$. Thus, the points of intersection are $(-5, -3)$ and $(1, 3)$.

Solution to Applied Problem

The center of mass, or center of gravity, of the triangle is its "balance point." It is located at the point of intersection of the medians of the triangle, the centroid.

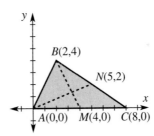

So we must find equations for two of the medians of $\triangle ABC$ and determine their simultaneous solution. The midpoint of $\overline{AC}$ is $M(4, 0)$ and the midpoint of $\overline{BC}$ is $N(5, 2)$. The slope of $\overline{MB}$ is $\dfrac{4 - 0}{2 - 4} = -2$. Thus, an equation of the line containing $\overline{MB}$ is $y - 0 = -2(x - 4)$, or $y = -2x + 8$. Similarly, the equation of the line containing $\overline{AN}$ is $y = \dfrac{2}{5}x$.

The center of mass is the simultaneous solution of $y = -2x + 8$ and $y = \dfrac{2}{5}x$. Solving this system of equations, we get $x = \dfrac{10}{3}$ and $y = \dfrac{4}{3}$.

GEOMETRY AROUND US

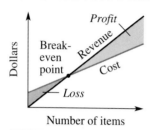

In business, a cost function and a revenue function are often plotted on the same graph. The cost function gives the total cost to produce x items, and the revenue function gives the total revenue from the sale of x items. The point at which these two graphs intersect is significant—it is called the break-even point. When the graph of revenue lies above the graph of cost, the company makes a profit. If the revenue graph lies below the cost graph, the company loses money.

PROBLEM SET 8.3

EXERCISES/PROBLEMS

1. Show that point P lies on the line with the given equation.
   (a) $y = 7x - 2; P(1, 5)$
   (b) $y = -2x + 5; P(-2, 9)$
   (c) $-2x = 6y + 3; P(7.5, -3)$
   (d) $3y = 4x + 2; P(4, 6)$

2. For each of the following equations find three points whose coordinates satisfy the equation.
   (a) $x - y = 4$ (b) $2x - 3y = 6$
   (c) $x = 3$ (d) $x + 4y = 0$

3. Identify the slope and the y-intercept for each line.
   (a) $y = -3x + 2$ (b) $y = 2x - 5$
   (c) $2y = 5x - 6$ (d) $2x - 7y = 8$

4. Write each equation in the form $y = mx + b$. Identify the slope and y-intercept.
   (a) $2y = 6x + 12$ (b) $4y + 3x = 0$
   (c) $8y - 2 = 0$ (d) $3x - 4y = 12$

5. Write the equation of the line, given its slope m and y-intercept b.
   (a) $m = 3, b = 7$ (b) $m = -1, b = -3$
   (c) $m = 2, b = 5$ (d) $m = -3, b = -\dfrac{1}{4}$

6. Write the equation in slope-intercept form for the line containing these pairs of points.
   (a) $(6, 3), (0, 2)$
   (b) $(-4, 8), (3, -6)$
   (c) $(-5, 2), (-3, 1)$
   (d) $(4, 7), (10, 7)$

7. Tell whether the line described by each equation rises to the right, falls to the right, is vertical, or is horizontal.
   (a) $y = 5x$
   (b) $y = -\frac{1}{3}x - 1$
   (c) $y = -10$
   (d) $x = -10$
   (e) $y + x = 0$
   (f) $2y - 3x = 0$

8. Write an equation of each line that satisfies the following.
   (a) Vertical through $(1, 3)$
   (b) Vertical through $(-5, -2)$
   (c) Horizontal through $(-3, 6)$
   (d) Horizontal through $(3, -3)$

9. Graph the line $3y = 4x + 3$, using the following steps:
   (a) Rewrite the equation in the $y = mx + b$ form.
   (b) What is its y-intercept? Plot the y-intercept.
   (c) What is its slope? Use the slope to find another point on the line and then graph the line.

10. Graph the line described by each equation.
    (a) $y = 2x - 1$
    (b) $y = -3x + 2$
    (c) $y = 2x + 1$
    (d) $x = -2$
    (e) $y = 3$
    (f) $2x = -6y$

11. (a) Graph the following lines on the same pair of coordinate axes.
   (i) $y = 2x$ (ii) $y = 2x + 3$ (iii) $y = 2x - 5$
   (b) What is the relationship among these lines?
   (c) Describe all lines of the form $y = cx + d$, where c is a fixed real number and d can be any real number.

12. (a) Graph the following lines on the same pair of coordinate axes.
   (i) $y = 2x + 3$ (ii) $y = -3x + 3$
   (iii) $y = \frac{1}{2}x + 3$
   (b) What is the relationship between these lines?
   (c) Describe all lines of the form $y = cx + d$, where d is a fixed real number and c can be any real number.

13. Write the equation of the line that passes through the given point and is parallel to the line whose equation is given.
   (a) $(5, -1)$; $y = 2x - 3$
   (b) $(1, -2)$; $y = -x - 2$
   (c) $(2, -5)$; $3x + 5y = 1$
   (d) $(-1, 0)$; $3x + 2y = 6$

14. Write the equation of the line that passes through the given point and is perpendicular to the line whose equation is given.
   (a) $(-2, 5)$; $y = -2x + 1$
   (b) $(6, 0)$; $y = 3x - 1$
   (c) $(1, -3)$; $2x + 4y = 6$
   (d) $(-5, -4)$; $3x - 2y = 8$

15. Find the equation of the perpendicular bisector of the segment whose endpoints are $(-3, -1)$ and $(6, 2)$.

16. Find equations of the perpendicular bisectors of the sides of the triangle whose vertices are $(0, 0)$, $(4, 0)$, and $(0, 3)$.

17. Find the equation of the line containing the median $\overline{AD}$ of $\triangle ABC$ with vertices $A(3, 7)$, $B(11, 2)$, and $C(1, 4)$.

18. Find the equation of the line containing altitude $\overline{PT}$ of $\triangle PRS$ with vertices $P(3, 5)$, $R(-1, 1)$, and $S(7, -3)$.

19. Find the equations of the lines containing the diagonals of the rectangle $PQRS$ with vertices $P(-2, -1)$, $Q(-2, 4)$, $R(8, 4)$, and $S(8, -1)$.

20. Find the equation of the line containing the perpendicular bisector of side $\overline{AB}$ of $\triangle ABC$, with vertices $A(0, 0)$, $B(2, 5)$, and $C(10, 5)$.

21. Use the substitution method to find the simultaneous solution of each system of linear equations, if a solution exists. If no solution exists, explain why not.
   (a) $y = 2x - 4$
      $y = -5x + 17$
   (b) $4x + y = 8$
      $5x + 3y = 3$
   (c) $x - 2y = 1$
      $x = 2y + 3$
   (d) $x - y = 5$
      $2x - 4y = 7$

22. Use the substitution method to find the simultaneous solution of each system of linear equations, if a solution exists. If no solution exists, explain why not.
   (a) $y = x - 1$
      $y = -2x + 8$
   (b) $y = \frac{1}{2}x - 1$
      $y = \frac{2}{3}x + \frac{5}{3}$
   (c) $2x - 4y = 7$
      $5x + y = 1$
   (d) $y = \frac{2}{3}x + 2$
      $3x - 9y = 14$

23. One method of solving a system of linear equations is called the **graphical method**. The lines are graphed and, if they intersect, the coordinates of the intersection point are determined. Graph the following pairs of linear equations to determine their simultaneous solutions, if any exist.
   (a) $y = 2x + 1$
      $y = 2x + 4$
   (b) $x + 2y = 4$
      $x + 2y = -2$
   (c) $-2x + 3y = 9$
      $x + y = -2$
   (d) $-2x + 3y = 9$
      $4x - 6y = -18$

24. Use the graphical method to find the simultaneous solution of each system of linear equations, if a solution exists.
   (a) $y = x + 1$
      $y = 3x - 1$
   (b) $x - y = -8$
      $2x + y = 2$
   (c) $y = -\frac{1}{2}x + 2$
      $x + 2y = 7$
   (d) $4x - 6y = 19$
      $-x + 3y = -7$

25. Another algebraic method of solving systems of equations is called the **elimination method** and involves eliminating one variable by adding or subtracting equivalent expressions. Consider the system

$$2x + y = 7$$
$$3x - y = 3$$

   (a) Add the left-hand sides of the equations and the right-hand sides. Because the original expressions were equal, the resulting sums are also equal. Notice that the variable y is eliminated.
   (b) Solve the equation you obtained in part 1.
   (c) Substitute this value of the variable into one of the original equations to find the value of the other variable.

26. Use the elimination method to solve the following systems of equations. [HINT: One equation may have to be multiplied by some number to facilitate eliminating one of the variables.]
    (a) $5x + 3y = 17$
       $2x - 3y = -10$
    (b) $4x - 4y = -3$
       $7x + 2y = 6$

27. Show that point P lies on the circle with the given equation.
    (a) $P(3, 4)$; $x^2 + y^2 = 25$
    (b) $P(-3, 5)$; $x^2 + y^2 = 34$
    (c) $P(-3, 7)$; $(x + 1)^2 + (y - 2)^2 = 29$
    (d) $P(\sqrt{5}, -3)$; $x^2 + (y + 5)^2 = 9$

28. Identify the center and radius of each circle whose equation is given.
    (a) $(x - 2)^2 + (y + 5)^2 = 64$
    (b) $(x + 3)^2 + (y - 4)^2 = 20$

29. Write the equation of the circle given by each of the following descriptions.
    (a) Center $(-1, -2)$ and radius $\sqrt{5}$
    (b) Center $(2, -4)$ and passing through $(-2, 1)$
    (c) Endpoints of a diameter at $(-1, 6)$ and $(3, -2)$

30. Write the equation of the circle having center C and passing through point P.
    (a) $P(2, -6)$, $C(3, -4)$
    (b) $P(-4, -3)$, $C(-2, 5)$

31. Find the coordinates of the centroid of the triangle with vertices $A(-5, -1)$, $B(3, 3)$, and $C(5, -5)$.

32. Find the coordinates of the orthocenter of the triangle with vertices $A(-2, 0)$, $B(-1, 6)$, and $C(5, 0)$.

33. Find the coordinates of the circumcenter of the triangle with vertices $A(-5, -1)$, $B(3, 3)$, and $C(5, -5)$.

34. Find the equation of the circumscribed circle for the triangle whose vertices are $J(4, 5)$, $K(8, -3)$, and $L(-4, -3)$.

35. (a) Use the graphical method to predict solutions to the system of equations given. How many solutions do you expect?

    $$x^2 + y^2 = 1 \quad \text{and} \quad y = \frac{x}{2} + 1$$

    (b) Use the substitution method to solve the simultaneous equations.

36. Solve the following pairs of simultaneous equations.
    (a) $x^2 + y^2 = 4$
       $(x - 2)^2 + (y - 2)^2 = 4$
    (b) $x^2 + y^2 = 4$
       $y + x = 0$

37. Find the area of a triangle formed by the x-axis, y-axis, and the line $2x + 3y = 6$.

38. Find the area of the triangle formed by the x-axis, the line $y = 2x$, and the line $8x + y = 40$.

PROOF

39. Prove Theorem 8.8: The circle with center (a, b) and radius r has the equation $(x - a)^2 + (y - b)^2 = r^2$.

APPLICATIONS

40. A catering company will cater a reception for $3.50 per person plus fixed costs of $100.
    (a) Complete the following chart.

Number of People	1	30	50	75	100	n
Total Cost						

    (b) Write a linear equation representing the relationship between number of people (x) and total cost (y).
    (c) What is the slope of this line? What does it represent?

41. A cab company charges a fixed fee of $0.60 plus $0.50 per mile.
    (a) Find the cost of traveling 10 miles; of traveling 25 miles.
    (b) Write an expression for the cost, y, of a trip of x miles.

42. A barrel in the form of a right circular cylinder having a diameter of three feet is to be constructed.
    (a) Write a linear equation that describes the volume of the barrel y as a function of its height x.
    (b) What is the slope of the line whose equation you wrote in (a)? What does it represent?
    (c) Write a linear equation that describes the surface area of the barrel y as a function of its height x.
    (d) What is the slope of the line whose equation you wrote in (c)? What does it represent?

43. Three sides of a lot must be fenced as shown. One end requires 60 feet of fence, and it must be fenced. The lengths of the other two congruent sides depend

on the owner's wishes and finances. The type of fencing to be used costs $7.25 per linear foot.

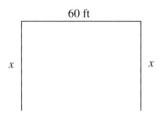

60 ft

x x

(a) Write a linear equation that describes the total cost, y, in terms of the length of a side, x.
(b) What is the slope of the line whose equation you wrote in (a)? What does it represent?
(c) What is the y-intercept of the line whose equation you wrote in (a)? What does it represent?
(d) If the owner has available $1700 for this fencing project, what is the maximum length for the side of the fenced area?

44. (a) A manufacturer can produce items at a cost of $0.65 per item with an operational overhead of $350. Let y represent the total production costs and x represent the number of items produced. Write an equation representing total production costs.
(b) The manufacturer sells the items for $1.15 per item. Let y represent the revenue and x represent the number of items sold. Write an equation representing total revenue.
(c) How many items does the manufacturer need to produce and sell to break even?

45. An organization is planning a dance. The band will cost $400 and advertising costs will be $100. Food will be supplied at $2.00 per person.
(a) How many people need to attend the dance in order for the organization to break even if tickets are sold at $7 each?
(b) How many people need to attend the dance in order for the organization to break even if tickets are sold at $6 each?
(c) How many people need to attend the dance if the organization wants to earn $400 profit and sells tickets for $6 each?

8.4

PROBLEM SOLVING USING COORDINATES

Applied Problem

Two lookouts posted in fire towers located 8 miles apart both spot the smoke from a fire. One observer determines the direction of the fire from her position as N 42° E, and the second sights the fire at N 30° W. Use the angles shown and the lines drawn to determine the location of the fire. [NOTE: This method of locating an object is called triangulation.]

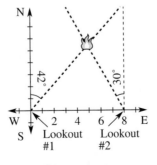

The most beautiful feature of coordinate geometry is that many common geometric shapes can be represented by equations. Thus, if we want to solve a problem in geometry, we may be able to represent the problem in algebraic terms, solve the problem using algebra, then interpret the result geometrically (Figure 8.27).

Problem Solving Using Coordinate Geometry

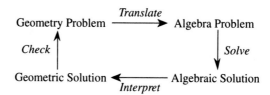

FIGURE 8.27

Some problems are more easily solved using coordinates instead of the triangle congruence and similarity techniques used in Chapters 2–6. The following examples use familiar theorems to illustrate coordinate proofs. In doing such proofs, geometric figures should be oriented as conveniently as possible in an effort to simplify computations.

ASSIGNING COORDINATES TO GEOMETRIC FIGURES

When writing a proof using coordinate geometry, we locate the coordinate axes conveniently to simplify the computations. For example, we may choose to put one vertex of a rectangle or parallelogram at the origin and another vertex somewhere else on the x-axis because this orientation makes several of the coordinates of the vertices zero. We use variables for some of the coordinates of the vertices to write our proof for a general figure. The next example shows how coordinates can be assigned to some common geometric figures.

EXAMPLE 8.18   Choose convenient coordinates for the vertices of each of the following.

(a) a rectangle   (b) a parallelogram

SOLUTION

(a) Because the coordinate axes are perpendicular, we can place two sides of the rectangle on the x-axis and y-axis [Figure 8.28(a)]. Placing three vertices on the coordinate axes also makes four coordinates of the vertices zero, which can simplify computations when slopes, distances, equations of sides, and so on are determined. These calculations are usually required when we write a coordinate proof. Notice that vertices B and D were given coordinates

(0, b) and (a, 0), respectively, to allow for the fact that sides $\overline{AB}$ and $\overline{AD}$ may not be congruent. Hence $ABCD$ is a general rectangle. If we had used vertices $A(0, 0)$, $B(0, a)$, $C(a, a)$, and $D(a, 0)$, the rectangle would have been a square. Any result we subsequently proved using those coordinates would be true only for squares and might not be true for all rectangles in general.

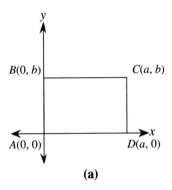

(a)

FIGURE 8.28

(b) To represent a parallelogram, we can place one side along the x-axis with one vertex at the origin [Figure 8.28(b)]. Now we must determine coordinates for vertices B and C. We choose two new variable coordinates for point B, (b, c) [Figure 8.28(c)]. The y-coordinate of point C must be c, the same as the y-coordinate of B, because $\overline{BC}$ is parallel to $\overline{AD}$. Moreover, because the slope of $\overline{AB}$ must equal the slope of $\overline{DC}$, the x-coordinate of C must be $a + b$ [Figure 8.28(c)].

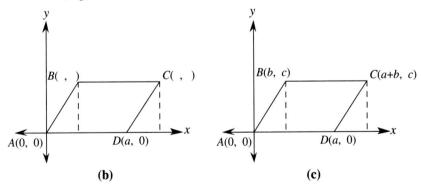

(b)   **(c)**

FIGURE 8.28

Keep in mind that the representations shown in Example 8.18 are not unique. Because the figures could be oriented differently on the coordinate axes, many other representations are possible. The ones shown, however, are convenient for many purposes.

COORDINATE PROOFS

When we use coordinates to prove a theorem about a geometric figure, there are two steps that we must complete.

1. *Assign appropriate coordinates to the figure.* As Example 8.18 illustrates, this means making certain that the figure is as general as necessary. For example, if we plan to prove a theorem about triangles in general, we should not draw a special case of a triangle such as a right triangle or an isosceles triangle. Placing sides on the coordinate axes and using 0 for as many coordinates as possible will often simplify calculations.

2. *Use the assigned coordinates to write a proof.* Some of the tools we have at hand to write coordinate proofs are
   (a) Slope (which is useful for showing that segments are parallel or perpendicular)
   (b) The midpoint formula (which is useful for showing when segments are bisected, for example)
   (c) The distance formula (which is useful for showing that segments are congruent)
   (d) The equation of a line (which is useful for finding points of intersection)

The following examples show how these two steps are applied to construct a coordinate proof of a theorem. Each of the theorems was proved earlier in the text using standard Euclidean techniques.

EXAMPLE 8.19   Use coordinates to prove that the diagonals of a rectangle are congruent.

SOLUTION   Using the steps mentioned in the preceding discussion, we must (1) assign coordinates to the vertices of a rectangle *ABCD* (Figure 8.29), and (2) show that its diagonals are congruent, or that $\overline{AC} \cong \overline{BD}$ in Figure 8.29.

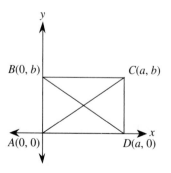

FIGURE 8.29

Using the distance formula, we have

$$AC = \sqrt{(a - 0)^2 + (b - 0)^2} = \sqrt{a^2 + b^2}$$

and

$$DB = \sqrt{(a - 0)^2 + (0 - b)^2} = \sqrt{a^2 + b^2}$$

Thus, $AC = BD$. Therefore, the diagonals are congruent.

EXAMPLE 8.20   Use coordinates to prove that the diagonals of a rhombus are perpendicular.

SOLUTION
(1) First we must assign coordinates to a rhombus $ABCD$. Since a rhombus is a parallelogram, we can use the vertices for a parallelogram obtained in Example 8.18 and shown in Figure 8.30(a). The fact that this figure is a rhombus means that $AB = AD$, or $\sqrt{a^2 + b^2} = c$, or $a^2 + b^2 = c^2$. We will use this relationship in part (2).

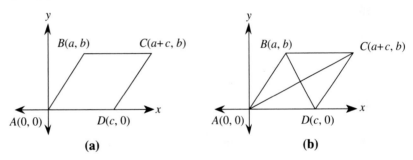

FIGURE 8.30

(2) We must show that $\overline{AC} \perp \overline{BD}$ [Figure 8.30(b)]. We will show that the product of their slopes is -1.

$$\text{Slope of } \overline{AC} = \frac{b - 0}{(a + c) - 0} = \frac{b}{a + c}$$

$$\text{Slope of } \overline{BD} = \frac{b - 0}{a - c} = \frac{b}{a - c}$$

Thus the product of their slopes is

$$\text{(i)} \quad \left(\frac{b}{a + c}\right)\left(\frac{b}{a - c}\right) = \frac{b^2}{a^2 - c^2}$$

Recall that in (1) we found that $a^2 + b^2 = c^2$, or $b^2 = c^2 - a^2$. Substituting this expression in place of b^2 in equation (i) we have that the product of the slopes is $\dfrac{b^2}{a^2 - c^2} = \dfrac{c^2 - a^2}{a^2 - c^2} = -1$.   Therefore, $\overline{AC} \perp \overline{BD}$, that is, the diagonals of this general rhombus are perpendicular.

EXAMPLE 8.21   Use coordinates to prove that for any quadrilateral *ABCD* whose sides have midpoints *W*, *X*, *Y*, and *Z* as shown in Figure 8.31(a), *WXYZ* is a parallelogram.

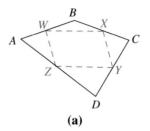

(a)

FIGURE 8.31

SOLUTION

(1) First we place *ABCD* on the coordinate plane. Point *A* is placed at the origin and $\overline{AD}$ is on the *x*-axis [Figure 8.31(b)].

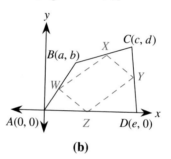

(b)

FIGURE 8.31

(2) Now we must show that *WXYZ* is a parallelogram. We will show $\overline{WX} \parallel \overline{ZY}$ and $\overline{WZ} \parallel \overline{XY}$. The coordinates of *WXYZ* are found using the midpoint formula as follows:

$$W = \left(\frac{a+0}{2}, \frac{b+0}{2}\right) = \left(\frac{a}{2}, \frac{b}{2}\right) \qquad X = \left(\frac{a+c}{2}, \frac{b+d}{2}\right)$$

$$Y = \left(\frac{c+e}{2}, \frac{d+0}{2}\right) = \left(\frac{c+e}{2}, \frac{d}{2}\right) \qquad Z = \left(\frac{e}{2}, 0\right)$$

$$\text{Slope of } \overline{WX} = \frac{\dfrac{b}{2} - \dfrac{b+d}{2}}{\dfrac{a}{2} - \dfrac{a+c}{2}} = \frac{d}{c}$$

$$\text{Slope of } \overline{ZY} = \frac{\dfrac{d}{2} - 0}{\dfrac{c+e}{2} - \dfrac{e}{2}} = \frac{d}{c}$$

Thus, $\overline{WX} \parallel \overline{ZY}$ because they have the same slope. It can be shown similarly that $\overline{WZ} \parallel \overline{XY}$ (details are left for the problem set). Because the opposite sides of WXYZ are parallel, it is a parallelogram.

Solution to Applied Problem

Find the equation of each line in slope-intercept form. The slope of line 1 is $m = \tan 48° \approx 1.1106$. Line 1 has a y-intercept of 0. So line 1 has equation $y = 1.1106x$. The slope of line 2 is $m = -\tan 60° \approx -1.7321$. Using this slope and the point (8, 0), we have

$$y - 0 = -1.7321 (x - 8) \quad \text{or} \quad y = -1.7321x + 13.856$$

Solving these equations simultaneously, we find $x \approx 4.9$ and $y \approx 5.4$. Hence the fire is at the point (4.9, 5.4).

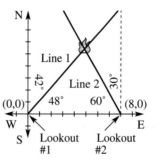

GEOMETRY AROUND US

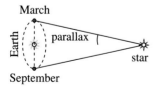

If you hold up one finger and look at it it with your left eye closed, then open your left eye and close your right eye, the finger will seem to have moved. This apparent change in location is called parallax, and it can be used to calculate the distance to nearby stars. By observing the star from two opposite points in Earth's orbit (6 months apart), astronomers can determine the angle between these two lines of sight. This angle and the known distance between the two observation points can be used to calculate the distance to the star. Astronomers have defined a parsec, a unit of distance, in terms of parallax. One parsec is defined to be the distance at which a star would have a parallax of one second of arc.

PROBLEM SET 8.4

PROOFS

1. Two vertices of a trapezoid have been assigned coordinates in the following figure.

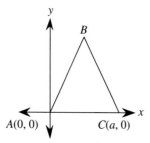

Determine appropriate coordinates for the remaining two vertices. Use the coordinates to verify that the resulting figure is a trapezoid and that it is not a special case.

2. In each of the following figures, two vertices of an isosceles trapezoid have been assigned coordinates.

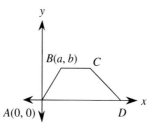

Determine appropriate coordinates for the remaining two vertices. Use the coordinates to verify that the resulting figure is an isosceles trapezoid in each case.

3. A theorem is to be proved about an isosceles triangle. Two vertices have been assigned coordinates as shown.

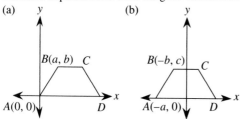

Determine appropriate coordinates for the third vertex. Use the coordinates to verify that the triangle is isosceles, but not necessarily equilateral.

4. A theorem involving right triangles is to be proved.
 (a) The sides have been positioned along the coordinate axes as shown.

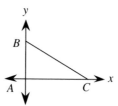

Label the coordinates of the vertices. Why is △ABC a right triangle?
 (b) Suppose that the vertices of the right triangle were assigned the coordinates $(0, 0)$, $(0, a)$, and $(a, 0)$. Would these coordinates be appropriate? Why or why not?

5. (a) A kite has been placed on the coordinate system as shown next, and two vertices have been assigned coordinates.

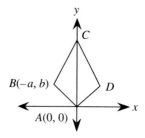

Determine appropriate coordinates for the remaining two vertices. Use the coordinates to verify that the resulting figure is a kite.
 (b) Suppose that vertices for a kite were given as $A(0, 0)$, $B(-a, a)$, $C(0, 2a)$, and $D(a, a)$. Would these coordinates be appropriate?

6. A student asked to draw a general quadrilateral and label the vertices drew the following figure.

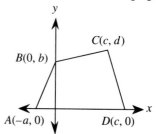

(a) Why is this representation of a general quadrilateral not appropriate?

(b) How would you change one vertex of the figure so that it is acceptable?

7. Complete the proof in Example 8.21 by showing that $\overline{WZ} \parallel \overline{XY}$.

8. For $\triangle ABC$ shown, write the equation of the median from vertex A.

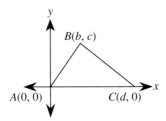

Use coordinates to write the following proofs.

9. Prove: If $ABCD$ is a square, then $\overline{AC} \perp \overline{BD}$.

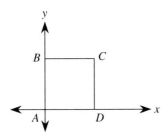

10. Prove: If $ABCD$ is a parallelogram, then $\overline{AC}$ and $\overline{BD}$ bisect each other.

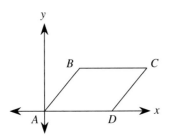

11. Prove: The opposite sides of a parallelogram are congruent.

12. Prove: The median from the vertex of an isosceles triangle is perpendicular to the base.

13. Prove: If M is the midpoint of $\overline{BC}$, where $\triangle ABC$ is a right triangle with $\angle A = 90°$, then $AM = BM = CM$.

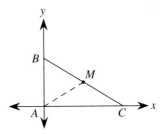

14. Prove: If M and N are midpoints of $\overline{AB}$ and $\overline{BC}$ respectively, then $\overline{MN} \parallel \overline{AC}$ and $MN = \frac{1}{2}AC$.

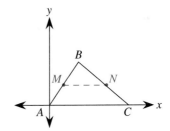

15. Prove: If $ABCD$ is an isosceles trapezoid, then $AC = BD$.

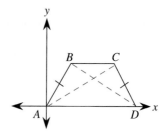

16. Prove: If M and N are midpoints of the legs of trapezoid $ABCD$, then $MN = \frac{1}{2}(AD + BC)$.

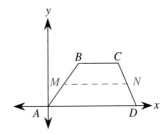

17. Prove: If *ABCD* is a rectangle and *M*, *N*, *O*, and *P* are midpoints of the sides as indicated, then *MNOP* is a rhombus.

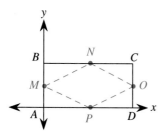

18. Prove: If *ABCD* is a kite and *M*, *N*, *O*, and *P* are midpoints of the sides as indicated, then *MNOP* is a rectangle.

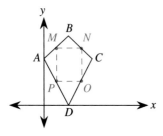

19. Prove: The segments connecting consecutive midpoints of the sides of a rhombus form a rectangle.

20. Prove: The segments connecting the midpoints of opposite sides of a quadrilateral bisect each other.

21. Prove: If *M* and *N* are midpoints of the congruent sides of an isosceles triangle as indicated, then *AN* = *CM*.

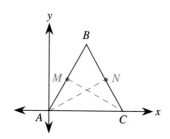

22. Prove: If *AN* and *CM* are altitudes to the congruent sides of an isosceles triangle, then *AN* = *CM*.

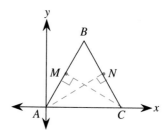

23. Prove: If the diagonals of a parallelogram are congruent, then it is a rectangle.

24. Prove: If the diagonals of a parallelogram are perpendicular, then it is a rhombus.

25. Prove: The perpendicular bisectors of the sides of a triangle are concurrent. [HINT: Write an equation for each of the three perpendicular bisectors. Then determine the point of intersection for each pair of lines.]

26. Prove: The altitudes of a triangle are concurrent. [HINT: Write an equation for each of the three altitudes. Then determine the point of intersection for each pair of lines.]

27. Prove: The medians of a triangle are concurrent. [HINT: Write an equation for each of the three medians. Then determine the point of intersection for each pair of lines.]

Solution to Initial Problem

A treasure has been hidden in a desert location. A map shows $\triangle ABC$, where B is 50 miles NE of A and C is 80 miles east of A [Figure 8.32(a)]. You are told that the treasure is at the midpoint of the median from A. Locate the treasure.

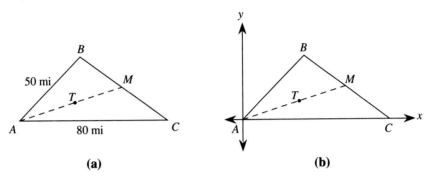

FIGURE 8.32

Place A at $(0, 0)$ and $\overline{AC}$ on the x-axis [Figure 8.32(b)]. The coordinates of C are $(80, 0)$. Because B is 50 mi NE, its coordinates are $(25\sqrt{2}, 25\sqrt{2})$. The midpoint M of $\overline{BC}$ is $\left(\dfrac{80 + 25\sqrt{2}}{2}, \dfrac{25\sqrt{2}}{2}\right)$. Finally, point T, the midpoint of $\overline{AM}$, has coordinates $\left(\dfrac{80 + 25\sqrt{2}}{4}, \dfrac{25\sqrt{2}}{4}\right)$ [Figure 8.32(c)].

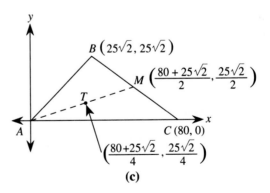

(c)

FIGURE 8.32

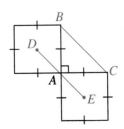

Additional Problems Where the Strategy "Use Coordinates" Is Useful

1. An explorer leaves base camp and travels 4 km north, 3 km west, and 2 km south. How far is he from the base camp?
2. Prove that $\overline{BC}$ is parallel to $\overline{DE}$ where D and E are the centers of the squares. [HINT: Use slopes.]

Writing for Understanding

1. A student claims that it doesn't make sense to say that a vertical line has "no slope." He says it seems more reasonable to say that a horizontal line has no slope, because it is flat. How would you explain the distinction, in terms of slope, between horizontal and vertical lines so that it makes sense to this student?

2. Give a geometric interpretation of what happens when you solve a system of two linear equations simultaneously. Be sure to include all possible outcomes. How do the possibilities change when there are more than two equations? When at least one of the equations is not linear?

3. Select two problems in Section 8.4, one whose proof is easier to do using coordinates than using triangle congruence principles as in Chapter 5 and a second that is easier to do using triangle congruence than coordinates. Based on these two examples, state which method you prefer and why.

4. The "Geometry Around Us" at the end of Section 8.2 relates slope to the pitch of a roof. Do some research on the pitch of a roof, focussing on the advantages of one pitch over another, taking into account such factors as space, construction materials, the climate, and so on. Summarize your findings.

CHAPTER REVIEW

Following is a list of key vocabulary, notation, and ideas for this chapter. Mentally review these items and, where appropriate, write down the meaning of each term. Then restudy the material that you are unsure of before proceeding to take the chapter test.

Section 8.1—Coordinates and Distance in the Plane

Vocabulary/Notation

origin 371	ordered pair 372
x-axis 371	x-coordinate 372
y-axis 372	y-coordinate 372
coordinate plane 372	projection of a point 372

Main Idea/Result

1. The distance formula and the midpoint formula can be used to verify properties of geometric figures.

Section 8.2—Slope

Vocabulary/Notation

slope of a line 381	run 381
rise 381	slope of a line segment 381

Main Ideas/Results

1. Parallel lines that are not vertical have the same slope.

2. If two lines are perpendicular and neither line is vertical, the product of their slopes is -1.

3. Slope can be used to verify properties of geometric figures.

Section 8.3—Equations of Lines and Circles

Vocabulary/Notation

point-slope form 392   standard form 392
slope-intercept form 392   system of equations 394

Main Ideas/Results

1. The equation of a line can be written in several forms.
2. Given (i) two points on a line or (ii) one point and the slope of the line, an equation of a line can be written.
3. A system of two linear equations has zero, one, or infinitely many solutions.
4. The simultaneous solution(s) to a system of equations can be determined in several ways.
5. Given the center and radius of a circle, an equation of the circle can be determined.

Section 8.4—Problem Solving Using Coordinates

Main Idea/Result

1. Coordinates can be used to solve problems and prove properties of geometric figures.

PEOPLE IN GEOMETRY

Maria Agnesi's (1718–1799) mathematical fame rests on the highly regarded *Instituzioni Analitiche*, a 1020-page, two-volume presentation of algebra, analytic geometry, and calculus. Published in 1748, it brought order and clarity to the mathematics invented by Descartes, Newton, Leibniz and others in the seventeenth century. Maria was the eldest of 21 children in a wealthy Italian family. She was a gifted child, with an extraordinary talent for languages. Her parents encouraged her to excel, and she received the best schooling available. Maria began work on *Instituzioni Analitiche* at age twenty and finished it ten years later, supervising its printing on presses installed in her home. After its publication, she was appointed honorary professor at the University of Bologna. Instead, Maria decided to dedicate her life to charity and religious devotion, and she spent the last 45 years of her life caring for the sick, aged, and indigent.

CHAPTER 8 TEST

TRUE-FALSE

Mark as true any statement that is always true. Mark as false any statement that is never true or that is not necessarily true. Be able to justify your answers.

1. If the slope of a line is 2, then the slope of any line perpendicular to it is -2.

2. If points A, B, and C are collinear, then $AB + BC = AC$.

3. If points A, B, and C are collinear and lie on a line that is not vertical, then the slope of $\overline{AB}$ equals the slope of $\overline{AC}$.

4. If two line segments bisect each other, then they intersect at their midpoints.

5. Given the coordinates of the vertices of $ABCD$, you can use the slope formula to determine whether it is a rhombus.

6. A system of two linear equations has exactly one simultaneous solution.

7. A horizontal line has a slope of 0.

8. The line $ax + by = c$ (where $b \neq 0$) has a slope of $\dfrac{a}{b}$.

9. If each coordinate of each vertex of $\triangle ABC$ is doubled, the area of the resulting triangle will be twice the area of $\triangle ABC$.

10. The graph of the line $y = 2 - 5x$ has y-intercept 2.

EXERCISES/PROBLEMS

11. For the points $A(3, -2)$ and $B(9, 4)$, find the following:
    (a) The midpoint of $\overline{AB}$
    (b) The distance $\overline{AB}$
    (c) The slope of $\overline{AB}$
    (d) The equation of the line containing $\overline{AB}$

12. Determine whether the three points $A(-30, 400)$, $B(-16, 232)$, and $C(20, -200)$ are collinear. Explain.

13. Sketch the graphs of the following equations.
    (a) $4x - 3y = 6$ (b) $y = -4x$
    (c) $x = 2$ (d) $x^2 + y^2 = 9$
    (e) $(x + 5)^2 + (y - 2)^2 = 36$

14. Solve each of the following systems of equations. If there is no solution, explain why not.
    (a) $3x - y = 11$ and $5x + 4y = 24$
    (b) $y = -x + 5$ and $3x + 3y = 13$
    (c) $5x + 3y = 15$ and $(x - 4)^2 + (y - 3)^2 = 4$

15. Write an equation of the perpendicular bisector of $\overline{AB}$ for $A(1, -6)$ and $B(5, 2)$.

16. The coordinates of three vertices of parallelogram $ABCD$ are shown. Determine the coordinates of B.

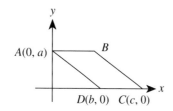

17. Quadrilateral $ABCD$ has vertices $A(-1, 1)$, $B(-2, 6)$, $C(3, 7)$, and $D(4, 2)$. Use these coordinates to verify that the diagonals of $ABCD$ are congruent.

18. A circle has center $O(-1, 4)$ and passes through the point $(5, 0)$. Write an equation of the circle.

19. Quadrilateral $ABCD$ has vertices $A(-5, 4)$, $B(8, 6)$, $C(12, 0)$, and $D(-1, -2)$. What type of quadrilateral is $ABCD$? Justify your answer.

20. The coordinates of the midpoints of the sides of a triangle are $M_1(6, 5)$, $M_2(5, -1)$, and $M_3(-1, 5)$. Find the coordinates of the vertices of the triangle.

PROOFS

21. Use coordinates to prove that if M_1, M_2, and M_3 are midpoints of the sides of equilateral $\triangle ABC$, then $\triangle M_1 M_2 M_3$ is an equilateral triangle.

22. Use coordinates to prove that if the diagonals of a parallelogram are perpendicular, then the parallelogram is a rhombus.

APPLICATIONS

23. A highway rises 220 feet in 0.8 mile. What is the slope of the road? What is the percent grade?

24. A copper wire has a diameter of 0.2 cm.
    (a) If x represents the length of the wire in centimeters and y represents the total volume of the wire, write a linear equation expressing y in terms of x.
    (b) What is the slope of the line whose equation you wrote in (a)? What does it mean?
    (c) The density of copper is 8.94 g/cm³. Again, let x represent the length of the wire and now let y represent the total mass of the wire. Write a linear equation expressing y in terms of x.
    (d) What is the slope of the line whose equation you wrote in (c)? What does it mean?

9

Transformation Geometry

Coordinate geometry provides an alternate way to verify relationships in geometry. Another more modern method, transformation geometry, involves a more dynamic approach. Here properties of geometric figures are studied with respect to translations, rotations, reflections, and size transformations. The art of Maurits Escher utilizes the concepts associated with transformation geometry. In his *Circle Limit 1* shown below, Escher pictures flying figures that can be rotated to each other, others that can be reflected and reduced in size to another, and so on. Many objects in nature can also be described using the language of transformation geometry. For example, the silhouette of a fir tree has reflection symmetry with respect to a vertical line through the center of its trunk, a starfish has 108° rotational symmetry because it rotates to itself after a turn equal to one fifth of 360°, and so on.

STRATEGY 14: USE SYMMETRY

In Chapter 2 we observed symmetry in many geometric shapes. If a figure has reflection symmetry, we know that angles and sides that reflect onto corresponding angles and sides are congruent. Similar information can be deduced from figures with rotation symmetry. By using a reflection or rotation, the following problem can be transformed into a simpler equivalent problem using symmetry.

INITIAL PROBLEM

Houses A and B are to be connected to a television cable line, l, at a transformer at some point P. Where should P be located so that the sum of the distances, $AP + PB$, is as small as possible?

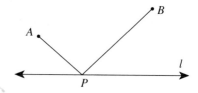

CLUES

The Use Symmetry strategy may be appropriate when

· Geometry problems involve transformations.
· Interchanging values does not change the representation of the problem.
· Symmetry limits the number of cases that need to be considered.
· Pictures or algebraic procedures appear to be symmetric.

A solution for the initial problem above is on page 456.

INTRODUCTION

Reflection and rotation symmetries were studied in Chapter 2. In this chapter we extend the notion of symmetry to a study of geometric figures using transformations. In Section 9.1 we study translations (slides), rotations (turns), and reflections (flips) and combinations of them. In Section 9.2, similitudes, transformations that preserve shapes of objects, are studied. Section 9.3 is devoted to solving problems using translations, rotations, reflections, and glide reflections.

9.1
ISOMETRIES AND CONGRUENCE

Applied Problem

An astronomer is constructing a star map to show the position of the constellation Cassiopeia at different times during one evening. The following map shows its position with respect to the North Pole at 9:00 P.M.

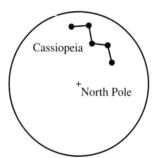

Cassiopeia

+ North Pole

What is the position of the constellation at 3:00 A.M. the next morning?

TRANSFORMATIONS

A translation, or slide, is one example of a transformation. To think of a slide, imagine a 'directed' line segment, that is, a line segment where one end of the segment has been designated by an arrowhead. We call such a directed line segment a **vector**. Associated with each vector is a distance (the length of the vector) and a direction (the measure of the angle the vector makes with the positive x-axis) [Figure 9.1(a)]. Now imagine moving every point in a plane the same distance and in the same direction as indicated by the vector v in Figure 9.1(a) [Figure 9.1(b)].

Distance — Direction — v

(a)

FIGURE 9.1

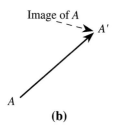

Image of A

(b)

FIGURE 9.1

A 1-1 correspondence that assigns every point A in a plane to another point A' in the plane is called a **transformation**. Point A' is called the **image** of A. In the case of a transformation described by a vector, the image of an A point is at the end of the arrow's tip if the other end of the vector is on the point A [Figure 9.1(b)]. We can denote such a vector as $\overrightarrow{AA'}$. [NOTE: Although the arrow written above two letters has represented a ray thus far, in this chapter this notation will represent a vector of finite length unless stated otherwise. In this context, an arrow such as the one in Figure 9.1(b) also has the *finite* length AA' whereas a ray has infinite length.] The point A is called the **preimage** of A'. One or many of the points may be assigned to themselves under a transformation. For example, in a rotation we will show that the center corresponds to itself.

We will study several types of transformations in this chapter, including translations, rotations, and reflections. Our emphasis will be on their properties and how they can be used to solve problems.

TRANSLATIONS

Every vector v defines a transformation in that it assigns to every point A in a plane, a point A' with the distance and direction determined by v [Figure 9.1(b)]. This type of transformation is called a **translation**. Informally, a translation of the plane can be thought of as a slide, sliding all the points of the plane the same distance and same direction.

EXAMPLE 9.1   Describe the image of $\triangle ABC$ under the translation determined by the vector v in Figure 9.2(a).

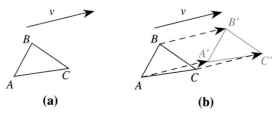

(a)   **(b)**

FIGURE 9.2

SOLUTION   Because the vector v takes all points the same distance and direction, $\triangle ABC$ is assigned to $\triangle A'B'C'$ as shown in Figure 9.2(b). Notice that vectors $\overrightarrow{AA'}$, $\overrightarrow{BB'}$, and $\overrightarrow{CC'}$ all have the same length and direction as the vector v.   •

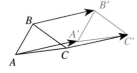

FIGURE 9.3

In Figure 9.2(b), it appears that $\triangle ABC \cong \triangle A'B'C'$. We can prove this is true by studying Figure 9.3. The translation that is determined by vector v forms the parallelogram $ABB'A'$ because $\overrightarrow{AA'}$ and $\overrightarrow{BB'}$ are congruent and parallel. Therefore, $\overline{AB} \cong \overline{A'B'}$ because they are opposite sides of a parallelogram. Similarly, $\overline{BC} \cong \overline{B'C'}$ and $\overline{AC} \cong \overline{A'C'}$. Thus, $\triangle ABC \cong \triangle A'B'C'$ by SSS Congruence.

By corresponding parts, $\angle ABC \cong \angle A'B'C'$, $\angle BAC \cong \angle B'C'A'$, and $\angle ACB \cong \angle A'C'B'$.

Theorem 9.1

Properties of Translations

1. Translations take lines to lines, rays to rays, and line segments to line segments.
2. Translations preserve distance.

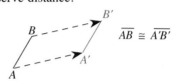
$$\overline{AB} \cong \overline{A'B'}$$

3. Translations preserve angle measure.

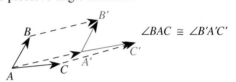
$$\angle BAC \cong \angle B'A'C'$$

4. Translations preserve perpendicularity.
5. Translations preserve parallelism.

Proof

Part 1 Because the proof of this result requires sophisticated ideas that are beyond the scope of this book, we will accept this result as we do postulates.

Part 2 This result was proved in the paragraph preceding Theorem 9.1.

Part 3 This proof also was given in the paragraph preceding Theorem 9.1.

Part 4 Because perpendicular lines form right angles and angle measure is preserved by part 3, it follows that perpendicularity is preserved.

Part 5 If two parallel lines are cut by a transversal they form congruent alternate interior angles. By part 1, the lines go to lines under translations. By part 3, angle measure is preserved. Therefore, the alternate interior angles of the new set of lines formed by the transversal are congruent. So the lines are parallel by Theorem 5.1.

Part 5 of Theorem 9.1 states that transformations preserve parallelism. That is, if two lines are parallel, their images are also parallel. However, as suggested in Figure 9.3, under a translation, the image of any line is parallel to the original line. For example, $\overline{AB}$ and $\overline{A'B'}$ in Figure 9.3 are parallel.

Another property of transformations that is important is orientation. Triangle $\triangle ABC$ is said to have **clockwise orientation** if its vertices when read "A, B, C"

are pictured in a clockwise orientation. Otherwise, the triangle is said to have a **counterclockwise orientation**. In Figure 9.4, $\triangle ABC$ has clockwise orientation and $\triangle DEF$ has counterclockwise orientation.

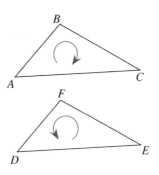

FIGURE 9.4

Check the orientation of a triangle and its image under several translations and you will find that translations preserve orientation of triangles (and all other shapes). Later in this section we will study transformations that do not preserve orientation.

Coordinates can be used to illustrate the effects of a translation as shown next.

EXAMPLE 9.2   A vector v, which represents a translation, is graphed on the coordinate system in Figure 9.5(a).

(a) A triangle has vertices $A(-1, 1)$, $B(4, -2)$, and $C(3, 6)$. Find the image of $\triangle ABC$ under the translation described by v. That is, find $\triangle A'B'C'$.

(b) Verify that $\triangle ABC \cong \triangle A'B'C'$ and that the translation preserves orientation.

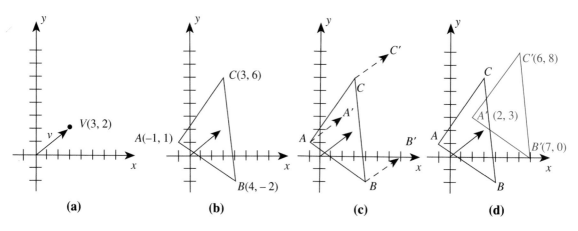

FIGURE 9.5

SOLUTION

(a) △ABC is graphed in Figure 9.5(b). Note that △ABC has a counterclockwise orientation. The effect of vector v will be to translate each point of △ABC three units to the right and two units up. If we apply the translation to each vertex of △ABC we get the three points A', B', and C' shown in Figure 9.5(c). Finally, connecting the vertices, we obtain △$A'B'C'$, the image of △ABC in Figure 9.5(d).

(b) We can use the distance formula and SSS congruence to verify that △$ABC \cong △A'B'C'$.

In △ABC,

$$AB = \sqrt{(4 + 1)^2 + (-2 - 1)^2} = \sqrt{34}$$

$$BC = \sqrt{(3 - 4)^2 + (6 + 2)^2} = \sqrt{65}$$

$$AC = \sqrt{(3 + 1)^2 + (6 - 1)^2} = \sqrt{41}$$

In △$A'B'C'$,

$$A'B' = \sqrt{(7 - 2)^2 + (0 - 3)^2} = \sqrt{34}$$

$$B'C' = \sqrt{(6 - 7)^2 + (8 - 0)^2} = \sqrt{65}$$

$$A'C' = \sqrt{(6 - 2)^2 + (8 - 3)^2} = \sqrt{41}$$

Because $AB = A'B'$, $BC = B'C'$, and $AC = A'C'$, we have △$ABC \cong △A'B'C'$ by SSS congruence. We have shown that the translation determined by the vector v takes △ABC to congruent triangle △$A'B'C'$. Also, because both triangles have counterclockwise orientation, the translation preserves orientation.

ROTATIONS

Another type of transformation is a rotation, or a turn. A rotation is determined by a point, O, and a "directed angle." A **directed angle** is an angle where one side is identified as the initial side and a second side is the terminal side. An angle can be directed either clockwise or counterclockwise. In Figure 9.6(a), A is rotated counterclockwise 60° around O to A'. Here $\overline{OA}$ is the initial side of $\angle AOA'$ and $\overline{OA'}$ is its terminal side. Angles that are directed counterclockwise are assigned a positive measure. Thus, we say that the measure of directed angle $\angle AOA'$ is 60. In Figure 9.6(b), B is rotated clockwise 90° around O to B' and so directed $\angle BOB'$ has measure $-90°$ to indicate its clockwise rotation. The point O is called the **center** of the rotation in each case.

Notice that under these rotations, $OA = OA'$ and $OB = OB'$. Also, the orientation of a directed angle is indicated by the way the angle is *named*. For example, the directed angle $\angle AOA'$ in Figure 9.6(a) is directed counterclockwise, and hence is positive, whereas the directed angle $\angle A'OA$ has a clockwise direction, hence its measure is $-60°$.

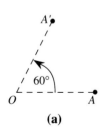

(a)

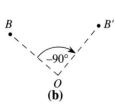

(b)

FIGURE 9.6

Based on the directed angles in Figure 9.6 we can state the following formal definition of a **rotation**: The image of a point X under the rotation determined by the directed angle $\angle AOB$ in Figure 9.7(a) is the point X' where

1. $OX = OX'$
2. $\angle XOX' = \angle AOB$ as *directed* angles [Figure 9.7(b)]

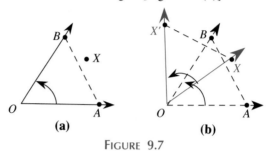

(a)   **(b)**

FIGURE 9.7

Once a rotation is defined by its center and its directed angle (whose vertex is the **center of rotation**), the image of every point in the plane is determined. That is, we can find the image A' of any point A in the plane.

The center, O, of a rotation always corresponds to itself, thus we call it a fixed point. In general, a point A is called a **fixed point** under a transformation if A and its image A' are the same point. Thus, the center of any rotation is a fixed point. The transformation that leaves all points fixed is called the **identity transformation**. The translation determined by the zero vector is the identity transformation. All other translations have no fixed points. The rotation whose directed angle is 360° is also the identity transformation since each point corresponds to itself.

Next we consider the effect a rotation has on a triangle.

EXAMPLE 9.3   Describe the image of $\triangle ABC$ under the rotation with center O and directed angle $\angle XOX'$ as shown in Figure 9.8(a).

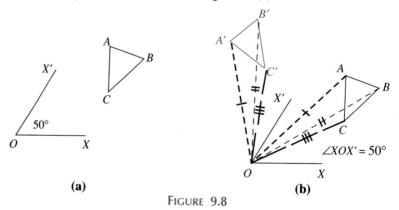

(a)   **(b)**

FIGURE 9.8

SOLUTION   The respective images A', B', C' are shown in Figure 9.8(b). Notice that $OA = OA'$, $OB = OB'$, and $OC = OC'$. Also, $\angle AOA' = \angle BOB' = \angle COC' = 50°$.

It appears that $\triangle A'B'C' \cong \triangle ABC$ but it has been rotated 50° counterclockwise around O. Due to the nature of the rotation in Figure 9.8(b), we can observe the following:

1. $OA = OA'$
2. $OB = OB'$
3. $\angle AOA' = \angle BOB' = 50°$

If we subtract $\angle AOB'$ from $\angle AOA'$ and $\angle BOB'$, we have

4. $\angle A'OB' = \angle AOB$

Thus, by SAS Congruence, we can conclude that $\triangle AOB \cong \triangle A'OB'$. Therefore, $AB = A'B'$. Similarly, $AC = A'C'$ and $BC = B'C'$. Using SSS congruence we have that $\triangle A'B'C' \cong \triangle ABC$. This shows, by example, that a rotation takes a triangle to a congruent triangle. Using methods analogous to those we used for translations, we can prove the following theorem.

Theorem 9.2

Properties of Rotations

1. Rotations take lines to lines, rays to rays, and line segments to line segments.
2. Rotations preserve distance.

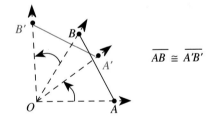

$$\overline{AB} \cong \overline{A'B'}$$

3. Rotations preserve angle measure.

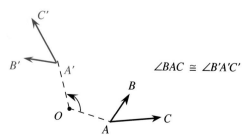

$$\angle BAC \cong \angle B'A'C'$$

4. Rotations preserve perpendicularity.
5. Rotations preserve parallelism.

The verification of properties 2–5 is left for the problem set. Several examples should make it clear that rotations preserve orientation just as translations do.

Although Theorems 9.1 and 9.2 show that translations and rotations have many similar properties, they differ in some ways. Except for the zero translation, translations have no fixed points. However, rotations always have at least one fixed point, namely the center. Also, the image of a line under a translation is parallel to the original line. But this is not generally true in a rotation. For example, in Figure 9.8(b), the image of $\overline{AB}$, namely $\overline{A'B'}$, is not parallel to $\overline{AB}$.

REFLECTIONS

In Section 2.2 we observed that some geometric figures were symmetric with respect to a line. This idea is associated with a type of transformation called a reflection. A **reflection with respect to line** l is defined by describing the location of the image of each point of the plane as follows:

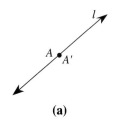

(a)

1. If A is a point on l, then $A = A'$ (that is, a point on the line of reflection is its own image) [Figure 9.9(a)].

2. If A is not on l, then l is the perpendicular bisector of $\overline{AA'}$ [Figure 9.9(b)].

Next we consider the effect of a reflection on a triangle.

EXAMPLE 9.4 Describe the image of $\triangle ABC$ under the reflection with respect to line l as shown in Figure 9.10(a).

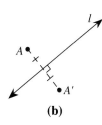

(b)

FIGURE 9.9

SOLUTION Line l will be the perpendicular bisector of $\overline{AA'}$. So we construct a line through A that is perpendicular to l, then mark off $\overline{A'P} \cong \overline{AP}$ [Figure 9.10(b)]. Then we find points B' and C' in the same way. The respective images A', B', C' of A, B, and C are shown in Figure 9.10(c). Here, again, it appears that $\triangle A'B'C' \cong \triangle ABC$. But notice that while the orientation of $\triangle ABC$ is clockwise, the orientation of $\triangle A'B'C'$ is counterclockwise.   •

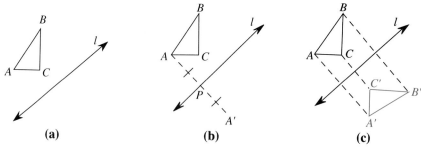

(a)   **(b)**   **(c)**

FIGURE 9.10

It can be proved that $\triangle A'B'C' \cong \triangle ABC$. This result is a consequence of the next theorem whose proof is left for the problem set.

Theorem 9.3

Properties of Reflections

1. Reflections take lines to lines, rays to rays, and line segments to line segments.
2. Reflections preserve distance.

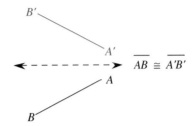

$$\overline{AB} \cong \overline{A'B'}$$

3. Reflections preserve angle measure.

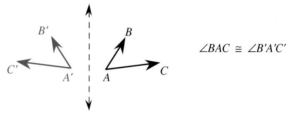

$$\angle BAC \cong \angle B'A'C'$$

4. Reflections preserve perpendicularity.
5. Reflections preserve parallelism.

Notice that $\overline{AB}$ is *not* parallel to its image $\overline{A'B'}$ in Figure 9.10(c). Therefore, reflections are similar to rotations in that the image of a line is not necessarily parallel to the original line. In Figure 9.10(c), we observed that $\triangle ABC$ has clockwise orientation and $\triangle A'B'C'$ has counterclockwise orientation. Thus, reflections differ from both translations and rotations since orientation is not preserved.

GLIDE REFLECTIONS

To motivate the final type of transformation in this section, consider the footprints in the sand pictured in Figure 9.11.

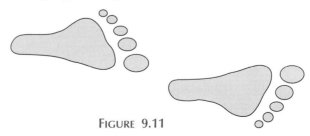

FIGURE 9.11

It is impossible to translate or rotate the left foot to the right foot since the feet have a different orientation. Also, no reflection line can be found to reflect one foot onto the other. But, as Figure 9.12 shows, one foot can be obtained from the other by performing a reflection followed by a translation.

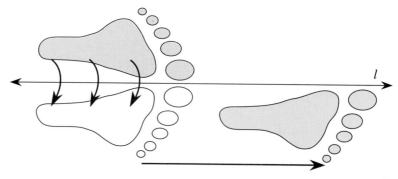

FIGURE 9.12

Notice that the left foot also could have been transformed to the right foot by first translating along *l*, then reflecting with respect to line *l*.

The transformation described in Figure 9.12 is called a **glide reflection**, which means a reflection followed by a translation that is not the identity translation. For simplicity, we will assume that the translation is in a direction parallel to the line of reflection. However, it can be shown that this assumption is not necessary. Because a glide reflection is composed of a reflection and a translation, it has all the properties common to both of these as stated in the next theorem.

Theorem 9.4

Properties of Glide Reflections

1. Glide reflections take lines to lines, rays to rays, and line segments to line segments.
2. Glide reflections preserve distance.
3. Glide reflections preserve angle measure.
4. Glide reflections preserve perpendicularity.
5. Glide reflections preserve parallelism.

Notice that a glide reflection reverses the orientation of geometric figures because the reflection reverses orientation, but the translation which we used in combination with the reflection does not. The results listed in Theorem 9.4 follow immediately due to the fact that glide reflections are combinations of a translation and a reflection, each of which satisfies 1 to 5 in the theorem.

We can use coordinates to describe the effect of a glide reflection as shown in the next example.

EXAMPLE 9.5 $\triangle ABC$ with vertices $A(1, 0)$, $B(2, 3)$, and $C(5, 1)$ is shown in Figure 9.13(a).

(a) Find the image of $\triangle ABC$ under the glide reflection defined by the translation from $(0, 0)$ to $(0, -2)$, followed by a reflection with respect to the y-axis.

(b) If a point has coordinates (s, t), what are the coordinates of its image under this glide reflection?

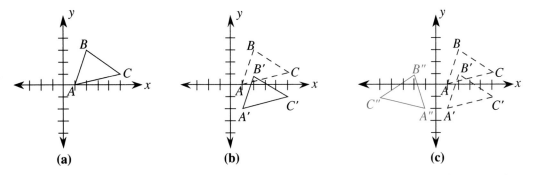

(a)   (b)   (c)

FIGURE 9.13

SOLUTION

(a) The translation from $(0, 0)$ to $(0, -2)$ shifts each point of the plane down two units. The image $\triangle A'B'C'$ of $\triangle ABC$ under this translation is shown in Figure 9.13(b). A reflection with respect to the y-axis flips $\triangle A'B'C'$ across the y-axis to yield $\triangle A''B''C''$ [Figure 9.13(c)]. Thus $\triangle A''B''C''$ is the image of $\triangle ABC$ under this glide reflection.

(b) Consider the images of points A, B, and C under this glide reflection. The points and their images are listed in Table 9.1.

Point	Image
$A(1, 0)$	$A''(-1, -2)$
$B(2, 3)$	$B''(-2, 1)$
$C(5, 1)$	$C''(-5, -1)$
(s, t)	$(?, ?)$

TABLE 9.1

In each case, the y-coordinate of the original point is decreased by 2 because of the downward translation. The x-coordinate of each image point is the opposite of the x-coordinate of its preimage because of the reflection with respect to the y-axis. Therefore, the image of the point (s, t) is $(-s, t - 2)$.

ISOMETRIES

All of the transformations we have discussed share the property that they preserve distance. A transformation that preserves distance is called an **isometry**.

That is, for any pair of points A and B and their images A' and B', we have $AB = A'B'$. Translations, rotations, reflections, and glide reflections are isometries because they all preserve distance. Surprisingly, although we will not prove it in this book, it can be proven that these are the *only* isometries. That is, if there is a transformation that preserves distance, then it must be one of the four types that we have studied.

Because isometries preserve distance, they also preserve angle measure, perpendicularity, and parallelism, and take lines to lines, and so on. In fact, we could have started this chapter by proving these properties about isometries. Then, when we showed that translations, rotations, and reflections preserved distance, they would automatically satisfy all the other properties of isometries.

In Chapter 4 we studied the concept of congruence of triangles. Because isometries preserve distance, the image of one triangle must be congruent to the original triangle by the SSS congruence postulate.

Theorem 9.5

Congruent Triangles and Isometries

Two triangles are congruent if and only if there is an isometry that takes one triangle to the other.

In general, two polygons are congruent if there is a one-to-one correspondence between the vertices of one polygon and the vertices of the other such that the corresponding angles and the corresponding sides are congruent. Because polygons can be subdivided into triangles by connecting vertices with diagonals, congruence can be extended to polygons by using triangle congruence. This result is summarized next.

Theorem 9.6

Congruent Polygons and Isometries

Two polygons are congruent if and only if there is an isometry that takes one polygon to the other.

The value in studying isometries is that we can use isometries to decide if *any* shapes, including those that are not polygons, are congruent. In general, we say that two shapes are **congruent** if and only if there is an isometry that takes one to the other. For example, the shaded footprints in Figure 9.11 are congruent because a glide reflection can be used to take one footprint to the other.

Solution to Applied Problem

In six hours the constellation will make one quarter turn about the North Pole. So the astronomer can draw the constellation's new position by finding the image of Cassiopeia after a 90° rotation. The new position is shown next.

GEOMETRY AROUND US

The four isometries described in this section can be applied again and again to create a repeated pattern that is pleasing to the eye. For example, patterns on wallpaper, pottery, and fabric often exhibit translations, reflections, rotations, or glide reflections. The pot and strip of wallpaper shown utilize a translation and glide reflection in their patterns.

PROBLEM SET 9.1

EXERCISES/PROBLEMS

1. Draw $\overrightarrow{PQ}$ and $\overline{AB}$ as illustrated next.

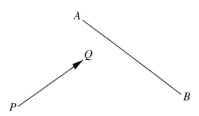

With a sheet of tracing paper on top, trace segment $\overline{AB}$ and point P. Slide the tracing paper (without turning) so that the traced point P moves to point Q. Make impressions of points A and B by pushing your pencil tip down. Label these impressions A' and B'. Draw segment $A'B'$, the translation image of segment $\overline{AB}$. Does $\overline{A'B'}$ appear to be parallel to $\overline{AB}$? Should it be? Explain.

2. Using tracing paper, find the image, $\triangle A'B'C'$, of $\triangle ABC$ under the translation $\overrightarrow{PQ}$.

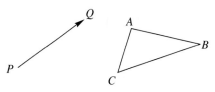

Does $\triangle A'B'C'$ appear to be congruent to $\triangle ABC$? Should it be? Explain.

3. (a) On graph paper, draw three other vectors that describe the translation $\overrightarrow{AB}$ as shown next.

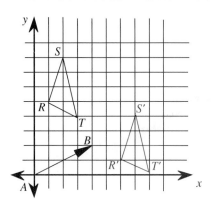

(b) Draw vectors that describe a translation that moves points 3 units down and 4 units to the right.

(c) Draw vectors that describe the translation that maps $\triangle RST$ to $\triangle R'S'T'$.

4. Find the coordinates of A' and B' that are the images of A and B under the translations listed in parts (a), (b), (c), and (d).

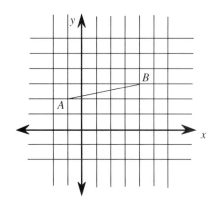

(a)

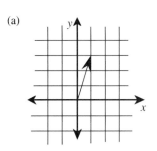

(b)

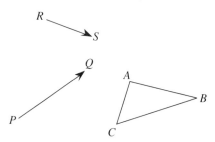

(c)

(d)

5. Using tracing paper, find the image $\triangle A'B'C'$ of $\triangle ABC$ under the translation $\overrightarrow{PQ}$.

Then find the image $\triangle A''B''C''$ of $\triangle A'B'C'$ under the translation $\overrightarrow{RS}$. Does $\triangle A''B''C''$ appear to be congruent to $\triangle ABC$? Should it be? Explain.

6. Draw any triangle, say $\triangle ABC$. Find its image $\triangle A'B'C'$ under some translation $\overrightarrow{PQ}$. Then find the image $\triangle A''B''C''$ of $\triangle A'B'C'$ under the translation $\overrightarrow{QP}$. What did you find? What general statement can you make based on this example?

7. Use a compass and straightedge to construct the image of $\overline{AB}$ under the translation $\overrightarrow{PQ}$. [HINT: Use the fact that $AA'B'B$ must be a parallelogram.]

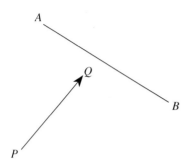

8. Use a compass and straightedge to construct the image of $\triangle ABC$ under the translation $\overrightarrow{PQ}$.

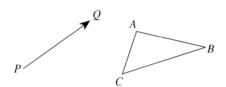

9. Find the 90° counterclockwise rotation of point P around point O.

(a)

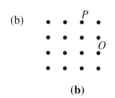

(a)

(b)

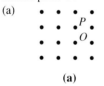

(b)

(c)

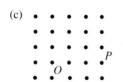

10. Find the rotation image of $\overline{AB}$ around point O for each of the following directed angles.

(a) $-90°$

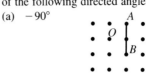

(b) 90°

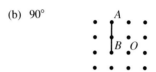

(c) 180°

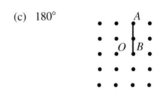

11. Find the measure of each of the following directed angles $\angle ABC$.

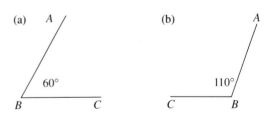

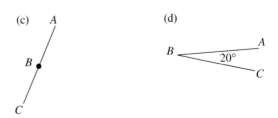

12. Using a protractor, draw and label ∠ABC to represent the following directed angles.
    (a) 40°   (b) −60°   (c) 110°   (d) −150°

13. A protractor and tracing paper may be used to find rotation images. For example, find the image of point A under a rotation of −50° about center O by following the steps below.

A.

O.

    (i) Draw a ray from O through A.
    (ii) With ray $\overrightarrow{OA}$ as the initial side, use your protractor to draw a directed angle ∠AOB of −50°.
    (iii) Place tracing paper on top and trace A.
    (iv) Keep O fixed and turn the tracing paper until A is on ray $\overrightarrow{OB}$. Make an imprint with your pencil for A′.

14. Using a protractor and tracing paper, find the rotation image of A about O for the following directed angles.
    (a) 75°   (b) −90°   (c) −130°   (d) 180°

A.

O.

15. Using a protractor, find the image, △A′B′C′, of △ABC under the rotation about point O of 90°. Are the sides of △ABC parallel to the corresponding sides of △A′B′C′? Should they be? Explain.

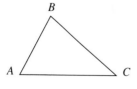

16. Using a protractor, find the image, △A′B′C′, of △ABC under the rotation around O of 70°. Then find the rotation image △A″B″C″ of △A′B′C′ under the rotation of 110° around O. Is there a rotation around O that takes △ABC to △A″B″C″? If so, describe it. Are the sides of △ABC parallel to the corresponding sides of △A″B″C″? Should they be? Explain.

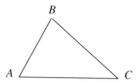

17. Using a compass and straightedge, find the −60° rotation image of △ABC around O.

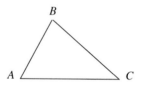

18. Using a compass and straightedge, find the image, △A′B′C′, of △ABC under the rotation of 90° about point O. Then, using tracing paper, find the translation image △A″B″C″ of △A′B′C′ under the translation from P to Q. [NOTE: Use the location of P and Q as given, not as they would be after the rotation.]

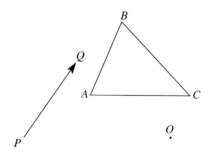

19. Find the reflection of point A with respect to each of the given lines.

(a)

(b)

(c)

20. Find the image $\overline{A'B'}$ of the segment $\overline{AB}$ under a reflection with respect to each of the given lines l.

(a)

(b)

(c)

21. The reflection image of P can be found using tracing paper.

(i) Choose a point Q on line l.

(ii) Trace line l and points P and Q on your tracing paper.

(iii) Flip your tracing paper over, matching line l and point Q.

(iv) Make an impression for P'.

(a) What kind of triangle is $\triangle PP'Q$? Explain.

(b) If M is the point where $\overline{PP'}$ intersects l, what kind of triangle is $\triangle PMQ$? Explain.

22. Find the reflection image of Q and P with respect to line l. Describe the figure $PP'Q'Q$ as completely as possible. Explain.

23. Using a compass and straightedge, find the reflection image of $\triangle ABC$ with respect to line l. Describe $ABCB'$ as completely as possible. Explain.

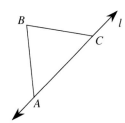

24. Using a compass and straightedge, find the reflection image of $\triangle ABC$ with respect to line l. Does the image appear to be congruent to $\triangle ABC$? Does it have the same orientation?

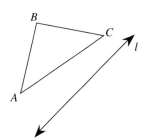

25. The reflection image of *A* in line *l* can be constructed with compass and straightedge. Use the fact that line *l* is the perpendicular bisector of $\overline{AA'}$. Recall that if point *P* is the intersection of $\overline{AA'}$ and line *l*, then $\overline{AA'} \perp l$ and $AP = PA'$. Using compass and straightedge, find *A'*.

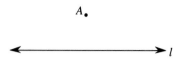

26. Using a compass and straightedge, find the image of $\overline{AB}$ under the glide reflection composed of the translation from *X* to *Y* followed by a reflection with respect to the line *l*.

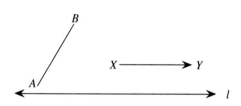

27. Find the coordinates of points *A'* and *B'* that are the images of points *A* and *B* under the following translations.

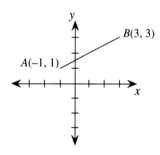

Translation that takes
(a) *P*(0, 0) to *Q*(1, 3)
(b) *P*(0, −1) to *Q*(4, −2)
(c) *P*(−1, −2) to *Q*(0, 0)
(d) *P*(3, −2) to *Q*(2, −3)

28. Give the coordinates of the images of the following points under the rotation of 90° around the origin.
(a) (2, 3)   (b) (−1, 3)   (c) (−1, 4)
(d) (−4, −2)   (e) (2, −4)   (f) (x, y)

29. Give the coordinates of the images of the following points under the −90° rotation around the origin.
(a) (1, 5)   (b) (−1, 3)   (c) (−2, 4)
(d) (−3, −1)   (e) (5, −2)   (f) (x, y)

30. Draw the image of quadrilateral *ABCD* shown next under the rotation around the origin of 180° for *A*(−2, 0), *B*(0, 2), *C*(3, 0), and *D*(−1, −2). What are the coordinates of *A'*, *B'*, *C'*, and *D'*?

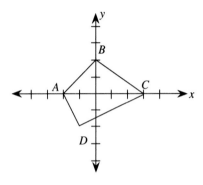

31. Give the coordinates of the images of the following points under the rotation of 180° around the origin.
(a) (3, −1) (b) (−6, −3) (c) (−4, 2) (d) (x, y)

32. We could express the results of the rotation of 90° around the origin applied to (4, 2) in the following way: (4, 2) rotates 90° around the origin to (−2, 4). Complete the following statements.
(a) (x, y) rotates 180° around the origin to (?, ?).
(b) (x, y) rotates 270° around the origin to (?, ?).
(c) (x, y) rotates 360° around the origin to (?, ?).

33. (a) Graph the triangle with vertices *A*(1, 2), *B*(3, 5), and *C*(6, 1) and its image under the reflection with respect to the *x*-axis.
(b) What are the coordinates of the images of points *A*, *B*, and *C* under the reflection with respect to the *x*-axis?
(c) If point *P* has coordinates (*a*, *b*), what are the coordinates of its image under the reflection with respect to the *x*-axis?

34. (a) Graph △*ABC* with *A*(−2, 1), *B*(0, 3), and *C*(3, −2) and its image under the glide reflection described by the translation from *P*(1, 1) to *Q*(1, 6) followed by the reflection with respect to the *y*-axis.

(b) What are the coordinates of the images of points *A*, *B*, and *C* under this glide reflection?

(c) If a point has coordinates (*a*, *b*), what are the coordinates of its image under this glide reflection?

35. In each part, identify the type of transformation that maps rectangle *A'B'C'D'* onto rectangle *A'B'C'D'*.

(a)

(b)

36. In each part, identify the type of transformation that maps rectangle *ABCD* onto rectangle *A'B'C'D'*.

(a)

(b)

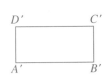

37. Trace each of the following pairs of figures. In each case, determine if there is a reflection that takes △*ABC* to △*A'B'C'*.

(a)

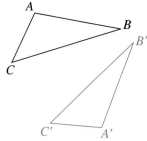

(b)

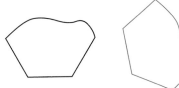

38. Determine the type of isometry that maps the shape on the left onto the shape on the right.

(a)

(b)

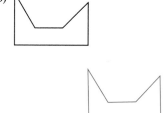

(c)

39. Several reflections are described below. In each part draw figures and their images to find the reflection line. The notation (*x*, *y*) → (*x*, 2 − *y*) means that the image of (*x*, *y*) is (*x*, 2 − *y*).

(a) (*x*, *y*) → (*x*, 2 − *y*)

(b) (*x*, *y*) → (*x*, 8 − *y*)

(c) (*x*, *y*) → (*x*, −4 − *y*)

40. (a) Find the coordinates of the images of the vertices of quadrilateral $ABCD$ for $A(-2, 0)$, $B(0, 2)$, $C(3, 0)$, and $D(-1, -2)$ under the translation $\overrightarrow{OD}$.

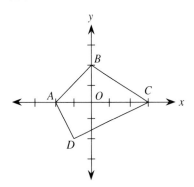

(b) Given a point (x, y), what are the coordinates of its image under this translation?

41. (a) Graph $\triangle ABC$ with $A(2, 1)$, $B(3, -5)$, and $C(6, 3)$ and its image with respect to the reflection in the y-axis.
   (b) What are the coordinates of the points A, B, and C under the reflection with respect to the y-axis?
   (c) If point P has coordinates (a, b), what are the coordinates of its image under the reflection with respect to the y-axis?

42. (a) Graph $\triangle ABC$, with $A(3, 1)$, $B(4, 3)$, and $C(5, -2)$, and its image under the reflection with respect to the line $y = x$.
   (b) What are the coordinates of the images of points A, B, and C under the reflection with respect to the line $y = x$?
   (c) If a point P has coordinates (a, b), what are the coordinates of its image under reflection with respect to the line $y = x$?

PROOFS

43. Prove property 2 of Theorem 9.2: Rotations preserve distance.

44. Prove property 3 of Theorem 9.2: Rotations preserve angle measure.

45. Prove property 4 of Theorem 9.2: Rotations preserve perpendicularity.

46. Prove property 5 of Theorem 9.2: Rotations preserve parallelism.

47. Prove property 2 of Theorem 9.3: Reflections preserve distance.

48. Prove property 3 of Theorem 9.3: Reflections preserve angle measure.

49. Prove property 4 of Theorem 9.3: Reflections preserve perpendicularity.

50. Prove property 5 of Theorem 9.3: Reflections preserve parallelism.

51. Prove property 2 of Theorem 9.4: Glide reflections preserve distance.

52. Prove property 3 of Theorem 9.4: Glide reflections preserve angle measure.

APPLICATIONS

53. Each wallpaper border shown illustrates at least one of the isometries in this section. Identify several isometries that take one design to another.
(a)

(b)

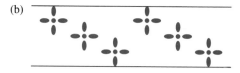

(c)

(d)

54. Each wallpaper border shown illustrates at least one of the isometries in this section. Identify several isometries that take one design to another.
(a)

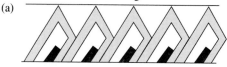

(b)

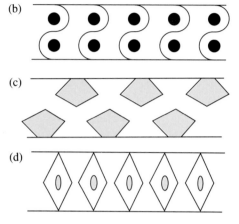

(c)

(d)

square at a right angle to the line of the first move. Given the knight represented by the dot in the position shown, find all possible moves for that piece and describe those moves in terms of transformations.

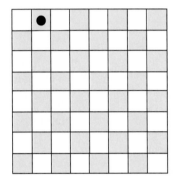

55. In the game of chess, the knight's move is two squares away in any row or column and then one

9.2

SIMILITUDES AND SIMILARITY

Applied Problem

The scale drawing shown represents the basic floor plan in the design of a home.

A draftsperson needs to create an enlargement of this floor plan that has four times the area of the plan shown but retains the same shape. How can this be done?

Just as congruence was generalized using the concept of isometry, similarity of triangles can be generalized to any geometric figures.

SIZE TRANSFORMATIONS

Congruence is related to the sizes and shapes of geometric figures, but similarity is related to only the shapes of figures. We begin our study of similarity using transformations by enlarging and reducing the sizes of figures proportionally.

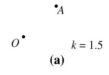

(a)

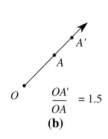

(b)

FIGURE 9.14

A **size transformation** of the plane is defined as follows. Let O be any point (called the **center of the size transformation**) and k be a positive real number (called the **scale factor**). The image of any point A under the size transformation with center O and scale factor k is the point A' where

1. A' is on $\overrightarrow{OA}$.

2. $\dfrac{OA'}{OA} = k$ [Figure 9.14(b)].

If $k = 1.5$ and O and A are given as in Figure 9.14(a), then A' is found as shown in Figure 9.14(b).

EXAMPLE 9.6   In Figure 9.15(a), find the image of point A for the following size transformations.

(a) The size transformation with center O and scale factor 3
(b) The size transformation with center O and $k = 0.5$
(c) The size transformation with center O and scale factor 1

SOLUTION

(a) To find the image of A, we extend $\overline{OA}$ through A and mark off point A' where $OA' = 3(OA)$ [Figure 9.15(b)].

(b) The image is A' because $OA' = 0.5(OA)$ and A' is on ray $\overrightarrow{OA}$ [Figure 9.15(c)].

(c) The image is A [Figure 9.15(d)] because $OA = OA'$. This is called the **identity size transformation**.   •

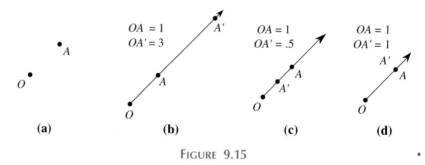

| (a) | (b) | (c) | (d) |

FIGURE 9.15   •

Size transformations have several properties similar to those of isometries. The next example illustrates some of them.

EXAMPLE 9.7   In Figure 9.16(a), find the image of $\triangle ABC$ for the size transformation with center O and scale factor 2. Is $\triangle ABC \sim \triangle A'B'C'$?

SOLUTION   The image is $\triangle A'B'C'$ [Figure 9.16(b)]. By the definition of a size transformation, we know that $\dfrac{OA'}{OA} = \dfrac{OB'}{OB} = 2$. Also $\angle AOB \cong \angle A'OB'$ because A and A' are collinear, as are B and B'. So by SAS Similarity

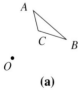

(a)

FIGURE 9.16

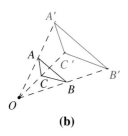

(b)

FIGURE 9.16

$\triangle OAB \sim \triangle OA'B'$. Thus $\dfrac{A'B'}{AB} = 2$ also. By the same reasoning, we can show that $\triangle OAC \sim \triangle OA'C'$ and $\triangle OCB \sim \triangle OC'B'$. Therefore, we have $\dfrac{A'C'}{AC} = 2$ and $\dfrac{B'C'}{BC} = 2$ respectively. Consequently, $\triangle ABC \sim \triangle A'B'C'$ by SSS Similarity. It follows that $\angle A \cong \angle A'$, $\angle B \cong \angle B'$, and $\angle C \cong \angle C'$. Moreover, it can be shown that the respective sides of the two triangles are parallel. •

This example suggests some of the results stated in the following theorem. The proofs of properties 3–6 are left for the problem set.

Theorem 9.7

Properties of Size Transformations

1. Size transformations take lines to lines, rays to rays, and line segments to line segments.
2. Size transformations preserve orientation.
3. Size transformations preserve ratios of distances.
4. Size transformations preserve angle measure.
5. Size transformations preserve perpendicularity.
6. Size transformations take lines to parallel lines.

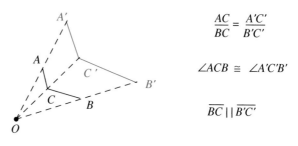

$$\frac{AC}{BC} = \frac{A'C'}{B'C'}$$

$$\angle ACB \cong \angle A'C'B'$$

$$\overline{BC} \,||\, \overline{B'C'}$$

EXAMPLE 9.8   Is there a size transformation that takes $\triangle RST$ in Figure 9.17(a) to $\triangle R'S'T'$? If so, find its center and scale factor. If not, explain why not.

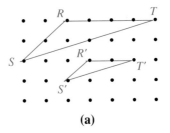

(a)

FIGURE 9.17

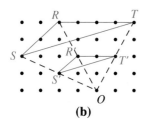

(b)

FIGURE 9.17

SOLUTION   If such a size transformation exists, then $\triangle RST$ as $\triangle R'S'T'$. By comparing lengths of the sides, we have

$$\frac{R'T'}{RT} = \frac{S'R'}{SR} = \frac{S'T'}{ST} = \frac{1}{2}$$

So $\triangle RST \sim \triangle R'S'T'$ by SSS Similarity. Therefore, we *may* be able to find a size transformation that takes $\triangle RST$ to $\triangle R'S'T'$. Since the corresponding sides are parallel, such a size transformation should exist. The center of the size transformation is collinear with S and S', with R and R', and with T and T'. The lines determined by $\overline{SS'}$, $\overline{RR'}$, and $\overline{TT'}$ intersect at the center O [Figure 9.17(b)]. Thus, the size transformation with center O and scale factor $\frac{1}{2}$ takes $\triangle RST$ to $\triangle R'S'T'$. •

SIMILITUDES

In Section 9.1 we found that congruence can be defined in terms of isometries. As suggested by Examples 9.7 and 9.8, a size transformation takes a figure to a similar figure. However, consider $\triangle ABC$ and $\triangle DEF$ [Figure 9.18(a)].

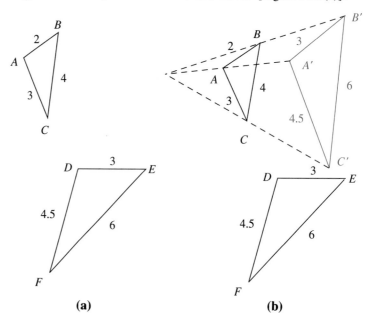

(a)   **(b)**

FIGURE 9.18

Although the two triangles are similar by SSS Similarity, no size transformation will take $\triangle ABC$ to $\triangle DEF$ because their corresponding sides are not parallel as required by part 6 of Theorem 9.7. But a size transformation with scale factor 1.5 will take $\triangle ABC$ to $\triangle A'B'C'$ where $\triangle A'B'C' \cong \triangle DEF$ [Figure 9.18(b)]. Then an isometry (here a rotation) can be found to take $\triangle A'B'C'$ to $\triangle DEF$ since they are congruent. A **similitude** is defined to be a combination of a size transformation and an isometry. This discussion motivates the following result.

Theorem 9.8

Similar Triangles and Similitudes

$\triangle ABC \sim \triangle A'B'C'$ if and only if there is a similitude that takes $\triangle ABC$ to $\triangle A'B'C'$.

EXAMPLE 9.9   Suppose that $\triangle ABC \sim \triangle DBE$ [Figure 9.19(a)]. Find the similitude that takes $\triangle ABC$ to $\triangle DBE$. [NOTE: Any similitude can be expressed as the composition of a size transformation followed by an isometry in infinitely many ways depending what point is chosen as the center of the size transformation. Thus, although there is only one similitude, there are many different ways to describe it.]

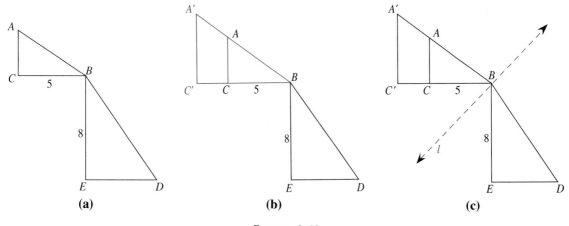

FIGURE 9.19

SOLUTION   Because the ratio $\dfrac{BE}{BC}$ is $\dfrac{8}{5}$, the scale factor of our size transformation must be $\dfrac{8}{5}$. For convenience, we choose the center of the size transformation to be B (however, we could have used any other point in the plane). Then the image of $\triangle ABC$ under the size transformation with center B and scale factor $\dfrac{8}{5}$ is $\triangle A'BC'$ [Figure 9.19(b)]. Because $\triangle A'BC'$ and $\triangle DBE$ have opposite orientation, the isometry that takes $\triangle A'BC'$ to $\triangle DBE$ must be a reflection or a glide reflection. Notice that the reflection with respect to the line l that bisects $\angle C'BE$ takes $\triangle A'BC'$ to $\triangle DBE$ [Figure 9.19(c)]. Thus the size transformation with center B and with scale factor $\dfrac{8}{5}$ followed by the reflection with respect to line l is a similitude that takes $\triangle ABC$ to $\triangle DBE$.   •

Two polygons are defined to be **similar** if and only if there is a correspondence between the polygons such that the corresponding angles are congruent and the corresponding sides are proportional.

EXAMPLE 9.10   Prove that any two squares are similar.

SOLUTION   Let *ABCD* and *EFGH* be two squares with sides of length *s* and *t* [Figure 9.20(a)].

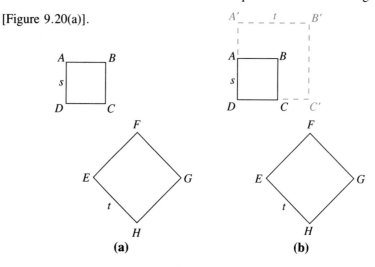

FIGURE 9.20

Any size transformation with scale factor $\frac{t}{s}$ takes *ABCD* to a square *A'B'C'D'* where $A'B'C'D' \cong EFGH$ [one such size transformation with center *D* is shown in Figure 9.20(b)]. Because the squares *A'B'C'D'* and *EFGH* are congruent, there must be an isometry that takes *A'B'C'D'* to *EFGH*. Thus, the combination of the size transformation and the isometry produces a similitude that takes *ABCD* to *EFGH*. Since similitudes preserve angle measure and proportionality of side lengths, the squares are similar.   •

This example suggests that all equilateral triangles are similar also. This suggests the following general result about polygons.

Theorem 9.9

Similar Polygons and Similitudes

Two polygons are similar if and only if there is a similitude that takes one to the other.

Just as with isometries, the value in studying similitudes is that we can use similitudes to decide if *any* shapes, including those that are not polygons, are similar. In general, we say that two shapes are **similar** if and only if there is a similitude that takes one to the other.

Solution to Applied Problem

If the ratio of the sides of two similar figures is $a:b$, then the ratio of their areas is $a^2:b^2$. Thus if the dimensions of the sides of the floor plan are doubled, that is, $b = 2a$, the area of the new floor plan would be four times the area of the original plan since $b^2 = (2a)^2 = 4a^2$.

GEOMETRY AROUND US

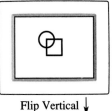

Flip Vertical ↓

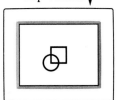

Computer graphics programs allow the user to manipulate a geometric figure drawn on the screen using transformations such as those described in this chapter. One popular graphics package has options such as Flip Horizontal, Flip Vertical, and Free Rotate. Users can also perform a size transformation on a figure by using a Magnifying Glass tool and can perform a translation by moving a figure using the mouse. An example of the use of the Flip Vertical option is shown here.

PROBLEM SET 9.2

EXERCISES/PROBLEMS

1. Trace O and A. Using a ruler, find A' for a size transformation with center O having the following scale factors.

(a) 4   (b) 3.75   (c) $2\frac{5}{8}$ (d) $\frac{2}{3}$

.A

O.

2. Trace O and A. Using a compass and straightedge, find A' for a size transformation with center O having the following scale factors.

(a) 3.5   (b) $\frac{3}{4}$ (c) 1   (d) 2.25

.A

O.

3. On the square lattice portions shown, find the image of point P under the size transformation with center O and scale factor 3.

(a)

O P

(b)

O

P

(c)

.O

.P

4. On the square lattice portions shown, find the image of $\overline{AB}$ under the size transformation with center O and scale factor 2.

(a)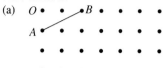

(b)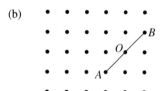

(c)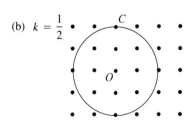

(a) $k = 3$

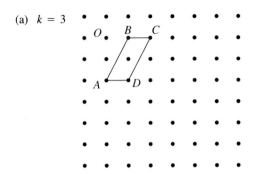

(b) $k = 0.5$

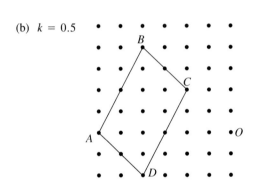

5. On the square lattice portions shown, find the image of circle C under a size transformation with center O and scale factor as given.

(a) $k = 2$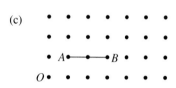

(b) $k = \dfrac{1}{2}$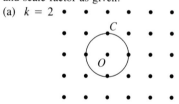

6. On the square lattice portions shown, find the image of parallelogram $ABCD$ under a size transformation with center O and scale factor as given.

7. (a) Find the image of $\angle ABC$ under the size transformation with center O and scale factor 2.

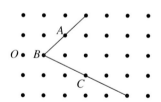

(b) How do the measures of $\angle ABC$ and $\angle A'B'C'$ compare?

8. (a) Find the image of $\triangle ABC$ under the size transformation with center O and scale factor 3.

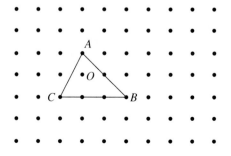

(b) Find the lengths of $\overline{AB}$ and $\overline{A'B'}$. What is the ratio $\dfrac{A'B'}{AB}$?

(c) Find the lengths of $\overline{BC}$ and $\overline{B'C'}$. What is the ratio $\dfrac{B'C'}{BC}$?

(d) Find the lengths of $\overline{AC}$ and $\overline{A'C'}$. What is the ratio $\dfrac{A'C'}{AC}$?

(e) In addition to the ratio of their lengths, what other relationship exists between a line segment and its image under the size transformation?

9. Given $\triangle ABC$ with vertices $A(0, 0)$, $B(4, 0)$, and $C(4, 3)$, find the following:
   (a) The image of $\triangle ABC$ after the size transformation with center A and with scale factor 2
   (b) The ratio of the area of $\triangle A'B'C'$ to the area of $\triangle ABC$

10. Given $\triangle ABC$ with vertices $A(0, 0)$, $B(4, 0)$, and $C(4, 3)$, find the following:
    (a) The image of $\triangle ABC$ after the size transformation with center C and scale factor $\dfrac{1}{2}$
    (b) The ratio of the perimeter of $\triangle A'B'C'$ to the perimeter of $\triangle ABC$

11. Trace the following circle. Then draw the image of circle C under a size transformation with center O having scale factor 3. Compare the circumference and area of C with those of its size transformation image.

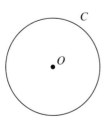

12. Trace the following circle. Then draw the image of circle C under a size transformation with center A having scale factor 0.5. Compare the circumference and area of C with those of its size transformation image.

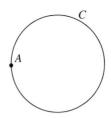

13. (a) Find the center of the size transformation that maps P to P' and Q to Q'.

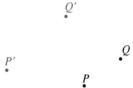

(b) Give an estimate for the scale factor of the size transformation.

14. (a) Find the center of the size transformation that maps P to P' and Q to Q'.

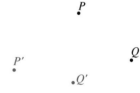

(b) Give an estimate for the scale factor of the size transformation.

15. Determine if $\triangle A'B'C'$ is a size transformation image of $\triangle ABC$ for $A'(0, 2)$, $B'(4, 4)$, $C'(8, 0)$, $A(0, 1)$, $B(2, 2)$, and $C(4, 0)$. If yes, find the center and scale factor. If no, explain why not.

16. Determine if $\triangle A'B'C'$ is a size transformation image of $\triangle ABC$ for $A'(5, 3)$, $B'(1, 5)$, $C'(9, 5)$, $A(5, 0)$, $B(3, 1)$, and $C(7, 2)$. If yes, find the center and scale factor. If no, explain why not.

17. A quadrilateral $ABCD$ has vertices $A(0, 0)$, $B(1, 3)$, $C(4, 2)$, and $D(5, 0)$.
    (a) Sketch the image of $ABCD$ under the size transformation with center A and scale factor 2 followed by a reflection with respect to the y-axis.
    (b) What are the coordinates of A', B', C', and D'?
    (c) If a point has coordinates (a, b), what are the coordinates of its image under this similitude?

18. A quadrilateral *ABCD* has vertices *A*(0, 0), *B*(1, 3), *C*(4, 2), and *D*(5, 0).
    (a) Sketch the image of *ABCD* under the size transformation with center *A* and scale factor 3 followed by a translation of 2 units to the right and 1 unit up.
    (b) What are the coordinates of *A'*, *B'*, *C'*, and *D'*?
    (c) If a point has coordinates (*a*, *b*), what are the coordinates of its image under this similitude?

19. A quadrilateral *ABCD* has vertices *A*(0, 0), *B*(1, 3), *C*(4, 2), and *D*(5, 0).
    (a) Sketch the image of *ABCD* under the rotation of 180° around *A* followed by the size transformation with center *A* and scale factor $\frac{1}{2}$.
    (b) What are the coordinates of *A'*, *B'*, *C'*, and *D'*?
    (c) If a point has coordinates (*a*, *b*), what are the coordinates of its image under this similitude?

20. A quadrilateral *ABCD* has vertices *A*(0, 0), *B*(1, 3), *C*(4, 2), and *D*(5, 0).
    (a) Sketch the image of *ABCD* under the clockwise rotation of 90° around point *A* followed by the size transformation with center *A* and scale factor 2.
    (b) What are the coordinates of *A'*, *B'*, *C'*, and *D'*?
    (c) If a point has coordinates (*a*, *b*), what are the coordinates of its image under this similitude?

21. Find the similitude that takes △*ABC* to △*A'B'C'* for *A*(0, 0), *B*(0, 2), *C*(2, 0), *A'*(−3, 0), *B'*(−3, 9), and *C'*(6, 0). [HINT: First find a size transformation that takes △*ABC* to some △*A''B''C''* that is congruent to △*A'B'C'*.]

22. Find the similitude that takes △*ABC* to △*A'B'C'* for *A*(1, 0), *B*(1, 3), *C*(2, 0), *A'*(−3, 0), *B'*(−3, 9), and *C'*(−6, 0).

23. Find the similitude that takes △*ABC* to △*A'B'C'*. Describe the similitude as completely as possible.

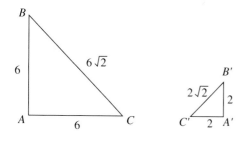

24. Find the similitude that takes △*ABC* to △*A'B'C'*. Describe the similitude as completely as possible.

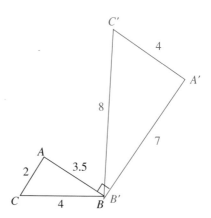

PROOFS

25. Describe in words how you would prove that any equilateral △*ABC* is similar to equilateral △*DEF*.

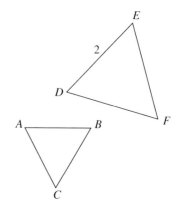

26. Prove that any circle is similar to any other circle.

27. Prove part 3 of Theorem 9.7: Size transformations preserve ratios of distances.

28. Prove part 4 of Theorem 9.7: Size transformations preserve angle measure.

29. Prove part 5 of Theorem 9.7: Size transformations preserve perpendicularity.

30. Prove part 6 of Theorem 9.7: Size transformations take lines to parallel lines.

9.3

PROBLEM SOLVING USING TRANSFORMATIONS

Applied Problem

Towns A and B are on opposite sides of a river. The towns are to be connected with a bridge, $\overline{CD}$, perpendicular to the river, so that the distance $AC + CD + DB$ is as small as possible. Where should the bridge be located?

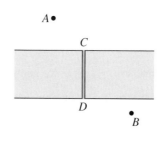

Our knowledge of transformations can be applied to solve problems. The following examples are of two types. First we examine proofs of some familiar results using transformations. These are presented to reinforce the idea that it is acceptable to prove a result using any of several techniques. Second, we look at examples of problems which may be very difficult to prove using techniques other than those of transformation geometry.

TRANSFORMATION PROOFS

In Chapter 5 we used congruent triangles to verify properties of quadrilaterals. The following example shows how one such property can be verified using transformations.

EXAMPLE 9.11   Using transformations, show that if the diagonals of a quadrilateral $ABCD$ bisect each other, then it is a parallelogram.

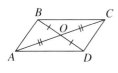

FIGURE 9.21

SOLUTION   Let $ABCD$ have diagonals $\overline{AC}$ and $\overline{BD}$ that bisect each other at O (Figure 9.21). Consider the rotation of $180°$ with center O that rotates A to C, B to D, C to A, and D to B. Because the rotation takes A to C and B to D, it must take $\overline{AB}$ to $\overline{CD}$. Therefore, $\overline{AB} \cong \overline{CD}$ because the rotation preserves length. Similarly, $\overline{AD} \cong \overline{CB}$. Thus $ABCD$ is a parallelogram because it has two pairs of opposite sides that are congruent.   •

The next example gives a transformation proof of a result proved in Chapter 6 using similar triangles.

EXAMPLE 9.12   In $\triangle ABC$ in Figure 9.22, $\overline{DE}$ divides $\overline{AB}$ and $\overline{CB}$ proportionally as indicated. Prove that $\overline{DE} \parallel \overline{AC}$.

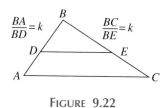

FIGURE 9.22

SOLUTION   Consider the size transformation with center B and scale factor k. Because $\dfrac{BA}{BD} = k$, we know that the size transformation takes D to A. Similarly, the size transformation takes E to C. Therefore, the size transformation takes $\overline{DE}$ to $\overline{AC}$. Because size transformations take line segments to parallel line segments, we have $\overline{DE} \parallel \overline{AC}$.   •

APPLICATIONS OF TRANSFORMATIONS

Transformations can also be useful in solving certain applied problems. Consider the following problem related to shots on a pool table.

EXAMPLE 9.13   In a pool game shown in Figure 9.23(a), ball A is blocking a direct shot from the cue ball C to the object ball B. Assuming there is no "spin" on the cue ball, where on the upper rail should a player aim ball C to hit ball B?

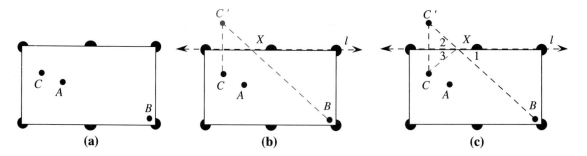

FIGURE 9.23

SOLUTION   Consider line l defined by the edge of the upper rail [Figure 9.23(b)]. Reflect C across l to C' and draw $\overline{C'B}$. The point X where $\overline{C'B}$ intersects l is the point where ball C should be hit [Figure 9.23(b)]. We will use angle measures to prove that X is the correct point [Figure 9.23(c)]. First, $\angle 1$ is the correct angle to hit B from X. Next, $\angle 1 = \angle 2$ because they are vertical angles. Finally, $\angle 2 = \angle 3$ because reflections preserve angle measure and $\angle 3$ was reflected to $\angle 2$. Thus, when ball C hits l at X forming $\angle 3$, it will rebound forming $\angle 1$ and hit ball B.

•

Other examples of pool shot paths caroming or "banking" off several cushions appear in the problem set.

Solution to Applied Problem

No matter where the bridge is located, CD, the width of the river, will be a constant in the sum $AC + CD + DB$. Hence we wish to minimize the sum $AC + DB$. Consider a translation in a direction from B toward the river and perpendicular to it for a distance equal to the width of the river, d. Let B' be the image of B. Then $\overline{AB'}$ is the shortest path from A to B'. Let C be the intersection of $\overline{AB'}$ and m, a line that represents one side of the river.

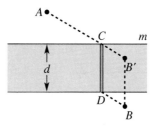

Let D be opposite C on the other side of the river, where $\overline{CD} \perp m$. That is, $\overline{CD} \parallel \overline{B'B}$ and $CD = B'B$. Hence, $BB'CD$ is a parallelogram since the opposite sides $\overline{CD}$ and $\overline{B'B}$ are congruent and parallel. Thus the sum $AC + CD + DB$ is as small as possible.

GEOMETRY AROUND US

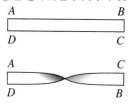

A surface with unique properties was created by the German mathematician, A. F. Moebius. To make a Moebius strip, start with a rectangular strip $ABCD$. Twist the strip once and tape the two ends together to form a "twisted loop." Then, for a surprising result, draw a continuous line down the middle of "one side" of your strip.

Moebius strips are among the shapes studied in a branch of mathematics called topology. Topology focuses on those features of geometric shapes that are not changed when the shape is stretched or deformed. One practical application of the Moebius strip is the manufacture of conveyor belts, fan belts, and so on, which wear more evenly.

PROBLEM SET 9.3

1. *ABCD* is a square and points *E*, *F*, *G*, and *H* are midpoints of its sides. List four reflections that map the square onto itself.

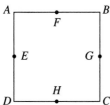

2. *ABCDEF* is a regular hexagon with center *O*. List three reflections and three rotations that map the hexagon onto itself.

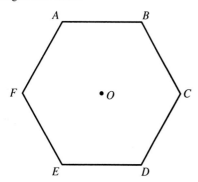

3. *ABCD* is a parallelogram. List all the isometries that map *ABCD* to itself.

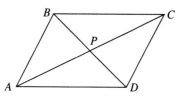

4. *ABCD* is a kite. List all the isometries that map *ABCD* onto itself.

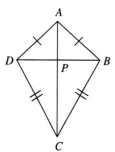

5. *ABCD* is a rhombus.

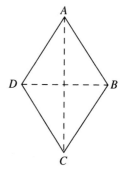

(a) List all the isometries that map *ABCD* onto itself.
(b) How is your solution related to Exercises 3 and 4?

6. *ABCD* is an isosceles trapezoid. Are there two isometries that map *ABCD* onto itself? Explain.

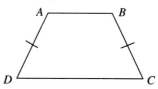

7. Suppose lines *l* and *m* as shown are parallel. Reflect △*ABC* over *l* and then reflect its image △*A'B'C'* over *m* to △*A″B″C″*.

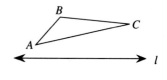

(a) Is △*ABC* ≅ △*A″B″C″*? Explain.
(b) Do △*ABC* and △*A″B″C″* have the same orientation?
(c) Will any points of the plane be fixed points after the two reflections have been performed?
(d) Is the result of two reflections an isometry? Explain.
(e) It was stated in Section 9.1 that an isometry must be either a translation, a rotation, a reflection, or a glide reflection. What type of isometry will be the result of two reflections across parallel lines? Explain.

8. Suppose lines *l* and *m* intersect at *O*. Reflect △*ABC* over *l* and then reflect its image △*A'B'C'* over *m* to △*A"B"C"*.

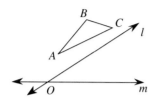

(a) Is △*ABC* ≅ △*A"B"C"*? Explain.
(b) Do △*ABC* and △*A"B"C"* have the same orientation?
(c) Will any points of the plane be fixed points after the two reflections have been performed? Explain.
(d) Is the result of two reflections an isometry? Explain.
(e) It was stated in Section 9.1 that an isometry must be either a translation, a rotation, a reflection, or a glide reflection. What type of isometry will be the result of two reflections over intersecting lines? Explain.

PROOFS

9. *ABCD* is a square. Point *P* is the intersection of the diagonals. Determine which of the transformations in (a)–(d) can be used to map the square onto itself.

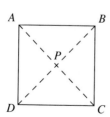

(a) The rotation around *P* of 90°
(b) The reflection with respect to $\overline{AC}$
(c) The reflection with respect to $\overline{BD}$
(d) The rotation around *P* of 270°
(e) Which of the transformations in (a)–(d) can be used to prove that the diagonals of a square are congruent? Explain.

10. *ABCD* is a kite and point *E* is the intersection of its diagonals. Which of the following transformations in (a)–(d) can be used to prove that △*ABC* ≅ △*ADC* using the transformation definition of congruence? Explain.

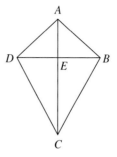

(a) The rotation around *E* of 180°
(b) The translation from *B* to *E*
(c) The rotation around *E* of 90°
(d) The reflection with respect to $\overleftrightarrow{AC}$

11. *ABCDE* is a regular pentagon with center *O*. Points *F*, *G*, *H*, *I*, and *J* are the midpoints of the sides. List all the reflections and all the rotations that map the pentagon onto itself. Then show how to prove that all the diagonals of a regular pentagon are congruent.

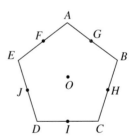

12. In *ABCD*, *AP* = *CP* and $\overline{AC} \perp \overline{BD}$.

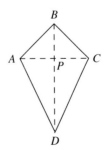

Use transformations to prove that
(a) *AB* = *BC*
(b) *AD* = *CD*
(c) *ABCD* is a kite.

13. In $ABCD$, $AP = CP$, $BP = DP$, and $\overline{AC} \perp \overline{BD}$. Using transformations, prove that $ABCD$ is a rhombus.

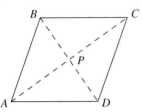

14. In $ABCD$, $AP = CP = BP = DP$ and $\overline{AC} \perp \overline{BD}$. Using transformations, prove that $ABCD$ is a square.

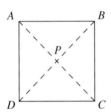

15. $ABCD$ is a parallelogram. Show that diagonal $\overline{AC}$ creates two congruent triangles using the transformation definition of congruence. [HINT: Use a rotation of $180°$ around the midpoint of $\overline{AC}$ and the fact that $180°$ rotations take a line to a parallel line.]

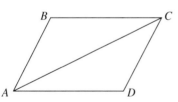

16. Use the results of problem 15 to show that the following statements are true.
  (a) Opposite angles of a parallelogram are congruent.
  (b) Opposite sides of a parallelogram are congruent.

17. Suppose that B is equidistant from A and C as shown. Using a transformation proof, show that B is on the perpendicular bisector of $\overline{AC}$. [HINT: Let P be a point on $\overline{AC}$ so that $\overline{BP}$ is the angle bisector of $\angle ABC$. Then use a reflection with respect to the line $\overleftrightarrow{BP}$.]

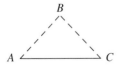

18. Let $ABCD$ be a kite with $AB = BC$ and $AD = DC$. Explain how problem 17 shows that the following statements are true.

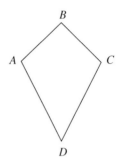

  (a) The diagonals of a kite are perpendicular.
  (b) A kite has reflection symmetry.

19. Suppose that B is on the perpendicular bisector l of $\overline{AC}$. Use the reflection across l to show that $AB = CB$ (i.e., that B is equidistant from A and C).

20. Suppose that P is on the angle bisector l of $\angle ABC$ where $\overline{AP} \perp \overline{AB}$ and $\overline{CP} \perp \overline{CB}$. Use the reflection across l to show that $AP = CP$ (i.e., that P is equidistant from A and C).

21. Suppose that $ABCD$ is a parallelogram. Using transformations, show that the diagonals $\overline{AC}$ and $\overline{BD}$ bisect each other. [HINT: Let P be the midpoint of $\overline{AC}$ and show that under a rotation of $180°$ about P, the image of B is D.]

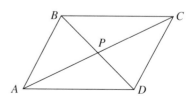

22. Suppose that *ABCD* is a parallelogram and *P* is the intersection of the diagonals. How does problem 21 show that a parallelogram has rotation symmetry?

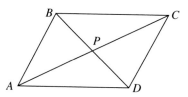

23. Points *P*, *Q*, and *R* are the midpoints of the sides of △*ABC*.

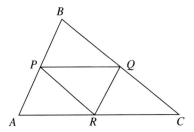

(a) Show that a size transformation with center *A* and scale factor 2 takes △*APR* to △*ABC*.

(b) What is the image of △*PBQ* under the size transformation with center *B* and scale factor 2?

(c) Find a size transformation that takes △*QCR* to △*BCA*.

24. Show how parts (a), (b), and (c) in problem 23 can be used to prove that *PBQR* is a parallelogram.

25. The following sequence of diagrams suggests a transformational proof of the Pythagorean theorem. Explain why this works.

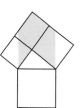

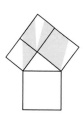

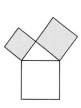

APPLICATIONS

26. On the billiards table, ball *A* is to carom off two of the rails (sides) and strike ball *B*. Draw a path for a successful shot.

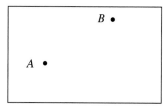

27. On the billiards table, ball *A* is to carom off three rails, then strike ball *B*. Draw a path for a successful shot.

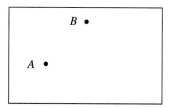

Solution to Initial Problem

Houses A and B are to be connected to a television cable line, l, at a transformer at some point P. Where should P be located so that the sum of the distances, $AP + PB$, is as small as possible?

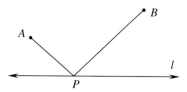

Reflect point B across line l to point B'. Then draw $\overline{AB'}$ and $\overline{BB'}$, forming points of intersection P and Q, respectively.

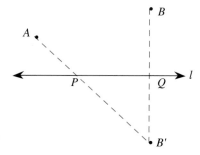

Next draw the perpendicular $\overline{AR}$ and the segment $\overline{PB}$.

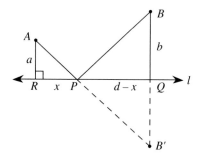

Because B' is the reflection image of B, we have $\angle BPQ \cong \angle B'PQ$. Also, $\angle APR \sim \angle B'PQ$ since they are vertical angles. Therefore, $\angle APR \cong \angle BPQ$. Thus, $\triangle APR \sim \triangle BPQ$ by AA Similarity. Let $RP = x$ and $RQ = d$. Thus, $PQ = d - x$. Because corresponding parts of similar triangles are proportional, we have

$$\frac{x}{d - x} = \frac{a}{b}$$

so that $bx = a(d - x)$, or

$$x = \frac{ad}{a + b}$$

where $d = RQ$. So the transformer P should be located at a distance $\dfrac{ad}{a + b}$ from R along $\overline{RQ}$.

Additional Problems Where the Strategy "Use Symmetry" Is Useful

1. A 4 by 4 quilt of squares shown has reflection symmetry.

Explain how to find all such quilts having reflection symmetry which are made up of 12 light squares and 4 dark squares.

2. Two vertices of an isosceles trapezoid are given. What are two possible sets of coordinates for the other vertices R and S?

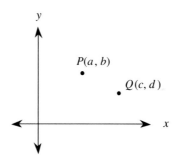

Writing for Understanding

1. Discuss all possible outcomes when two isometries are performed in sequence. For example, what is the result when a reflection is followed by a reflection?

2. The relationship of the size of a photograph of an object to the size of the object is an example of a size transformation. Describe several other applications of size transformations in everyday life.

3. Designs based on symmetries and patterns based on isometries and similitudes are found in many cultures as mentioned in the "Geometry Around Us" in Section 9.1. Find pictures of such designs in pottery, weavings, etc., from a culture other than your own. Describe how transformations are utilized in the artwork in the pictures.

4. Do some research on tessellations created by Maurits Escher. Select three of your favorite ones and explain how transformations play a role in the tessellation.

CHAPTER REVIEW

Following is a list of key vocabulary, notation, and ideas for this chapter. Mentally review these items and, where appropriate, write down the meaning of each term. Then restudy the material that you are unsure of before proceeding to take the chapter test.

Section 9.1—Isometries and Congruence

Vocabulary/Notation

Vector 419
Transformations 420
Image 420
Preimage 420
Translation (slide) 420
Clockwise orientation 421
Counterclockwise orientation 422
Directed angle 423

Rotation (turn) 424
Center of rotation 424
Fixed point 424
Identity transformation 424
Reflection (flip) 426
Glide reflection 428
Isometry 429
Congruent 430

Main Ideas/Results

1. There are several properties common to translations, rotations, reflections, and glide reflections but there are differences.
2. If a transformation is an isometry, then it must be a translation, rotation, reflection, or glide reflection.
3. An isometry is used to define congruence in general.

Section 9.2—Similitudes and Similarity

Vocabulary/Notation

Size transformation 440
Center of a size transformation 440

Scale factor 440
Similitude 442

Main Ideas/Results

1. Size transformations have several important properties.
2. A similitude is a combination of a size transformation and an isometry.
3. A similitude is used to define similarity in general.

Section 9.3—Problem Solving Using Transformations

Main Idea/Result

1. Isometries and similitudes are useful in solving problems in geometry.

PEOPLE IN GEOMETRY

H.S.M. Coxeter (b. 1907) is known for his researches and expositions of geometry. He is the author of 11 books, including *Introduction to Geometry*, *Projective Geometry*, *Non-Euclidean Geometry*, *The Fifty-Nine Icosahedra*, and *Mathematical Recreations and Essays*. He contends that Russia, Germany, and Austria are the countries that do the best job of teaching geometry in the schools, because they still regard it as a subject worth studying: "In English-speaking countries, there was a long tradition of dull teaching of geometry. People thought that the only thing to do in geometry was to build a system of axioms and see how you would go from there. So children got bogged down in this formal stuff and didn't get a lively feel for the subject. That did a lot of harm."

CHAPTER 9 TEST

TRUE-FALSE

Mark as true any statement that is always true. Mark as false any statement that is never true or that is not necessarily true. Be able to justify your answers.

1. Every rotation has at least one fixed point.

2. Every reflection has at least one fixed point.

3. A size transformation takes a line segment to a parallel line segment.

4. A glide reflection takes a line segment to a parallel line segment.

5. If the perimeter of $\triangle ABC$ is 20 cm and $\triangle A'B'C'$ is the image of $\triangle ABC$ under any isometry, then the perimeter of $\triangle A'B'C'$ is 20 cm.

6. A reflection preserves the orientation of a triangle.

7. The transformation that takes each point (a, b) to the point $(a - 3, b)$ is a translation.

8. The transformation that takes each point (a, b) to the point $(a, -b)$ is a reflection.

9. If two triangles are congruent and have the same orientation, then there is a translation that takes one triangle to the other.

10. If two polygons are similar and have the same orientation, then there is a size transformation that takes one polygon to the other.

EXERCISES/PROBLEMS

11. (a) Find coordinates of the vertices of $\triangle A'B'C'$, the image of $\triangle ABC$ under the translation $\overrightarrow{PQ}$.

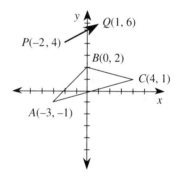

(b) If a point has coordinates (a, b), what are the coordinates of its image under the translation $\overrightarrow{PQ}$?

12. Use your protractor and a ruler to draw the image of the figure shown under a rotation of $-40°$ about center O.

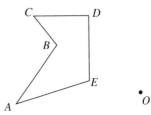

13. Using a compass and straightedge, find the reflection image of $\triangle ABC$ with respect to line l.

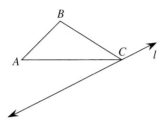

14. Using a protractor and ruler, find the image of $ABCD$ under the glide reflection that consists of the translation from A to B followed by the reflection with respect to line l.

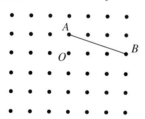

15. On the square lattice shown, draw the image of $\overline{AB}$ under a rotation of $270°$ about point O.

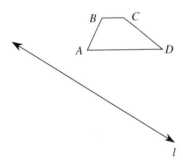

16. On the square lattice shown, draw the image of $\triangle ABC$ under a size transformation with center O and scale factor 2.

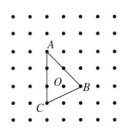

17. A triangle has vertices $A(1, 2)$, $B(4, 4)$, and $C(6, 3)$.
  (a) Sketch $\triangle A'B'C'$ the image of $\triangle ABC$ under the size transformation centered at the origin with scale factor 3 followed by a reflection with respect to the y-axis.
  (b) What are the coordinates of A', B', and C'?
  (c) If a point has coordinates (a, b), what are the coordinates of its image under this similitude?

18. Determine the type of isometry that takes the shape on the left to the shape on the right.

  (a)

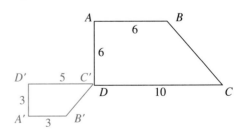

  (b)

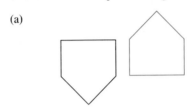

  (c)

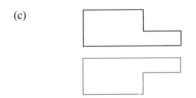

19. Describe the similitude that takes trapezoid $ABCD$ to trapezoid $A'B'C'D'$.

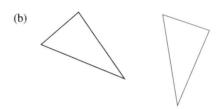

20. Suppose that a transformation is applied to all points in the plane and the image of AB is AB. Under what kind of transformation might this occur?

PROOFS

21. Show that a glide reflection can be performed by first reflecting and then translating or by first translating and then reflecting. That is, show that the image of an arbitrary point is the same in each case. [NOTE: Recall that we assume that the translation is in a direction parallel to the line of reflection.]

22. Suppose that *ABCD* is a rhombus. Use reflections with respect to lines $\overleftrightarrow{AC}$ and $\overleftrightarrow{BD}$ to show that *ABCD* is a parallelogram.

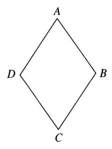

APPLICATIONS

23. On the billiards table shown, ball *A* is to carom off four of the rails and then strike ball *B*.

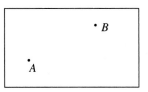

Draw a path for a successful shot.

24. A graphic designer working on a logo needs to put the design shown on a coordinate system and then double its size. How can she accomplish this?

1 Real Numbers and Solving Equations

The real numbers are used in the study of algebra and geometry, particularly in measurement. In this section, we describe the real numbers and their properties. Also, exponents and radicals are discussed because they arise in many computations. Finally, we review the skills needed to solve equations such as $3x + 4 = 5x - 3$ and $x^2 + 5x + 6 = 0$.

THE REAL NUMBERS

Before defining the real numbers, we define three other common sets of numbers. The set of **natural numbers** is the set $\{1, 2, 3, \ldots\}$. The set of integers includes the natural numbers and their opposites together with the number zero. That is, $\{\ldots, -3, -2, -1, 0, 1, 2, 3, \ldots\}$ is the set of **integers**. The set of **rational numbers** consists of numbers that can be expressed in the form $\dfrac{a}{b}$, where a and b are integers and $b \neq 0$. For example, $\dfrac{1}{2} = 0.5$, $\dfrac{1}{4} = 0.25$, and $\dfrac{-2}{5} = -0.4$ are examples of rational numbers in both their fraction and decimal representations. Notice that the rational numbers include positive numbers, negative numbers, and zero. The decimal form of $\dfrac{1}{3}$ is $0.33333\ldots$, or $0.\overline{3}$. The decimal $0.\overline{3}$ is an example of an **infinite repeating decimal**. Other examples of infinite repeating decimals are $0.2828\ldots = 0.\overline{28}$ and $1.401717\ldots = 1.40\overline{17}$. Because $\dfrac{1}{2} = 0.5$ can also be written as $0.5\overline{0}$, it can be thought of as an infinite repeating decimal. In summary, the rational numbers have decimal representations that are repeating. Those decimals that repeat only zero, such as $0.5\overline{0}$, are called **terminating decimals**. It can be shown that every repeating decimal represents a rational number. In fact, we could have defined the rational numbers as those numbers whose decimal representations repeat.

463

There are decimals that do not repeat. For example, 0.1010010001... is not a repeating decimal because the number of zeros increases as the decimal is continued. It can also be shown that $\sqrt{2}$, $\sqrt{3}$, π, and many other numbers do not have repeating decimal representations. Therefore, these numbers are not rational. Those decimals that do not have a repeating decimal representation are called the **irrational numbers**.

The **real numbers** are all numbers that can be represented as a decimal. In other words, the real numbers consist of all the rational numbers (repeating decimals) together with all the irrational numbers (nonrepeating decimals).

EXAMPLE T1.1   Which of the following descriptions fit each number given: real number, rational number, irrational number, integer, natural number?

(a) 13   (b) $-4.\overline{6}$ (c) $\dfrac{-11}{3}$ (d) 0.12131415 . . .

SOLUTION

(a) 13 is a natural number and thus an integer. Because it can be written as $\dfrac{13}{1}$, it is also a rational number. Because it is rational, it is a real number.

(b) $-4.\overline{6}$ is an infinite repeating decimal, and thus it is a rational number. Because it is rational, it is also a real number.

(c) $\dfrac{-11}{3}$ is a rational number since it is a ratio of two integers. It is also a real number.

(d) 0.12131415... is not an infinite repeating decimal because the digits in the pattern vary. Thus it is an irrational and also a real number.   •

The real numbers can be graphed on a **real number line** (Figure T1.1).

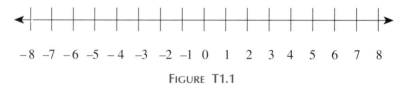

$$-8\ -7\ -6\ -5\ -4\ -3\ -2\ -1\ 0\ 1\ 2\ 3\ 4\ 5\ 6\ 7\ 8$$

FIGURE T1.1

Every point on this line represents a real number. With respect to the number line, a real number tells how far from 0 the point is on the number line and if it is located to the right of 0 (positive) or to the left of zero (negative). For example, the point representing -5 is located 5 units to the left of 0 on the real number line.

The **absolute value** of a real number r, written $|\,r\,|$, is its nonnegative (positive or zero) distance from 0. For example, $|\,3\,| = 3$ because the point representing 3 is 3 units away from 0 on the number line. Similarly, $|\,0\,| = 0$, and $|\,-7\,| = 7$. The absolute value can be used to find the distance between any two points on the real number line. If r and s are points representing any two real numbers, then the **distance** between r and s is $|r - s|$. For example, the distance between 7 and 4 is $|7 - 4| = |3| = 3$ (Figure T1.2). We can also write the distance between 7 and 4 as $|4 - 7| = |-3| = 3$.

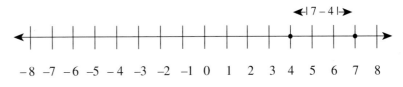

FIGURE T1.2

The distance between -3 and 8 is $|-3 - 8| = |-11| = 11$ (Figure T1.3).

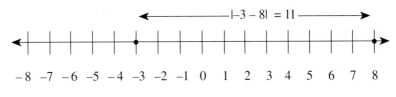

FIGURE T1.3

EXPONENTS

Exponents are used to denote repeated multiplication. That is, if r is a nonzero real number and m is a natural number, then r^m means $r \cdot r \cdots r$ where m factors of r are multiplied together. For example, 3^2 means $3 \cdot 3 = 9$, 5^3 means $5 \cdot 5 \cdot 5 = 125$, etc. Also, r^0 is defined to be 1. That is, $4^0 = 1$, $(-7)^0 = 1$, and so on. The following properties of exponents are useful in simplifying calculations.

Properties of Exponents

Let r and s be real numbers and m and n be nonnegative integers. Then the following properties of exponents hold.

1. $r^m \cdot r^n = r^{m+n}$

2. $(r^m)^n = r^{mn}$

3. $r^m s^m = (rs)^m$

4. $\dfrac{r^m}{s^m} = \left(\dfrac{r}{s}\right)^m$, for $s \neq 0$

5. $\dfrac{r^m}{r^n} = \begin{cases} r^{m-n}, & \text{for } m \geq n \\ \dfrac{1}{r^{n-m}}, & \text{for } m < n \end{cases}$

EXAMPLE T1.2   Use properties of exponents to do the following.

(a) Rewrite $2^3 \cdot 2^4$ using one exponent.

(b) Simplify the following.

(i) $3^4 \cdot 5^4$ (ii) $(3^2)^4$ (iii) $\dfrac{5^7}{5^4}$ (iv) $3^5 \cdot 3^6$ (v) $\dfrac{12^5}{4^5}$

SOLUTION

(a) $2^3 \cdot 2^4 = (2 \cdot 2 \cdot 2) \cdot (2 \cdot 2 \cdot 2 \cdot 2) = 2 \cdot 2 \cdot 2 \cdot 2 \cdot 2 \cdot 2 \cdot 2 = 2^7$. We could also apply property 1 and write $2^3 \cdot 2^4$ as $2^{3+4} = 2^7$.

(b) (i) $3^4 \cdot 5^4 = (3 \cdot 5)^4 = 15^4$ (ii) $(3^2)^4 = 3^{2 \cdot 4} = 3^8$

(iii) $\dfrac{5^7}{5^4} = 5^{7-4} = 5^3$ (iv) $3^5 \cdot 3^6 = 3^{5+6} = 3^{11}$

(v) $\dfrac{12^5}{4^5} = \left(\dfrac{12}{4}\right)^5 = 3^5$ •

SQUARE ROOTS AND CUBE ROOTS

The **principal square root** of a nonnegative number r, written $\sqrt{r}$, is defined to be the nonnegative number s, where $s^2 = r$. For example, $\sqrt{9} = 3$ because $3^2 = 9$, $\sqrt{25} = 5$ because $5^2 = 25$, and so on. Notice that when working with real numbers, the number r must be a nonnegative number because s^2 is always nonnegative. Also, although there are two numbers whose square is 4, namely 2 and -2, 2 is the *principal* square root of 4 since it is nonnegative. Similarly, the **cube root** of a number r, written $\sqrt[3]{r}$, is defined to be the real number t, where $t^3 = r$. For example, $\sqrt[3]{8} = 2$ because $2^3 = 8$ and $\sqrt[3]{-125} = -5$ since $(-5)^3 = -125$.

The square roots and cube roots of some numbers are irrational, but their decimal approximations can be determined by using a scientific calculator.

EXAMPLE T1.3   Use a calculator to find the following square roots and cube roots. Where answers are not exact, round to four decimal places.

(a) $\sqrt{2.25}$ (b) $\sqrt{65}$ (c) $\sqrt[3]{10}$

SOLUTION

(a) To evaluate $\sqrt{2.25}$ on a typical scientific calculator, use the sequence 2.25 $\boxed{\sqrt{}}$. The result here is 1.5. Thus $\sqrt{2.25} = 1.5$. Note that $1.5^2 = 2.25$. On some calculators, this sequence of keystrokes may be different.

(b) Using the same keystrokes as in (a), we get $\sqrt{65} \approx 8.062257748$ (on a ten-digit display calculator). Rounding to four decimal places yields 8.0623. Thus, $\sqrt{65} \approx 8.0623$.

(c) Some scientific calculators have a cube root key on them. In that case, the sequence of keystrokes to find $\sqrt[3]{10}$ is 10 $\boxed{\sqrt[3]{}}$. Thus $\sqrt[3]{10} \approx$

2.1544. Notice that $2.1544^3 \approx 9.999516957 \approx 10$. If your calculator does not have a cube root key, you can find the cube root of a number by using the $\boxed{y^x}$ key as follows:

$$10 \;\; \boxed{y^x} \;\; \boxed{(} \;\; 1 \;\; \boxed{\div} \;\; 3 \;\; \boxed{)} \;\; \boxed{=}$$

This method uses the fact that $x^{\frac{1}{3}} = \sqrt[3]{x}$ •

To evaluate products and quotients of square roots or to simplify computations, the following properties of square roots are often helpful.

Properties of Square Roots

$$1. \;\; \sqrt{rs} = \sqrt{r}\sqrt{s}$$

$$2. \;\; \sqrt{\frac{r}{s}} = \frac{\sqrt{r}}{\sqrt{s}}, \text{ where } s \neq 0.$$

Although the properties stated are for square roots, the same properties hold for cube roots and other roots as well.

EXAMPLE T1.4   Simplify each of the following using a property of radicals.

(a) $\sqrt{60}$ (b) $\sqrt{8}\sqrt{18}$ (c) $\sqrt{2^3 3^5}$ (d) $\sqrt[3]{3}\sqrt[3]{9}$ (e) $\sqrt{\frac{10}{49}}$

SOLUTION

(a) $\sqrt{60} = \sqrt{4 \cdot 15} = \sqrt{4}\sqrt{15} = 2\sqrt{15}$
  [NOTE: $2\sqrt{15}$ is the *exact* value of $\sqrt{60}$, and it is the simplified form of that number. For many purposes, the decimal approximation $\sqrt{60} \approx 7.746$ may be more appropriate.]

(b) $\sqrt{8}\sqrt{18} = \sqrt{8 \cdot 18} = \sqrt{144} = 12$

(c) $\sqrt{2^3 3^5} = \sqrt{2^2 \cdot 2 \cdot 3^4 \cdot 3} = \sqrt{2^2 \cdot 3^4}\sqrt{2 \cdot 3} = 2 \cdot 3^2\sqrt{6} = 18\sqrt{6}$

(d) $\sqrt[3]{3}\,\sqrt[3]{9} = \sqrt[3]{27} = 3.$

(e) $\sqrt{\frac{10}{49}} = \frac{\sqrt{10}}{\sqrt{49}} = \frac{\sqrt{10}}{7}$ •

Square roots and cube roots are added and subtracted in the same way that we combine like terms in algebraic expressions. For example, just as $3x + 2x = (3 + 2)x = 5x$, we have $3\sqrt{7} + 2\sqrt{7} = 5\sqrt{7}$. Also, just as we cannot combine $3x$ and $7y$ in $3x + 7y$, neither can we rewrite $3\sqrt{2} + 7\sqrt{5}$ in a simpler way because $\sqrt{2}$ and $\sqrt{7}$ are different.

ORDER OF OPERATIONS

Often formulas or other algebraic expressions that involve several operations and/or exponents and radicals must be evaluated. In these cases, we need to refer to the order of operations convention.

Order of Operations

1. First perform all operations within grouping symbols, such as parentheses, working from the inside out if more than one set of grouping symbols is present.
2. Then evaluate all powers and roots.
3. Next perform all multiplications and divisions in order from left to right.
4. Finally, perform all additions and subtractions in order from left to right.

EXAMPLE T1.5   Use the order of operations convention to evaluate each of the following.

(a) $3 + 2^3 \cdot 4 - 1$ (b) $(3 + 2^3) \cdot 4 - 1$ (c) $3 + 2^3 \cdot (4 - 1)$
(d) $(3 + 2^3) \cdot (4 - 1)$ (e) $3 + 2[15 - 4(2 + 1)]$

SOLUTION

(a) $3 + 2^3 \cdot 4 - 1 = 3 + 8 \cdot 4 - 1 = 3 + 32 - 1 = 35 - 1 = 34$
(b) $(3 + 2^3) \cdot 4 - 1 = (3 + 8) \cdot 4 - 1 = 11 \cdot 4 - 1 = 44 - 1 = 43$
(c) $3 + 2^3 \cdot (4 - 1) = 3 + 2^3 \cdot 3 = 3 + 8 \cdot 3 = 3 + 24 = 27$
(d) $(3 + 2^3) \cdot (4 - 1) = (3 + 8)(4 - 1) = 11 \cdot 3 = 33$
(e) $3 + 2[15 - 4(2 + 1)] = 3 + 2[15 - 4 \cdot 3] = 3 + 2[15 - 12] = 3 + 2 \cdot 3 = 9$ •

We often employ the order of operations convention when evaluating formulas as shown in the next example.

EXAMPLE T1.6   The formula for the surface area of a right circular cylinder, in square inches, is $S = 2\pi r^2 + 2\pi rh$ where r is the radius of the cylinder in inches and h is the height of the cylinder in inches. Evaluate this formula for a cylinder with radius 5 inches and height 12 inches.

SOLUTION   We must substitute 5 for r and 12 for h in the given formula and then evaluate it, following the order of operations convention.

$$S = 2\pi r^2 + 2\pi rh$$
$$= 2\pi(5)^2 + 2\pi \cdot 5 \cdot 12$$

$$= 2\pi \cdot 25 + 10\pi \cdot 12$$

$$= 50\pi + 120\pi$$

$$= 170\pi$$

$$\approx 534.07 \text{ in}^2$$

The approximation 534.07 was obtained by using the $\boxed{\pi}$ key on a scientific calculator and rounding to two decimal places.

EQUATIONS

Often problems arise in geometry that require you to draw upon your algebra skills to solve equations. The two types of equations that occur most frequently in this book are:

Linear equations Those of the form $ax + b = 0$, where a and b represent real numbers, $a \neq 0$, and x is real number variable.

Quadratic equations Those of the form $ax^2 + bx + c = 0$, where a, b, and c are real numbers, $a \neq 0$, and x is a real number variable.

There are two main principles associated with solving linear equations.

Principle 1 If a number is added to or subtracted from both sides of an equation, the solution of the resulting equation will be the same as the solution of the original equation.

Principle 2 If both sides of an equation are multiplied or divided by the same nonzero number, the solution of the resulting equation will be the same as the solution of the original equation.

The next example illustrates how these principles can be applied to solve a particular linear equation.

EXAMPLE T1.7   Solve for x: $3x + 4 = 5x - 6$

SOLUTION   To solve the equation for x we isolate x by subtracting the same numbers from both sides of the equation and by dividing both sides by the same number.

$$
\begin{array}{rcll}
3x + 4 &=& 5x - 6 & \\
\underline{-4} && \underline{-4} & \text{Principle 1} \\
3x &=& 5x - 10 & \\
\underline{-5x} && \underline{= -5x} & \text{Principle 1} \\
-2x &=& -10 & \\
\dfrac{-2x}{-2} &=& \dfrac{-10}{-2} & \text{Principle 2} \\
x &=& 5 &
\end{array}
$$

CHECK We can check the solution by substituting 5 for each x in the equation and verifying that a true statement results.

$$3(5) + 4 = \mathbf{19} \quad \text{and} \quad 5(5) - 6 = \mathbf{19} \checkmark$$

.

In addition to the two equation principles, there are specific techniques used to solve quadratic equations. The next example illustrates two of those methods.

EXAMPLE T1.8 Solve for x: $x^2 + 6x + 5 = 0$

SOLUTION

METHOD 1: FACTORING This technique applies the principle of zero products which says that if the product of two numbers is zero, then one of the numbers must be zero. That is, if $ab = 0$, then $a = 0$ or $b = 0$.

$$x^2 + 6x + 5 = 0$$
$$(x + 1)(x + 5) = 0$$
$$x + 1 = 0 \quad \text{or} \quad x + 5 = 0$$
$$x = -1 \quad \text{or} \quad x = -5$$

CHECK $(-1)^2 + 6(-1) + 5 = 1 + (-6) + 5 = 0 \checkmark$
$(-5)^2 + 6(-5) + 5 = 25 + (-30) + 5 = 0 \checkmark$

Thus, both solutions satisfy the quadratic equation.

METHOD 2: THE QUADRATIC FORMULA If the variable expression of a quadratic equation to be solved does not factor easily, we can use the quadratic formula to find the solution(s). In fact, the quadratic formula can be used to solve any quadratic equation, whether it factors or not.

Quadratic Formula

For a quadratic equation of the form $ax^2 + bx + c = 0$, for $a \neq 0$, the solutions of the equation are given by the formula

$$x = \frac{-b \pm \sqrt{b^2 - 4ac}}{2a}$$

For $x^2 + 6x + 5 = 0$, we have $a = 1$, $b = 6$, and $c = 5$, so

$$x = \frac{-6 \pm \sqrt{6^2 - 4(1)(5)}}{2(1)}$$

$$= \frac{-6 \pm \sqrt{36 - 20}}{2}$$

$$= \frac{-6 \pm \sqrt{16}}{2}$$

$$= \frac{-6 \pm 4}{2}$$

So,

$$x = \frac{-6 + 4}{2} \qquad \text{or} \qquad x = \frac{-6 - 4}{2}$$

$$x = \frac{-2}{2} \qquad \text{or} \qquad x = \frac{-10}{2}$$

$$x = -1 \qquad \text{or} \qquad x = -5$$

NOTE: We use order of operations to evaluate the original expression. Here the radical and the fraction bar act as grouping symbols. That is, we first evaluate the expression under the square root, then evaluate the numerator, and finally perform the division.

For quadratic equations in which $b = 0$, another method is usually faster. The next example illustrates this method which is called the **square root method**.

EXAMPLE T1.9   Solve for x: $3x^2 + 1 = 41$

SOLUTION   To solve for x, we isolate the x^2 term and then take the square root of both sides of the equation.

$$3x^2 + 1 = 41$$

$$\underline{\quad -1 \quad -1 \quad} \qquad \text{Principle 1}$$

$$3x^2 = 40$$

$$\frac{3x^2}{3} = \frac{40}{3} \qquad \text{Principle 2}$$

$$x^2 = \frac{40}{3}$$

$$x = \pm\sqrt{\frac{40}{3}} \qquad \text{NOTE: There are } \textit{two} \text{ square roots of } \frac{40}{3}.$$

$$x = \pm\frac{\sqrt{40}}{\sqrt{3}} \qquad \text{Property 2 of square roots}$$

$$x = \pm\frac{2\sqrt{10}}{\sqrt{3}} \qquad \text{because } \sqrt{40} = \sqrt{4 \cdot 10} = \sqrt{4}\sqrt{10} = 2\sqrt{10}$$

$$x \approx \pm 3.65 \qquad \text{using a scientific calculator and rounding to two places}$$

The answers $\pm\sqrt{\frac{40}{3}}$ and $\pm\frac{2\sqrt{10}}{\sqrt{3}}$ are acceptable if exact answers are requested and ± 3.65 is an acceptable approximate answer.

Equations that are neither linear nor quadratic sometimes arise. For example, an equation may contain a third-degree term as the next example shows.

EXAMPLE T1.10   The volume of a sphere is given by the formula $V = \frac{4}{3}\pi r^3$.

If the volume of a particular sphere is 20π, find the value of r.

SOLUTION   We have $20\pi = \frac{4}{3}\pi r^3$ and we must solve for r. First we isolate r^3 and then we take the cube root of both sides of the equation.

$$20\pi = \frac{4}{3}\pi r^3$$

$$\frac{3}{4} \cdot 20\pi = \frac{3}{4} \cdot \frac{4}{3}\pi r^3$$

$$15\pi = \pi r^3$$

$$\frac{15\pi}{\pi} = \frac{\pi r^3}{\pi}$$

$$15 = r^3$$

$$\sqrt[3]{15} = \sqrt[3]{r^3}$$

$$\sqrt[3]{15} = r$$

Therefore, $r = 2.466$ to three decimal places.   •

FORMULAS

The same techniques that were used to solve equations containing a single variable can also be used to solve a formula relating two or more variables for one of those variables. The next example illustrates an application of those techniques to a geometric formula.

EXAMPLE T1.11   The formula for the volume of a right circular cylinder is $V = \pi r^2 h$, where r is the radius of the cylinder and h is the height. Solve the formula for r.

SOLUTION   We must solve for r in the formula $V = \pi r^2 h$.

$$V = \pi r^2 h$$

$$\frac{V}{\pi h} = \frac{\pi r^2 h}{\pi h} \qquad \text{Principle 2}$$

$$\frac{V}{\pi h} = r^2$$

Therefore, $r = \sqrt{\dfrac{V}{\pi h}}$.

NOTE: We chose the nonnegative square root here since the radius of the cylinder cannot be a negative number.   •

PROBLEM SET T1

EXERCISES/PROBLEMS

1. Which of the following are rational numbers?

   (a) -7 (b) $\dfrac{2}{15}$ (c) 4.08

   (d) $\sqrt{6}$ (e) $91.7\overline{2}$

2. Which of the following are rational numbers?

   (a) $-\dfrac{10}{7}$ (b) $\sqrt[3]{35}$

   (c) $0.981981198111\ldots$ (d) 128

   (e) $0.00\overline{2}$

3. Evaluate each of the following.

   (a) $|-8|$ (b) $|-5-9|$

   (c) $|-5|-|9|$

4. Evaluate each of the following.

   (a) $|6|$ (b) $|-5-(-1)|$

   (c) $|-5|-|-1|$

5. Find the distance between the points -60 and 13 on the real number line.

6. Find the distance between the points -45 and -27 on the real number line.

7. Use properties of exponents to simplify each of the following.

   (a) $7^2 \cdot 7^4$ (b) $5^3 \cdot 6^3$

   (c) $(4^3)^5$ (d) $\dfrac{10^{12}}{10^2}$

8. Use properties of exponents to simplify each of the following.

   (a) $\dfrac{24^5}{8^5}$ (b) $7^2 \cdot 4^2$

   (c) $\dfrac{9^4}{9^6}$ (d) $(5^4)^7$

9. Use properties of exponents to express $\dfrac{2^4 \cdot 4^3}{8^2}$ using one exponent. [HINT: Try expressing all powers using the same base number.]

10. Use properties of exponents to express $\dfrac{81 \cdot 27^5}{9^3 \cdot 3^4}$ using one exponent.

11. Use your calculator to find the following square roots and cube roots. For approximate values, round your answers to four decimal places.

   (a) $\sqrt{95}$ (b) $\sqrt[3]{-16}$ (c) $\sqrt{5^2 + 4}$

   (d) $\sqrt{2\pi - 1}$ (e) $\sqrt[3]{12}$

12. Use your calculator to find the following square roots and cube roots. For approximate values, round your answers to four decimal places.

   (a) $\sqrt[3]{29}$ (b) $\sqrt{4^2 + 5^2}$ (c) $\sqrt{\dfrac{50}{7}}$

   (d) $2\sqrt{40}$ (e) $\sqrt[3]{3 \cdot 4^2 + 5}$

13. Use a property of radicals to simplify each of the following.

   (a) $\sqrt{45}$ (b) $\sqrt{12}\sqrt{27}$

   (c) $\sqrt[3]{4}\sqrt[3]{16}$ (d) $\sqrt{\dfrac{125}{9}}$

14. Use a property of radicals to simplify each of the following.

   (a) $\sqrt{48}$ (b) $\sqrt{6}\sqrt{15}$

   (c) $\sqrt[3]{25}\sqrt[3]{5}$ (d) $\sqrt{\dfrac{98}{25}}$

15. Perform each of the indicated additions and subtractions.

   (a) $10x + 3y + x$ (b) $\sqrt{5} + 3\sqrt{5} - \sqrt{7}$

   (c) $10\sqrt[3]{2} - 2\sqrt[3]{2}$ (d) $6\pi + 15 + 3\pi$

   (e) $2\pi + \dfrac{\pi}{2}$

16. Perform each of the indicated additions and subtractions.

   (a) $3a + 2a - 10b - a$

   (b) $3\sqrt{15} + 2\sqrt{15} - \sqrt{15}$

   (c) $2\sqrt[3]{2} + 7\sqrt[3]{2} + \sqrt[3]{2}$

   (d) $\pi + 8\pi$

   (e) $\dfrac{\pi}{2} + \dfrac{5\pi}{2} + 1$

17. Evaluate each of the following using the order of operations convention.

   (a) $3 \cdot 2^4 - 4 \div 2$ (b) $3(2^4 - 4) \div 2$

   (c) $2 \cdot 3^2 \cdot 5 - 4 \cdot 2$ (d) $\dfrac{10(8 - 5) + 2 \cdot 3}{2^3 + 1}$

18. Evaluate each of the following using the order of operations convention.

   (a) $(5 + 2)^2 - 8$ (b) $5 + 2^2 - 8$

   (c) $5 \cdot 2^3 \cdot 3^2 + 7 \cdot 4$ (d) $\dfrac{5^3 - 3^4 + 2^2}{1 + 3(2 + 3)}$

19. The volume of a right circular cone, in cubic inches, is given by the formula $V = \frac{1}{3}\pi r^2 h$, where r is the radius of the base in inches and h is the height of the cone in inches.
    (a) Find V if r is 4 inches and h is 9 inches.
    (b) Solve for h in the formula.

20. The area of a trapezoid, in square centimeters, is given by $A = \frac{1}{2}h(b_1 + b_2)$, where h is the height in cm and b_1 and b_2 are the lengths of the bases in cm.
    (a) Find A if h is 6 cm, b_1 is 5 cm, and b_2 is 2 cm.
    (b) Solve for b_1 in the formula.

21. The perimeter of a rectangle is given by $P = 2w + 2l$. Solve for w in the formula.

22. The area of a sector of a circle is given by the formula $A = \frac{x}{360}\pi r^2$ where x is the measure of the central angle of the sector. Solve for x in the formula.

23. Solve each of the following linear equations for x.
    (a) $3x - 5 = 19$
    (b) $2x = 3x - 4$
    (c) $2 + \frac{2}{5}x = \frac{1}{4}x + 5$
    (d) $2(x - 3) + 4x = x + 24$

24. Solve each of the following linear equations for x.
    (a) $2x + 7 = 25$
    (b) $7x + 2 = x + 32$
    (c) $\frac{1}{2}x - 6 = \frac{1}{3}x$
    (d) $3(1 - x) - 8 = 3x - (5x + 7)$

25. Use factoring to solve for x:
    $x^2 - 7x + 12 = 0$

26. Use factoring to solve for x:
    $2x^2 + x - 15 = 0$

27. Use the quadratic formula to solve for x:
    $x^2 + 7x + 8 = 0$

28. Use the quadratic formula to solve for x:
    $2x^2 - 8x + 5 = 0$

29. Use any method to solve each of the following quadratic equations.
    (a) $4x^2 = 25$
    (b) $x^2 - x - 6 = 0$
    (c) $x^2 + 3x + 1 = 0$

30. Use any method to solve each of the following quadratic equations.
    (a) $x^2 - 8 = 4$
    (b) $3x^2 + 11x - 4 = 0$
    (c) $2x^2 - x - 6 = 0$

31. Solve for r in the formula $A = \pi r^2$.

32. Solve for b in the formula $a^2 + b^2 = c^2$.

33. Solve for x: $2x^3 - 4 = 246$

34. Solve for r in the formula $V = \pi r^3$.

35. Solve for x to three decimal places:
    $2\pi x^2 = 239$

36. Solve for x to three decimal places:
    $\frac{2}{3}\pi x^2 + 12 = 70$

TOPIC REVIEW

Following is a list of key vocabulary, notation, and ideas for this topic. Mentally review these items and, where appropriate, write down the meaning of each term. Then restudy the material that you are unsure of before proceeding to take the topic test.

Vocabulary/Notation

Natural numbers 463
Integers 463
Rational numbers 463
Irrational numbers 464
Real numbers 464

Absolute value, $|r|$ 464
Distance between r and s, $|r - s|$ 465
Exponent 465
Principal square root, $\sqrt{r}$ 466
Cube root, $\sqrt[3]{r}$ 466

Main Ideas/Results

1. The real numbers are used in algebra and geometry.
2. Exponents and roots are used to express numbers and simplify computations.
3. The order of operations is used to evaluate expressions and formulas.
4. Solving linear, quadratic, and other equations is helpful in solving geometry problems.

TOPIC 1 TEST

TRUE-FALSE

Mark as true any statement that is always true. Mark as false any statement that is never true or that is not necessarily true. Be able to justify your answers.

1. The cube root of a real number is a nonnegative real number.

2. The real number 3.14 is an irrational number.

3. The product of $\sqrt{6}$ and $\sqrt{5}$ is $\sqrt{30}$; that is, $\sqrt{6} \cdot \sqrt{5} = \sqrt{30}$.

4. The sum of $\sqrt{6}$ and $\sqrt{5}$ is $\sqrt{11}$; that is, $\sqrt{6} + \sqrt{5} = \sqrt{11}$.

5. A quadratic equation has at most two solutions.

EXERCISES/PROBLEMS

6. Use properties of exponents to simplify each of the following.
   (a) $8^3 \cdot 5^3$
   (b) $\dfrac{8^3}{8^7}$
   (c) $(8^3)^6$
   (d) $8^3 \cdot 16^2$

7. Use your calculator to find the following square roots and cube roots. For approximate values, round your answers to four decimal places.
   (a) $\sqrt[3]{115}$ (b) $\sqrt{6^2 - 1.8^2}$ (c) $\sqrt{\dfrac{15}{28}}$

8. Use a property of radicals to simplify each of the following.
   (a) $\sqrt{75}$
   (b) $\sqrt{10} \cdot \sqrt{10}$
   (c) $\sqrt[3]{6}\sqrt[3]{36}$
   (d) $\sqrt{\dfrac{21}{2}}$

9. Perform each of the indicated additions and subtractions.
   (a) $9x - y + 7x + 6y$ (b) $2\pi + 3\pi$
   (c) $\sqrt{2} + 4\sqrt{2}$ (d) $5\sqrt{3} + 7 + \dfrac{1}{2}\sqrt{3}$

10. Evaluate each of the following using the order of operations convention.
    (a) $3 \cdot 5^2 - 4 \cdot 2^3$ (b) $\dfrac{2 + 3(8 - 5)^2}{6}$
    (c) $1 + 12 \div 4 \div 2$

11. The formula for the surface area of a right rectangular prism (a box), in square inches, is $S = 2lw + 2lh + 2hw$. The variables l, w, and h represent the length, width, and height, respectively, of the prism, in inches. Find S if l is 10 inches, w is 8 inches, and h is 5 inches.

12. Solve for x: $10x - 8 = 3(4 - x) + 6$

13. Use factoring to solve for x:
    $3x^2 - 26x + 35 = 0$

14. Use the quadratic formula to solve for x:
    $x^2 - 11x - 4 = 0$

15. Use the square root method to solve for x:
    $5x^2 - 7 = 42$

16. Solve for r in the formula $V = \dfrac{2}{3}\pi r^3$.

2 Elementary Logic

A knowledge of logic allows us to evaluate the validity of arguments, in and out of mathematics. The validity of an argument depends on its logical form and not on the particular meaning of the terms it contains. For example, the argument, "All Xs are Ys; all Ys are Zs; therefore all Xs are Zs" is valid no matter what X, Y, and Z are. In this topic section we will study how logic can be used to represent arguments symbolically and to analyze arguments using tables and diagrams.

STATEMENTS

Statements are the building blocks on which logic is built. A **statement** is a declarative sentence that is true or false but not both. Examples of statements are:

1. Geographically, Alaska is the largest state of the United States. (True)
2. Based on population, Texas is the largest state of the United States. (False)
3. $2 + 3 = 5$. (True)
4. $3 < 0$. (False)

The following are not statements as defined in logic.

1. Oregon is the best state. (Subjective)
2. Help! (An exclamation)
3. Where were you? (A question)
4. The rain in Spain (Not a sentence)
5. This sentence is false. (Neither true nor false!)

Often, ideas in mathematics can be made clearer by using variables and diagrams. For example, the equation $2m + 2n = 2(m + n)$, where the variables m and n are whole numbers, can be used to show that the sum of any two arbitrary even numbers, $2m$ and $2n$ here, is the even number $2(m + n)$. Figure T2.1 shows that $(x + y)^2 = x^2 + 2xy + y^2$, where each term in the expanded product is the area of the rectangular region so designated. In particular, symbols and diagrams can be used to clarify the logic of verbal arguments. Statements are represented symbolically by lowercase letters (e.g., p, q, r, and s).

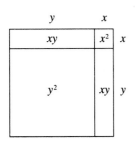

FIGURE T2.1

477

New statements can be created from existing statements in several ways. For example, if p represents the statement "The sun is shining," then the **negation** of p, written $\sim p$ and read "not p," is the statement "The sun is not shining." When a statement is true, its negation is false, and when a statement is false, its negation is true. That is, a statement and its negation have opposite truth values. We summarize this relationship between a statement and its negation using a **truth table**:

p	$\sim p$
T	F
F	T

This table shows that when the statement p is T, then $\sim p$ is F, and when p is F, then $\sim p$ is T.

LOGICAL CONNECTIVES

Two or more statements can be joined, or connected, to form **compound statements**. Next we study the four commonly used **logical connectives** "and," "or," "if-then," and "if and only if."

And

If p is the statement "It is raining" and q is the statement "The sun is shining," then the **conjunction** of p and q is the statement "It is raining *and* the sun is shining." The conjunction is represented symbolically as "$p \wedge q$." The conjunction of two statements p and q *is* true only when both p and q are true. That is, we say that the statement "It is raining and the sun is shining" is true only in the case where it is true that the sun is shining *and* it is raining. The next truth table displays the possible values of p and q with the corresponding values of $p \wedge q$.

p	q	$p \wedge q$
T	T	T
T	F	F
F	T	F
F	F	F

Notice that the two statements p and q each have two possible truth values, T and F. Hence there are four possible combinations of T and F to consider.

Or

The **disjunction** of statements p and q is the statement "p or q," which is represented symbolically as "$p \vee q$." In practice, there are two common uses of "or": the exclusive "or" and the inclusive "or." The statement "I will go or I will not go" is an example of the use of the **exclusive "or"** because either "I will go" is true or "I will not go" is true, but both cannot be true simultaneously. The **inclusive "or"** (called "and/or" in everyday language) allows for the situation in which both parts are true. For example, the statement "It will rain or the sun will shine" uses the inclusive "or." It is true if (1) it rains, (2) the sun shines, or (3) it rains and the sun shines. That is, the inclusive "or" in $p \lor q$ allows for both p and q to be true. In mathematics, we agree to use the inclusive "or" whose truth values are summarized in the next truth table.

p	q	$p \lor q$
T	T	T
T	F	T
F	T	T
F	F	F

EXAMPLE T2.1   Decide if the following statements are true or false, where p represents the statement "Rain is wet" and q represents the statement "Black is white."

(a) $\sim p$ (b) $p \land q$ (c) $(\sim p) \lor q$
(d) $p \land (\sim q)$ (e) $\sim (p \land q)$ (f) $\sim[p \lor (\sim q)]$

SOLUTION
(a) p is T, so $\sim p$ is F.
(b) p is T and q is F, so $p \land q$ is F.
(c) $\sim p$ is F and q is F, so $(\sim p) \lor q$ is F.
(d) p is T and $\sim q$ is T, so $p \land (\sim q)$ is T.
(e) p is T and q is F, so $p \land q$ is F and $\sim (p \land q)$ is T.
(f) p is T and $\sim q$ is T, so $p \lor (\sim q)$ is T and $\sim[p \lor (\sim q)]$ is F.   •

If-Then

One of the most important compound statements is the implication. The statement "If p, then q," denoted by "$p \Rightarrow q$," is called an **implication** or **conditional** statement. The statement p is called the **hypothesis** and q is called the **conclusion.** To construct the truth table for $p \Rightarrow q$, consider the following conditional promise given to a math class: "If you average at least 90% on all tests, then you will earn an A." Let p represent "Your average is at least 90% on all tests" and q represent "You earn an A." Then there are four possibilities:

Average at Least 90%	Earn an A	Promise Kept
Yes	Yes	Yes
Yes	No	No
No	Yes	Yes
No	No	Yes

Notice that the only way the promise can be broken is in line 2. In lines 3 and 4, the promise is not broken since an average of at least 90% was not attained. (In these cases, a student may still earn an A—it does not affect the promise either way.) This example suggests the following truth table for the conditional $p \Rightarrow q$.

p	q	$p \Rightarrow q$
T	T	T
T	F	F
F	T	T
F	F	T

The truth values for $p \wedge q$ and $q \wedge p$ are always the same. Also, the truth tables for $p \vee q$ and $q \vee p$ are identical (Why?). But the truth tables of $p \Rightarrow q$ and $q \Rightarrow p$ are not identical. Consider this example: Let p be "You live in New York City" and q be "You live in New York state." Then $p \Rightarrow q$ is true since if you live in New York City, then you must live in New York state. However, $q \Rightarrow p$ is not true because living in New York state does not necessarily imply that you live in New York City. For example, you may live in Albany. The conditional $q \Rightarrow p$ is called the converse of $p \Rightarrow q$. As the example shows, a conditional may be true and its converse false. On the other hand, a conditional and its converse may both be true. There are two other variants of a conditional, the contrapositive and the inverse, that are used in mathematics. A summary of these statements is given next.

Given conditional: $p \Rightarrow q$
The **converse** of $p \Rightarrow q$ is $q \Rightarrow p$.
The **inverse** of $p \Rightarrow q$ is $(\sim p) \Rightarrow (\sim q)$.
The **contrapositive** of $p \Rightarrow q$ is $(\sim q) \Rightarrow (\sim p)$.

The following truth table displays the various truth values for these four statements. Remember that the only case in which a conditional is false is when the hypothesis is true and the conclusion is false.

p	q	$\sim p$	$\sim q$	Conditional $p \Rightarrow q$	Contrapositive $\sim q \Rightarrow \sim p$	Converse $q \Rightarrow p$	Inverse $\sim p \Rightarrow \sim q$
T	T	F	F	T	T	T	T
T	F	F	T	F	F	T	T
F	T	T	F	T	T	F	F
F	F	T	T	T	T	T	T

Notice that the columns of truth values under the conditional $p \Rightarrow q$ and its contrapositive are the same. When this is the case, we say that the two statements are logically equivalent. In general, two statements are **logically equivalent** when they have the same truth tables. Similarly, the converse of $p \Rightarrow q$ and the inverse of $p \Rightarrow q$ have the same truth table. Therefore they, too, are logically equivalent. In mathematics, replacing a conditional with a logically equivalent conditional often facilitates the solution of a problem.

EXAMPLE T2.2   Prove that if x^2 is odd, then x is odd.

SOLUTION   Rather than trying to prove that the given conditional is true, consider its logically equivalent contrapositive: If x is not odd (i.e., x is even), then x^2 is not odd (i.e., x^2 is even). Even numbers are of the form $2m$, where m is a whole number. Thus the square of $2m$, $(2m)^2 = 4m^2 = 2(2m^2)$, is also an even number because it is of the form $2n$. Thus if x is even, then x^2 is even. Because this statement is true and it is logically equivalent to the original statement, the original statement is also true.   •

If and Only If

The connective "p if and only if q" is called the **biconditional** and is written $p \Leftrightarrow q$. It is the conjunction of $p \Rightarrow q$ and its converse $q \Rightarrow p$. That is, $p \Leftrightarrow q$ is logically equivalent to $(p \Rightarrow q) \wedge (q \Rightarrow p)$. The truth table of $p \Leftrightarrow q$ follows.

p	q	$p \Rightarrow q$	$q \Rightarrow p$	$(p \Rightarrow q) \wedge (q \Rightarrow p)$	$p \Leftrightarrow q$
T	T	T	T	T	T
T	F	F	T	F	F
F	T	T	F	F	F
F	F	T	T	T	T

Notice that the biconditional $p \Leftrightarrow q$ is true when p and q have the same truth values and false otherwise.

The words "necessary" and "sufficient" can be used to describe conditionals and biconditionals. For example, the statement "Water is *necessary* for the formation of ice" means "If there is ice, then there is water." On the other hand, the statement "Water is *sufficient* for the formation of ice" would be false, because it means that water guarantees the formation of ice. The statement "A rectangle with two adjacent sides the same length is a sufficient condition to define a square" means "If a rectangle has two adjacent sides the same length, then it is a square." Symbolically we have the following:

$p \Rightarrow q$ means q **is necessary for** p

$p \Rightarrow q$ means p **is sufficient for** q

$p \Leftrightarrow q$ means p **is necessary and sufficient for** q.

ARGUMENTS

Deductive or **direct reasoning** is a process of reaching a conclusion from one (or more) statements, called the hypothesis (or hypotheses). This somewhat informal definition can be rephrased using the language and symbolism in the preceding section. An **argument** is a set of statements where one of the statements is called the conclusion and the rest comprise the hypothesis.

A **valid argument** is an argument in which the conclusion must be true whenever the hypothesis is true. When an argument is valid, we say that the conclusion follows from the hypothesis. For example, consider the following argument:

If it is snowing, then it is cold.
It is snowing.
Therefore, it is cold.

In this argument, when the two statements in the hypothesis, namely "If it is snowing, then it is cold" and "It is snowing" are both true, then one can conclude that "It is cold." That is, this argument is valid because the conclusion follows from the hypothesis.

An argument is an **invalid argument** if its conclusion can be false when its hypothesis is true. An example of an invalid argument is the following:

If it is raining, then the streets are wet.
The streets are wet.
Therefore, it is raining.

For convenience, we will represent this argument as $[(p \Rightarrow q) \wedge q] \Rightarrow p$. This is an invalid argument because the streets could be wet from a variety of causes (e.g., a street cleaner, an open fire hydrant, etc.) without having had any rain. In this example, $p \Rightarrow q$ is true and q may be true, whereas p is false. The next truth table also shows that this argument is invalid, since it is possible to have the hypothesis $[(p \Rightarrow q) \wedge q]$ true with the conclusion p false.

p	q	$p \Rightarrow q$	$(p \Rightarrow q) \wedge q$	$[(p \Rightarrow q) \wedge q] \Rightarrow p$
T	T	T	T	T
T	F	F	F	T
F	T	T	T	*F*
F	F	T	F	T

The argument with hypothesis $[(p \Rightarrow q) \wedge \sim p]$ and conclusion $\sim q$ is another example of a common invalid argument form. Here, when p is F and q is T, $[(p \Rightarrow \sim q) \wedge p]$ is T and $\sim q$ is F.

There are three important valid argument forms that are used repeatedly in logic. The first two "laws" are discussed briefly in Chapter 4.

1. **Law of Detachment:** $[(p \Rightarrow q) \wedge p] \Rightarrow q$.

In words, the law of detachment says that whenever a conditional statement and its hypothesis are true, the conclusion is also true. That is, the conclusion can be "detached" from the conditional. An example of the use of this law follows.

If a number ends in zero, then it is a multiple of 10.
Forty is a number that ends in zero.
Therefore, 40 is a multiple of 10.

NOTE: Strictly speaking, the sentence "a quadrilateral has four right angles" is an "open" sentence because no particular quadrilateral is specified. Therefore the sentence is neither true nor false as given. The sentence "Quadrilateral *ABCD* in Figure T2.2 has four right angles" is a true statement. Because the use of open sentences is prevalent throughout mathematics, we will permit such "open" sentences in conditional statements without pursuing an in-depth study of such sentences.

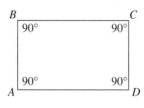

FIGURE T2.2

The following truth table verifies that the law of detachment is a valid argument form. Notice that the final column consists of all T's.

p	q	$p \Rightarrow q$	$(p \Rightarrow q) \wedge p$	$[(p \Rightarrow q) \wedge p] \Rightarrow q$
T	T	T	T	T
T	F	F	F	T
F	T	T	F	T
F	F	T	F	T

In line 1 of the preceding truth table, when the hypothesis $(p \Rightarrow q) \wedge p$ is true, the conclusion, q, is also true. This law of detachment is used in everyday language and thought.

Diagrams also can be used to decide the validity of arguments. Consider the following argument.

All mathematicians are logical.
Greta is a mathematician.
Therefore, Greta is logical.

This argument can be pictured using an **Euler diagram** (Figure T2.3). The "mathematician" circle within the "logical people" circle represents the statement "All mathematicians are logical." The point labeled "Greta" in the "mathematician" circle represents "Greta is a mathematician." Because the "Greta" point is within the "logical people" circle, we conclude that "Greta is logical."

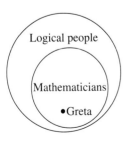

FIGURE T2.3

The second common valid argument form is the law of syllogism (or hypothetical syllogism).

2. **Law of Syllogism:** $[(p \Rightarrow q) \wedge (q \Rightarrow r)] \Rightarrow (p \Rightarrow r)$.

The following argument is an application of this law.

If a number is a multiple of eight, then it is a multiple of four.
If a number is a multiple of four, then it is a multiple of two.
Therefore, if a number is a multiple of eight, then it is a multiple of two.

This argument can be verified using an Euler diagram (Figure T2.4). The circle within the "multiples of 4" circle represents the "multiples of 8" circle. The "multiples of 4" circle is within the "multiples of 2" circle. Thus, from the diagram, it must follow that "If a number is a multiple of 8, then it is a multiple of 2" because all the multiples of 8 are within the "multiples of 2" circle.

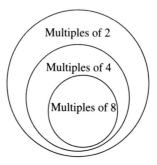

Multiples of 2

Multiples of 4

Multiples of 8

FIGURE T2.4

The following truth table also proves the validity of the law of syllogism.

p	q	r	$p \Rightarrow q$	$q \Rightarrow r$	$p \Rightarrow r$	$(p \Rightarrow q) \wedge (q \Rightarrow r)$	$[(p \Rightarrow q) \wedge (q \Rightarrow r)] \Rightarrow (p \Rightarrow r)$
T	T	T	T	T	T	T	T
T	T	F	T	F	F	F	T
T	F	T	F	T	T	F	T
T	F	F	F	T	F	F	T
F	T	T	T	T	T	T	T
F	T	F	T	F	T	F	T
F	F	T	T	T	T	T	T
F	F	F	T	T	T	T	T

Observe that in rows 1 and 5 the hypothesis $(p \Rightarrow q) \wedge (q \Rightarrow r)$ is true. In both cases, the conclusion, $p \Rightarrow r$, is also true. Thus the argument is valid.

The final valid argument we study here is used often in mathematical reasoning.

3. **Law of Contraposition:** $[(p \Rightarrow q) \wedge (\sim q)] \Rightarrow \sim p.$

Consider the following argument (Figure T2.5).

If a quadrilateral is a square, then it has four 90° angles.
ABCD does not have four 90° angles.
Therefore, *ABCD* is not a square.

This argument is an application of the law of contraposition. Figure T2.5(b) illustrates this argument. All points outside the larger circle represent quadrilaterals that do *not* have four congruent angles. Clearly, any point outside the larger circle must be outside the smaller circle. Thus, because *ABCD* would lie outside the circle labelled "Quadrilaterals with four 90° angles," it is not a square.

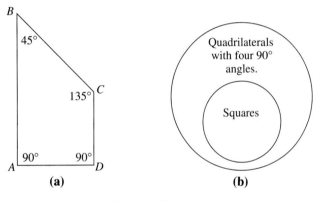

(a)   (b)

FIGURE T2.5

The next truth table provides a verification of the validity of this argument form.

p	q	$p \Rightarrow q$	$(p \Rightarrow q) \wedge \sim q$	$\sim p$	$[(p \Rightarrow q) \wedge (\sim q)] \Rightarrow \sim p$
T	T	T	F	F	T
T	F	F	F	F	T
F	T	T	F	T	T
F	F	T	T	T	T

Notice that row 4 is the only instance when the hypothesis $(p \Rightarrow q) \wedge \sim q$ is true. Here, the conclusion of $[(p \Rightarrow q) \wedge (\sim q)] \Rightarrow \sim p$, namely p, is also true. Therefore the argument is valid.

The three argument forms we have been studying also can be applied to statements that are modified by "quantifiers," that is, words such as "all," "some," "every," or their equivalents. Here, again, Euler diagrams can be used to decide the validity or invalidity of various arguments. Consider the following argument.

All logicians are mathematicians.
Some philosophers are not mathematicians.
Therefore, some philosophers are not logicians.

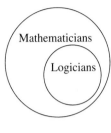

FIGURE T2.6

The first line of this argument is represented by the Euler diagram in Figure T2.6. But, because the second line guarantees that there are "some" philosophers outside the "mathematician" circle, a dot is used to represent at least one philosopher who is not a mathematician (Figure T2.7). Observe that, due to the dot, there is always at least one philosopher who is not a logician. Therefore the argument is valid. Note that the word "some" means "at least one."

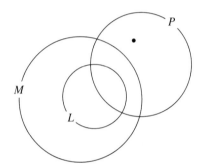

FIGURE T2.7

Next consider the following argument.

All rock stars have green hair.
No presidents of banks are rock stars.
Therefore, no presidents of banks have green hair.

An Euler diagram that represents this argument is shown in Figure T2.8, where G represents all people with green hair, R represents all rock stars, and P represents all bank presidents. Note that Figure T2.8 allows for presidents of banks to have green hair, because the circles G and P may have an element in common. Thus the argument, as stated, is invalid because the hypothesis can be true while the conclusion is false. The validity or invalidity of the arguments

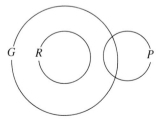

FIGURE T2.8

given in the problem set can be determined in this way using Euler diagrams. Be sure to consider *all* possible relationships among the sets before drawing any conclusions.

PROBLEM SET T2

EXERCISES/PROBLEMS

1. Determine which of the following are statements.
   (a) What's your name?
   (b) The rain in Spain falls mainly in the plain.
   (c) Happy New Year!
   (d) Five is an odd number.

2. Determine which of the following are statements.
   (a) Is $\triangle ABC$ isosceles?
   (b) Go away!
   (c) The sum of the angle measures
   (d) The area of a circle with radius r is $2\pi r$.

3. Let r, s, and t be the following statements:

   r: Roses are red.
   s: The sky is blue.
   t: Turtles are green.

   Translate the following statements into English.
   (a) $r \wedge s$ (b) $r \wedge (s \vee t)$
   (c) $s \Rightarrow (r \wedge t)$ (d) $(\sim t \wedge t) \Rightarrow \sim r$

4. Write the following in symbolic form using p, q, r, $\sim$, $\wedge$, $\vee$, $\Rightarrow$, $\Leftrightarrow$, where p, q, and r represent the following statements:

   p: The sun is shining.
   q: It is raining.
   r: The grass is green.

   (a) If it is raining, then the sun is not shining.
   (b) It is raining and the grass is green.
   (c) The grass is green if and only if it is raining and the sun is shining.
   (d) Either the sun is shining or it is raining.

5. Fill in the headings of the following truth table using p, q, $\wedge$, $\vee$, $\sim$, and $\Rightarrow$.

p	q		
T	T	T	F
T	F	F	T
F	T	T	T
F	F	T	T

6. Fill in the headings of the following truth table using p, q, $\wedge$, $\vee$, $\sim$, and $\Rightarrow$.

p	q		
T	T	T	T
T	F	F	T
F	T	F	F
F	F	F	T

7. If p is T, q is F, and r is T, find the truth values for the following:
   (a) $p \wedge \sim q$ (b) $\sim(p \vee q)$
   (c) $(\sim p) \Rightarrow r$ (d) $(\sim p \wedge r) \Leftrightarrow q$
   (e) $(\sim q \wedge p) \vee r$ (f) $p \vee (q \Leftrightarrow r)$
   (g) $(r \wedge \sim p) \vee (r \wedge \sim q)$
   (h) $(p \wedge q) \Rightarrow (q \vee \sim r)$

8. Suppose that $p \Rightarrow q$ is known to be false. Give the truth values for the following:
   (a) $p \vee q$ (b) $p \wedge q$
   (c) $q \Rightarrow p$ (d) $\sim q \Rightarrow p$

9. Construct one truth table that contains truth values for all of the following statements and determine which are logically equivalent.
   (a) $(\sim p) \vee (\sim q)$ (b) $(\sim p) \vee q$
   (c) $(\sim p) \wedge (\sim q)$ (d) $p \Rightarrow q$
   (e) $\sim(p \wedge q)$ (f) $\sim(p \vee q)$

10. Use a truth table to determine which of the following are always true.
    (a) $(p \Rightarrow q) \Rightarrow (q \Rightarrow p)$
    (b) $p \Rightarrow p$
    (c) $[p \wedge (p \Rightarrow q)] \Rightarrow q$
    (d) $(p \vee q) \Rightarrow (p \wedge q)$
    (e) $(p \wedge q) \Rightarrow p$

11. State the hypothesis (or hypotheses) and conclusion for each of the following arguments.
    (a) All football players are introverts. Tony is a football player, so Tony is an introvert.
    (b) Bob is taller than Jim, and Jim is taller than Sue. So Bob is taller than Sue.

12. State the hypothesis (or hypotheses) and conclusion for each of the following arguments.
    (a) All penguins are elegant swimmers. No elegant swimmers fly, so penguins don't fly.
    (b) $x < y$ and $y < z$ implies that $x < z$.

13. Determine the validity of the following arguments.
    (a) All professors are handsome.
        Some professors are tall.
        Therefore, some handsome people are tall.
    (b) If I can't go to the movie, then I'll go to the park.
        I can go to the movie.
        Therefore, I will not go to the park.
    (c) If you score at least 90%, then you'll earn an A.
        If you earn an A, then your parents will be proud.
        You have proud parents.
        Therefore, you scored at least 90%.
    (d) Some arps are bomps.
        All bomps are cirts.
        Therefore, some arps are cirts.
    (e) All equilateral triangles are equiangular.
        All equiangular triangles are isosceles.
        Therefore, all isosceles triangles are equilateral.
    (f) If you work hard, then you will succeed.
        You do not work hard.
        Therefore, you will not succeed.
    (g) Some girls are teachers.
        All teachers are college graduates.
        Therefore, all girls are college graduates.
    (h) If it doesn't rain, then the street won't be wet.
        The street is wet.
        Therefore, it rained.

14. Determine a valid conclusion that follows from each of the given statements and explain your reasoning.
    (a) If you study hard, then you will be popular. You will study hard.
    (b) If Scott is quick, then he is a basketball star. Scott is not a basketball star.
    (c) All friends are respectful. All respectful people are trustworthy.
    (d) Every square is a rectangle. Some parallelograms are rhombuses. Every rectangle is a parallelogram.

15. Using each pair of statements, determine whether (i) p is necessary for q, (ii) p is sufficient for q, or (iii) p is necessary and sufficient for q.
    (a) p: Bob has some water.
        q: Bob has some ice.

    (b) p: It is snowing.
        q: It is cold.
    (c) p: It is December.
        q: 31 days from today it is January.

16. Using each pair of statements, determine whether (i) p is necessary for q, (ii) p is sufficient for q, or (iii) p is necessary and sufficient for q.
    (a) p: Janet is at least 16 years old.
        q: Janet has her driver's license.
    (b) p: A square has a side that measures 5 inches.
        q: A square has an area of 25 square inches.
    (c) p: Ted walked to work this morning.
        q: Ted woke up this morning.

17. Which of the laws (detachment, syllogism, or contraposition) is being used in each of the following arguments?
    (a) If Joe is a professor, then he is learned. If you are learned, then you went to college. Joe is a professor, so he went to college.
    (b) If you have children, then you are an adult. Bob is not an adult, so he has no children.
    (c) If I am broke, I will ride the bus. When I ride the bus, I am always late. I'm broke, so I am going to be late.

18. Which of the laws (detachment, syllogism, or contraposition) is being used in each of the following arguments?
    (a) All women are smart. Helen of Troy was a woman. So Helen of Troy was smart.
    (b) If today is Tuesday, then tomorrow is Wednesday. Tomorrow is Saturday, so today is not Tuesday.
    (c) If I don't eat breakfast, I'll be hungry by 10:00. If I'm hungry before noon, I always snack before lunch. I skipped breakfast, so I will snack before lunch.

19. If possible, determine the truth value of each statement. Assume that a and b are true, p and q are false, and x and y have unknown truth values. If a value can't be determined, write "unknown."
    (a) $p \Rightarrow (a \lor b)$ (b) $b \Rightarrow (p \lor a)$
    (c) $x \Rightarrow p$ (d) $a \lor p$
    (e) $b \Rightarrow q$ (f) $b \Rightarrow x$
    (g) $a \land (b \lor x)$ (h) $(y \lor x) \Rightarrow a$
    (i) $(y \land b) \Rightarrow p$ (j) $(a \lor x) \Rightarrow (b \land q)$
    (k) $x \Rightarrow a$ (l) $x \lor p$
    (m) $\sim x \Rightarrow x$ (n) $x \lor (\sim x)$
    (o) $\sim[y \land (\sim y)]$

20. Decide the truth value of each statement.
  (a) Alexander Hamilton was once president of the United States.
  (b) The world is flat.
  (c) If dogs are cats, then the sky is blue.
  (d) If Tuesday follows Monday, then the sun is hot.
  (e) If Christmas Day is December 25, then Texas is the largest state in the United States.

21. Rewrite each argument in symbolic form, then check the validity of the argument.
  (a) If today is Wednesday (*w*), then yesterday was Tuesday (*t*). Yesterday was Tuesday, so today is Wednesday.
  (b) The plane is late (*l*) if it snows (*s*). It is not snowing. Therefore, the plane is not late.

22. Rewrite each argument in symbolic form, then check the validity of the argument.
  (a) If I do not study (*s*), then I will eat (*e*). I will not eat if I am worried (*w*). Hence, if I am worried, I will study.
  (b) Meg is married (*m*) and Sarah is single (*s*). If Bob has a job (*j*), then Meg is married. Hence Bob has a job.

23. Draw an Euler diagram to represent the following argument and decide if it is valid.

  All timid creatures (*T*) are bunnies (*B*).
  All timid creatures are furry (*F*).
  Some cows (*C*) are furry.
  Therefore, all cows are timid creatures.

24. Use the following Euler diagram to determine which of the following statements are true. (Assume that there is at least one person in every region within the circles.)

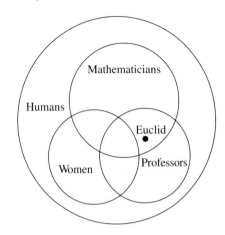

  (a) All women are mathematicians.
  (b) Euclid was a woman.
  (c) All mathematicians are men.
  (d) All professors are humans.
  (e) Some professors are mathematicians.
  (f) Euclid was a mathematician and human.

PROOF

25. Prove that the conditional $p \Rightarrow q$ is logically equivalent to $\sim p \lor q$.

TOPIC REVIEW

Following is a list of key vocabulary, notation, and ideas for this topic. Mentally review these items and, where appropriate, write down the meaning of each term. Then restudy the material that you are unsure of before proceeding to take the topic test.

Vocabulary/Notation

Statement, *p* 477
Negation, $\sim p$ 478
Truth table 478
Compound statements 478
Logical connective 478
Conjunction (and), $p \land q$ 478
Disjunction (or), $p \lor q$ 479
Exclusive "or" 479

Inclusive "or" 479
Implication/conditional
  (if . . . , then . . .), $p \Rightarrow q$ 479
Hypothesis 479
Conclusion 479
Converse 479
Inverse 479
Contrapositive 479

Main Ideas/Results

1. Logic is used to determine if arguments are valid or invalid.

2. Statements and arguments can be represented symbolically and pictorially.

3. Valid arguments can be identified using truth tables and Euler diagrams.

TOPIC 2 TEST

TRUE-FALSE

Mark as true any statement that is always true. Mark as false any statement that is never true or that is not necessarily true. Be able to justify your answers.

1. The disjunction of p and q is true whenever p is true and q is false.

2. If $p \Rightarrow q$ is true, then $\sim p \Rightarrow \sim q$ is true.

3. "I am older than 20 or younger than 30" is an example of an exclusive "or."

4. The converse of $p \Rightarrow q$ is $q \Rightarrow p$.

5. $p \Rightarrow q$ means p is necessary for q.

EXERCISES/PROBLEMS

6. Find the converse, inverse, and contrapositive of each statement.
   (a) $p \Rightarrow \sim q$ (b) $\sim p \Rightarrow q$ (c) $\sim q \Rightarrow \sim p$

7. Decide the truth value of each statement.
   (a) $4 + 7 = 11$ and $1 + 5 = 6$.
   (b) $2 + 5 = 7 \Leftrightarrow 4 + 2 = 8$.
   (c) $3 \cdot 5 = 12$ or $2 \cdot 6 = 11$.
   (d) If $2 + 3 = 5$, then $1 + 2 = 4$.
   (e) If $3 + 4 = 6$, then $8 \cdot 4 = 31$.
   (f) If 7 is even, then 8 is even.

8. Use Euler diagrams to check the validity of each argument.
   (a) Some men are teachers.
       Sam Jones is a teacher.
       Therefore, Sam Jones is a man.

   (b) Gold is heavy.
       Nothing but gold will satisfy Amy.
       Hence nothing that is light will satisfy Amy.
   (c) No cats are dogs.
       All poodles are dogs.
       So some poodles are not cats.
   (d) Some cows eat hay.
       All horses eat hay.
       Only cows and horses eat hay.
       Frank eats hay, so Frank is a horse.
   (e) All chimpanzees are monkeys.
       All monkeys are animals.
       Some animals have two legs.
       So some chimpanzees have two legs.

9. Complete the following truth table.

p	q	$p \wedge q$	$p \vee q$	$p \Rightarrow q$	$\sim q \Leftrightarrow p$	$\sim q \Rightarrow \sim p$
T					T	
	T	T	F			
F				F		T
				T	F	

10. Using an Euler diagram, display an example of an invalid argument. Explain.

11. For the following pair of statements, determine whether p is necessary for q, p is sufficient for q, or p is both necessary and sufficient for q.

    p: $\triangle ABC$ has exactly two congruent sides.
    q: $\triangle ABC$ has exactly two congruent angles.

12. Determine a valid conclusion that follows from each pair of statements, if possible. Explain your reasoning.
    (a) All polyhedra have edges. A tetrahedron is a polyhedron.
    (b) If it is 5:00 P.M., then it is quitting time at work. It is 4:15 P.M.

13. Suppose that $\sim p \Rightarrow q$ is known to be false. Give truth values for each of the following.
    (a) $p \vee q$ (b) $\sim p \Leftrightarrow q$ (c) $\sim p \wedge \sim q$

14. Is it ever the case that the conjunction, disjunction, and implication of two statements are all true at the same time? All false? If so, what are the truth values of each statement? If not, explain why not.

3 Inequalities in Algebra and Geometry

In Chapter 4 we studied triangle congruence and equations arising from corresponding parts of congruent triangles. In Chapter 7 we studied congruence relationships among central angles, arcs, and chords of circles. In this section, we study inequality relationships in triangles and circles that are proved using results from Chapters 4 and 7. We also review the skills needed to solve inequalities such as $5x - 8 > 37$.

INEQUALITIES

Sometimes we need to solve problems in geometry that involve inequalities. That is, we may have a situation in which the measure of one angle is larger than the measure of another, one line segment is shorter than another, and so on. We use the symbol ">" to indicate that one number is greater than another and "<" to indicate less than. These symbols are often combined with the equals sign to form the symbol "≤," which means "is less than or equal to" and "≥," which means "is greater than or equal to." For example, $n \leq 5$ means "n is less than 5 *or* n is equal to 5."

We can interpret the statement "$3 < 5$" in several ways. Using a number line, we see that $3 < 5$ and 3 is to the left of 5 (and 5 is to the right of 3) (Figure T3.1).

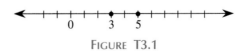

FIGURE T3.1

In general, for real numbers a and b we say that a is less than b (and write $a < b$) if and only if a is to the left of b on the real number line.

We can also relate inequalities to equations. To say that $3 < 5$ means that there is a positive real number, 2 in this case, that we can add to 3 to obtain 5.

In other words, we say that $3 < 5$ because $3 + 2 = 5$. In general, we say that a **is less than** b if and only if there is a positive real number p such that $a + p = b$. We say that b **is greater than** a whenever a is less than b.

If a and b are real numbers represented on the real number line, either they share the same position (hence $a = b$), or a is to the left of b ($a < b$) or b is to the left of a ($a > b$). This relationship is stated next.

Trichotomy Property

For any real numbers a and b, exactly one of the following statements is true:

$$a = b \quad \text{or} \quad a > b \quad \text{or} \quad a < b$$

Often more than two numbers are compared. For example, if we know that $\angle A < \angle B$ and $\angle B < 90°$, we can conclude that $\angle A < 90°$. This is an example of the following theorem.

Theorem T3.1

Transitive Property for Less Than

Let a, b, and c represent three real numbers. If $a < b$ and $b < c$, then $a < c$.

Although this property is stated for less than, similar results hold for ">," "≤," and "≥." That is, if $a > b$ and $b > c$, then $a > c$. The proof of the transitive property is left for the problem set.

SOLVING INEQUALITIES

In geometry we sometimes need to use algebra to solve an inequality such as $5x - 8 > 37$ for x. That is, we need to find all real number values for x so that the inequality is true. We can accomplish this by applying the following principles and using techniques similar to those we use when solving equations.

Inequality Principle 1: If a real number is added to or subtracted from both sides of an inequality, the direction of the inequality is unchanged.

If $a < b$, then $a + c < b + c$.

Inequality Principle 2: If both sides of an inequality are multiplied or divided by the same *positive* real number, the direction of the inequality is unchanged.

If $a < b$ and $c > 0$, then $ac < bc$.

Inequality Principle 3: If both sides of an inequality are multiplied or divided by the same *negative* real number, the direction of the inequality is *reversed*.

If $a < b$ and $c < 0$, then $ac > bc$.

In the case of Principles 1 and 2, we say that the inequality is preserved.

EXAMPLE T3.1   Solve each of the following inequalities for x.
(a) $5x - 8 > 37$ (b) $5x + 4 \geq 9x - 52$

SOLUTION

(a) To solve this inequality for x, we isolate x. First we simplify the inequality by using Inequality Principle 1 and adding 8 to both sides.

$$
\begin{array}{rl}
5x - 8 & > 37 \\
\underline{+ 8 \quad + 8} & \qquad \text{Inequality Principle 1} \\
5x & > 45
\end{array}
$$

Next we apply Inequality Principle 2 and divide both sides by 5.

$$
\dfrac{5x}{5} > \dfrac{45}{5} \qquad \text{Inequality Principle 2}
$$

$$
x > 9
$$

Notice that the direction of the inequality did not change in the second step since we divided by a *positive* number.

CHECK   There are infinitely many solutions to the inequality $x > 9$, all of which will satisfy the original inequality $5x - 8 > 37$. We can check such solutions by substituting some values into the original inequality. For $x = 12$, we have $5(12) - 8 = $ **52** and $52 > 37$. Next we check one value that should *not* satisfy the original inequality. For $x = 7$, we have $5(7) - 8 = $ **27** and $27 < 37$, rather than "greater than." Thus, 7 is not one of the solutions of $5x - 8 > 37$. Although it is impossible to check all values, checking one that works and one that doesn't often helps catch errors.

(b) As we did in (a), we will isolate x by first subtracting and then dividing.

$$
\begin{array}{rl}
5x + 4 \geq \quad 9x - 52 & \\
\underline{-9x \qquad\qquad -9x} & \qquad \text{Inequality Principle 1} \\
-4x + 4 \geq -52 & \\
\underline{\quad -4 \qquad -4 \quad} & \qquad \text{Inequality Principle 1} \\
-4x \geq -56 & \\
\dfrac{-4x}{-4} \leq \dfrac{-56}{-4} & \qquad \text{Inequality Principle 3} \\
 & \qquad \text{(Notice the inequality} \\
 & \qquad \text{is reversed.)} \\
x \leq \quad 14 &
\end{array}
$$

CHECK   In the original inequality, try $x = 13$ (which should be a solution) and $x = 15$ (which should *not* be a solution).

$$5(13) + 4 = \textbf{69} \text{ and } 9(13) - 52 = \textbf{65}, \text{ and } 69 \geq 65. \ \checkmark$$
$$5(15) + 4 = \textbf{79} \text{ and } 9(15) - 52 = \textbf{83}, \text{ and } 79 < 83. \ \checkmark$$

•

TRIANGLES

Consider $\triangle ABC$ in Figure T3.2(a) where $AC > AB$. It appears that $\angle B > \angle C$. To prove this, find point D on $\overline{AC}$ where $AB = AD$ [Figure T3.2(b)].

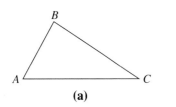
(a)

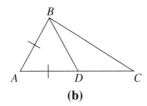
(b)

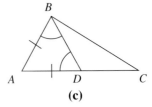
(c)

FIGURE T3.2

Because $AB = AD$, we have $\angle ABD = \angle ADB$ [Figure T3.2(c)]. We also know $\angle ABC = \angle ABD + \angle DBC$. Therefore $\angle ABC > \angle ABD = \angle ADB > \angle C$. Justify this inequality. [HINT: Notice $\angle ADB$ is an exterior angle of $\triangle BDC$.] In other words, $\angle ABC > \angle C$. This result is summarized next.

Theorem T3.2

If two sides of a triangle are not congruent, the angle opposite the longer side is greater than the angle opposite the shorter side.

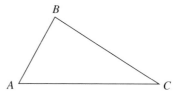

If $AC > AB$, then $\angle B > \angle C$.

The converse of Theorem T3.2 is also true. In fact, we can use Theorem T3.2 to help make the proof of its converse, which follows.

Theorem T3.3

If two angles of a triangle are not congruent, then the side opposite the larger angle is longer than the side opposite the smaller angle.

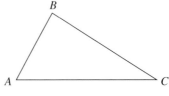

If $\angle B > \angle C$, then $AC > AB$.

Given $\angle B > \angle C$ (Figure T3.3)

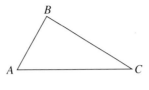

FIGURE T3.3

Prove $AC > AB$

Plan We will use indirect reasoning. First we will assume that $\angle B > \angle C$ and $AC = AB$ and show that this is impossible. Then we will assume that $\angle B > \angle C$ and $AC < AB$ and show that this, too, is impossible. Thus we will conclude that $AC > AB$.

Proof First, assume that $AC = AB$. Then $\triangle ABC$ is isosceles and $\angle B = \angle C$, which contradicts our hypothesis $\angle B > \angle C$. Next assume that $AC < AB$. Then we could also write $AB > AC$. It follows from Theorem T3.2 that $\angle C > \angle B$, which contradicts our hypothesis. Therefore, because $AC = AB$ and $AC < AB$ are not possible when $\angle B > \angle C$, it follows that $AC > AB$.   •

EXAMPLE T3.2   Show that the shortest distance from a point to a line is the length of the perpendicular from the point to the line. That is, given line l and point P in Figure T3.4(a), prove that PA is the shortest distance from P to l.

SOLUTION   Let B be any other point on line l. Draw $\overline{PB}$. We will show that $PA < PB$, for any point B [Figure T3.4(b)]. Because $\triangle PAB$ is a right triangle, we have $\angle PAB > \angle PBA$. Thus, by Theorem 3.3, we know that $PB > PA$. So the shorter of the distances from P to l is PA, the perpendicular distance.   •

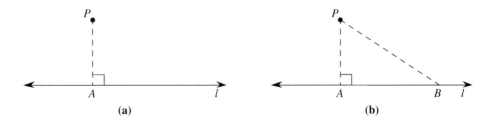

(a)   (b)

FIGURE T3.4

The next theorem, called the Triangle Inequality, shows that there is a relationship among the lengths of the sides of a triangle. For example, consider $\triangle ABC$ in Figure T3.5. If you were asked to walk from A to C, which would be the shorter route: (i) along $\overline{AC}$ or (ii) from A to B along $\overline{AB}$ and then to C along $\overline{BC}$? The next theorem answers this question. It states that the sum of the lengths of any two sides of a triangle is greater than the length of the third side, as appears to be the case in Figure T3.5.

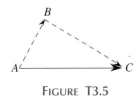

FIGURE T3.5

Theorem T3.4

The Triangle Inequality

The sum of the lengths of any two sides of a triangle is greater than the length of the third side.

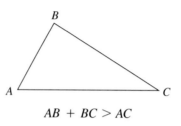

$$AB + BC > AC$$

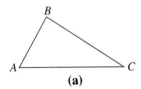

(a)

Given $\triangle ABC$ [Figure T3.6(a)]

Prove $AB + BC > AC$

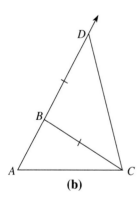

(b)

FIGURE T3.6

Proof On $\overrightarrow{AB}$, find point D where B is between A and D, and $BD = BC$ [Figure T3.6(b)]. Because $\triangle BCD$ is isosceles, $\angle D = \angle BCD$. Because $\angle ACD > \angle BCD$, we have $\angle ACD > \angle D$ by substitution. By the previous theorem we conclude that $AD > AC$. But $AD = AB + BD$. So we have $AB + BD > AC$, and because $BD = BC$, we substitute BC for BD to have $AB + BC > AC$.   •

EXAMPLE T3.3   Given $\triangle ABC$ as shown, where $AB = 5$m and $AC = 6$m, what lengths are possible for side $\overline{BC}$ (Figure T3.7)?

SOLUTION   By the Triangle Inequality, we know that the sum of the lengths of any two sides must be greater than the length of the third side. So we can write

(i) $5 + 6 > BC$,   (ii) $6 + BC > 5$,   and   (iii) $5 + BC > 6$

From inequality (i), we know that $11 > BC$, or $BC < 11$.
From inequality (ii), we know that $BC > -1$, but because BC must be nonnegative, this gives us no new information.
From inequality (iii), we know that $BC > 1$.
Therefore, we have $BC > 1$ and $BC < 11$, which we write as $1 < BC < 11$.
That is, BC must be between 1 m and 11 m.   •

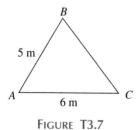

FIGURE T3.7

The next two theorems are consequences of the preceding ones. Due to the complexity of the proof of the first theorem, we simply state it without proof here.

Theorem T3.5

SAS Inequality

If two sides of one triangle are congruent to two sides of another, but the included angle of the first triangle is less than the included angle of the other triangle, then the side opposite the included angle in the first triangle is shorter than the corresponding side of the second triangle.

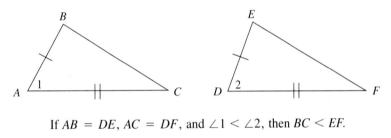

If $AB = DE$, $AC = DF$, and $\angle 1 < \angle 2$, then $BC < EF$.

The proof of the next theorem, which uses the SAS Inequality Theorem, is left for the problem set.

Theorem T3.6

SSS Inequality

If two sides of one triangle are congruent to two sides of another, but the third side of the first triangle is shorter than the third side of the other triangle, then the angle opposite the third side in the first triangle is less than the corresponding angle in the second triangle.

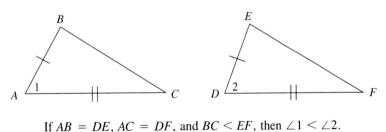

If $AB = DE$, $AC = DF$, and $BC < EF$, then $\angle 1 < \angle 2$.

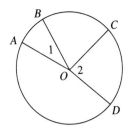

CIRCLES

Recall that congruent central angles intercept congruent arcs and chords. Similar inequality relationships also hold for circles. For example, $\angle 1 < \angle 2$ in circle O (Figure T3.8). Because $\angle 1 = \widehat{AB}$ and $\angle 2 = \widehat{CD}$, we have $\widehat{AB} < \widehat{CD}$. The converse of this argument is also true. We summarize these two results as follows.

Theorem T3.7

In a circle or in circles with the same radius, one central angle is less than another if and only if the measure of the intercepted arc of the first angle is less than the measure of the intercepted arc of the second angle.

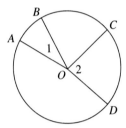

$\angle 1 < \angle 2$ if and only if $\widehat{AB} < \widehat{CD}$.

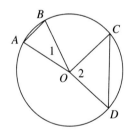

FIGURE T3.9

Next consider chords $\overline{AB}$ and $\overline{CD}$ in circle O of Figure T3.9. Suppose that $\angle AOB < \angle COD$. Because $OA = OB = OC = OD$ and $\angle 1 < \angle 2$, by the SAS Inequality Theorem, $AB < CD$. Conversely, if $AB < CD$, then by the SSS Inequality Theorem, $\angle 1 < \angle 2$. We summarize this result in the next theorem.

Theorem T3.8

In a circle or in circles with the same radius, one central angle is less than another if and only if the length of the chord intercepted by the first angle is less than the length of the chord intercepted by the second.

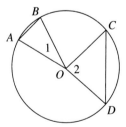

$\angle 1 < \angle 2$ if and only if $AB < CD$.

The two preceding theorems can be combined to form the next theorem.

Theorem T3.9

In a circle or in circles with the same radius, the length of one chord is less than the length of a second chord if and only if the measure of the arc intercepted by the first chord is less than the measure of the arc intercepted by the second.

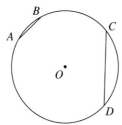

$AB < CD$ if and only if $\overset{\frown}{AB} < \overset{\frown}{CD}$.

In Figure T3.9 it appears that the chord nearer the center of the circle is longer than the one farther from the center. Next we prove this result.

Let $x < y$ in Figure T3.10(a) where $\overline{OE}$ and $\overline{OF}$ are perpendicular to $\overline{AB}$ and $\overline{CD}$ respectively.

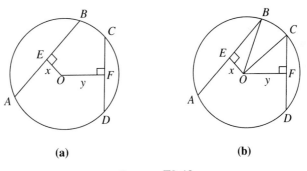

(a)   (b)

FIGURE T3.10

By Problem 32 of Section 7.2, E and F are midpoints of $\overline{AB}$ and $\overline{CD}$ respectively. By the Pythagorean Theorem, $x^2 + (BE)^2 = (BO)^2$ and $y^2 + (CF)^2 = (CO)^2$ [Figure T3.10(b)]. Because $(BO)^2 = (CO)^2$ we have $x^2 + (BE)^2 = y^2 + (CF)^2$. Because $x < y$, we have $x^2 < y^2$ and $x^2 + (BE)^2 < y^2 + (BE)^2$. Finally, $y^2 + (CF)^2 < y^2 + (BE)^2$ by substitution, so $(CF)^2 < (BE)^2$, or $CF < BE$. This implies that $2CF < 2BE$, or that $CD < AB$, which was what we were trying to prove. This argument can be reversed, leading to the following result.

Theorem T3.10

In a circle or in circles with the same radius, the length of one chord is greater than the length of a second chord if and only if the first chord is nearer to the center than the second.

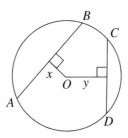

$AB > CD$ if and only if $x < y$.

Notice that an immediate consequence of this theorem is the fact that diameters of a circle are the longest chords.

PROBLEM SET T3

EXERCISES/PROBLEMS

1. Solve each inequality for x.
   (a) $4x - 1 < 23$ (b) $9 + 6x \leq -6$
   (c) $3x - 8 > 7x + 12$

2. Solve each inequality for x.
   (a) $3x - 7 > 11$ (b) $12 - x < 4$
   (c) $2x + 6 \leq 7x - 34$

3. Solve each inequality for x.
   (a) $2 - 3(x - 4) \leq 2(1 - x)$
   (b) $\dfrac{2x - 1}{7} \leq -3$
   (c) $x(x - 5) \leq x^2 - 3(x - 4)$

4. Solve each inequality for x.
   (a) $3 - (x - 1) > 3x + 4$
   (b) $\dfrac{6x - 5}{2} \geq -25$
   (c) $2x(x - 1) \geq x^2 + x(x + 2) + 8$

5. Given $\triangle ABC$ as shown, explain why $\angle B > \angle C$.

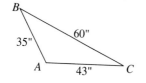

6. Given $\triangle ABC$ as shown, explain why $AB < BC$.

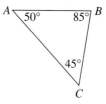

7. In $\triangle PQR$, $\angle P = 41°$ and $\angle Q = 80°$.
   (a) What is the measure of $\angle R$?
   (b) Which of the three sides is the longest?
   (c) Which of the three sides is the shortest?

8. In $\triangle VWX$, $\angle W = 26°$ and $\angle X = 105°$.
   (a) What is the measure of $\angle V$?
   (b) Which of the three sides is the longest?
   (c) Which of the three sides is the shortest?

9. In circle O, $\angle AOB = 95°$, $\angle BOC = 42°$, and $\angle COD = 60°$.

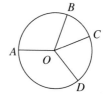

(a) What is the measure of ∠AOD?
(b) Which of $\widehat{AB}$ or $\widehat{BC}$ is larger? Why?
(c) Which of $\widehat{AD}$ or $\widehat{AC}$ is larger? Why?

10. In circle $\underline{O}$, $\widehat{AD}$ has measure 112°, $\widehat{BC}$ has measure 34°, and $\overline{BD}$ is a diameter.

(a) What is the measure of $\widehat{CD}$?
(b) Which of ∠AOC or ∠AOD is larger? Why?
(c) Which of ∠DOA or ∠DOC is larger? Why?

11. Answer the following questions about the circle O as shown.

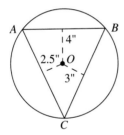

(a) Which side of △ABC is longest? Explain.
(b) Which angle of △ABC is the largest? Explain.
(c) Which of arcs $\widehat{AB}$, $\widehat{BC}$, or $\widehat{AC}$ has the smallest measure? Explain.

12. Answer the following questions about circle O where $AB = 5''$, $BC = 2''$, and $DE = 9''$.

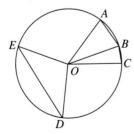

(a) Which of ∠AOB, ∠BOC, or ∠EOD is the largest? Explain.
(b) Which of $\widehat{AB}$, $\widehat{BC}$, or $\widehat{DE}$ has the smallest measure? Explain.
(c) Which of ∠EOD or ∠AOC is larger? Explain.

13. Assuming that the following figure is drawn precisely, write three valid inequality statements relating pairs of AB, BD, BE, or BC. For example, $AB < BC$.

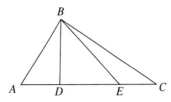

14. Using the figure in problem 13, write three valid inequality statements relating pairs of ∠A, ∠ADB, ∠AEB, or ∠C. For example, ∠ADB > ∠AEB.

15. Given parallelogram $ABCD$ with dimensions as shown, what lengths are possible for side $\overline{BC}$?

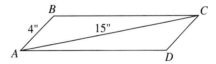

16. Given quadrilateral $PQRS$ with dimensions as shown, what lengths are possible for diagonal $\overline{QS}$?

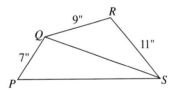

PROOFS

17. Prove that the hypotenuse is longer than either of the legs of a right triangle.

18. △ABC is a right triangle with D on $\overline{AC}$. Prove that $BD < BC$ using two different methods.

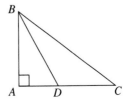

19. In △ABC, AB = BC and D is on $\overline{AC}$. Prove that AB > BD.

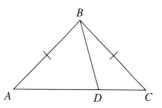

20. ABCD is a kite with AB = BC, AD = CD, and AB < AD. Prove: ∠D < ∠B.

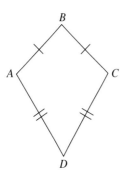

21. For rectangle ABCD, prove that the diagonal AC is longer than either of the sides using two different methods.

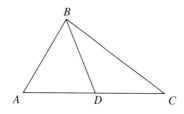

22. Let △ABC be any triangle with $\overline{BD}$ the median to $\overline{AC}$. Prove that BD < ½(AB + BC).

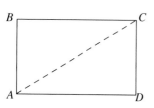

23. In isosceles trapezoid ABCD, AB = BC = CD and BC < AD. Prove that ∠A < ∠C.

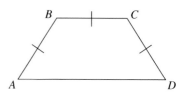

24. Let ABCD be any quadrilateral. Prove AC + BD < AB + BC + CD + AD. Restate this result in words.

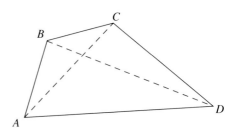

25. In the circle O as shown, $\overset{\frown}{AB}$ = 90° and $\overset{\frown}{CD}$ > 90°, and D, O, and B are collinear. Prove that the area of △CDO is less than the area of △ABO.

26. Let A be a point in the exterior of circle O. Prove that the shortest distance from A to the circle is AB where B is the point of intersection of $\overline{OA}$ and circle O.

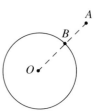

27. Let A be a point in the interior of circle O. Prove that the shortest distance from A to the circle is AB where B is the point of intersection of ray $\overrightarrow{OA}$ and circle O.

28. Let P be any point in the interior of circle O except O itself. Show that the shortest chord containing P must be the one where $\overline{OP}$ is perpendicular to the chord.

[HINT: Draw chord $\overline{AB}$ where P is on $\overline{AB}$ and $\overline{OP}$ is perpendicular to $\overline{AB}$. Then draw any other chord $\overline{CD}$

through point P. Then consider the perpendicular segment $\overline{OQ}$ from O to $\overline{CD}$.]

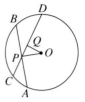

29. Prove the transitive property for less than. That is, prove that if $a < b$, and $b < c$, then $a < c$. [HINT: Use the definition of less than: $a < b$ if there is a positive p such that $a + p = b$.]

30. Prove Theorem T3.5: SSS Inequality—If two sides of one triangle are congruent to two sides of another, but the third side of the first triangle is shorter than the third side of the other triangle, then the angle opposite the third side in the first triangle is less than the corresponding angle in the second triangle.

TOPIC REVIEW

Following is a list of key vocabulary, notation, and ideas for this topic. Mentally review these items and, where appropriate, write down the meaning of each term. Then restudy the material that you are unsure of before proceeding to take the topic test.

Vocabulary/Notation

Less then 494	Triangle Inequality 498
Greater than 494	SAS Inequality 499
Trichotomy Property 494	SSS Inequality 499

Main Ideas/Results

1. In triangles, there are inequality relationships between the lengths of the sides and the measures of angles opposite those sides.

2. In circles, there are inequality relationships among the measures of the central angles and the measures of their intercepted arcs and lengths of their chords.

3. The lengths of chords are related to their distances from the center of the circle.

TOPIC 3 TEST

TRUE-FALSE

Mark as true any statement that is always true. Mark as false any statement that is never true or that is not necessarily true. Be able to justify your answers.

1. If a, b, and c are real numbers such that $a < b$ and $b < c$, then $a < c$.

2. If a, b, c, and d are real numbers such that $a < b$ and $c < d$, then $ac < bd$.

3. If $\angle A \geq \angle B$ and $\angle B < 20°$, then $\angle A > 20°$.

4. If the three angles of a triangle have different measures, then the triangle is scalene.

5. A triangle cannot have sides of lengths 8 cm, 11 cm, and 20 cm.

EXERCISES/PROBLEMS

6. Solve for x: $17 - 3x > 20$

7. Solve for x: $\dfrac{x - 7}{2} < 3x + 4$

8. In $\triangle RST$, the sides have lengths as follows: $RS = 4$ cm, $ST = 12$ cm, $RT = 10$ cm.
   (a) Which angle of $\triangle RST$ is the largest? Explain.
   (b) Which angle of $\triangle RST$ is the smallest? Explain.

9. In $\triangle XYZ$, the angles have measures as follows: $\angle X = 32°$, $\angle Z = 84°$.
   (a) Which side of $\triangle XYZ$ is the longest? Explain.
   (b) Which side of $\triangle XYZ$ is the shortest? Explain.

10. In circle O, $\angle AOB = 115°$ and $\angle BOC = 80°$.

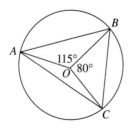

   (a) Which has the greater measure: $\overset{\frown}{AB}$ or $\overset{\frown}{AC}$?
   (b) Which is longer: $\overline{AO}$ or $\overline{BC}$?

11. Given $\triangle ABC$ with dimensions as shown, what lengths are possible for side $\overline{AC}$?

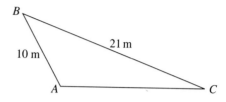

12. In circle O, $\overline{AC}$ is a diameter, $\overline{BO} \perp \overline{DO}$, $\angle AOD = 140°$, and $\overline{OE}$ bisects $\angle AOD$.

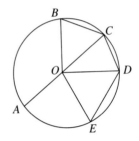

   (a) Which chord is closest to O: $\overline{BC}$, $\overline{CD}$, or $\overline{DE}$? Explain.
   (b) Which angle is the smallest: $\angle OBC$, $\angle OCD$, or $\angle OED$? Explain.

13. Let $\triangle ABC$ be any triangle with $\overline{BD}$ the angle bisector of $\angle B$. Prove that (i) $\angle ADB > \angle ABD$ and (ii) $\angle CDB > \angle CBD$.

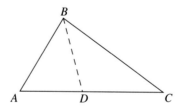

14. Prove that the sum of the lengths of the diagonals of a pentagon is less than twice its perimeter.

4 Non-Euclidean Geometry

In Euclid's famous work, *The Elements*, many definitions, five postulates of a geometric nature, and five common notions that could be applied to both numbers and geometry were given. The five postulates of a geometric nature were as follows:

Postulate 1. A straight line can be drawn from any point to any point.

Postulate 2. It is possible to extend a finite straight line indefinitely.

Postulate 3. A circle can be drawn with any point as center and any distance as radius.

Postulate 4. All right angles are equal.

Postulate 5. If a straight line meets two straight lines, so as to make the two interior angles on the same side of it taken together less than two right angles, these straight lines, being continually produced, shall at length meet on the side on which are the angles that are less than two right angles.

Postulate 5, commonly called the **Parallel Postulate**, was the center of much debate. For over 2000(!) years there were many attempts to prove the Parallel Postulate from the other postulates. However, it has been shown that it is impossible to prove it in this way.

Exploration of the Parallel Postulate stimulated interest in this area of geometry. It was demonstrated that, when assuming the first four postulates, some theorems that were proved using the Parallel Postulate could, in fact, be used to prove the Parallel Postulate. Thus, they were equivalent formulations of the Parallel Postulate. Four such results are listed next.

1. Playfair's Axiom: Given a line l and a point P not on l, there is only one line m containing P that is parallel to l.

2. Parallel lines are everywhere equidistant.

3. The sum of the measures of the angles in a triangle is 180°.

4. Proclus' Axiom: If a line intersects one of two parallel lines, then it must also intersect the other line.

In addition to motivating these equivalent formulations of the Parallel Postulate, the attempt to prove it led to the discovery of other geometries. In this section we provide a look at several of these geometries.

ABSOLUTE GEOMETRY

In *The Elements*, Euclid proved many theorems before actually employing the Parallel Postulate. The body of knowledge that follows from the postulates that preceded the Parallel Postulate became known as **absolute geometry**. In the 17th century, Girolamo Saccheri (1676–1733), an Italian Jesuit, approached the dilemma of trying to prove the Parallel Postulate by first proving additional theorems that didn't rely on the Parallel Postulate. In his study, Saccheri introduced some quadrilaterals, since named **Saccheri quadrilaterals**, that had a pair of opposite sides congruent and two right angles as shown in Figure T4.1. The other two angles were called **summit angles**. He was hoping that he could infer that these quadrilaterals were rectangles without using the Parallel Postulate. But he could not. Later it was shown that Saccheri quadrilaterals were not rectangles unless the Parallel Postulate was assumed.

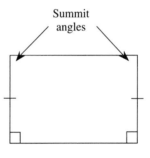

FIGURE T4.1

Although the Parallel Postulate seems "obviously" true, this is only because of our world of reference. That is, because we are generally limited to a small portion of the earth, it seems as if we are living on a plane. But, suppose that we wanted to study the geometry of a sphere. The sphere, itself, would be our plane and points on the sphere our points. The shortest distance between two points on a sphere is an arc of a **great circle**, that is, a circle of the sphere that passes through both endpoints of a diameter of a sphere. The equator of the earth is one example of a great circle. In the geometry of a sphere, these great circles are defined to be the lines. Figure T4.2(a) shows two examples of lines on a sphere. Figure T4.2(b) shows two examples of curves that would not be considered lines in this geometry.

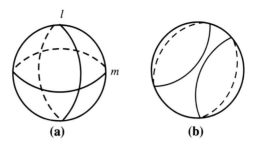

(a)   **(b)**

FIGURE T4.2

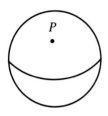

FIGURE T4.3

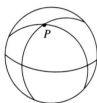

FIGURE T4.4

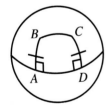

FIGURE T4.5

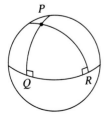

FIGURE T4.6

Using the points of a sphere as our points, the great circles of a sphere as our lines, and the sphere as our plane, we can develop a geometry. It is interesting to note, however, that the Parallel Postulate does not hold in this geometry. Using Playfair's Axiom, consider any great circle (such as the equator) and any point P not on the equator (Figure T4.3). Then any other great circle containing P also will intersect the equator. That is, there is no "line" through P that is parallel to the equator. Figure T4.4 shows two such great circles. This model also presents a Saccheri quadrilateral that is not a rectangle because the angles at B and C are greater than 90° (Figure T4.5).

ELLIPTIC GEOMETRY

In 1854, Bernhard Riemann (1826–1866) formalized this idea of a spherical geometry by replacing the Parallel Postulate with the following.

Elliptic Parallel Postulate

Given a line l and a point P not on l, there is no line containing P that is parallel to l.

It can be shown in absolute geometry that parallel lines do *exist* (only the uniqueness is guaranteed by the Parallel Postulate). Thus, Euclid's Postulate 1 must be modified as follows to be consistent with the Elliptic Parallel Postulate: Given any two points, there is *at least* one line containing the points.

Some of the results that follow from the Elliptic Parallel Postulate are quite surprising in that they differ so much from or actually contradict results derived in Euclidean geometry. For example, consider Figure T4.6. The great circles through P and Q and through P and R are perpendicular to the equator. So $\triangle PQR$ has two right angles, a property not shared by triangles in Euclidean geometry. This means that the sum of the angle measures in $\triangle PQR$ exceeds 180°. Some results of the Elliptic Parallel Postulate are listed on the next page.

1. There are no parallel lines.
2. The sum of the measures of the angles in any triangle exceeds 180°.
3. There are no rectangles.
4. If two triangles are similar, they must be congruent.
5. Two lines intersect in two points.

HYPERBOLIC GEOMETRY

Even before Riemann developed his alternative to Euclidean geometry, Joano Bolyai (1802–1860), a Hungarian army officer, and Nichollas Lobachevsky (1793–1856), a Russian mathematician, each independently developed a geometry based on absolute geometry using the following postulate in place of the Parallel Postulate.

Hyperbolic Parallel Postulate

Given a line l and a point P not on l, there exist at least two lines containing P that are parallel to l.

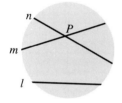

FIGURE T4.7

The following model can be used to picture a geometry that satisfies this parallel postulate. Consider a plane to be the interior of a circle, points to be the points of this interior, and lines to be chords of the circle (without the endpoints). Lines l, m, and n and point P are shown in Figure T4.7. Notice how m and n contain P, yet don't intersect l. Thus, m and n are parallel to l. In fact, there are infinitely many lines through P in this plane that are parallel to l. The usual angle measure does not hold in this geometry. Instead, there is a formula to determine angle measure that is based on the usual Euclidean angle measure.

Some theorems that arise from the Hyperbolic Parallel Postulate follow.

1. The sum of the measures of the angles of any triangle is less than 180°.
2. Different triangles may have different angle sums.
3. There are no rectangles.
4. If two triangles are similar, they must be congruent.
5. Parallel lines are not everywhere equidistant.

Thus far we have seen three ways in which Euclidean geometry has been studied based on the Parallel Postulate. First, Saccheri developed absolute geometry, that is, Euclidean geometry without the parallel postulate. Next, Bolyai and Lobachevsky developed an alternative geometry that allowed for more than one line to be parallel to a given line. Finally, elliptic geometry allowed for no parallel lines. The development of these latter two geometries had a profound impact on mathematics. It encouraged mathematicians to alter axioms and to study the resulting systems. These systems yielded important mathematical results even if they didn't model any known physical situation as Euclid's system modeled the plane, which was considered to represent portions of the Earth.

A summary and comparison of three of these geometries is provided in the following table.

	Euclidean Geometry	Elliptic Geometry	Hyperbolic Geometry
Model	Plane	Sphere	Interior of a circle
Parallels to a line through a point	One	None	Infinitely many
Angle sum in a triangle	180°	> 180°	< 180°

The postulates of Euclidean geometry may be altered in other ways to produce other very different geometries. We will complete our treatment of non-Euclidean geometry by studying two other systems, finite geometries and taxicab geometry.

FINITE GEOMETRIES

Finite geometries are composed of a finite set of points. For example, consider points A, B, C, and D in Figure T4.8(a).

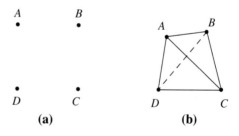

| (a) | (b) |

FIGURE T4.8

We define the set $\{A, B, C, D\}$ to be our only plane. The six possible pairs of points will be our lines. So the "line" AB in this geometry consists of only the points A and B. One way of modeling this geometry is to think of the points as vertices of a triangular pyramid with a pair of points forming a "line" if they are endpoints of the same side of the pyramid [Figure T4.8(b)]. Keep in mind that the pairs of points are the "lines" in this geometry, not the segments connecting them as shown in Figure T4.8(b).

Next we list three postulates that this example models.

Postulate 1. There are exactly four points.

Postulate 2. Every pair of points is on a line.

Postulate 3. Every line contains a pair of points.

The next several theorems can be deduced from these three postulates.

Theorem 1. If two lines have a point in common, they have exactly one point in common.

Theorem 2. There are exactly six lines.

Theorem 3. Each point contains exactly three lines.

Theorem 4. Given any line l and point P *not* on the line, there is one and only one line parallel to l that contains P.

These theorems can be proved by reasoning deductively as we do in Euclidean geometry. On the other hand, because this geometry contains a small finite number of points, we can also prove them by checking all possible cases. Although the pyramid model may be used to check cases, an alternate method is to assign letters to the points and to check the theorems using the letters. In the four-point geometry described by $\{A, B, C, D\}$, let the pairs AB, AC, AD, BC, BD, and CD be the lines. These points and "lines" satisfy Postulates 1, 2, and 3, so they represent a model for the four-point geometry. This model, using pairs of letters as lines, can be used to verify Theorems 1–4. We prove Theorem 3 in the following example. The proofs of the other three theorems are left for the problem set.

EXAMPLE T4.1   Prove that in the four-point geometry, each point contains exactly three lines.

SOLUTION   The four points in this geometry are A, B, C, and D. The lines in this geometry are the pairs AB, AC, AD, BC, BD, and CD. We must show that every point appears in exactly three of these pairs.

A appears in AB, AC, and AD. B appears in AB, BC, and BD.

C appears in AC, BC, and CD. D appears in AD, BD, and CD.

Thus, every point does appear in three pairs and so contains exactly three lines.   •

TAXICAB GEOMETRY

Euclidean geometry was studied from a coordinate point of view in Chapter 8. There the distance between two points (or the length of a segment) was found using the Pythagorean theorem (Figure T4.9). Given a coordinate system, there

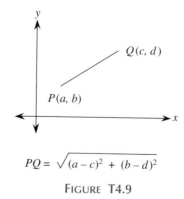

$$PQ = \sqrt{(a-c)^2 + (b-d)^2}$$

FIGURE T4.9

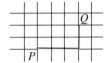

FIGURE T4.10

is another way to measure distance. Imagine a coordinate grid as intersecting streets, North/South and East/West (Figure T4.10). In taxicab geometry, the **taxicab distance** from P to Q is the shortest distance a taxi would travel along this grid of intersecting streets. In Figure T4.10, $PQ = 3 + 2 = 5$, as indicated by the colored line segments. If $P(a, b)$ and $Q(c, d)$ are two points in the plane, then, using absolute value, the distance from P to Q is given by $|a - c| + |b - d|$. In this geometry, the points, lines, and plane are the same as in Euclidean geometry. Just the distance formula has been changed. Moreover, taxicab geometry satisfies many of the same postulates as Euclidean geometry, even the Parallel Postulate.

In this book we study a set of postulates that includes Euclid's postulates as well as triangle congruence postulates. Within this set of postulates, there is one that taxicab geometry does not satisfy. Consider the right triangles $\triangle ABC$ and $\triangle DEF$ in Figure T4.11. Notice that *using taxicab distance $AB = 2 = AC$ and $DF = 2 = DE$.* Because the triangles are right triangles, they should be congruent by the SAS Congruence Postulate. But notice that they are not the same size (although they are the same *shape*); therefore they are *not* congruent. Thus, the SAS Congruence Postulate does not hold in taxicab geometry.

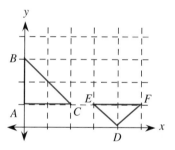

FIGURE T4.11

There are other surprising consequences of the altered distance formula in taxicab geometry. The next example illustrates one such result.

EXAMPLE T4.2   Find the "circle" with center $(0, 0)$ and radius 2 in taxicab geometry.

SOLUTION   In Figure T4.12(a), points $A(-2, 0)$, $B(0, 2)$, $C(2, 0)$, and $D(0, -2)$ are on the circle since their taxicab distance from the origin is 2.

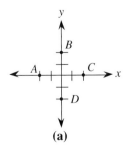

(a)

FIGURE T4.12

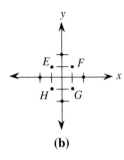

(b)

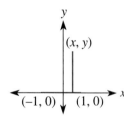

(c)

FIGURE T4.12

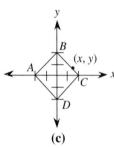

FIGURE T4.13

Also, $E(-1, 1)$, $F(1, 1)$, $G(1, -1)$, and $H(-1, -1)$ are 2 taxicab units from the origin [Figure T4.12(b)]. In Figure T4.12(c), let (x, y) be a point on $\overline{BC}$. The equation of $\overline{BC}$ in the Euclidean sense is $y = -x + 2$. By substitution, the point $(x, -x + 2)$ represents an arbitrary point of $\overline{BC}$ for $0 \leq x \leq 2$. Thus, the taxicab distance from $(0, 0)$ to $(x, -x + 2)$ is $x + (-x + 2) = 2$. This shows that any point on $\overline{BC}$ is on the "circle" of points at a distance of 2 from $(0, 0)$. Continuing in this manner, we can show that the "circle" with center $(0, 0)$ and radius 2 in taxicab geometry is actually the square $ABCD$!   •

In Euclidean geometry, a point is equidistant from the endpoints of a segment if and only if it is on the perpendicular bisector of the segment. The next example explores this idea in taxicab geometry.

EXAMPLE T4.3

(a) Find all points equidistant from $(-1, 0)$ and $(1, 0)$.

(b) Find all points equidistant from $(0, 0)$ and $(1, 3)$.

SOLUTION

(a) Let (x, y) be any point equidistant from $(-1, 0)$ and $(1, 0)$ (Figure T4.13). Then the distance from $(-1, 0)$ to (x, y) is $(x + 1) + y$, and the distance from $(1, 0)$ to (x, y) is $1 - x + y$. Because these distances are equal, we have $(x + 1) + y = 1 - x + y$, or $2x = 0$, or $x = 0$. Thus, any point equidistant from $(-1, 0)$ and $(1, 0)$ must be on the y-axis. That is, $(0, y)$ is equidistant from $(-1, 0)$ and $(1, 0)$ where the distance from either point to $(0, y)$ is $y + 1$.

(b) First, observe that $(1, 1)$ is 2 units from both $(0, 0)$ and $(1, 3)$ [Figure T4.14(a)].

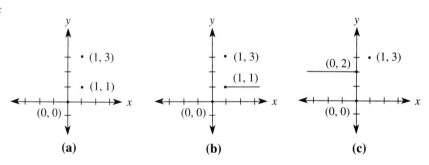

FIGURE T4.14

Also, in general, $(x, 1)$, for $x > 1$, is $x + 1$ units from both points [Figure T4.14(b)]. Similarly, the point $(x, 2)$, for $x < 0$, is equidistant from $(0, 0)$ and $(1, 3)$ [Figure T4.14(c)]. Finally, the points on the segment whose endpoints are $(0, 2)$ and $(1, 1)$ are equidistant from $(0, 0)$ and $(1, 3)$ [Figure T4.14(d)]. The verification of this result is left for the problem set.

Thus, the jagged line shown in Figure T4.14(e) is the set of points that are equidistant from (0, 0) and (1, 3).

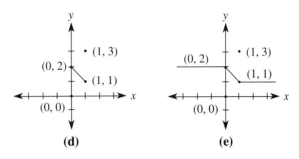

(d)   **(e)**

FIGURE **T4.14**

Other interesting properties of taxicab geometry are revealed in the problem set. A very nice treatment of this subject is found in the Dover Publications book *Taxicab Geometry* by Eugene F. Krause.

PROBLEM SET T4

EXERCISES/PROBLEMS

1. In Euclidean geometry, two distinct lines in a plane are either parallel or intersect in exactly one point. Discuss the intersection of two distinct lines in the spherical geometry of this section. Include a picture with your explanation.

2. In Euclidean geometry, given a line *l* and a point *P* not on the line, there is exactly one line through *P* that is perpendicular to *l*. Discuss the number of lines through *P* that are perpendicular to *l* in the geometry of the sphere. Include a picture with your explanation.

3. In the spherical geometry of this section, a "line" is a great circle. What will a "circle" look like in this model? [HINT: Think of a circle as the set of points equidistant from a fixed point.]

Problems 4–12 deal with taxicab geometry. Let $d_T(A, B)$ represent the taxicab distance between *A* and *B* and $d_E(A, B)$ represent the usual Euclidean distance.

4. Find $d_T(A, B)$ and $d_E(A, B)$ for the following pairs.

   (a) $A(0, 0)$, $B(0, 10)$
   (b) $A(1, 1)$, $B(1, 8)$
   (c) $A(-2, -3)$, $B(7, -3)$
   (d) $A(5, 4)$, $B(5, -2)$

   (e) Using parts (a)–(d), describe, in general, pairs of points where the taxicab distance equals the Euclidean distance.

5. Find $d_T(A, B)$ and $d_E(A, B)$ for the following pairs.
   (a) $A(2, 3)$, $B(-1, 6)$
   (b) $A(-1, -2)$, $B(3, 4)$
   (c) $A(3, 5)$, $B(9, -2)$
   (d) It is claimed that $d_T(A, B) \geq d_E(A, B)$ for all pairs *A*, *B*. Determine if this claim is correct for the pairs in parts (a)–(c). If it is, do you think that it true for any pair of points?

6. Is the Pythagorean Theorem true in taxicab geometry? Explain.

7. (a) The length of the side of a square is 4 and the diameter of a circle is 4. Compare their areas using the Euclidean distance.
   (b) Repeat part (a) using the taxicab distance and Example T4.1.

8. Determine if the following statement is true or false for the given points:

   $$d_T(A, B) + d_T(B, C) \geq d_T(A, C)$$

   (a) $A(1, 1)$, $B(2, 3)$, $C(5, 9)$
   (b) $A(-1, 4)$, $B(4, 1)$, $C(0, -5)$

9. Recall that in Euclidean geometry, P is between A and B if $AP + PB = AB$. Plot each of the following triples of points to see if one appears to be between the other two. Then check to see if

$$d_T(A, P) + d_T(P, B) = d_T(A, B)$$

   (a) $A(2, 3)$, $B(3, 6)$, $C(4, 9)$
   (b) $A(-1, -3)$, $B(2, 0)$, $C(5, 3)$

10. Complete Example T4.3 by verifying that every point on the segment with endpoints $(0, 2)$ and $(1, 1)$ is equidistant from the two points $(0, 0)$ and $(1, 3)$.

11. Figure T4.11 shows two triangles that satisfy SAS using taxicab distance but are not congruent. Find two triangles that satisfy ASA using taxicab distance, yet are not congruent.

12. Figure T4.11 shows two triangles that satisfy SAS using taxicab distance but are not congruent. Find two triangles that satisfy SSS using taxicab distance yet are not congruent.

Problems 13–15 apply to the four-point finite geometry discussed in this section. Use the points and lines associated with $\{A, B, C, D\}$ to prove the following theorems.

13. Prove Theorem 1: If two lines have a point in common, they have exactly one point in common.

14. Prove Theorem 2: There are exactly six lines.

15. Prove Theorem 4: Given any line l and point P *not* on the line, there is one and only one line parallel to l that contains P.

Let $\{A, B, C, D, E, F, G\}$ be a set of points and the following triples be "lines": ABC, AGF, AED, BGD, CGE, CFD, and EBF. Determine if the statements in problems 16–20 are true or false in this seven-point geometry. If a statement is true, prove that it is true.

16. Two distinct lines intersect in exactly one point.

17. Each point is on exactly three lines.

18. Each point is on at least one line with every other point.

19. For every pair of points, there are exactly two lines containing neither point.

20. If any three lines are not concurrent (that is, do not all share the same point), there is exactly one point that is not on any of the three lines.

Problems 21–23 deal with Saccheri quadrilaterals. You may use results about triangle congruence, but you may not use Euclid's Parallel Postulate or any of its equivalents.

21. Prove that the diagonals of a Saccheri quadrilateral are congruent.

22. Prove that the summit angles of a Saccheri quadrilateral are congruent.

23. A **Lambert quadrilateral** is one that has three right angles. It can be shown that the fourth angle is not obtuse. In the Lambert quadrilateral $ABCD$ shown prove that $CD \le AB$. [HINT: Suppose that $AB < CD$. Let E be a point on CD, consider $\angle BED$, and observe that $ABED$ is a Saccheri quadrilateral.]

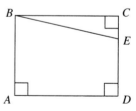

TOPIC REVIEW

Following is a list of key vocabulary, notation, and ideas for this topic. Mentally review these items and, where appropriate, write down the meaning of each term. Then restudy the material that you are unsure of before proceeding to take the topic test.

Vocabulary/Notation

Main Ideas/Results

1. The Parallel Postulate was the focus of much attention in Euclidean geometry as well as the stimulus for the formation of many new geometries.

2. Saccheri developed absolute geometry in his attempt to prove the Parallel Postulate from the other postulates.

3. There are non-Euclidean geometries whose parallel postulates differ from the standard Euclidean parallel postulate in two ways:
   (i) Hyperbolic geometry is an example of a non-Euclidean geometry where given a line l and a point P not on the line, there is more than one line through P parallel to l.
   (ii) Elliptic geometry is a non-Euclidean geometry where given a line l and a point P not on the line, there are no lines through P parallel to l.

4. There are finite geometries that model many aspects of Euclidean geometry.

5. Taxicab geometry is a non-Euclidean geometry that satisfies all of Euclid's postulates. However, it does not satisfy the SAS Congruence Postulate.

TOPIC 4 TEST

TRUE-FALSE

Mark as true any statement that is always true. Mark as false any statement that is never true or that is not necessarily true. Be able to justify your answers.

1. Two points determine a unique line in the elliptic geometry on the sphere as described in this section.

2. Two points determine a unique line in the interior of a circle model of hyperbolic geometry as described in this section.

3. In the geometry on a sphere as described in this section, a triangle can have two right angles.

4. The taxicab distance from $A(-1, 4)$ to $B(2, 3)$ is greater than the Euclidean distance AB.

5. A 3-4-5 triangle in taxicab geometry is not necessarily a right triangle.

EXERCISES/PROBLEMS

6. In Euclidean geometry, the measure of an exterior angle of a triangle is equal to the sum of the measures of the two nonadjacent interior angles. Discuss whether this relationship holds in the geometry of the sphere. Include a picture with your explanation.

7. A three-point geometry can be defined by letting the set of points $\{A, B, C\}$ represent the plane. Then the lines will be pairs of points. The following three postulates will be accepted.

Postulate 1: There are exactly three points.
Postulate 2: Every pair of points is on a line.
Postulate 3: Every line contains exactly two points.

Answer the following questions about this three-point geometry. Justify your answers.
(a) How many lines are in this geometry?
(b) If two lines intersect, how many points do they have in common?
(c) Are there parallel lines in this geometry?

8. The length of a side of an equilateral triangle is 8.
(a) Draw a sketch of the triangle and find the area of the triangle using Euclidean distance.
(b) Draw a sketch of the triangle in taxicab geometry. (Be careful! More than one triangle is possible.)
(c) Find the area of the triangle using taxicab distance.

9. Suppose that two triangles satisfy the SSS Similarity Theorem using taxicab distance. Must the two triangles be similar? Justify your answer or provide a counterexample.

10. Prove that if a quadrilateral is both a Saccheri and a Lambert quadrilateral, then it is a rectangle.

1 Getting Started: Making Proofs

HOW TO USE THIS SUMMARY:

Suppose that you need to show that two segments are congruent. Check the list of suggestions under "Congruent Segments." The first suggestion indicates that you may be able to show that the segments are congruent by showing that they are corresponding parts of congruent triangles. Another reminder is that opposite sides of a parallelogram are congruent. Can you show that the two segments are opposite sides of a parallelogram?

Be sure to note the page number after each result so that you don't use a result that has not yet been proved in your course.

CONGRUENT ANGLES

CONGRUENT SEGMENTS

CONGRUENT TRIANGLES

PERPENDICULAR LINES AND RIGHT ANGLES

PARALLEL LINES

PARALLELOGRAM

SAS Similarity   284
LL Similarity   285
SSS Similarity   286
Side-Splitting Theorem   294
Image under a similitude   443

CONGRUENT CENTRAL ANGLES

Intercept the same or congruent arcs   320
Intercept the same or congruent chords   336

CONGRUENT ARCS

Intercepted by congruent central angles   320
Intercepted by congruent chords   336
Intercepted by parallel lines   336

CONGRUENT INSCRIBED ANGLES

Intercept the same or congruent arcs   324
Intercept the same or congruent chords   336

CONGRUENT CHORDS

Intercept the same or congruent central angles   336
Intercept the same or congruent arcs   336
Equidistant from the center   336

TANGENT LINES

Perpendicular to a radius at a point of the circle   349

2 Summary of Postulates, Theorems, and Corollaries

Thm 3.1   **Perimeters of Common Quadrilaterals**

Figure	Perimeter
Square with sides of length s	$4s$
Rectangle with sides of lengths a and b	$2a + 2b$
Parallelogram with sides of lengths a and b	$2a + 2b$
Rhombus with sides of length s	$4s$
Kite with sides of lengths a and b	$2a + 2b$

96

Post 3.1   **Circumference of a Circle:** The circumference of a circle is the product of π and the diameter of the circle.
$C = \pi d = 2\pi r$ 98

Post 3.2   **Area of a Region Enclosed by a Simple Closed Curve:**
(a) For every simple closed curve and unit square, there is a positive real number that gives the number of unit squares (and parts of unit squares) that exactly tessellate the region enclosed by the simple closed curve.
(b) The area of a region enclosed by a simple closed curve is the sum of the areas of the smaller regions into which the region can be subdivided.   99

Post 3.3   Any two triangles whose corresponding angles and sides are congruent have the same area.   100

Thm 3.2   **Area of a Right Triangle:** The area of a right triangle is one-half the product of the lengths of its legs.
$A = \dfrac{1}{2}ab$ 101

Thm 3.3   **Area of a Triangle:** The area of a triangle is one-half the product of the length of one side and the height to that side.
$A = \dfrac{1}{2}bh$ 101

Thm 3.4   **Area of a Parallelogram:** The area of a parallelogram is equal to the product of the length of one of its sides and the height to that side.
$A = bh$ 111

Thm 3.5   **Area of a Trapezoid:** The area of a trapezoid is equal to one-half the product of the height and the sum of the lengths of its two bases.
$A = \dfrac{1}{2}h\,(b_1 + b_2)$ 112

Thm 3.6   **Area of a Regular Polygon:** The area of a regular polygon is equal to one-half the product of the perimeter of the polygon and the perpendicular distance from its center to one of its sides.
$A = \dfrac{1}{2}hP$ 113

Thm 3.7   **Area of a Circle:** The area of a circle is equal to the product of pi and the square of the radius.
$A = \pi r^2$ 114

Thm 3.8   **Pythagorean Theorem:** The sum of the squares of the lengths of the legs of a right triangle is equal to the square of the length of the hypotenuse.   121

Thm 3.9   **30-60 Right Triangle:** In a 30-60 right triangle the length of the leg opposite the 30° angle is one-half the length of the hypotenuse. The length of the leg opposite the 60° angle is $\frac{\sqrt{3}}{2}$ times the length of the hypotenuse.   125

Thm 3.10   **45-45 Right Triangle:** In a 45-45 right triangle, the length of the hypotenuse is $\sqrt{2}$ times the length of a leg.   127

Thm 3.11   **Surface Area of a Right Prism:** The surface area of a right prism is the sum of twice the area of its base plus the product of the perimeter of the base and the height of the prism.
$SA = 2A + hP$ 134

Thm 3.12   **Surface Area of a Right Regular Pyramid:** The surface area of a right pyramid is the sum of the area of its base plus one-half the product of the perimeter of the base and the slant height.
$SA = A + \frac{1}{2}lP$ 135

Thm 3.13   **Surface Area of a Right Circular Cylinder:** The surface area of a right circular cylinder is twice the area of its base plus the circumference of its base times its height.
$SA = 2A + hC = 2\pi r^2 + 2\pi rh$ 136

Thm 3.14   **Surface Area of a Right Circular Cone:** The surface area of a right circular cone is the sum of the area of its base and one-half the circumference of its base times its slant height.
$SA = A + \frac{1}{2}lC = \pi r^2 + \frac{1}{2}l(2\pi r) = \pi r(r + l)$ 137

Thm 3.15   **Surface Area of a Sphere:** The surface area of a sphere is 4π times the square of its radius.
$SA = 4\pi r^2$ 138

Post 3.4   **Volume Postulate**
(a) For every polyhedron and unit cube, there is a real number that gives the number of unit cubes (and parts of unit cubes) that exactly fill the region enclosed by the polyhedron.
(b) The volume of the region enclosed by a polyhedron is the sum of the volumes of the smaller regions into which the region can be subdivided.   143

Thm 3.16   **Volume of a Right Rectangular Prism:** The volume of a right rectangular prism is the product of its length, width, and height.
$V = lwh$ 145

Cor 3.17   **Volume of a Cube:** The volume of a cube is the cube of the length of its side.
$V = s^3$ 145

Thm 3.18   **Volume of a Pyramid:** The volume of a pyramid is one-third the product of the area of its base and its height.
$$V = \frac{1}{3}Ah$$ 146

Thm 3.19   **Volume of a Right Circular Cylinder:** The volume of a right circular cylinder is the product of the area of its base and its height.
$V = Ah = \pi r^2 h$ 147

Thm 3.20   **Volume of a Right Circular Cone:** The volume of a right circular cone is one-third the product of the area of its base and its height.
$$V = \frac{1}{3}Ah = \frac{1}{3}\pi r^2 h$$ 147

Thm 3.21   **Volume of a Sphere:** The volume of a sphere is $\frac{4}{3}\pi$ times the cube of its radius.
$$V = \frac{4}{3}\pi r^3$$ 148

Thm 4.1   If x and y are the measures of a pair of vertical angles, then $x = y$.   170

Post 4.1   **SAS Congruence Postulate:** If two sides and the included angle of one triangle are congruent respectively to two sides and the included angle of another triangle, then the two triangles are congruent.   177

Thm 4.2   **LL Congruence Theorem:** If two legs of one right triangle are congruent respectively to two legs of another right triangle, then the two triangles are congruent.   178

Thm 4.3   **HL Congruence Theorem:** If the hypotenuse and a leg of one right triangle are congruent respectively to the hypotenuse and a leg of another right triangle, then the two triangles are congruent.   178

Post 4.2   **ASA Congruence Postulate:** If two angles and the included side of one triangle are congruent respectively to two angles and the included side of another triangle, then the two triangles are congruent.   180

Post 4.3   **SSS Congruence Postulate:** If three sides of one triangle are congruent respectively to three sides of another triangle, then the two triangles are congruent.   181

Thm 4.4   **Converse of the Pythagorean Theorem:** If the sum of the squares of the lengths of two legs of a triangle equals the square of the third side, then the triangle is a right triangle.   182

Thm 4.5   In an isosceles triangle, the angles opposite the congruent sides are congruent.   188

Cor 4.6   Every equilateral triangle is equiangular.   189

Thm 4.7   If two angles of a triangle are congruent, then the sides opposite those angles are congruent.   189

Cor 4.8   Every equiangular triangle is equilateral.   190

Thm 4.9   In an isosceles triangle, the ray that bisects the vertex angle bisects the base and is perpendicular to it.   190

Thm 4.10   **Perpendicular Bisector Theorem:** A point is on the perpendicular bisector of a line segment if and only if it is equidistant from the endpoints of the segment.   191

Thm 4.11   The measure of an exterior angle of a triangle is greater than the measure of either of the nonadjacent interior angles.   193

Thm 5.1   If two lines are cut by a transversal to form a pair of congruent alternate interior angles, then the lines are parallel.   219

Cor 5.2   If two lines are both perpendicular to a transversal, then the lines are parallel.   220

Cor 5.3   If two lines cut by a transversal form a pair of congruent corresponding angles, then the lines are parallel.   220

Cor 5.4   If two lines cut by a transversal form a pair of supplementary interior angles on the same side of the transversal, then the lines are parallel.   221

Post 5.1   **The Parallel Postulate:** Given a line, l, and a point, P, not on l, there is only one line, m, containing P, such that $l \parallel m$.   222

Thm 5.5   If a pair of parallel lines is cut by a transversal, then the alternate interior angles formed are congruent.   222

Cor 5.6   If two lines are parallel and a line is perpendicular to one of the two lines, then it is perpendicular to the other line.   224

Cor 5.7   If two parallel lines are cut by a transversal, then each pair of corresponding angles formed is congruent.   224

Cor 5.8   If two parallel lines are cut by a transversal, then both pairs of interior angles on the same side of the transversal are supplementary.   224

Thm 5.9   **Angle Sum in a Triangle Theorem:** The angle sum in a triangle is 180°.   228

Cor 5.10   **The Exterior Angle Theorem:** An exterior angle of a triangle is equal to the sum of the two nonadjacent interior angles.   229

Cor 5.11   **AAS Congruence Theorem:** If two angles and a side of one triangle are congruent respectively to two angles and a side of another triangle, then the two triangles are congruent.   230

Cor 5.12   **HA Congruence Theorem:** If the hypotenuse and an acute angle of one right triangle are congruent respectively to the hypotenuse and an acute angle of another right triangle, then the two triangles are congruent.   230

Thm 5.13   **Angle Bisector Theorem:** A point is on the bisector of an angle if and only if it is equidistant from the sides of the angle.   231

Thm 5.14   A diagonal of a parallelogram forms two congruent triangles.   235

Cor 5.15   In a parallelogram, the opposite sides are congruent and the opposite angles are congruent.   235

Cor 5.16   Parallel lines are everywhere equidistant.   236

Thm 5.17   If the opposite sides of a quadrilateral are congruent, then the quadrilateral is a parallelogram.   236

Thm 5.18   If the opposite angles of a quadrilateral are congruent, then the quadrilateral is a parallelogram.   236

Thm 5.19   If a quadrilateral has two sides that are parallel and congruent, then it is a parallelogram.   236

Thm 5.20   The diagonals of a parallelogram bisect each other.   238

Thm 5.21   If the diagonals of a quadrilateral bisect each other, then it is a parallelogram.   239

Thm 5.22   Every rhombus is a parallelogram.   240

Thm 5.23   The diagonals of a rhombus are perpendicular to each other.   240

Thm 5.24   If the diagonals of a parallelogram are perpendicular to each other, then the parallelogram is a rhombus.   240

Thm 5.25   A parallelogram is a rhombus if and only if its diagonals bisect the opposite angles.   242

Thm 5.26   Every rectangle is a parallelogram.   246

Thm 5.27   A parallelogram with one right angle is a rectangle.   247

Thm 5.28   The diagonals of a rectangle are congruent.   247

Thm 5.29   If a parallelogram has congruent diagonals, then it is a rectangle.   248

Thm 5.30   The base angles of an isosceles trapezoid are congruent.   249

Thm 5.31   If the base angles of a trapezoid are congruent, then it is isosceles.   250

Thm 5.32   If three parallel lines form congruent segments on one transversal, then they form congruent segments on any transversal.   258

Thm 6.1   **Cross Multiplication Theorem:** $\dfrac{a}{b} = \dfrac{c}{d}$ if and only if $ad = bc$.

273

Thm 6.2 $\dfrac{a}{b} = \dfrac{c}{d}$ if and only if $\dfrac{d}{b} = \dfrac{c}{a}$.

$\dfrac{a}{b} = \dfrac{c}{d}$ if and only if $\dfrac{a}{c} = \dfrac{b}{d}$.

$\dfrac{a}{b} = \dfrac{c}{d}$ if and only if $\dfrac{b}{a} = \dfrac{d}{c}$.

274

Thm 6.3   If $\dfrac{a}{b} = \dfrac{b}{c}$, then b is the geometric mean of a and c.   276

Post 6.1   **AAA Similarity Postulate:** Two triangles are similar if and only if three angles of one triangle are congruent respectively to three angles of the other triangle.   282

Thm 6.4   **AA Similarity Theorem:** Two triangles are similar if two angles of one triangle are congruent respectively to two angles of the other triangle.   283

Cor 6.5   Two right triangles are similar if an acute angle of one triangle is congruent to an acute angle of the other triangle.   283

Thm 6.6   **SAS Similarity Theorem:** Two triangles are similar if two sides of one triangle are proportional to two sides of another triangle and the angles included between the sides are congruent.   284

Cor 6.7   **LL Similarity:** Two right triangles are similar if the legs of one triangle are proportional respectively to the legs of the other triangle.   285

Thm 6.8   **SSS Similarity Theorem:** Two triangles are similar if three sides of one triangle are proportional to three sides of another triangle.   286

Thm 6.9   In a right triangle, the altitude to the hypotenuse is the mean proportional between the two segments formed by the altitude on the hypotenuse.   293

Thm 6.10   **Side-Splitting Theorem:** A line parallel to one side of a triangle forms a triangle similar to the original triangle and divides the other two sides of the triangle into proportional corresponding segments.   294

Thm 6.11   **Midsegment Theorem:** A midsegment of a triangle is parallel to the third side and is half as long.   295

Cor 6.12   **Midquad Theorem:** If $ABCD$ is *any* quadrilateral and E, F, G, H are midpoints the sides $\overline{AB}$, $\overline{BC}$, $\overline{CD}$, and $\overline{AD}$ respectively, then $EFGH$ is a parallelogram.   296

Thm 6.13   In a right triangle with acute angle A, $\dfrac{\sin A}{\cos A}$ is $\tan A$.   307

Thm 6.14   In a right triangle with acute angle A, $\sin^2 A + \cos^2 A = 1$.   307

Post 7.1   **Length of an Arc:** The ratio of the length of an arc of a circle to the circumference of the circle equals the ratio of the central angle of the arc to $360°$.   321

Post 7.2   **Area of a Sector:** The ratio of the area of a sector of a circle to the area of the circle is equal to the ratio of the central angle of the sector to $360°$.   322

Thm 7.1   **Inscribed Angle Theorem:** The measure of an inscribed angle in a circle is equal to half the measure of its intercepted arc.   323

Cor 7.2   Inscribed angles that intercept the same arcs (or congruent arcs) are congruent.   324

Cor 7.3   An angle is inscribed in a semicircle if and only if it is a right angle.   325

Thm 7.4   The perpendicular bisector of a chord contains the center of the circle.   331

Cor 7.5   The intersection of the perpendicular bisectors of any two nonparallel chords of a circle is the center of the circle.   331

Cor 7.6   If two circles, O and O', intersect in two points A and B, then the line containing O and O' is the perpendicular bisector of $\overline{AB}$.   332

Thm 7.7   If two chords intersect, then the measure of any one of the vertical angles formed is equal to half the sum of the two arcs intercepted by the angle and its vertical angle.   333

Thm 7.8   If two chords of a circle intersect, the product of the lengths of the two segments formed on one chord is equal to the product of the lengths of the two segments formed on the other chord.   334

Thm 7.9   If two secants intersect outside a circle, the measure of the angle of intersection is half the difference of the measures of the intercepted arcs.   338

Thm 7.10   If two secants intersect outside a circle, then the product of the lengths of the segments formed on one secant is equal to the product of the lengths of the segments on the other secant.   339

Thm 7.11   A radius or diameter of a circle is perpendicular to a tangent line at its point of tangency.   340

Thm 7.12   The measure of an angle formed when a chord intersects a tangent line at the point of tangency is half the measure of the arc intercepted by the chord and the tangent line.   341

Thm 7.13   If a secant and a tangent line intersect outside a circle, the measure of the angle formed is half the difference of the measures of the intercepted arcs.   342

Thm 7.14   If two tangent lines intersect outside a circle, the measure of the angle formed is half the difference of the measures of the intercepted arcs.   342

Cor 7.15   If two tangents are drawn to a circle from the same point in the exterior of the circle, the distances from the common point to the points of tangency are equal.   344

Thm 7.16   If tangent and secant lines are drawn to a circle from the same point in the exterior of the circle, the length of the tangent segment is the mean proportional between the length of the external secant segment and the length of the secant.   345

Thm 7.17 **Circumcenter of a Triangle:** The perpendicular bisectors of a triangle intersect in a single point, the circumcenter.   351

Thm 7.18 **Orthocenter of a Triangle:** The altitudes of a triangle intersect in a single point, the orthocenter.   352

Thm 7.19 **Incenter of a Triangle:** The angle bisectors of a triangle intersect in a single point, the incenter.   354

Thm 7.20 **Centroid of a Triangle:** The medians of a triangle intersect in a single point, the centroid, which is two thirds of the way from any vertex to the other endpoint of the median from that vertex.   355

Thm 8.1 **Distance Formula:** If $P(a, b)$ and $Q(c, d)$ are two points in the coordinate plane, then $PQ = \sqrt{(c - a)^2 + (d - b)^2}$.   374

Thm 8.2 **Test for Collinearity:** The points P, Q, and R are collinear, with Q between P and R, if and only if $PQ + QR = PR$.   375

Thm 8.3 **Midpoint Formula:** If $P(a, b)$ and $Q(c, d)$ are endpoints of a line segment, then the midpoint of $\overline{PQ}$ is $M\left(\dfrac{a + c}{2}, \dfrac{b + d}{2}\right)$.   376

Thm 8.4 **Slopes of Parallel Lines:** Two lines in a coordinate plane are parallel if and only if
(a) their slopes are equal, or
(b) their slopes are undefined.   384

Thm 8.5 **Slopes of Perpendicular Lines:** Two lines in a coordinate plane are perpendicular if, and only if, (a) one line is horizontal and the other is vertical, or (b) the product of their slopes equals -1.   386

Thm 8.6 **Forms of Equations of Lines:** Every nonvertical line in the plane can be expressed in either of the following two forms where m represents the slope of the line, b is the y-intercept of the line, and (r, s) is a point on the line.
$$\text{Point-Slope Form: } y - s = m(x - r)$$
$$\text{Slope-Intercept Form: } y = mx + b \quad 392$$

Thm 8.7 **Solutions of a System of Two Linear Equations:** Two linear equations have 0, 1, or infinitely many simultaneous solutions if, and only if, the lines they represent are parallel, intersect in one point, or coincide, respectively.   395

Thm 8.8 **Equation of a Circle:** The circle with center (a, b) and radius r has the equation $(x - a)^2 + (y - b)^2 = r^2$.   398

Thm 9.1 **Properties of Translations**
1. Translations take lines to lines, rays to rays, and line segments to line segments.
2. Translations preserve distance.
3. Translations preserve angle measure.
4. Translations preserve perpendicularity.
5. Translations preserve parallelism.   421

Thm 9.2   **Properties of Rotations**
1. Rotations take lines to lines, rays to rays, and line segments to line segments.
2. Rotations preserve distance.
3. Rotations preserve angle measure.
4. Rotations preserve perpendicularity.
5. Rotations preserve parallelism.   425

Thm 9.3   **Properties of Reflections**
1. Reflections take lines to lines, rays to rays, and line segments to line segments.
2. Reflections preserve distance.
3. Reflections preserve angle measure.
4. Reflections preserve perpendicularity.
5. Reflections preserve parallelism.   427

Thm 9.4   **Properties of Glide Reflections**
1. Glide reflections take lines to lines, rays to rays, and line segments to line segments.
2. Glide reflections preserve distance.
3. Glide reflections preserve angle measure.
4. Glide reflections preserve perpendicularity.
5. Glide reflections preserve parallelism.   428

Thm 9.5   **Congruent Triangles and Isometries:** Two triangles are congruent if and only if there is an isometry that takes one triangle to the other.   430

Thm 9.6   **Congruent Polygons and Isometries:** Two polygons are congruent if and only if there is an isometry that takes one polygon to the other.   430

Thm 9.7   **Properties of Size Transformations**
1. Size transformations take lines to lines, rays to rays, and line segments to line segments.
2. Size transformations preserve orientation.
3. Size transformations preserve ratios of distances.
4. Size transformations preserve angle measure.
5. Size transformations preserve perpendicularity.
6. Size transformations take lines to parallel lines.   441

Thm 9.8   **Similar Triangles and Similitudes:** $\triangle ABC \sim \triangle A'B'C'$ if and only if there is a similitude that takes $\triangle ABC$ to $\triangle A'B'C'$.   443

Thm 9.9   **Similar Polygons and Similitudes:** Two polygons are similar if and only if there is a similitude that takes one to the other.   444

Answers to Odd-Numbered Problems, Chapter Tests, and Topics Sections

Section 1.1

1. 13

3. 12

5. One solution is shown. Many solutions are possible.

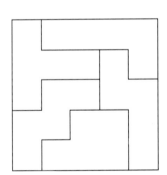

7.

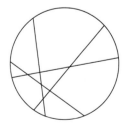

9. 40

11. (a) (b)

13. It cannot be done unless he uses fewer than all of the blocks. Because the sum of all ten heights is 55 cm, two equal towers would each be 27.5 cm tall, which is impossible.

15. (a)

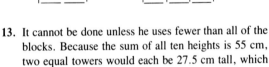

(b)

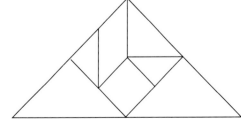

17. (a)

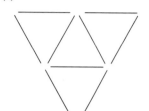

(b)

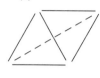

(b) Yes, one solution is shown next.

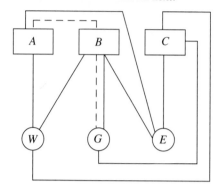

19. (a) No, there will always be at least one house not connected to one of the utilities.

21. 15 inches

Section 1.2

1. (a)

X	X	X	X	X
X	X	X	X	X
X	X	X	X	X
X	X	X	X	X

(b) $1 + 2 + 3 + 4 = \dfrac{1}{2}(4 \times 5)$

(c) 1275, 2850

(d) $\dfrac{n(n + 1)}{2}$

3. $\dfrac{n(n + 1)}{2}$

5. (a) ii   (b) i   (c) i   (d) ii

7. (a) 36, 49, 144, n^2

(b) $-12, -14, -24, -2n$

(c) 243, 729, 177147, 3^{n-1}

(d) 13, 15, 25, $2n + 1$

9. 15

11. (a) 19, 31, 50

(b) 6, 13, 33

(c) 8, 11, 30

(d) 13, 23, 36, 59

13. 268

15. 100 squares, $1 + 3 + 5 + \cdots + (2n - 1) = n^2$ squares

17. 220

19. (a) 27   (b) "D" column

21. $\dfrac{1}{3}$

23. 7

Chapter 1 Test

1. Draw a Picture, Guess and Test, Use a Variable, Look for a Pattern, Make a Table, Solve a Simpler Problem

2. Use a Variable

3.

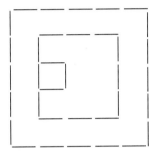

4.

5. 37°, 37°, 106°

6.

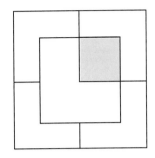

7. (a) 7 pieces   (b) 11 pieces   (c) 16 pieces

8. 109 squares, $1 + 2(2) + 2(3) + \cdots + 2(n) = n^2 + n - 1$

9. (a) 85, $a_n = 9n - 5$ (b) 1025, $a_n = 2^n + 1$

10. (a) 9   (b) 36   (c) 100   (d) $1 + 8 + 27 + \ldots + n^3$

11.

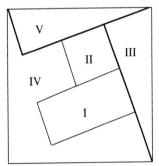

Section 2.1

1. 10 line segments: $\overline{AB}, \overline{AC}, \overline{AD}, \overline{AE}, \overline{BC}, \overline{BD}, \overline{BE},$ $\overline{CD}, \overline{CE}, \overline{DE}$

3. 8 rays: $\overrightarrow{AB}, \overrightarrow{BC}, \overrightarrow{CD}, \overrightarrow{DE}, \overrightarrow{ED}, \overrightarrow{DC}, \overrightarrow{CB}, \overrightarrow{BA}$

5. 9 ways: $\overleftrightarrow{AB}, \overleftrightarrow{AC}, \overleftrightarrow{AD}, \overleftrightarrow{BC}, \overleftrightarrow{BD}, \overleftrightarrow{BE}, \overleftrightarrow{CD}, \overleftrightarrow{CE}, \overleftrightarrow{DE}$

7. (a) 4.34   (b) 1.91   (c) 6.65   (d) 4.22

9. (a) ∠BAC or ∠CAB, ∠BAD or ∠DAB, ∠CAD or ∠DAC
   (b) ∠BAC and ∠CAD

11. (a) 10   (b) 4   (c) 6

13. (a) 60° acute   (b) 114° obtuse   (c) 90° right

15. ∠AFB = 60°, ∠CFD = 35°

17. ∠1 = 54°, ∠2 = 126°

19. ∠A = 72°

21. 3 lines: 7 regions, 4 lines: 11 regions, 5 lines: 16 regions, 10 lines: 56 regions, n lines: $\dfrac{n(n+1)}{2} + 1$ regions

23. (a) 41.4°   (b) 84.25°   (c) 25.504°   (d) 123.58°

25. (a) 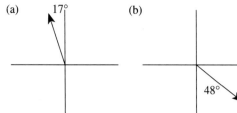   (b)

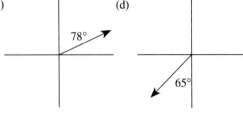

27. (a) 40°   (b) 320°   (c) 277°43′   (d) 155°25′

29. 109°6′

Section 2.2

1. (a) △DEF, △GHI, △KJL
  (b) △GHI
  (c) △ABC, △MNO
  (d) △ABC, △JKL
  (e) △DEF

3. (a) Yes   (b) No–not all line segments
  (c) No–part is retraced   (d) No–part is crossed

5. (a) OFHQ
  (b) ADPK or AEGJ
  (c) MNOS, MOQK, KQHI
  (d) △ACK, △IJK, △FGH, △OCD, △CEF, △OPQ,
    △ABL, △MCO
  (e) △MNO, △MSO, △LBC
  (f) MNOS, KQHI
  (g) COSM
  (h) △ROQ, △LKC
  (i) △ROP, △ACL
  (j) KCOQ, KQHJ, CDOM, FODE, KMOP
  (k) LBCK

7. (a) Four triangles: 2-9-9, 4-8-8, 6-7-7, 8-6-6
  (b) Four triangles: 3-8-9, 4-7-9, 5-6-9, 5-7-8
  (c) None

9. (a)

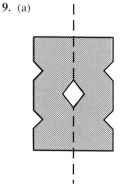

(b)

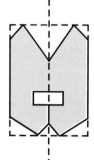

(c)

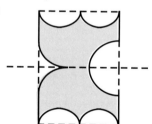

11. (a)

(b)

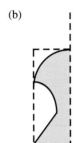

(c)

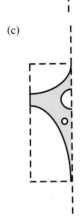

13. (a) 20   (b) 6   (c) 10   (d) 2   (e) 8

15. Assuming congruent ones are left out:
(a) 11   (b) 1   (c) 2   (d) 1

17. (a) 56   (b) 24   (c) 24

19. (a)

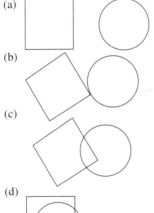

(b)

(c)

(d)

21. (a)

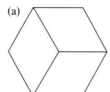

(b)

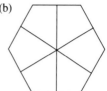

23. (a) Yes, 2   (b) Yes, 1

25. (a) Yes, 1   (b) No

27. (a) Draw the lines of symmetry in the regular n-gons in (i)—(iii). How many does each have?
(i) Five

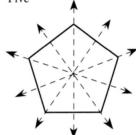

(ii) Six

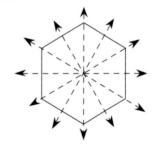

(iii) Eight

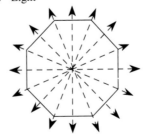

(b) n

29. (a) Reflection symmetry only
   (b) Both
   (c) Reflection symmetry only
   (d) Reflection symmetry only
   (e) Both
   (f) Reflection symmetry only

Section 2.3

1. (a) $\angle BAC = 54°$, $\angle ABC = 90°$, $\angle ACB = 36°$–the angle sum should be $180°$
   (b) $\angle HDE = 100°$, $\angle DEF = 126°$, $\angle EFG = 89°$, $\angle FGH = 115°$, $\angle GHD = 110°$–the angle sum should be $540°$

3. (a) One of many correct examples follows.

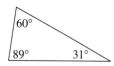

   (b) One of many correct examples follows.

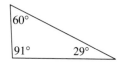

   (c) Not possible. If $a = b = c$, then we have that $a^2 + b^2 = c^2 + c^2 = 2c^2$ rather than c^2, so the Pythagorean Theorem fails.
   (d) Not possible. If a triangle is equiangular then it is equilateral so all sides would have the same length. Thus, it cannot be scalene.

5. (a) $30°$ (b) $20°$ (c) $62°$ (d) $34°$

7. (a) $98°$
   (b) each $125°$
   (c) $120°$
   (d) $95°$, $100°$, $80°$, $85°$

9. (a)

   (b)

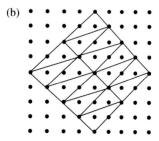

11. (a) $150°$ (b) $157.5°$ (c) $144°$ (d) $162°$
    (e) $160°$ (f) $170°$

13. (a)   (b)

15. 11 sides

17. (a) $70°$ (b) $130°$ (c) $120°$ (d) $20°$ (e) $20°$
    (f) $80°$ (g) $60°$ (h) $100°$

19. Sum $= 180°$

21.

Number of Sides and Angles in a Polygon	Number of Triangles into Which the Polygon Can Be Divided	Sum of Angle Measures in the Polygon	Measure of One Angle in the Polygon
3	1	$180°$	$180°/3 = 60°$
4	2	$2(180°) = 360°$	$360°/4 = 90°$
5	3	$3(180°) = 540°$	$540°/5 = 108°$
6	4	$4(180°) = 720°$	$720°/6 = 120°$
8	6	$6(180°) = 1080°$	$1080°/8 = 135°$
10	8	$8(180°) = 1440°$	$1440°/10 = 144°$
n	$n - 2$	$(n - 2)(180°)$	$\dfrac{(n - 2)180°}{n}$

23. (a) Measures of $\angle A$, $\angle B$, and $\angle C$ add up to $180°$ since they form a straight angle.
    (b) Rectangle, $\frac{1}{2}AC$
    (c) Yes

25. $\angle ACB = 69°24'10''$

27. (a) $540°$, no error   (b) $718°5'$, error $= 1°55'$

Section 2.4

1. Either upper left-hand circle or lower right-hand circle

3. (a)

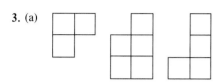

    Top   Front   Right

(b)

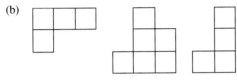

    Top   Front   Right

(c)

    Top   Front   Right

(d)

    Top   Front   Right

5.

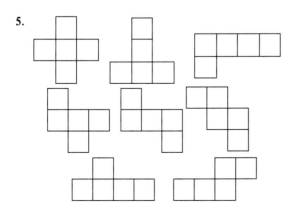

7. a, b

9.

11. (iii)

13. (a) *ABCDE, FGHIJ*
    (b) *ABGF, BCHG, CDIH, EDIJ, AEJF*
    (c) *ABGF, BCHG, CDIH, FGHIJ*
    (d) Right pentagonal prism

15. (a) Square pyramid
    (b) Triangular pyramid
    (c) Octagonal pyramid

17. (a) Oblique circular cone
    (b) Oblique circular cylinder
    (c) Sphere

19.

Base	F	V	F + V	E
Triangle	5	6	11	9
Quadrilateral	6	8	14	12
Hexagon	8	12	20	18
n-gon	$n + 2$	$2n$	$3n + 2$	$3n$

Yes, Euler's formula holds for these figures.

21.

Figure	F	V	F + V	E
(i)	7	10	17	15
(ii)	9	9	18	16
(iii)	8	12	20	18

Yes, Euler's formula holds for these figures.

23. 12 vertices–one possible answer is a right hexagonal prism

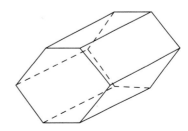

25. (a) Circle   (b) Rectangle

27. (a) 3   (b) 6   (c) 9

29. (a) 6   (b) 4   (c) 6   (d) Infinitely many

31. (a) Equilateral Triangle   (b) Square

33. Refer to the following figure.
Cut 1: Through *ROM*
Cut 2: Through *RON*

Cut 3: Through *MON*
Cut 4: Through *MRN*

Section 2.5

1. (a) $\dfrac{1 \text{ ft}}{12 \text{ in.}}$ (b) $\dfrac{5280 \text{ ft}}{1 \text{ mi}}$ (c) $\dfrac{16 \text{ oz}}{1 \text{ lb}}$ (d) $\dfrac{1 \text{ gal}}{4 \text{ qt}}$

3. (a) $\dfrac{60 \text{ min}}{1 \text{ hr}}$ (b) $\dfrac{1 \text{ min}}{60 \text{ sec}}$

   (c) $\dfrac{1 \text{ yd}}{36 \text{ in.}}$ (d) $\dfrac{1 \text{ man-day}}{24 \text{ man-hours}}$

5. (a) 72 oz   (b) 62.4 degrees   (c) 1,584,000 in.
   (d) 0.0105 kL

7. (a) 66 ft/sec   (b) $8\frac{1}{3}$ dollars/hour
   (c) 2.5 ft/century   (d) 16 m/mL

9. (a) 15.24 cm   (b) 91.44 m   (c) 128.016 m
   (d) 0.621 mi

11. (a) 30.48 m/sec   (b) 0.807 lb/yd

13. 104.61 km/hr

15. 22.49 km/hr

17. 0.000235 mL/m

19. 1.17×10^{-8} mph

21. Light is 1,073,000 times faster than sound.

23. (a) 5.88×10^{12} mi/yr   (b) 4.47×10^{14}
   (c) 0.716 hours

Answers for the Additional Problems

1. 199 days

2. 4950 angles

Chapter 2 Test

1. F–Three points may be collinear.

2. T

3. F–An isosceles triangle may also be equilateral.

4. F–The vertex angle of a *regular* hexagon measures 120°.

5. F–The diameter of a circle is twice its radius.

6. T

7. T

8. T

9. T

10. F–A polyhedron has faces that are polygonal regions.

11. Equiangular–all right angles

12. $a = 76°$, $b = 28°$, $c = 50°$, $d = 24°$, $e = 156°$, $f = 26°$, $g = 50°$

13. 165°

14. (a) *IBMG, IABR, HIRG*
   (b) *HIDF, GRDF, ABGH*
   (c) *IGED, PQJO*
   (d) $\triangle GHI, \triangle GRI, \triangle IAB, \triangle IRB, \triangle BRM, \triangle MRG, \triangle EFD$
   (e) $\triangle ICG, \triangle QJO, \triangle JOP, \triangle OPQ, \triangle QPJ, \triangle BMG, \triangle MGI, \triangle GIB, \triangle IBM, \triangle LMN, \triangle CKL, \triangle BKJ$
   (f) *PQJO*
   (g) *GHIC, RNCJ*
   (h) $\triangle ICD, \triangle NCD, \triangle GNM, \triangle GJC, \triangle GJI, \triangle IJQ, \triangle IJB, \triangle BCK, \triangle IKM, \triangle IJO$
   (i) $\triangle LMG, \triangle IJR, \triangle QRP, \triangle PRO, \triangle JRO, \triangle JRQ, \triangle GRN, \triangle IBK$

(j) *IABM*, *IACD*, *HACE*, *HIDE*, *HIMG*, *GNDE*, *GMDF*, *GNDF*, *INGH*, *KCDM*, *LCDM*, *AINC*

(k) *IBCD*, *GMDE*

15. (a) $\angle 1$ and $\angle 2$, $\angle 3$ and $\angle 4$, $\angle 1$ and $\angle 3$, $\angle 2$ and $\angle 4$, $\angle 6$ and $\angle 9$

(b) $\angle 7$ and $\angle 10$, $\angle 5$ and $\angle 8$, $\angle 5$ and $\angle 7$, $\angle 8$ and $\angle 10$

(c) $\angle 9$ or $\angle 6$

(d) $\angle 1$ and $\angle 2$, $\angle 2$ and $\angle 4$, $\angle 4$ and $\angle 3$, $\angle 3$ and $\angle 1$, $\angle 5$ and $\angle 6$, $\angle 6$ and $\angle 7$, $\angle 7$ and $\angle 10$, $\angle 10$ and $\angle 9$, $\angle 9$ and $\angle 8$, $\angle 8$ and $\angle 5$

(e) $\angle 1$, $\angle 4$, $\angle 7$, $\angle 10$, $\angle 8$, $\angle 5$

(f) $\angle 2$, $\angle 3$

16. (a) rotation 5, reflection 6

(b) rotation 1, reflection 2

(c) rotation and reflection–both infinitely many

17.

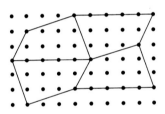

18. (a) 24   (b) 3   (c) 15   (d) 6

19. 11,700,000 milliseconds

20. (a) One possible answer is shown.

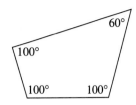

(b) One possible answer is shown.

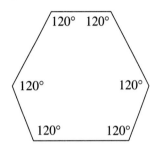

21. 6.04 g/L

22. 540°10′–the traverse does not close; error = 10′

23. $\dfrac{45\ \cancel{\text{sec}}}{1} \cdot \dfrac{1\ \cancel{\text{min}}}{60\ \cancel{\text{sec}}} \cdot \dfrac{33\frac{1}{3}\ \text{rev}}{1\ \cancel{\text{min}}} = 25$ revolutions

24. Right trapezoidal prism

Section 3.1

1. (a) 7 square units   (b) 11.5 square units

(c) 12 square units

3.

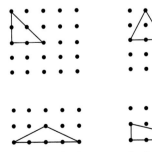

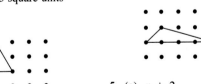

5. (a) $n + 2$ (b) $2n + 2$

(c) $3n + 2$ (d) $4n + 2$

(e) $6n + 2$ (f) $(m - 2)n + 2$

7. (a) $P = 28$ in., $A = 40$ in^2

(b) $P = 22.8$ cm, $A = 32.49$ cm^2

9. (a) $P = 24$ m, $A = 24$ m^2

(b) $P = 15.6$ cm, $A = 9.34$ cm^2

(c) $P = 17.07$ in., $A = 12.5$ in^2

11. (a) $C = 25.13$ in. (b) $C = 33.93$ m

13. (a) $P = 22$ in. (b) $P = 53.2$ m

15. (a) $P = 20$ cm (b) $P = 60$ in. (c) $P = ns$

17. (a) 0.35 ft² (b) $20,000,000$ cm² (c) 2.93 m²

19. (a) $A = 38.5$ ft² (b) $A = 3.58$ m²

21. Width $= 14$ in., length $= 22$ in.

23. The perimeter is 2 times the original perimeter, 3 times the original perimeter, 10 times the original perimeter, and n times the original perimeter, respectively.

25. No. Counterexample: If length is 15 m and width is 5 m, then the perimeter is 40 m and the area is 75 m². But if length is 10 m and width is 10 m, the perimeter is also 40 m, but the area is 100 m².

27.

	side	height	area
$a =$	52 mm	45 mm	1170 mm²
$b =$	48 mm	50 mm	1200 mm²
$c =$	59 mm	40 mm	1180 mm²

29. (a) One tetromino with the maximum perimeter of 10 is shown. Others are possible.

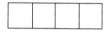

(b) The following tetromino has the minimum perimeter of 8.

31. The gap is 4.77 ft. You could crawl under it, but not walk under it unless you were less than 4.77 ft tall.

33. width $= 4$, length $= 4$; width $= 3$, length $= 6$

35. (a) $A = \frac{1}{2}b_1h$

(b) $A = \frac{1}{2}b_2h$.

(c) $A = \frac{1}{2}b_1h + \frac{1}{2}b_2h$

(d) $A = \frac{1}{2}(b_1 + b_2)h$, $b_1 + b_2 = b$, $A = \frac{1}{2}bh$

(e) $A = \frac{1}{2}bh$; yes

37. No. Even if the basketballs were both only 29.5″ in circumference, their diameters would be 9.39″ each. So, side by side, the width would be 18.78″. Thus, in order for the balls to fit through the ring side by side, the ring must have a diameter greater than 18.78″.

39. (a) 77 ft (b) 42 yd²

41. 8 rolls, $127.92

43. 0.56 ft by 0.56 ft, 1.68 ft by 1.68 ft

45. 458.3 revolutions per minute

47. (a) 1323.19 ft, 1/4 mi
(b) 18.85 ft
(c) 18.85 ft
(d) 18.85 ft in the second lane and 37.7 ft in the third lane

Section 3.2

1. (a) $A = 28$ cm² (b) $A = 20$ in²
(c) $A = 56.55$ cm² (d) $A = 215.35$ mm²

3.

	Radius	Diameter	Circumference	Area
(a)	12	24	75.4	452.4
(b)	22.5	45	141.37	1590.43
(c)	$\frac{\sqrt{6}}{2}$	$\sqrt{6}$	7.7	4.71
(d)	5.43	10.86	34.12	92.8

5.

	Side	Height	Area
$\overline{AD}$	35 mm	22 mm	770 mm²
$\overline{DC}$	27 mm	29 mm	783 mm²

7. (a) $A = 412.44$ in² (b) $A = 93.38$ cm²

9. 6 in.

11. The area of the circle is 4 times the original area, 25 times the original area, and n^2 times the original area respectively.

13. (a) $A = 21.46$ in² (b) $A = 0.86$ in²

15. Approximately 1310 mm²

17. $A = 32\pi - 64$ in² ≈ 36.53 in²

19. Circle

21. (a) $A = \frac{1}{2}bh$ (b) Congruent, $A = \frac{1}{2}bh$

    (c) $A = \frac{1}{2}bh + \frac{1}{2}bh$ (d) $A = bh$; yes

23. Large

25. (a) 125.66 acres   (b) 34.34 acres

27. 70.3 oz = 4 lb 6.3 oz

29. (a) 19.6%   (b) 16.0 ft

31. Probability = 0.497

Section 3.3

1. (a) 30 in.   (b) 8.5 cm   (c) $\sqrt{63} \approx 7.94$ m

3. (a) $x = 11$ cm, $y = 10.2$ cm
    (b) $x = 23.3$ in., $y = 33$ in., $z = 22.6$

5. (a) $A = 180$ m²   (b) $A \approx 93.53$ in²

7. (a) Yes   (b) No   (c) Yes

9. (a) Leg $= (7\sqrt{3})/3 \approx 4.04$ cm
       Hyp. $= (14\sqrt{3})/3 \approx 8.08$ cm
    (b) Legs $= 6\sqrt{2} \approx 8.49$ in. each

11. Height $= 8.7\sqrt{3} \approx 15.1$ cm; Area ≈ 425.1 cm²

13. $A = 120$ cm²

15. $A = 44.18\sqrt{2} \approx 62.48$ m²

17. Length $= 8$ in., width $= 3$ in.

19. $PQ = 8.5$ in., $PR = 16.4$ in.

21. (a) $a^2 + b^2 = c^2$
    (b) $a^2 + b^2 > c^2$
    (c) $a^2 + b^2 < c^2$

23. Area of the large square is c^2.
Area of the four triangles is $4(\frac{1}{2}ab) = 2ab$.
Area of the small square is $(b - a)^2 = b^2 - 2ab + b^2$.
Area of the large square equals the area of the small square plus the area of the four triangles. So, $c^2 = b^2 - 2ab + b^2 + 2ab$, or $c^2 = a^2 + b^2$.

25. 12.56 in.

27. 15.5 ft

29. 550 ft

31. 82 ft

33. 189 ft

35. 2.23 cm

37. 120 ft

39. 45 in. of green fabric
22.5 in. of white fabric
22.5 in. of yellow fabric
90 in. of blue fabric

41. Yes. If the tennis racket is placed from one corner to the farthest (diagonally opposite) corner, the length of the diagonal is 33.54 in.

43. 0.60 in.

Section 3.4

1. (a) $SA = 800$ in²   (b) $SA = 216$ ft²
    (c) $SA = 240$ cm²

3. (a) $SA = 1141$ cm²   (b) $SA = 275$ cm²
    (c) $SA = 765$ cm²   (d) $SA = 865$ cm²

5. (a) $SA = 43$ cm²   (b) $SA = 166$ cm²
    (c) $SA = 1247$ cm²

7. (a) $SA = 1206$ units²   (b) $SA = 4825$ units²
    (c) $SA = 1612$ units²

9. (a) $SA = 168$ cm²   (b) $SA = 366$ in²

11. $SA = 17.44$ cm²

13. $d \approx 8.2$ cm

15. The surface area of the sphere with radius 12 cm is 9 times the surface area of the sphere with radius 4 cm.

17. 12 in. × 8 in. × 6 in.

19. (a) $6 \times 6 \times 1$
    (b) $2 \times 9 \times 2$
    (c) $3 \times 3 \times 4$, $SA = 66$ square units
    (d) $1 \times 1 \times 36$, $SA = 146$ square units

21. 43.4 ft^2

23. 4.7 liters

25. $SA = 1.77$ in^2

27. \$32.76

29. 11.85 m (to two decimal places)

31. One third, or about 2094 cm^2

Section 3.5

1. (a) $V = 1500$ in^3 (b) $V = 144$ in^3
   (c) $V = 200$ cm^3

3. (a) $V = 2958$ cm^3 (b) $V = 342$ cm^3
   (c) $V = 1562$ cm^3 (d) $V = 1951$ cm^3

5. (a) $V = 14$, $SA = 46$ (b) $V = 7$, $SA = 30$
   (c) $V = 21$, $SA = 54$

7. (a) $V = 905$ in^3 (b) $V = 51$ in^3
   (c) $V = 7238$ m^3 (d) $V = 157$ km^3

9. $V \approx 478$ cm^3, $SA \approx 404$ cm^3

11. (a) 0.23 ft^3 (b) 1,200,000 cm^3
    (c) 11,326,739 mm^3

13. The new volume is 8 times the original volume.

15. (a) The area of the larger square is 9 times the area of the smaller square.
    (b) The volume of the larger cube is 27 times the volume of the smaller cube.
    (c) The new volume is 8 times the original volume.

17. $2 \times 9 \times 18$

19. The cylinder with the short side of the rectangle as its height has the greater volume. The cylinder with the short side of the rectangle as its height has volume 81.78 in^3, and the cylinder with the long side of the rectangle as its height has volume 62.98 in^3.

21. (a) $V \approx 31,700,000$ ft^3
    (b) $SA \approx 458,000$ ft^2

23. \$50

25. 318 cm^3

27. (a) 6,272,640 in^3, 3630 ft^3
    (b) 112.53 tons
    (c) 27,116 gallons

29. (a) 90 in^3 (b) No

31. Approximately 36 kg

33. 78 cm^3

35. 646 gallons

37. (a) 120,000 cm^3 (b) 31,000 cm^2

39. 4969 in^3

41. 325,828.8 gallons

43. 0.22 in^3, 97.25%

45. 369 ft^3

47. 42.65 hr

49. (a) $h = 40$ ft
    (b) Approximately 848,000 gal
    (c) The sphere has the smaller surface area.

Answers for the Additional Problems

1. $11^2 + 10^2 + \cdots + 1^2 = 506$

2. (a) Cut through two opposite edges.

   (b) Cut with a plane through one vertex and at equal distances from an adjacent vertex. Other solutions are possible.

   (c) Cut with a plane through two opposite vertices and through the midpoints of adjacent faces as shown.

Chapter 3 Test

1. F–Perimeter is $2l + 2w$.

2. F–Only when the triangle is a right triangle

3. T

4. F–For example, rectangles 5 cm by 10 cm and 7 cm by 8 cm.

5. F–The shape could be a pyramid.

6. F–The surface area of the cylinder is larger.

7. F–The altitude depends on which side is used as a base.

8. T

9. T

10. F–The area will be four times as large as the original area.

11. (a) $\dfrac{\pi d^2}{4}$ (b) $\dfrac{c^2}{4\pi}$

12. 8 cm

13. (a) 162 in^3
   (b) Volume is multiplied by 4.
   (c) Volume is multiplied by 9.
   (d) 216 in^2

14. $A = \dfrac{3\sqrt{3}}{2}x^2$

15.

	D	R	SA	C	V
(a)	56	28	3136π	56π	$\dfrac{87808\pi}{3}$
(b)	10	5	100π	10π	$\dfrac{500\pi}{3}$
(c)	18	9	324π	18π	972π
(d)	6	3	36π	6π	36π

16. 105.86 in^2, 49.42 in.

17. 1364 kg, 13.1 ft, 1902 g/cm^2

18. 13 square units

19. 432 in^2, 545 in^3

20. Approximately 108 ft^2

21. Approximately 575 revolutions

22. 24,086 ft^3

23. 147,000,000 yd^3

24. Approximately 53 ft

Section 4.1

1. The vertices lie in one and only one plane.

3. *ABCDEFGH* has an edge length of 3 and/or a surface area of 54 in^2.

5. Lines *l* and *m* intersect.

7. The lines are perpendicular if and only if they form right angles.

9. $\triangle ABC$ has no two sides that are congruent.

11. No deduction possible

13. $\triangle QRS$ is acute.

15. *ABCD* is a quadrilateral and/or *ABCD* is a rhombus.

17. T. Converse: If the length of a side of a square is 5 cm, then its area is 25 cm^2. T. Biconditional: A square has area 25 cm^2 if and only if the length of a side is 5 cm.

19. T. Converse: If two angles share a common side, they are adjacent. T. Biconditional: Two angles are adjacent if and only if they share a common side.

21. T. Converse: If the diagonal of a square measures $10\sqrt{2}$ in., then the side of the square measures 10 in. T. Biconditional: The diagonal of a square measures $10\sqrt{2}$ if and only if the sides measure 10 in.

23. T. Converse: If $a^2 + b^2 = c^2$, $\triangle ABC$ has a right angle at $\angle C$. T. Biconditional: $\triangle ABC$ has a right angle at $\angle C$ if and only if $a^2 + b^2 = c^2$.

25. T. Converse: If a triangle is a right triangle, then it has sides lengths 3, 4, and 5. F.

27. T. Converse: If a triangle has rotation symmetry, it is equilateral. T. Biconditional: A triangle has rotation symmetry if and only if it is equilateral.

29. T. Converse: If the triangle is equiangular, then it has three equal sides. T. Biconditional: A triangle is equiangular if and only if it has three equal sides.

31. T. Converse: If a pyramid has $n + 1$ vertices, then it has a base with n sides. T. A pyramid has a base with n sides if and only if it has $n + 1$ vertices.

33. s

35. Any one of $r \Rightarrow t, r \Rightarrow p, s \Rightarrow t$

37. Any one of $p \Rightarrow t, p \Rightarrow$ not r, not $q \Rightarrow t$

39. If *ABCD* is a parallelogram, then *ABCD* is a quadrilateral.

41. If a prism has n sides, then it has $2n$ vertices.

43. If a polyhedron is regular, then its faces are regular polygons.

45. (1) Given
(2) Angle Sum in a Triangle Theorem
(3) Definition of a Right Angle
(4) Substitution [(3) into (2)]
(5) Subtraction
(6) Definition

Section 4.2

1. $\overline{PQ} \cong \overline{XY}, \overline{QR} \cong \overline{YZ}, \overline{PR} \cong \overline{XZ}, \angle P \cong \angle X,$
$\angle Q \cong \angle Y, \angle R \cong \angle Z$

3. $\triangle ABC \cong \triangle VWU$

5. $\triangle ABC \cong \triangle DEF$, SAS

7. Not necessarily congruent

9. $\triangle QRS \cong \triangle VTU$, SAS

11. $\triangle ABE \cong \triangle DBC$, ASA

13. $\triangle WXZ \cong \triangle YXZ$, LL

15. $\triangle RYV \cong \triangle TWX$, LL

17. $\triangle CBD \cong \triangle ABE$, SAS
$\triangle CBE \cong \triangle ABD$, SAS

19. $\triangle QSR \cong \triangle QTP$, SSS
$\triangle STP \cong \triangle TSR$, SSS

21. $\overline{DE}$

23. $\angle B$

25. $\overline{FE}$

27. $\angle D, \angle BCD, \angle DEF, \angle EFA, \angle FAB$

29. $\overline{EC}, \triangle ACB \cong \triangle ECD$ by SAS and corresponding parts of congruent triangles are congruent.

31. $\angle FAC$

33. (a)

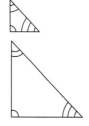

(b)

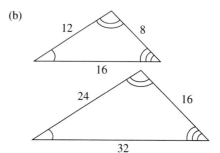

(c)

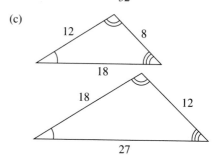

35. Given: $\overline{UW} \cong \overline{YW}$ and $\overline{VW} \cong \overline{XW}$

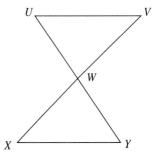

Prove: $\angle U \cong \angle Y$

Statement	Reason
1. $\overline{UW} \cong \overline{YW}$	1. Given
2. $\overline{VW} \cong \overline{XW}$	2. Given
3. $\angle UWV \cong \angle YWX$	3. Vertical angles
4. $\triangle UWV \cong \triangle YWX$	4. SAS Congruence
5. $\angle U \cong \angle Y$	5. C.P.

37. Consider rhombus $ABCD$ as shown.

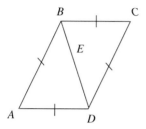

The four sides are congruent and the diagonal is congruent to itself. Thus $\triangle ABD \cong \triangle CDB$ by SSS Congruence.

39. Outline: Use congruent segments formed by intersecting diagonals to show $\triangle BEC \cong \triangle DEA$. Then $\angle DBC \cong \angle BDA$ by C.P.

41. We have $\overline{BE} \cong \overline{DE}$ and $\overline{AE} \cong \overline{CE}$ since E is the midpoint of both diagonals. Also $\angle BEC \cong \angle DEA$ since they are vertical angles. Thus, $\triangle BEC \cong \triangle DEA$ by SAS. Then $\overline{BC} \cong \overline{DA}$ and $\angle DBC \cong \angle BDA$ by C.P. So $\triangle DBC \cong \triangle BDA$ by ASA.

43. Outline: Show that $\triangle ACF \cong \triangle AEB$ by HL and then use C.P.

Section 4.3

1. $\angle Z = 70°$ because $\triangle XPY \cong \triangle ZPY$ by LL.

3. (a) 76° (b) 62° (c) 14° (d) 56°
(e) 34° (f) 42°

5. (a) Theorem 4.9
(b) $\triangle EBA \cong \triangle EBC$ by SAS and C.P.

7. Outline: Use Theorem 4.7 to show $\triangle WVY$ and $\triangle YWX$ are isosceles. Then show that they are congruent. The result follows by C.P.

9. We have $\overline{PX} \cong \overline{RY}$. Because $\overline{PY}$ and $\overline{RX}$ are altitudes, we have $\angle PXR = 90° = \angle RYP$. Also, $\overline{PR} \cong \overline{PR}$. So $\triangle PXR \cong \triangle RYP$.

11. Outline: Since $\overline{BD}$ is an altitude, it is perpendicular to $\overline{AC}$. Show that $\overline{BD}$ bisects $\overline{AC}$ by using $\triangle ABD$, $\triangle CBD$, and C.P.

13. Outline: Use the fact that the median bisects the base to show that it divides the triangle into two congruent triangles. Then two congruent, adjacent, supplementary angles are formed.

15. Outline: If $\triangle ABC$ is equilateral, show that $\triangle ABC \cong \triangle CBA$. Then the angles are congruent by C.P.

17. Outline: If $\triangle ABC$ is equiangular, show that $\triangle ABC \cong \triangle CBA$.

19. Each of the pictures must be centered on the perpendicular bisector of the base (or top) of the wall. That will be the case if and only if the distances from the hanger for a picture to points A and D (or B and C) are the same.

Section 4.4

23. The perpendicular bisector of $\overline{AC}$ also bisects $\angle B$ in $\triangle ABC$.

25. The perpendicular bisectors of the three sides of a triangle intersect in a single point.

27. The three altitudes of a triangle, if extended, intersect in a single point.

29. Since all sides of the triangle constructed are the same length, then triangle is equilateral.

31. Since two sides and their included angle are used to construct the triangle, then any triangles constructed in this way must be congruent by SAS.

33. Outline: The altitude from B is perpendicular to its base $\overline{AC}$. Use construction 5 and its justification.

35. $\triangle ABC$ is a right isosceles triangle. If $\overline{CD}$ is perpendicular to $\overline{AB}$, then $\angle DCB = 45°$. This means CD bisects $\angle C$. By Theorem 4.9, the bisector of the vertex angle in an isosceles triangle bisects the base, so $AD = DB$.

37. (a) The well should be located somewhere along the perpendicular bisector of $\overline{AB}$.
(b) The well should be located somewhere along the perpendicular bisector of $\overline{AC}$.
(c) If the well is located at the point of intersection of the perpendicular bisectors of $\overline{AB}$ and $\overline{AC}$, then it will be equidistant from A, B, and C.

Answers for the Additional Problems

1. $EC = \sqrt{52}$ in.

2. Approximately 0.025 in.

Chapter 4 Test

1. T

2. T

3. F–A biconditional can be written when a conditional and its *converse* are both true.

4. T

5. T

6. F–It is called the hypothesis.

7. F–There is no AAA Congruence. At least one pair of corresponding sides must be congruent.

8. T

9. T

10. F–The congruent angles must be the *included* angles.

11. (a) If a number is an integer, then it is a real number.
(b) If a number is a real number, then it is an integer.
(c) No, since the converse is not true.

12. 20.1 cm

13. She gains weight. Other valid conclusions include: She carries her work shoes; She doesn't carry a lunch; etc.

14. $\triangle VWZ \cong \triangle YWX$ by SAS.

15. $\triangle ABC \cong \triangle ADC$ by HL.

16. $\triangle PQR \cong \triangle PTS$ by SAS. Also, $\triangle PTR \cong \triangle PQS$ by SAS.

17. Construct a pair of perpendicular lines to form a 90° angle. Bisect one of the 90° angles to form two 45° angles. Then bisect one of the 45° angles to form two 22.5° angles.

21. $\angle PQT \cong \angle RQS$ because they are vertical angles. Therefore, $\triangle PQT \cong \triangle RQS$ by SAS. By C.P., $\overline{PT} \cong \overline{RS}$. Thus $\triangle PST \cong \triangle RTS$ by SSS.

22. (a) Because $ABCD$ is isoceles, $\angle BAD \cong \angle CDA$. Because $\overline{QA}$ and $\overline{PD}$ bisect $\angle BAD$ and $\angle CDA$ respectively, $\angle QAD \cong \angle PDA$. Therefore, $\triangle ARD$ is isosceles because $\overline{AR}$ and $\overline{DR}$ are opposite congruent angles.
(b) Since $\angle C \cong \angle B$, $\overline{CD} \cong \overline{AB}$, and $\angle CDP \cong \angle BAQ$, $\triangle PCD \cong \triangle QBA$ by ASA.
(c) From (a), $AR = DR$, and from (b), $AQ = DP$. By subtraction, $RQ = RP$, so $\triangle PQR$ is isosceles.

23. Approximately 1138 ft²

24. He should travel along the path that is the perpendicular bisector of the line segment joining his home and his friend's house.

Section 5.1

1. $\angle 3$ and $\angle 5$, $\angle 9$ and $\angle 11$

3. $\angle 3$ and $\angle 8$, $\angle 11$ and $\angle 6$, $\angle 1$ and $\angle 5$, $\angle 9$ and $\angle 13$

5. 180°

7. $\angle 3 = 65°$, $\angle 8 = 65°$

9. $\angle 1 = 60°$, $\angle 2 = 120°$, $\angle 3 = 60°$, $\angle 4 = 120°$, $\angle 5 = 60°$, $\angle 6 = 45°$, $\angle 7 = 75°$, $\angle 8 = 60°$, $\angle 9 = 45°$, $\angle 10 = 75°$, $\angle 11 = 45°$, $\angle 12 = 135°$, $\angle 13 = 45°$, $\angle 14 = 135°$

11. $\angle 11$

13. $\angle 17$

15. $\angle 13 = 140°$, $\angle 5 = 40°$

17. $\angle 1 = 80°$, $\angle 2 = 100°$, $\angle 3 = 135°$, $\angle 4 = 125°$, $\angle 5 = 55°$, $\angle 6 = 125°$, $\angle 7 = 55°$, $\angle 8 = 45°$, $\angle 9 = 135°$, $\angle 10 = 45°$, $\angle 11 = 80°$, $\angle 12 = 100°$, $\angle 13 = 125°$, $\angle 14 = 55°$, $\angle 15 = 100°$, $\angle 16 = 80°$, $\angle 17 = 100°$, $\angle 18 = 80°$, $\angle 19 = 125°$, $\angle 20 = 55°$

19. F

21. T–by Corollary 5.3

23. $\angle B = 80°$–by Corollary 5.7

25. $\angle ADF = 80°$–by Corollary 5.7

27. $\angle DFE = 80°$–by Theorem 5.5

29. $\angle 1 = 71°$, $\angle 2 = 67°$, $\angle 3 = 71°$, $\angle 4 = 42°$

31. $\angle 1 = 45°$, $\angle 2 = 135°$, $\angle 3 = 55°$, $\angle 4 = 100°$

33. Given: Lines l and m each perpendicular to line t

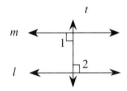

Prove: $l \parallel m$
Proof:

1. $l \perp t$ and $m \perp t$	1. Given
2. $\angle 1 = 90°$ and $\angle 2 = 90°$	2. Perpendicular lines form 90° angles
3. $\angle 1 = \angle 2$	3. Substitution
4. $l \parallel m$	4. Theorem 5.1

35. Let l and m be two lines forming supplementary interior angles with transversal t as shown.

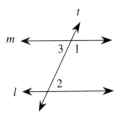

So $\angle 1 + \angle 2 = 180°$. Also $\angle 1 + \angle 3 = 180°$ since they form a straight angle. Then by substitution we have $\angle 1 + \angle 2 = \angle 1 + \angle 3$, so $\angle 2 = \angle 3$. Since these are alternate interior angles, lines l and m are parallel by Theorem 5.1.

37. Let l and m be parallel and $t \perp l$.

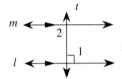

We know $\angle 1 = 90°$. By Theorem 5.5, $\angle 1 = \angle 2$. So $\angle 2 = 90°$. Thus, $m \perp t$.

39. Outline: Use Theorem 5.5 to show that a pair of alternate interior angles are congruent. Then use adjacent supplementary angles to show interior angles are supplementary.

41. Outline: Use the result proved in problem 40 to show a pair of alternate exterior angles are congruent. Then use adjacent supplementary angles to show that a pair of exterior angles on the same side of the transversal are supplementary.

Section 5.2

1. T–Corollary 5.10

3. T–$\angle 3 = \angle 6$ by Theorem 5.5 and the sum of supplementary angles is 180°.

5. F

7. T–Corollary 5.10

9. T–$\angle 2 + \angle 3 = \angle 5$ by Corollary 5.10 and $\angle 10 + \angle 11 = \angle 5$ by Corollary 5.10, so by substitution $\angle 2 + \angle 3 = \angle 10 + \angle 11$.

11. T–Sum of angles in a triangle is 180°.

13. F

15. T–$\angle 15 = \angle 8 + \angle 9 + \angle 11$ by Corollary 5.10.

17. T–$\angle 12 = \angle 8 + \angle 14$ by Corollary 5.10, and $\angle 14 = \angle 8$ because base angles of isosceles triangles are congruent, so by substitution $\angle 12 = 2\angle 8$.

19. T–$\angle 13 + \angle 12 = 180°$ because the sum of supplementary angles is 180°; $\angle 11 = \angle 12$ because base angles of isosceles triangles are congruent, so by substitution $\angle 13 + \angle 11 = 180°$.

21. $\angle 1 = 130°$, $\angle 2 = 35°$, $\angle 3 = 65°$, $\angle 4 = 115°$

23. Area $= 69$ in^2

25. Given: Congruent angles as marked in the following figure.

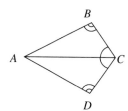

Prove: $\overline{CD} \cong \overline{CB}$

Proof:

1. $\angle ACD \cong \angle ACB$ and $\angle ADC \cong \angle ABC$	1. Given
2. $\overline{AC} \cong \overline{AC}$	2. Identity
3. $\triangle ACB \cong \triangle ACD$	3. AAS
4. $CD \cong CB$	4. C.P.

27. Given $\triangle ABC$ and $\triangle DEF$ as shown.

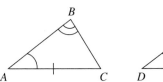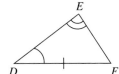

It is given that $\angle A \cong \angle D$, $\angle B \cong \angle E$, and $\overline{AC} \cong \overline{DF}$. Since $\angle A + \angle B + \angle C = 180°$ and $\angle D + \angle E + \angle F = 180°$, we have $\angle A + \angle B + \angle C = \angle D + \angle E + \angle F$. Subtracting equals from both sides we get $\angle C = \angle F$. Thus $\triangle ABC \cong \triangle DEF$ by ASA.

29. Outline: Use a picture similar to the one used to prove subgoal 1 and let $BP = PC$ and $\angle B = \angle C = 90°$. Use HL to show the triangles congruent and thus $\angle A$ is bisected.

Section 5.3

1. F

3. T–Corollary 5.15

5. F

7. F

9. F

11. T–Theorem 5.14

13. T–Theorem 5.22, Theorem 5.20

15. T–$AB = AD$, $BC = DC$, $CA = CA$, so $\triangle ADC \cong \triangle ABC$ by SSS Congruence.

17. T–Theorem 5.23, Theorem 5.9 (Angle Sum in a Triangle Theorem)

19. T–Opposite angles of a parallelogram are congruent.

21. $\angle X = 115°$, $\angle Y = 65°$, $\angle W = 65°$, $\angle Z = 115°$

23. $\angle 1 = 30°$, $\angle 2 = 60°$, $\angle 3 = 60°$

25. $\angle 1 = 35°$, $\angle 2 = 80°$, $\angle 3 = 125°$

27. Side $= 2\sqrt{12} \approx 6.92$

29. $\angle 1 = 140°$, $\angle 2 = 30°$, $\angle 3 = 110°$, $\angle 4 = 40°$

31. Height $= 8$ cm; area $= 160$ cm²

33. Diagonal $= 16$ cm

35. 72°, 72°, 108°, 108°

37. Given: $ABCD$ as shown

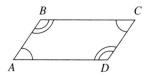

Prove: $ABCD$ is a parallelogram

Proof:

1. $\angle A = \angle C$ and $\angle B = \angle D$	1. Given
2. $\angle A + \angle B + \angle C + \angle D = 360°$	2. Angle sum in a rectangle is 360°.
3. $2(\angle A) + 2(\angle B) = 360°$	3. Substitution
4. $\angle A + \angle B = 180°$	4. Divide both sides by 2.
5. $\overline{BC} \parallel \overline{AD}$	5. Corollary 5.4
6. $\angle A + \angle D = 180°$	6. $\angle B = \angle D$
7. $\overline{AB} \parallel \overline{CD}$	7. Corollary 5.4
8. $ABCD$ is a parallelogram.	8. Definition

39. Given is rhombus $ABCD$ with diagonals $\overline{AC}$ and $\overline{BD}$.

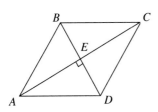

By Theorem 5.23, we know that $\overline{AC} \perp \overline{BD}$, so $\angle AED = 90°$. Since the sum of the angles in $\triangle AED$ is 180°, we know $\angle CAD + \angle BDA + 90° = 180°$, or $\angle CAD + \angle BDA = 90°$. Thus, $\angle CAD$ and $\angle BDA$ are complementary.

41. Outline: Draw in bisectors of two consecutive angles. Use the fact that any two consecutive angles of a parallelogram are supplementary and that the bisectors divide each angle into two congruent angles.

43. Since $\overline{AE}$ bisects $\angle A$, we have $\angle BAE = \angle EAF = \frac{1}{2}(\angle A)$. Similarly $\angle ECF = \angle DCF = \frac{1}{2}(\angle C)$. Since $ABCD$ is a parallelogram, $\angle A = \angle C$, so $\frac{1}{2}(\angle A) = \frac{1}{2}(\angle C)$. Thus, $\angle EAF = \angle ECF$. Also, $\angle BEA = \angle EAF$ and $\angle ECF = \angle DFC$ since each is a pair of alternate interior angles formed by parallel lines. Thus, by substitution, $\angle BEA = \angle DFC$. Since these angles are equal, so are $\angle CEA$ and $\angle AFC$. Therefore, by Theorem 5.18, $ABCD$ is a parallelogram since two pairs of opposite angles are congruent.

45. By Theorem 5.21, if the diagonals bisect each other, the quadrilateral is a parallelogram. Because the legs act as diagonals of a quadrilateral, the board will always be parallel to the ground.

47. 11 times

Section 5.4

1. F

3. T–Theorem 5.26, Theorem 5.20

5. T–Theorem 5.28

7. F

9. T–$AB = BA$, $AC = BD$ by Theorem 5.28 and $BC = AD$ so $\triangle ABD \cong \triangle BAC$ by SSS Congruence.

11. T–definition of rhombus, Theorem 5.23

13. T–Theorem 5.26, Theorem 5.20

15. T–Theorem 5.5

17. T–$AB = BC$, $BE = CE$, $EA = EB$ by Theorem 5.28, Theorem 5.20 so $\triangle ABE \cong \triangle BCE$ by SSS.

19. T–Theorem 5.25, definition of rectangle

21. T–Theorem 5.30

23. T–$AB = CD$, $AD = AD$, $\angle BAD \cong \angle CDA$ by Theorem 5.30. So $\triangle BAD \cong \triangle CDA$ by SAS.

25. T–Problem 23; corresponding parts of congruent triangles are congruent.

27. T–$\angle ABD = \angle DCA$ by Problem 23; corresponding parts of congruent triangles are congruent; $\angle ABD = \angle DCA$ and $\angle ABC = \angle DCB$ by Theorem 5.30. So $\angle DBC = \angle ACB$, which implies that $\triangle BEC$ is isosceles. So $BE = EC$.

29. Area $= 108$ cm²

31. Side $= 8\sqrt{2} \approx 11.31$ mm

33. Legs $= \sqrt{20} \approx 4.47$ in.

35. $EH = 45$ in.

37. $WZ = \sqrt{468}$ in. ≈ 21.63 in.

39. $CF = 6$ cm, $CB = 6$ cm

41. Let $ABCD$ be a rhombus with one right angle. Since a rhombus has four congruent sides, $ABCD$ must be a square.

43. Let *ABCD* be a rhombus with congruent diagonals. By Theorems 5.22 and 5.29, *ABCD* is a rectangle. A rhombus that is a rectangle is a square.

45. Given: *ABCD* is a parallelogram with $\angle A = 90°$.

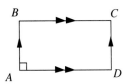

Prove: *ABCD* is a rectangle.
Proof:

1. *ABCD* is a parallelo-   1. Given
   gram with $\angle A = 90°$.
2. $\angle C = 90°$ 2. Opposite angles of a
                                parallelogram are
                                equal.
3. $\angle A + \angle B = 180°$ 3. Interior angles on
   and $\angle A + \angle D =$ the same side of the
   $180°$ transversal
4. $90° + \angle B = 180°$ 4. Substitution
   and
   $90° + \angle D = 180°$
5. $\angle B = 90°$ and   5. Subtracting equals
   $\angle D = 90°$ from both sides
6. *ABCD* is a rectangle.   6. Definition

47. Let *ABCD* be an trapezoid with congruent diagonals as shown.

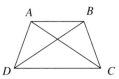

Extend $\overline{AB}$ and $\overline{DC}$ as shown to form trapezoids *EADH* and *BFGC* congruent to *ABCD*.

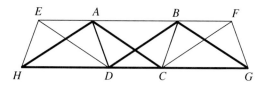

Then $\triangle HAC \cong \triangle GBD$ by SSS. Therefore, $\angle ACH \cong \angle BDG$ by C.P. So $\triangle ACD \cong \triangle BCD$ by SAS. Thus, $\overline{AD} \cong \overline{BC}$ by C.P.

49. Outline: Draw in the diagonals of kite *ABCD*. Use congruent triangles to show angles are congruent. Use the fact that adjacent angles are congruent or that two isosceles triangles are formed by a diagonal.

Section 5.5

21. $A = 42.92$ cm²

23. $M_1M_3 = M_2M_4$ and $\overline{M_1M_3} \perp \overline{M_2M_4}$

25. Let *ABCD* represent quadrilateral constructed. By construction, one angle should be congruent to $\angle A$ and the opposite sides should be parallel.

27. The same construction used in Exercise 6 could be used with $\angle A = 90°$. The resulting parallelogram will have four right angles, so it will be a rectangle.

29. One construction could be to first construct the 60° angle by constructing an equilateral triangle. Then bisect the angle making the length of the bisector *d*.

Next construct a perpendicular bisector *l* of *d*. The points where *l* intersects the sides of the 60° angle will be two vertices of the rhombus. The fourth vertex will be at the end of *d*.

31. Copy one of the diagonals, say *b*. Then find the midpoint of *b*. Bisect the other diagonal of length *c*. Using SSS, form a triangle using lengths half of *b*, half of *c*, and *a*. Extending the diagonal sides of this triangle to lengths *b* and *c* will give the other two vertices of the parallelogram. Theorem 5.21 can be used to prove that the resulting figure is a parallelogram.

33. $CU = UQ$

Answers for the Additional Problems

1. Suppose c^2 is a square and a and b are both odd. Then $a = 2m + 1$ and $b = 2n + 1$, for some whole numbers m and n. So

$$a^2 + b^2 = (2m + 1)^2 + (2n + 1)^2$$
$$= 4m^2 + 4m + 1 + 4n^2 + 4n + 1$$
$$= 4(m^2 + m + n^2 + n) + 2$$

which is not a perfect square because it has a factor of 2 but not a factor of 4. Therefore, it must be true that a and b cannot *both* be odd.

2. Suppose that $\overline{BD}$ and $\overline{BE}$ are altitudes to side $\overline{AC}$ as shown.

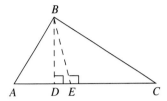

Then $\triangle BDE$ is a triangle with two right angles, which is impossible. Thus, $\overline{BD}$ is the only altitude to $\overline{AC}$.

Chapter 5 Test

1. T

2. F–Corresponding angles must be congruent.

3. F–Angle sum is 360°.

4. F–Bases of the trapezoid are not congruent.

5. F–Any parallelogram has opposite angles congruent.

6. T

7. T

8. F–The diagonals of an isosceles trapezoid are congruent.

9. T

10. F–Any point on the angle bisector is equidistant from the sides of the angle, but not from just any two points on the sides.

11. 20°, 160°, 160°

12. $\angle CBD = 30°$, $BD = 2$, $AC = 2$

13. *ABCD* is a rhombus.

14. 24 yd

15. $\angle 6$ and $\angle 11$, $\angle 7$ and $\angle 10$, $\angle 8$ and $\angle 13$, $\angle 9$ and $\angle 12$

16. $\angle 6$ and $\angle 14$, $\angle 7$ and $\angle 15$, $\angle 8$ and $\angle 16$, $\angle 4$ and $\angle 12$, $\angle 5$ and $\angle 13$, $\angle 9$ and $\angle 17$

17. $\angle 1 = 30°$, $\angle 5 = 40°$, $\angle 10 = 75°$

21. The diagonals of a square are congruent and bisect each other. Thus, $\triangle ADE \cong \triangle ABE$ by SSS.

22. Because E and F are midpoints of $\overline{AB}$ and $\overline{CD}$, respectively, and *ABCD* is a parallelogram, $AE = BE = CF = DF$. Thus, $\triangle EBC \cong \triangle FDA$ by SAS. (Also $\angle B \cong \angle D$, and $\overline{BC} \cong \overline{AD}$.) By C.P., $\overline{CE} \cong \overline{AF}$. Because $\overline{CE} \cong \overline{AF}$ and $\overline{AE} \cong \overline{CF}$, *AECF* is a parallelogram.

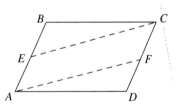

23. Approximately 34 ft³

24. The two crossbars are the same length, as are the distances between them on the vertical supports. Thus, these four segments form a parallelogram. When one support bar descends, the other rises, but they remain parallel and vertical because of the parallel crossbars. The pans are perpendicular to the supports, so they remain horizontal.

Section 6.1

1. $\dfrac{1}{4}$

3. $\dfrac{5}{1}$

5. $n = 30$

7. $n = 20$

9. $s = 30$

11. $d = \dfrac{5}{2}$

13. $x = 320$

15. $x = 15$

17. 6

19. $\sqrt{10\pi} \approx 5.6$

21. No

23. For a cylinder with a height, h, and a base with radius r, its volume is $\pi r^2 h$. The new cylinder has volume $\pi(2r)^2(3h) = 12\pi r^2 h$. Thus, the ratio of the volume of the original cylinder to the volume of the new cylinder is $12\pi r^2 h : \pi r^2 h$, or 12:1.

25. If $\dfrac{a}{b} = \dfrac{c}{d}$, then $ad = bc$ (by cross multiplication) and $da = bc$. Thus $\dfrac{d}{b} = \dfrac{c}{a}$ (by cross multiplication). All of the other proofs follow similarly.

27. Given: $\dfrac{a}{b} = \dfrac{c}{d}$

Prove: $\dfrac{a + b}{b} = \dfrac{c + d}{d}$

Proof:

1. $\dfrac{a}{b} = \dfrac{c}{d}$ 1. Given
2. $ad = bc$ 2. Cross multiplication
3. $ad + bd = bc + bd$ 3. Addition
4. $(a + b)d = b(c + d)$ 4. Distributive property
5. $\dfrac{a + b}{b} = \dfrac{c + d}{d}$ 5. Cross multiplication

29. If $\dfrac{a}{b} = \dfrac{c}{b}$, then $ab = bc$ by cross multiplication. Dividing both sides by b yields $a = c$.

31. Outline: To show that $\dfrac{a}{b} = \dfrac{a + c}{b + d}$, cross multiply and simplify. Then work backward starting with $\dfrac{a}{b} = \dfrac{c}{d}$.

33. Outline: Show that if $\dfrac{a_1}{b_1} = \dfrac{a_2}{b_2} = \cdots = \dfrac{a_n}{b_n}$, then $\dfrac{a_1 + a_2 + \cdots + a_n}{b_1 + b_2 + \cdots + b_n} = \dfrac{a}{b}$. To prove, extend the proof for problem 32.

35. (a) 105 mi, 350 mi, $35n$ mi   (b) 29 in.

37. 24,530 mi

39. (a) 115 people
(b) 15 fewer people

41. 499.2 lb/ft^2

43. 29 in.

45. 17 in.

47. 206 mph

49. Suppose you take the ratio of the larger dimension to the smaller dimension.
(a) greater than the golden ratio
(b) less than the golden ratio
(c) greater than the golden ratio
(d) less than the golden ratio
(e) less than the golden ratio

Section 6.2

1. $\triangle ABC \sim \triangle DEC$ by AA Similarity

3. $\triangle LMN \sim \triangle OPQ$ by LL Similarity

5. $DE = 10$, $DF = 14$

7. $XY = \dfrac{27}{8}$ cm, $VW = \dfrac{128}{9}$ cm

9. $EG = 4$, $HI = 3\sqrt{3}$, $FH = 2\sqrt{3}$

11. $AC = \dfrac{56}{9}$ in., $BC = \dfrac{65}{4}$ in.

13. (a) Bases: 1:2, Heights: 1:2, Areas: 1:4
(b) Bases: 2:3, Heights: 2:3, Areas: 4:9
(c) Lengths: 1:3, Widths: 1:3, Areas: 1:9
(d) The ratio of the areas is equal to the square of the ratios of the linear dimensions.

15. Outline: Use SSS Similarity.

17. Outline: Since base angles in isosceles triangles are congruent, all *four* base angles are congruent. Use AA Similarity.

19. Outline: In trapezoid *ABCD,* show that transversals $\overline{AC}$ and $\overline{BD}$ form congruent alternate interior angles. Use AA Similarity.

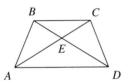

21. Given: $\triangle ABC \sim \triangle A'B'C'$ with medians $\overline{AM}$ and $\overline{A'M'}$

Prove: $\dfrac{AM}{A'M'} = \dfrac{AB}{A'B'}.$

Proof:

1. $\triangle ABC \sim \triangle A'B'C'$ 1. Given
2. $\angle B \cong \angle B'$ 2. Definition of similar triangles
3. $\dfrac{BC}{AB} = \dfrac{B'C'}{A'B'}$ 3. Definition of similar triangles
4. $\dfrac{BC}{2AB} = \dfrac{B'C'}{2A'B'}$ 4. Dividing both sides by 2
5. $\dfrac{BM}{AB} = \dfrac{B'M'}{A'B'}$ 5. Substitution
6. $\triangle ABM \sim \triangle A'B'M'$ 6. SAS Similarity
7. $\dfrac{AM}{A'M'} = \dfrac{AB}{A'B'}$ 7. Definition of similar triangles

23. Outline: Let $\triangle ABC$ be similar to $\triangle A'B'C'$ where $\overline{BM}$ and $\overline{B'M'}$ are altitudes.

 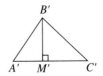

Use the result proved in problem 22.

25. Outline: The volumes of the cones will be $\dfrac{1}{3}\pi(AC)^2(BC)$ and $\dfrac{1}{3}\pi(A'C')^2(B'C')$. Using the fact that $\dfrac{AC}{BC} = \dfrac{A'C'}{B'C'}$, show that $\dfrac{\dfrac{1}{3}\pi(AC)^2(BC)}{\dfrac{1}{3}\pi(A'C')^2(B'C')} = \left(\dfrac{AC}{A'C'}\right)^3.$

27. Outline: Show that $\angle A \cong \angle A'$, and $\angle B \cong \angle B'$, and use AA Similarity.

29. Given: $\triangle ABC$ and $\triangle A'B'C'$ with parallel sides as shown. Extend $B'C'$ to intersect $\overline{AB}$ at *D.*

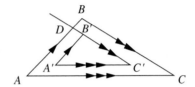

$\angle A'B'C' \cong \angle ADB'$ and $\angle ADB' \cong \angle ABC$ since they are pairs of corresponding angles. Therefore, $\angle A'B'C' \cong \angle ABC$. Similarly, $\angle B'C'A' \cong \angle BCA$. So $\triangle ABC \sim \triangle A'B'C'$ by AA Similarity.

31. 22 m

33. $\triangle AEB \sim \triangle CDE$ by SAS Similarity, therefore $\angle DCE = \angle EAB$. Also, $\triangle ABE$ is isosceles, so $\angle EAB = \angle EBA$. This implies that $\overline{AB} \parallel \overline{DC}$ because alternate interior angles are congruent. So the board will always be parallel to the ground.

35. 6.53 ft

37. (a) $V \approx 357$ cm^3 (b) $d \approx 9.5$ cm

39. Carol's shadow $= 6\frac{2}{11}$ ft $\approx 6'10''$
    Tom's shadow $= 8\frac{9}{11}$ ft $\approx 8'2''$

41. (a) $P = 3, A = \dfrac{\sqrt{3}}{4}$ (b) $P = \dfrac{9}{2}, A = \dfrac{3\sqrt{3}}{16}$
    (c) $P = \dfrac{27}{4}, A = \dfrac{9\sqrt{3}}{64}$ (d) $P = \dfrac{81}{8}, A = \dfrac{27\sqrt{3}}{256}$
    (e) $P = 3\left(\dfrac{3}{2}\right)^n, A = \dfrac{3^{n-1}\sqrt{3}}{4^n}$

Section 6.3

1. $AC = \dfrac{15}{4}$

3. $AD = \dfrac{15}{4}$

5. $AB = \dfrac{28}{5}$

7. $6\sqrt{6}$

9. $\dfrac{8}{3}$

11. 2

13. $EF = \dfrac{9}{4}, BC = \dfrac{8}{3}$

15. $AB = 6, AG = 4$

17. $MN = 6$ cm

19. 105 cm²

21. Area $= 118.79$ m²

23. (a) The midquad is a rectangle when diagonals $\overline{PR}$ and $\overline{QS}$ are perpendicular. That is, when $PQRS$ is a rhombus or a kite.
   (b) The midquad is a rhombus when diagonals $\overline{PR}$ and $\overline{QS}$ are congruent. That is, when $PQRS$ is a rectangle or an isosceles trapezoid.
   (c) The midquad is a square when conditions in (a) and (b) both hold. That is, when $PQRS$ is a square.

25. Observe that x is the mean proportion between a and 1. Thus, $\dfrac{a}{x} = \dfrac{x}{1}$, or $a = x^2$.

27. Given: M is the midpoint of $\overline{AB}$ in right $\triangle ABC$.

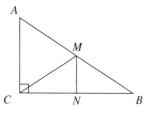

Prove: $MA = MC = MB$
Proof:
1. $MA = MB$ 1. Definition of midpoint
2. Construct $\overline{MN}$ per-   2. Construction
   pendicular to $\overline{CB}$.

3. $CN = NB$ 3. Theorem 6.10
4. $\triangle MCN \cong \triangle MBN$ 4. LL
5. $MC = MB$ 5. C.P.

29. Given: $\overline{BE}$ bisects $\angle ABC$ and $\dfrac{AB}{BE} = \dfrac{BD}{BC}$.

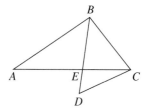

Prove: $\angle A = \angle D$
Proof:
1. $\angle ABC \cong \angle DBC$ 1. Definition of angle
                                    bisector
2. $\dfrac{AB}{BE} = \dfrac{BD}{BC}$ 2. Given

3. $\triangle ABE \sim \triangle DBC$ 3. SAS Similarity
4. $\angle A = \angle D$ 4. Definition of similar
                                    triangles

31. Given: $\triangle ABC$ as shown.

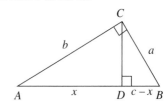

$\dfrac{c}{a} = \dfrac{a}{c-x}$ and $\dfrac{c}{b} = \dfrac{b}{x}$. Therefore, $c(c-x) = a^2$ and $cx = b^2$. Thus $c^2 - cx = a^2$ and $cx = b^2$ leads to $c^2 - b^2 = a^2$ and $a^2 + b^2 = c^2$.

33. Given: Equilateral $\triangle ABC$ with midsegment $\overline{MN}$
   Prove: $\triangle MBN$ is equilateral.
   Proof:
1. $AB = BC = AC$ 1. Given
   and $\overline{MN}$ a midsegment
2. $MN = \frac{1}{2}AC$ 2. Midsegment Thm
3. $MB = \frac{1}{2}AB$ and   3. Definition of midpoint
   $NB = \frac{1}{2}BC$
4. $MB = NB = MN$ 4. Substitution
5. $\triangle MBN$ is   5. Definition of
   equilateral.   equilateral triangle

35. Outline: Draw in the diagonals of the rhombus. Then use the Midsegment Theorem and the fact that the diagonals of a rhombus are perpendicular.

37. Outline: Use the Midsegment Theorem and SSS Similarity.

39. Outline: (c) Show that $\triangle AEC \sim \triangle EBC$, so $\dfrac{a}{x} = \dfrac{x}{b}$.

Section 6.4

1. $\sin A = \dfrac{5}{13}$

3. $\tan A = \dfrac{5}{12}$

5. $\cos B = \dfrac{5}{13}$

7. $\tan A = \dfrac{7}{5}$

9. $\cos A = \dfrac{7}{11}$

11. $\cos B = \dfrac{\sqrt{2}}{2}$

13. $a = 5.03$

15. $c = 8\frac{1}{3}$

17. $\angle A = 35.5°$

19. $\angle A = 46.7°$

21. $\cos B = 0.5$

23. $c \approx 0.85$

25. $\angle A = 19.5°, \angle B = 70.5°, CA = 8\sqrt{2}\ ''$

27. $AB = 13.7$ m, $BC = 7.8$ m, $\angle A = 35°$

29. Theorem 6.13: $\dfrac{\sin 45°}{\cos 45°} = \dfrac{0.7071}{0.7071} = 1 = \tan 45°$
Theorem 6.14: $\sin^2 45° + \cos^2 45° = (0.7071)^2 + (0.7071)^2 = 1$

31. Theorem 6.13: $\dfrac{\sin 30°}{\cos 30°} = \dfrac{0.5}{0.866} = 0.5774 = \tan 30°$
Theorem 6.14: $\sin^2 30° + \cos^2 30° = (0.5)^2 + (0.866)^2 = 1$

33. $a = 5, b = 5\sqrt{3}, \angle B = 60°$

35. $c = 26, \angle A = 67.4°, \angle B = 22.6°$

37. $\angle B = 67°, a = 1.7, c = 4.3$

39. $\angle A = 78°, a = 10.1, b = 2.2$

41. Area ≈ 390.1 ft²

43. Area ≈ 66.5 in²

45. Area ≈ 337.4 cm²

47. $(\cos A)(\tan A) = \dfrac{b}{c} \cdot \dfrac{a}{b} = \dfrac{a}{c} = \sin A$

49. Let $\triangle ABC$ be an isosceles right triangle.

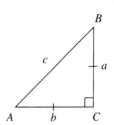

$\sin A = \dfrac{a}{c} = \dfrac{b}{c} = \cos A$

51. Let $\triangle ABC$ be as shown.

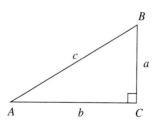

$\sin A = \dfrac{a}{c} = \cos B = \cos(180° - 90° - A) = \cos(90° - A)$

53. Let △ABC be as shown.

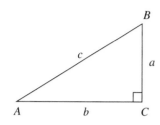

$$\frac{1}{\cos A} = \frac{1}{b/c} = \frac{c}{b} = \sec A$$

55. Let △ABC be as shown.

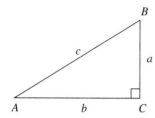

$$\frac{\cos A}{\sin A} = \frac{\frac{b}{c}}{\frac{a}{c}} = \frac{b}{a} = \cot A$$

57. Let △ABC be as shown.

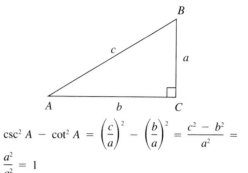

$$\csc^2 A - \cot^2 A = \left(\frac{c}{a}\right)^2 - \left(\frac{b}{a}\right)^2 = \frac{c^2 - b^2}{a^2} = \frac{a^2}{a^2} = 1$$

59. 92 ft

61. 42.89 ft

63. 80.3 ft

65. (a) 50 ft   (b) 46.9 ft

67. 42.3 ft

69. 126 m

Answers for the Additional Problems

1. 16 cm, 30 cm, 34 cm

2. $AC = 7.66$ in.

Chapter 6 Test

1. F–An equivalent proportion is $\dfrac{a}{c} = \dfrac{b}{d}$.

2. T

3. F–The included angles must also be congruent.

4. F–A midsegment is parallel to one side of the triangle.

5. T

6. T

7. F–For example, some rectangles are not similar to a square.

8. T

9. T

10. F–The cosecant ratio is the reciprocal of the sine ratio.

11. $x = 7.5$

12. $\sqrt{91} \approx 9.54$

13. $EF = 63$

14. (a) $5^2 + 12^2 = 25 + 144 = 169 = 132$. So $a^2 + b^2 = c^2$.

(b) For one angle, the sine is $\dfrac{5}{13}$ and the cosine is $\dfrac{12}{13}$. For the other angle the sine is $\dfrac{12}{13}$ and the cosine is $\dfrac{5}{13}$.

(c) $\sin^2 A + \cos^2 A = \left(\dfrac{5}{13}\right)^2 + \left(\dfrac{12}{13}\right)^2 = \dfrac{25}{169} + \dfrac{144}{169} = \dfrac{169}{169} = 1.$

15. Perimeter $= 40$ cm

16. Hypotenuse $= 13.5$

17. $x = 6$

18. $\triangle BAC \sim \triangle BDE$ by AA Similarity, because $\angle A = \angle BDE$ and $\angle B = \angle B$.
$\triangle BAC \sim \triangle BFD$ by AA Similarity, because $\angle BFD = \angle BAC$ and $\angle DBF = \angle ABC$.

19. Approximately 1023 m²

20. $a = 48$, $\angle A = 41.1°$, $\angle B = 48.9°$

21. The volume of the original cone is $\frac{1}{3}\pi r^2 h$. The volume of the new cone is $\frac{1}{3}\pi\left(\dfrac{r}{2}\right)^2(2h) = \frac{1}{2}(\frac{1}{3}\pi r^2 h)$. There-fore, the ratio of the volume of the new cone to the volume of the original cone is 1:2.

22. $\angle 1 = \angle 3$ since they are alternate interior angles and $\angle 2 = \angle 4$ since they are vertical angles. Therefore, the triangles are similar by AA Similarity.

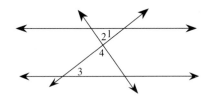

23. Approximately 241 ft

24. Length $= 35.4$ ft

Section 7.1

1. Central angle

3. $\angle BAC = 50°$

5. $\overset{\frown}{DC} = 80°$

7. $\overset{\frown}{AD} = 100°$

9. $\angle DAB = 90°$

11. $\angle C = 40°$

13. $\angle B = 35°$

15. $\angle CEB = 105°$

17. $\overset{\frown}{AD} = 110°$

19. $\angle A = 30°$, $\angle B = 30°$, $\angle C = 30°$

21. $\angle AEB = 120°$

23. We cannot necessarily conclude that $\overset{\frown}{AB} = 120°$ because we do not know that E is the center of the circle.

25. $\angle BOC = 80°$

27. Radius $= 19.5''$

29. $\angle B = 85°$

31. $PS = \sqrt{251} \approx 15.84$ in.

33. $\angle ABC = 111°$, $\angle BCD = 120°$, $\angle CDE = 129°$, $\angle DEA = 90°$, $\angle EAB = 90°$

35. Given: Rectangle $ABCD$ inscribed in a circle.

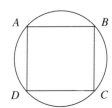

Prove: $\overline{AC}$ and $\overline{BD}$ are diameters.
Proof:

1. $\angle ADC$ and $\angle ABC$ are inscribed in the circle.   1. Given

2. $\angle ADC = \angle ABC = 90°$ 2. $ABCD$ is a rectangle

3. $\overset{\frown}{ABC} = \overset{\frown}{ADC} = 180°$ 3. Inscribed Angle Theorem

4. $\overline{AC}$ and $\overline{BD}$ are diameters.   4. Diameters have endpoints on the endpoints of semicircles.

37. $\overline{BD}$ is a diameter.

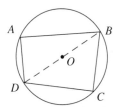

Thus, $\angle BAD = \angle BCD = 90°$. Since their sum is 180°, the sum of the measures of the remaining two angles is $360° - 180° = 180°$. Hence $\angle ADC$ and $\angle ABC$ are supplementary.

39. Outline: Use the fact that $\overset{\frown}{AC} = \angle AOC$ and $\triangle AOB$ is isosceles to show that $\angle ABC = \frac{1}{2}\angle AOC$.

41. Arc length $= \dfrac{16\pi}{9} \approx 5.59$ in.

43. Volume $= 59.13$ in³

45. Distance $= 2971.95$ mi

47. Distance $= 691.15$ mi

49. Distance $= 4354.25$ mi

Section 7.2

1. $\angle BEC = 149°$

3. $\angle CED = 70°$

5. $DE = 1.6$ cm

7. $CE = 5$ in.

9. $AF = 8.4$ in.

11. $\angle AFE = 104.5°$

13. $AE = 12$ cm

15. $AD = 16$ ft

17. $3, 6, 10, \dfrac{n(n-1)}{2}$

19. Given: Chords $\overline{AB}$ and $\overline{CD}$ are congruent.

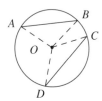

Prove: $\overset{\frown}{AB} = \overset{\frown}{CD}$
Proof:

1. $AB = CD$	1. Given
2. $AO = BO = CO = DO$	2. Radii of the same circle
3. $\triangle ABO \cong \triangle DCO$	3. SSS Congruence
4. $\angle AOB \cong \angle DOC$	4. C.P.
5. $\overset{\frown}{AB} = \overset{\frown}{CD}$	5. Definition of the measure of an arc

21. Outline: Use the fact that $\angle AOB = \angle COD$ to show $\triangle AOB \cong \triangle COD$ by SAS. Thus $\overline{AB} \cong \overline{CD}$.

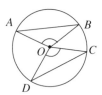

23. Let O and P be congruent circles. Then $AO = BO = CP = DP$. Since $\angle AOB = \angle CPD$, the measures of their arcs are the same.

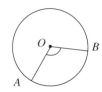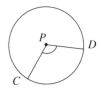

25. Given: Circle O with $\overline{AB} \cong \overline{CD}$

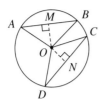

Prove: $OM = ON$

Proof:
1. $\overline{AB} \cong \overline{CD}$	1. Given
2. $\triangle AOB \cong \triangle COD$	2. SSS Congruence
3. $\angle B \cong \angle D$	3. C.P.
4. $\triangle BOM \cong \triangle DON$	4. H.A.
5. $OM = ON$	5. C.P.

27. $ABCD$ is a kite inscribed in a circle.

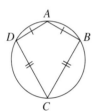

Since $\overset{\frown}{ABC} + \overset{\frown}{ADC} = 360°$, $\angle B + \angle D = 180°$. Since $\angle B = \angle D$, they must be right angles.

29. Outline: Let M be the midpoint of $\overline{AB}$.

Show that $\triangle AMO \cong \triangle BMO$ by SSS Congruence. Then $\angle AMO \cong \angle BMO$.

31. Let l be the perpendicular bisector of $\overline{AB}$.

Then $\triangle AMO \cong \triangle BMO$ by LL Congruence. Therefore, $\angle AOM \cong \angle BOM$ and $\overset{\frown}{AC} \cong \overset{\frown}{BC}$. Thus l bisects $\overset{\frown}{ACB}$.

33. The wastepaper basket should be placed at the point of intersection of the perpendicular bisectors of any two of the segments $\overline{JW}$, $\overline{WS}$, or $\overline{JS}$. This would be the center of the circle containing J, S, and W so each would be an equal distance from the wastepaper basket.

Section 7.3

1. $\angle C = 26°$

3. $EC = \dfrac{\sqrt{249} + 3}{2}$ cm ≈ 9.39 cm

5. $AB = 2.46$ in.

7. $\angle SPQ = 65°$

9. $\overset{\frown}{PQ} = 216°$

11. $\overset{\frown}{PQR} = 88°$

13. $\overset{\frown}{AE} = 90°$

15. $\overset{\frown}{BD} = 45°$

17. $DE = \dfrac{47}{7} \approx 6.71$

19. $\angle CBO = 90°$

21. $AC = BC$ by Corollary 7.15 and $OC = OC$, so $\triangle ACO \cong \triangle BCO$ by HL.

23. $\overset{\frown}{AE} = 150°$

25. $AB = 8\sqrt{2} \approx 11.31$ cm

27. $AP = 3.5$ ft, $BQ = 8.5$ ft

29. $\overset{\frown}{EF} = 28°$, $\overset{\frown}{GH} = 35°$, $\overset{\frown}{DI} = 27°$

31. Let $\overline{AB}$ and $\overline{AC}$ be tangent to a circle as shown.

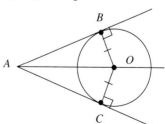

Since $BO = CO$ and $AO = AO$, we have $\triangle AOB \cong \triangle AOC$ by HL. Therefore, $\overline{AB} \cong \overline{AC}$ by C.P., so $AB = AC$.

33. Let A be on circle O and $\overleftrightarrow{AB} \perp \overline{AO}$.

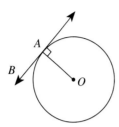

We will use an indirect proof to show that $\overleftrightarrow{AB}$ is tangent to circle O. Suppose $\overleftrightarrow{AB}$ is not tangent to circle O. Then it must intersect circle O in a second point C.

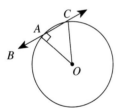

But $\triangle ACO$ is isosceles since $AO = CO$, and $\angle OAC = \angle OCA$. Thus $\triangle ACO$ has two right angles, which is impossible. Therefore our assumption is false and $\overleftrightarrow{AB}$ is tangent to circle O at A.

35. Let $\overleftrightarrow{AD}$ and $\overleftrightarrow{BC}$ be two parallel secants of circle O. There are two cases to consider: (i) O is on one of the lines and (ii) O is not on one of the lines.

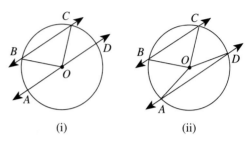

(i)   (ii)

In case (i) $\angle OBC = \angle OCB$, $\angle BCO = \angle COD$, $\angle CBO = \angle BOA$. Thus, $\angle BOA = \angle COD$, so $\overparen{AB} = \overparen{CD}$. To prove (ii), construct line l through O parallel to $\overleftrightarrow{BC}$ and $\overleftrightarrow{AD}$ and repeat the argument in (i) two times.

37. Outline: Draw tangent line $\overleftrightarrow{AB}$ with $\overleftrightarrow{AC} \perp \overleftrightarrow{AB}$.

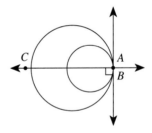

Use the fact that $\overleftrightarrow{AC} \perp \overleftrightarrow{AB}$, to show that the centers of both circles and A are collinear.

39. Diameter = 2 ft

Section 7.4

29. Let $\overline{AE}$ and $\overline{CD}$ be angle bisectors.

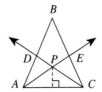

Each point on $\overline{AE}$ is equidistant from $\overline{AB}$ and $\overline{AC}$ and each point on $\overline{CD}$ is equidistant from $\overline{AC}$ and $\overline{BC}$. Thus, point P, the intersection of $\overline{AE}$ and $\overline{CD}$, must be equidistant from all three sides. Since it is equidistant from $\overline{BA}$ and $\overline{BC}$, it must also be on the bisector of $\angle ABC$. Thus all three angle bisectors intersect in P.

31. All the points on an angle bisector are equidistant from the sides of the angle. Thus the point of intersection of the three angle bisectors of a triangle is the same distance from all three sides. This distance is the radius of the inscribed circle.

33. Outline: Draw the chord and the two radii determined when making the first arc.

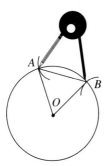

Show that △*AOB* is equilateral. Thus ∠*AOB* = 60°. Each subsequent arc will form another 60° angle.

35. Outline: Show that the two tangent lines form right angles with the diameter and that these are congruent alternate interior angles.

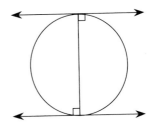

Answers for the Additional Problems

1. Consider the following three cases: (i) acute triangles, (ii) right triangles, and (iii) obtuse triangles. In each case the centroid lies inside the triangle because all three medians lie inside the triangle.

2. Consider the following two cases: (i) the first integer is odd and (ii) the first integer is even. Given that the three integers are consecutive integers, in case (i) we must have odd + even + odd and in case (ii) even + odd + even. Because the sum of the integers must be 180°, an even integer, only case (i) satisfies the conditions.

Chapter 7 Test

1. T

2. F–The incenter of a triangle always lies inside the triangle.

3. T

4. F–The diameter (4 inches) is the longest chord.

5. F–Its measure is half the measure of the intercepted arc.

6. T

7. T

8. T

9. T

10. F–∠*AOB* is larger than ∠*POQ*.

11. (a) $\overleftrightarrow{AC}$ (b) ∠*ECD*, ∠*BCE*
    (c) $\overset{\frown}{ECD}$ (d) ∠*OEA*   (e) $\overline{BC}, \overline{CD}, \overline{CE}$

12. 56°

13. 47°

14. 105°

15. (a) Approximately 63.7%
    (b) Approximately 78.5%
    (c) Circular plug

16. (a) *AB* = 30°   (b) ∠*EDF* = 34°

17. *TS* = 15 ft and *QT* = 4 ft

18. $\dfrac{40}{3}$

21. Because $\overset{\frown}{AB} = \overset{\frown}{CB}$, we have $\angle AOB \cong \angle XOB$. Therefore, $\triangle AOD \cong \triangle COD$ by SAS ($\overline{OA} \cong \overline{OC}$, $\angle AOB \cong \angle COB$, $\overline{OP} \cong \overline{OP}$). By C.P., $\angle ADO \cong \angle CDO$ and $AD = CD$. Because they form a straight angle, they are 90° angles. Thus, $\overline{OB}$ is the perpendicular bisector of $\overline{AC}$.

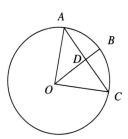

22. Because $\triangle ABO$, $\triangle BCO$, and $\triangle CDO$ are isosceles triangles, they have pairs of congruent angles as shown. We are given that $\angle 2 = 2(\angle 3)$.

(a) Prove $\angle AOB = 4(\angle 3)$: Because $\angle AOB$ is an exterior angle of $\triangle BCO$, $\angle AOB = \angle 2 + \angle 2$,

or $2(\angle 2)$. Because $\angle 2 = 2(\angle 3)$, we have by substitution $\angle AOB = 2[2(\angle 3)] = 4(\angle 3)$.

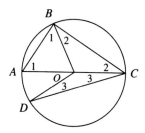

(b) Prove $\angle AOD + \angle ABO = 90°$: $\angle AOD + \angle COD = 180°$. $\angle COD = 180° - 2(\angle 3) = 180° - \angle 2$. Therefore, $\angle AOD + 180° - \angle 2 = 180°$, so $\angle AOD = \angle 2$. Because $\angle ABC$ is inscribed in a semicircle, $\angle ABC = 90°$. Thus, $\angle 1 + \angle 2 = 90°$, so $\angle ABO = 90° - \angle 2$. Therefore, $\angle AOD + \angle ABO = \angle 2 + (90° - \angle 2) = 90°$.

23. $r \approx 278'$

24. 146,200 ft²

Section 8.1

1. (a)

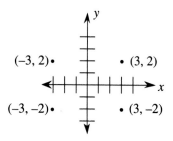

(b)

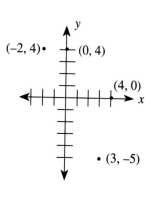

3. (a) 5   (b) 5   (c) 4   (d) 9

5. (a) Yes   (b) No   (c) Yes

7. $n = 3$ or $n = \dfrac{9}{5}$

9. (a) $(-\frac{3}{2}, 1)$ (b) $(-1, 2)$ (c) $(-\frac{5}{2}, \frac{9}{2})$
(d) $(3, 1)$ (e) $(2, 7)$ (f) $(\frac{3}{2}, \frac{3}{2})$

11. (a) Right isosceles   (b) Right scalene
(c) Obtuse isosceles   (d) Acute scalene

13. $AB = CD = \sqrt{18}$ or $3\sqrt{2}$, $BC = AD = 5$. Because opposite sides are congruent, $ABCD$ is a parallelogram.

15. Not a rectangle

17. $C(5\sqrt{3}, 5)$ or $C(-5\sqrt{3}, 5)$

19. Sides of $A'B'C'D'$ are one third as long as sides of $ABCD$.

21. (a) The perimeter of $A'B'C'D'$ is 3 times the perimeter of $ABCD$.
   (b) The area of $A'B'C'D'$ is 9 times the area of $ABCD$.

23. (a)

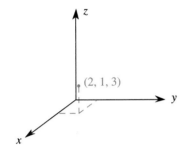

(b)

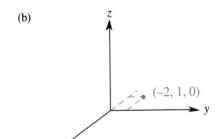

(c)

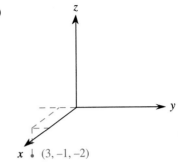

$x \downarrow (3, -1, -2)$

25. Octant 2: $(-, +, +)$ Octant 3: $(-, -, +)$
Octant 4: $(+, -, +)$ Octant 5: $(+, +, -)$
Octant 6: $(-, +, -)$ Octant 7: $(-, -, -)$
Octant 8: $(+, -, -)$

27. (a) $\sqrt{6}$ (b) $\sqrt{89}$

29. Outline: Use coordinates and the distance formula to show that $PM + MQ = PQ$. Then P, M, and Q must be collinear.

31. Let $P(x, y)$ be a point in the plane. Choose two values of x such that $r < s < x$. Then $A(r, mr + b)$ and $B(s, ms + b)$ are points on line l.
$AB = \sqrt{(s - r)^2 + [(ms + b) - (mr + b)]^2} = \sqrt{(s - r)^2 + m^2(s - r)^2} = \sqrt{(1 + m^2)(s - r)^2} = (s - r)\sqrt{1 + m^2}$. In a similar way, $BP = (x - s)\sqrt{1 + m^2}$ and $AP = (x - r)\sqrt{1 + m^2}$. So $AB + BP = (s - r + x - s)\sqrt{1 + m^2} = (x - r)\sqrt{1 + m^2} = AP$. Thus, A, B, and P are collinear.

Section 8.2

1. (a) $\dfrac{1}{2}$ (b) $\dfrac{4}{3}$ (c) 0   (d) Undefined

3. (a) Positive   (b) 0   (c) Undefined
   (d) Negative

5. (a) Yes   (b) Yes

7. (a) No   (b) No   (c) Yes   (d) Yes

9. (a) $(6, 6)$ or $(10, -4)$ (b) $(9, 4)$ or $(5, 0)$

11. Slope of $\overline{AB}$ = slope of $\overline{CD}$ = 1
Slope of $\overline{BC}$ = slope of $\overline{AD}$ = $-\frac{3}{4}$
So $ABCD$ is a parallelogram.

13. (a) Yes   (b) No

15. (a) 1   (b) $-\frac{2}{3}$ (c) 0   (d) undefined

17. $a = 10$

19. 11

21. (a) No   (b) No   (c) Yes

23. (a) No   (b) Yes   (c) No   (d) No

25. The midpoints of $\overline{AB}$ and $\overline{BC}$ are $M_1\left(-\dfrac{3}{2}, 2\right)$ and $M_2(3, 3)$, respectively. Using the distance formula

we have $M_1M_2 = \sqrt{\left(3 + \dfrac{3}{2}\right)^2 + (3 - 2)^2} =$

$\dfrac{\sqrt{85}}{2}$ and $AC = \sqrt{85}$. Thus, $M_1M_2 = \dfrac{1}{2}AC$. Also,

the slope of $M_1M_2 = \dfrac{3 - 2}{3 + \dfrac{3}{2}} = \dfrac{2}{9}$ and the slope of $\overline{AC}$

is $\dfrac{2}{9}$. So $M_1M_2 \parallel \overline{AC}$.

27. Outline: Determine the slopes of the diagonals of the quadrilateral. Use the fact that a rhombus is a parallelogram that has perpendicular diagonals.

29. (a) $\dfrac{PO}{OQ} = \dfrac{RS}{PS}$

(b) Since $\angle POQ = \angle RSP = 90°$ and $\dfrac{PO}{OQ} = \dfrac{RS}{PS}$, $\triangle QOP \sim \triangle PSR$ by SAS Similarity.

(c) $\dfrac{PO}{RS} = \dfrac{QO}{PS} = \dfrac{QP}{PR}$

(d) $\dfrac{QO + PS}{PS} = \dfrac{PO + RS}{RS}$ by problem 27 in Section 6.1

(e) $QO + PS = QO + OT = QT$ and $PO + RS = ST + RS = RT$. So from (d) we have $\dfrac{QT}{RS} = \dfrac{RT}{RS}$ and $\triangle PSR \sim \triangle QTR$ by SAS Similarity.

(f) $\dfrac{QT}{PS} = \dfrac{RT}{RS} = \dfrac{QR}{PR}$

(g) Since $\dfrac{QT}{PS} = \dfrac{QR}{PR}$, we have $QR = PR \cdot \dfrac{QT}{PS}$.

(h) $\dfrac{QP}{RS} = \dfrac{QO}{PS} \Rightarrow \dfrac{QP + PR}{PR} + \dfrac{QO + PS}{PS}$

$\Rightarrow QP + PR = \dfrac{PR}{PS}(QO + PS)$

$\Rightarrow QP + PR = PR\left(\dfrac{PO + RS}{RS}\right)$ from (c)

$\Rightarrow QP + PR = \dfrac{PR \cdot RT}{RS}$

$\Rightarrow QP + PR = \dfrac{PR \cdot QT}{PS}$ from (f)

(i) P, Q, and R are collinear since $QP + PR = QR$.

31. 0.08, approximately 4.6°

33. (a) 792 ft
(b) 28,512 ft
(c) Percent grade = slope. For example, a grade of 6% equals a slope of $\dfrac{6}{100}$.

Section 8.3

1. (a) $5 = 7(1) - 2$
(b) $9 = -2(-2) + 5$
(c) $-2(7.5) = 6(7.5) - 3$
(d) $3(6) = 4(4) + 2$

3. (a) $m = -3, b = 2$ (b) $m = 2, b = -5$
(c) $m = \frac{5}{2}, b = -3$ (d) $m = \frac{2}{7}, b = -\frac{8}{7}$

5. (a) $y = 3x + 7$ (b) $y = -x - 3$
(c) $y = 2x + 5$ (d) $y = -3x - \frac{1}{4}$

7. (a) Rises to the right   (b) Falls to the right
(c) Horizontal   (d) Vertical
(e) Falls to the right   (f) Rises to the right

9. (a) $y = \frac{4}{3}x + 1$
(b) $b = 1$
(c) $m = \frac{4}{3}$

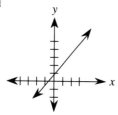

11. (a)

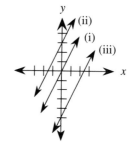

(b) The lines are parallel with slope 2.
(c) The lines all have the same slope c and so are parallel.

13. (a) $y = 2x - 11$
(b) $y = -x - 1$
(c) $y = -\dfrac{3}{5}x - \dfrac{19}{5}$
(d) $y = -\dfrac{3}{2}x - \dfrac{3}{2}$

15. $y = -3x + 5$

17. $y = -\frac{4}{3}x + 11$

19. $y = \frac{1}{2}x$, $y = -\frac{1}{2}x + 3$

21. (a) $(3, 2)$
   (b) $(3, -4)$
   (c) No solution–parallel lines
   (d) $\left(\dfrac{13}{2}, \dfrac{3}{2}\right)$

23. (a) No solution–parallel lines
   (b) No solution–parallel lines
   (c) $(-3, 1)$
   (d) All points on the line $-2x + 3y = 9$

25. (a) $5x = 10$
   (b) $x = 2$
   (c) $y = 3$

27. (a) $3^2 + 4^2 = 9 + 16 = 25$
   (b) $(-3)^2 + 5^2 = 9 + 25 = 34$
   (c) $(-3 + 1)^2 + (7 - 2)^2 = 4 + 25 = 29$
   (d) $(\sqrt{5})^2 + (-3 + 5)^2 = 5 + 4 = 9$

29. (a) $(x + 1)^2 + (y + 2)^2 = 5$
   (b) $(x - 2)^2 + (y + 4)^2 = 41$
   (c) $(x - 1)^2 + (y - 2)^2 = 20$

31. $(1, -1)$

33. $\left(\dfrac{4}{9}, -\dfrac{17}{9}\right)$

35. (a) 2 solutions   (b) $(0, 1)$ and $(-\frac{4}{5}, \frac{3}{5})$

37. 3 square units

39. Outline: Use the distance formula and the fact that the distance from any point (x, y) on the circle to the center (a, b) of the circle is r.

41. (a) \$5.60, \$13.10   (b) $y = 0.5x + 0.6$

43. (a) $y = 435 + 14.5x$
   (b) 14.5–For each foot added to the length of the side fence, the cost increases by \$14.50.
   (c) 435–A fixed cost, meaning that the owner will pay this cost even if the side fence has length zero feet.
   (d) Approximately 87 feet

45. (a) 100   (b) 125   (c) 225

Section 8.4

1. Because slope of $\overline{AD}$ = slope of $\overline{BC}$ = 0, we have $AD \parallel BC$. Thus, C and D may be labelled as $C(c, b)$ and $D(d, 0)$. Also, $AB = \sqrt{a^2 + b^2}$ and $CD = \sqrt{(d - c)^2 + b^2}$. Because, in general, these two lengths are not the same, $ABCD$ is not isosceles.

3. $B\left(\dfrac{a}{2}, b\right)$

$$AB = \sqrt{\left(\dfrac{a}{2} - 0\right)^2 + (b - 0)^2} = \sqrt{\dfrac{a^2}{4} + b^2}$$

$$BC = \sqrt{\left(a - \dfrac{a}{2}\right)^2 + (0 - b)^2} = \sqrt{\dfrac{a^2}{4} + b^2}$$

Because $AB = BC$, $\triangle ABC$ is isosceles. However,

because $\sqrt{\dfrac{a^2}{4} + b^2}$ is not necessarily equal to a,

$\triangle ABC$ is not necessarily equilateral.

5. (a) $C(0, c)$ and $D(a, b)$
$$AB = AD = \sqrt{a^2 + b^2}$$
$$BC = CD = \sqrt{(0 - a)^2 + (c - b)^2}$$
$$= \sqrt{a^2 + (c - b)^2}$$
So $ABCD$ is a kite since two pairs of adjacent nonoverlapping sides are congruent.
   (b) No. In that case, $ABCD$ is a rhombus, a special case of a kite, since $AB = BC = CD = AD = \sqrt{a^2 + a^2} = a\sqrt{2}$. In fact, because $\overline{AB} \perp \overline{BC}$, $ABCD$ is a square.

7. Slope of $\overline{WZ} = \dfrac{\dfrac{b}{2} - 0}{\dfrac{a}{2} - \dfrac{e}{2}} = \dfrac{b}{a - e}$

Slope of $\overline{YX} = \dfrac{\dfrac{b + d}{2} - \dfrac{d}{2}}{\dfrac{a + c}{2} - \dfrac{c + e}{2}} =$

$\dfrac{b + d - d}{a + c - c - e} = \dfrac{b}{a - e}$

Because the slope of $\overline{WZ}$ equals the slope of $\overline{YX}$, we know that $\overline{WZ} \parallel \overline{YX}$.

9. Outline: Given square *ABCD*

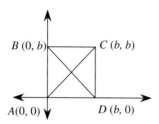

Use coordinates to show that the product of the slopes of $\overline{AC}$ and $\overline{BD}$ is -1.

11. Given parallelogram *ABCD*

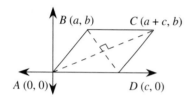

With coordinates shown, *ABCD* is a parallelogram, because slope $\overline{AD}$ = slope $\overline{BC}$ = 0 and slope $\overline{AB}$ = slope $\overline{DC} = \dfrac{b}{a}$.

$AD = c - 0 = c$

$BC = (a + c) - a = c$

$CD = \sqrt{(a + c - c)^2 + (b - 0)^2} = \sqrt{a^2 + b^2}$

$AB = \sqrt{(a - 0)^2 + (b - 0)^2} = \sqrt{a^2 + b^2}$

Thus, $\overline{AD} \cong \overline{BC}$ and $\overline{AB} \cong \overline{DC}$.

13. Given that *M* is the midpoint of $\overline{BC}$

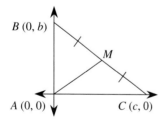

M, the midpoint of $\overline{BC}$, has coordinates $\left(\dfrac{c}{2}, \dfrac{b}{2}\right)$. By definition, $BM = CM$.

$CM = \sqrt{\left(\dfrac{c}{2} - c\right)^2 + \left(\dfrac{b}{2} - 0\right)^2} = \sqrt{\left(\dfrac{c}{2}\right)^2 + \left(\dfrac{b}{2}\right)^2}$

$AM = \sqrt{\left(\dfrac{c}{2} - 0\right)^2 + \left(\dfrac{b}{2} - 0\right)^2} = \sqrt{\left(\dfrac{c}{2}\right)^2 + \left(\dfrac{b}{2}\right)^2}$

Therefore, $AM = CM = BM$.

15. Outline: Given isosceles trapezoid *ABCD*

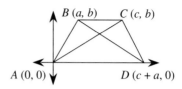

Use coordinates and the distance formula to show that $AC = BD$.

17. Outline: Given rectangle *ABCD* with midpoints *M, N, O,* and *P*.

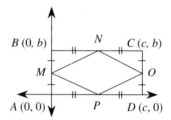

Use coordinates and the midpoint formula to find the coordinates of *M, N, O,* and *P*. Then use the distance formula to show $MN = NO = OP = PM$.

19. Given: *M, N, O, P* midpoints of sides of rhombus *ABCD*

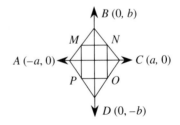

Prove: *MNOP* is a rectangle.

Proof:

1. $AB = BC = CD = DA = \sqrt{a^2 + b^2}$ 1. Distance formula

2. *ABCD* is a rhombus   2. Definition

3. Midpoints are $M\left(-\dfrac{a}{2}, \dfrac{b}{2}\right), N\left(\dfrac{a}{2}, \dfrac{b}{2}\right),$ $O\left(\dfrac{a}{2}, -\dfrac{b}{2}\right), P\left(-\dfrac{a}{2}, -\dfrac{b}{2}\right)$ 3. Midpoint formula

4. Slope of $\overline{MN}$ = slope of $\overline{OP}$ = 0   4. Slope formula

5. $\overline{MP}$ and $\overline{NO}$ are both vertical   5. Slope definition

6. $\overline{MN} \perp \overline{NO}$, $\overline{NO} \perp \overline{OP}$,   6. Theorem 8.5
$\overline{OP} \perp \overline{PM}$, and
$\overline{PM} \perp \overline{MN}$
7. $MNOP$ is a rectangle.   7. Definition

21. Outline: Given isosceles $\triangle ABC$ with M, N midpoints of $\overline{AB}$, $\overline{CB}$ respectively

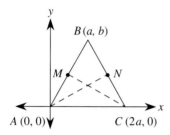

Use coordinates and the midpoint formula to find coordinates of M and N. Then show $AN = CM$ using the distance formula.

23. Given parallelogram $ABCD$ with $AC = BD$.

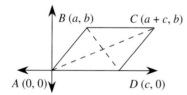

$BD = \sqrt{(a - c)^2 + b^2}$ and $AC = \sqrt{(a + c)^2 + b^2}$. Since $AC = BD$, $(a - c)^2 = (a + c)^2$. Therefore, $a^2 - 2ac + c^2 = a^2 + 2ac + c^2$, so $-2ac = 2ac$, or $4ac = 0$. Thus either $a = 0$ or $c = 0$. If $c = 0$, then $A = D$ and $ABCD$ is not a parallelogram. Hence, $a = 0$. Thus B and C have coordinates $B(0, b)$ and $C(c, b)$. So, $\angle A = 90°$ and $ABCD$ is a rectangle.

25. Given that l_1, l_2, l_3 are the perpendicular bisectors of the sides of $\triangle ABC$

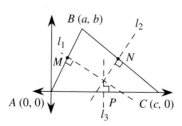

M, N, P have coordinates: $M\left(\dfrac{a}{2}, \dfrac{b}{2}\right)$, $N\left(\dfrac{a + c}{2}, \dfrac{b}{2}\right)$, $P\left(\dfrac{c}{2}, 0\right)$. Equation of l_3 is $x = \dfrac{c}{2}$.

Slope of l_1 is $-\dfrac{a}{b}$, so its equation is $y - \dfrac{b}{2} = -\dfrac{a}{b}\left(x - \dfrac{a}{2}\right)$, or $y = -\dfrac{a}{b}x + \dfrac{a^2 + b^2}{2b}$. Slope of l_2 is $\dfrac{c - a}{b}$, so its equation is

$$y - \dfrac{b}{2} = -\left(\dfrac{c - a}{b}\right)\left(x - \dfrac{a + c}{2}\right), \text{ or}$$

$$y = \dfrac{c - a}{b}x + \dfrac{a^2 + b^2 - c^2}{2b}.$$

l_1 and l_3 intersect where $x = \dfrac{c}{2}$ and

$$y = -\dfrac{a}{b}\left(\dfrac{c}{2}\right) + \dfrac{a^2 + b^2}{2b} = \dfrac{a^2 + b^2 - ac}{2b}.$$

l_2 and l_3 intersect where $x = \dfrac{c}{2}$

$$y = \left(\dfrac{c - a}{b}\right)\left(\dfrac{c}{2}\right) + \dfrac{a^2 + b^2 - c^2}{2b} = \dfrac{a^2 + b^2 - ac}{2b}.$$

Thus l_1, l_2, l_3 are concurrent.

27. Outline: Given M, N, P midpoints of $\overline{AB}$, $\overline{BC}$, $\overline{AC}$

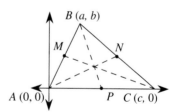

Using coordinates, find each midpoint. Using the coordinates of the vertices and the midpoints, write the equation of each median. Show that each pair of medians intersect in the same point.

Answers for the Additional Problems

1. $\sqrt{20} \approx 4.47$ km

2. Place the figure on a coordinate system with $B = (0, 2a)$ and $C = (2a, 0)$. Then $D = (-a, a)$ and $E = (a, -a)$. Using these coordinates, the slope of $\overline{BC}$ is -1 and the slope of $\overline{DE}$ is -1. Therefore, $\overline{BC} \parallel \overline{DE}$.

Chapter 8 Test

1. F–slope of a perpendicular line is $-\frac{1}{2}$.

2. F–only true if B is between A and C

3. T

4. T

5. T

6. F–It might have zero or infinitely many solutions.

7. T

8. F–The slope of the line is $-\dfrac{a}{b}$.

9. F–Area of new triangle will be 4 times the area of $\triangle ABC$.

10. T

11. (a) $(6, 1)$ (b) $\sqrt{72}$ or $6\sqrt{2}$
    (c) 1 (d) $y = x - 5$

12. Yes, they are collinear.
    Slope of $\overline{AB}$ = slope of $\overline{AC}$ = slope of $\overline{BC}$. Also, $AB + BC = AC$.

13. (a)

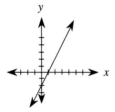

(b)

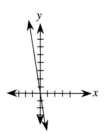

(c)

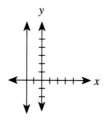

(d)

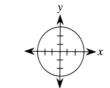

(e)

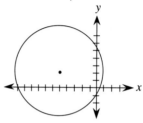

14. (a) $(4, 1)$ (b) No solution—parallel lines
    (c) No solution—Line does not intersect the circle.

15. $y = -\frac{1}{2}x - \frac{1}{2}$ or $x + 2y = -1$

16. $(c - b, a)$

17. $AC = \sqrt{(3 + 1)^2 + (7 - 1)^2}$
$= \sqrt{16 + 36} = \sqrt{52}$ or $2\sqrt{13}$
$BD = \sqrt{(4 + 2)^2 + (2 - 6)^2}$
$= \sqrt{36 + 16} = \sqrt{52}$ or $2\sqrt{13}$

18. $(x + 1)^2 + (y - 4)^2 = 52$

19. $ABCD$ is a parallelogram because $\overline{AB}$ and $\overline{CD}$ have the same slope and $\overline{AD}$ and $\overline{BC}$ have the same slope. Thus, opposite sides are parallel.

20. $(7, 11), (5, -1), (-1, 5)$

21.

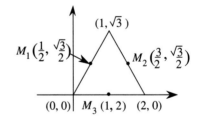

$$M_1M_2 = \frac{3}{2} - \frac{1}{2} = 1$$

$$M_1M_3 = \sqrt{\left(\frac{1}{2}\right)^2 + \left(\frac{\sqrt{3}}{2}\right)^2} = 1$$

$$M_2M_3 = \sqrt{\left(\frac{1}{2}\right)^2 + \left(\frac{\sqrt{3}}{2}\right)^2} = 1$$

Therefore, $\triangle M_1M_2M_3$ is equilateral.

22. If the diagonals are perpendicular, then the product of their slopes is -1.

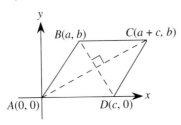

$$\left(\frac{b}{a-c}\right) \cdot \left(\frac{b}{a+c}\right) = -1$$

$$\frac{b^2}{a^2 - c^2} = -1$$

$b^2 = c^2 - a^2$

$a^2 + b^2 = c^2$

This shows that $AB = AD$. Because opposite sides of a parallelogram are congruent and $AB = AD$, all sides must be congruent. Therefore $ABCD$ is a rhombus.

23. Slope is $\dfrac{5}{96}$, 5.2% grade.

24. (a) $y = (0.01)\pi x$ or $y \approx 0.0314x$

(b) $(0.01)\pi$–For each cm added to the length of the wire, the volume increases by $(0.01)\pi$ cm³.

(c) $y = (0.0894)\pi x$

(d) $(0.0894)\pi$–For each cm added to the length of the wire, the mass increases by $(0.0894)\pi$ g.

Section 9.1

1. Yes. Yes. Translations take lines to parallel lines.

3. (a)

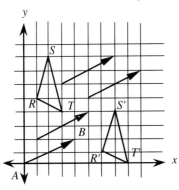

Many other answers are possible.

(b)

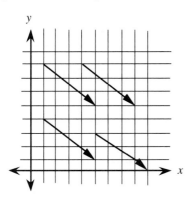

Many other answers are possible.

(c)

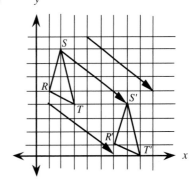

5. Yes. Yes. Each translation takes a triangle to a congruent triangle.

7.

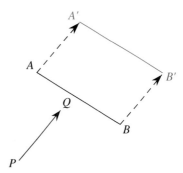

9. (a)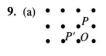

(b)

(c)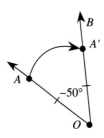

11. (a) $-60°$ (b) $110°$ (c) $180°$ (or $-180°$) (d) $-20°$

13.

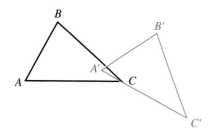

15. No. Rotations don't necessarily take lines to parallel lines.

17.

19. (a) 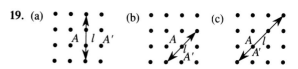 (b) (c)

21. (a) Isosceles because $QP = QP'$
(b) A right triangle because $\overline{PP'} \perp l$

23. $ABCB'$ is a kite because $CB = CB'$ and $AB = AB'$.

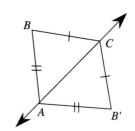

25.

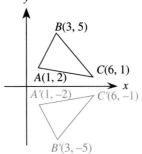

27. (a) $A'(0, 4)$, $B'(4, 6)$
(b) $A'(3, 0)$, $B'(7, 2)$
(c) $A'(0, 3)$, $B'(4, 5)$
(d) $A'(-2, 0)$, $B'(2, 2)$

29. (a) $(5, -1)$ (b) $(3, 1)$ (c) $(4, 2)$ (d) $(-1, 3)$
(e) $(-2, -5)$ (f) $(y, -x)$

31. (a) $(-3, 1)$ (b) $(6, 3)$ (c) $(4, -2)$ (d) $(-x, -y)$

33. (a)

(b) $A'(1, -2)$, $B'(3, -5)$, $C'(6, -1)$
(c) $(x, -y)$

35. (a) Translation   (b) Reflection

37. (a) Yes (b) No

39. (a)

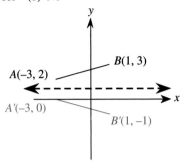

Other figures could be drawn to determine the line of reflection, which is the line $y = 1$.

(b)

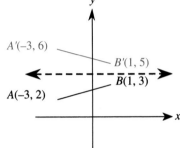

Other figures could be drawn to determine the line of reflection, which is the line $y = 4$.

(c)

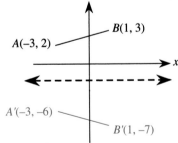

Other figures could be drawn to determine the line of reflection, which is the line $y = -2$.

41. (a)

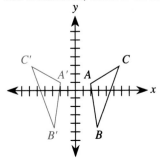

(b) $A'(-2, 1)$, $B'(-3, -5)$, $C'(-6, 3)$

(c) $(-a, b)$

43. Let $\overline{A'B'}$ be the image of $\overline{AB}$ under the rotation through θ about O.

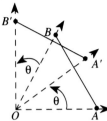

By definition of a rotation, $OA = OA'$ and $OB = OB'$. $m\angle A'OB' = m\angle AOB + \theta - \theta = m\angle AOB$. Thus $\triangle OA'B' \cong \triangle OAB$ (SAS) and $AB = A'B'$.

45. Outline: Suppose $\overline{AB} \perp \overline{CD}$ intersect at P. Let $\overline{A'B'}$ and $\overline{C'D'}$ be their images under the rotation about O. Then apply property 3 of Theorem 9.2.

47. Show that reflections preserve distance.
Given: $\overline{A'B'}$ is the reflection of $\overline{AB}$ with respect to line l, $\overline{AA'}$ intersects l at P, and $\overline{BB'}$ intersects l at Q.

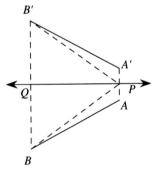

Prove: $\overline{AB} \cong \overline{A'B'}$
Proof:
1. $\overline{B'Q} \cong \overline{BQ}$ and $\angle BQP \cong \angle B'QP$ — 1. Definition of reflection
2. $\triangle BQP \cong \triangle B'QP$ — 2. SAS Congruence
3. $BP = B'P$ and $\angle BPQ \cong \angle B'PQ$ — 3. C.P.
4. $\overline{AP} \cong \overline{A'P}$ and $\angle APQ \cong \angle A'PQ = 90°$ — 4. Definition of reflection
5. $\angle APB \cong \angle APB'$ — 5. Complements of congruent angles
6. $\triangle APB \cong \triangle A'PB'$ — 6. SAS Congruence
7. $\overline{AB} \cong \overline{A'B'}$ — 7. C.P.

49. Outline: Use the fact that angle measure is preserved.

51. Outline: Let $\overline{A''B''}$ be the image of $\overline{AB}$ under the glide reflection consisting of reflection with respect to line l and translation $\overrightarrow{PQ}$.

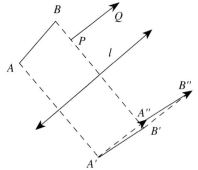

Use the fact that both reflections and translations preserve distance.

53. (a) Translation and rotation
(b) Translation and rotation
(c) Translation
(d) Translation, rotation, reflection, glide reflection

55. The three possible moves are shown. Each of these moves is equivalent to a glide reflection because it involves either a horizontal or vertical translation followed by a reflection.

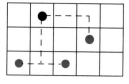

Section 9.2

1. (a) $\dfrac{OA'}{OA} = 4$

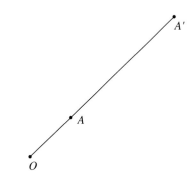

(b) $\dfrac{OA'}{OA} = 3.75$

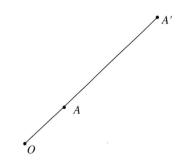

(c) $\dfrac{OA'}{OA} = 2\dfrac{5}{8} = 2.625$

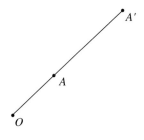

(d) $\dfrac{OA'}{OA} = \dfrac{2}{3}$

3. (a)

(b)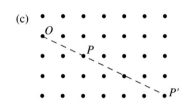

(c)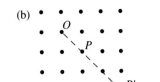

5. (a)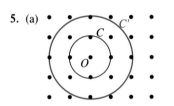

(b)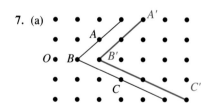

7. (a)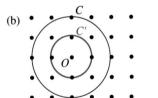

(b) $\angle ABC \cong \angle A'B'C'$

9. (a)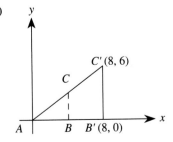

(b) $\dfrac{\text{Area of } \triangle A'B'C'}{\text{Area } \triangle ABC} = \dfrac{24}{6} = 4$

11. The circumference of circle C' is 3 times the circumference of circle C, and the area of circle C' is 9 times the area of circle C.

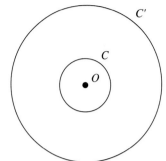

13. (a)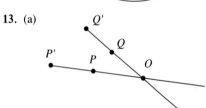

(b) Scale factor is approximately 2.

15.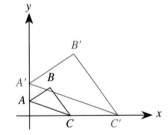

Yes, $\triangle A'B'C'$ is a size transformation image of $\triangle ABC$ with center (0, 0) and scale factor 2.

17. (a)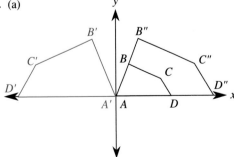

(b) $A'(0, 0)$, $B'(-2, 6)$, $C'(-8, 4)$, $D'(-10, 0)$

(c) $(-2a, 2b)$

19. (a)

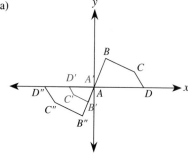

(b) $A'(0, 0)$, $B'(-\frac{1}{2}, -\frac{3}{2})$, $C'(-2, -1)$, $D'(-\frac{5}{2}, 0)$

(c) $\left(-\frac{a}{2}, -\frac{b}{2}\right)$

21. First perform the size transformation with center A and scale factor $\frac{9}{2}$. Then translate 3 units to the left.

23. First apply the size transformation with center A and scale factor $\frac{1}{3}$ to $\triangle ABC$. Call the image $\triangle A''B''C''$. Then reflect $\triangle A''B''C''$ across the line that is the perpendicular bisector of $\overline{C''C'}$.

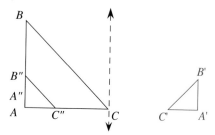

25. There is a size transformation with scale factor of 2 (in this case) that takes $\triangle ABC$ to $\triangle A'B'C'$. Since $\triangle A'B'C' \cong \triangle DEF$, there is an isometry which takes $\triangle A'B'C'$ to $\triangle DEF$. This size transformation and isometry constitute a similitude, so $\triangle ABC \cong \triangle DEF$.

27. Consider a size transformation with center O that takes A to A', B to B', C to C'. Then $\frac{OA'}{OA} = \frac{OB'}{OB} = \frac{OC'}{OC}$. We have $\angle AOB = \angle A'OB'$, $\angle BOC = \angle B'OC'$ since A and A', B and B', C and C' are collinear. Thus $\triangle OAB \sim \triangle OA'B'$ and $\triangle OBC \sim \triangle OB'C'$ by SAS Similarity. Then $\frac{OA'}{OA} = \frac{A'B'}{AB}$ and $\frac{OB'}{OB} = \frac{B'C'}{BC}$. So $\frac{A'B'}{AB} = \frac{B'C'}{BC}$. Thus, $BC \cdot A'B' = AB \cdot B'C'$ and $\frac{A'B'}{B'C'} = \frac{AB}{BC}$.

29. Outline: Consider the size transformation with center O that takes A to A', B to B', C to C', with $\overline{AB} \perp \overline{BC}$. Use the fact that size transformations preserve angle measure.

Section 9.3

1. Reflections with respect to the lines containing the following segments: $\overline{AC}, \overline{BD}, \overline{EG}, \overline{FH}$

3. Rotations of 180° and 360° around P

5. (a) The reflections with respect to lines $\overleftrightarrow{AC}$ and $\overleftrightarrow{BD}$ and 180° and 360° rotations around the point of intersection of the diagonals

(b) Because a rhombus is both a kite and a parallelogram, the rhombus must have at least the same isometries as the kite and parallelogram.

7. (a) Yes. Reflections take triangles to congruent triangles.

(b) Yes

(c) No

(d) Yes. Both reflections preserve distance, thus one reflection followed by another will preserve distance.

(e) A translation in the direction perpendicular to the parallel lines and twice the distance between the lines.

9. All. Parts (a) and (d) because they rotate one diagonal to the other.

11. Reflections with respect to $\overleftrightarrow{AI}, \overleftrightarrow{BJ}, \overleftrightarrow{CF}, \overleftrightarrow{DG}$, and $\overleftrightarrow{EH}$. Rotations of 72°, 144°, 216°, and 288° with respect to the center of the pentagon. Any of the rotations listed will rotate a diagonal to another diagonal. For example, for the rotation of 72°, $\overline{AC}$ rotates to $\overline{EB}$.

Thus they are congruent. Similarly, $\overline{EB}$ rotates to $\overline{DA}$, and so on.

13. The reflection with respect to $\overleftrightarrow{AC}$ takes B to D. Therefore, $AB = AD$ and $BC = DC$. The reflection with respect to $\overleftrightarrow{BD}$ takes A to C. Therefore, $AB = CB$ and $AD = CD$. Combining these four equations we have $AB = AD = DC = BC$. Thus $ABCD$ is a rhombus.

15. Let M be the midpoint of $\overline{AC}$. Then the image of A under the $180°$ rotation with center M is C and of C is A. If this same rotation is applied to B, the image is D since $ABCD$ is a parallelogram. Thus, this rotation takes $\triangle ABC$ to $\triangle CDA$, so they are congruent.

17. Outline: Use the fact that $\overline{AB}$ goes to $\overline{BC}$ under a reflection with respect to $\overleftrightarrow{BP}$. Then show $\triangle ABP \cong \triangle CBP$.

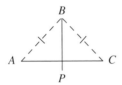

19. Given: Point B is on the perpendicular bisector l of $\overline{AC}$

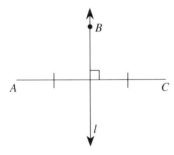

Prove: $AB = CB$
Proof:

1. l is the perpendicular bisector of $\overline{AC}$	1. Given
2. Reflection image of A with respect to l is C and reflection image of B with respect to l is B	2. Definition of reflection
3. Image of $\overline{AB}$ is $\overline{CB}$	3. Reflections take lines to lines.
4. $\overline{AB} \cong \overline{CB}$	4. Reflections preserve congruence.
5. $AB = CB$	5. Definition of congruence

21. Outline: Use the fact that A goes to C under the rotation and rotations take parallel lines to parallel lines to show that B must be on $\overleftrightarrow{DC}$ after the rotation. Show B must also be on $\overleftrightarrow{AD}$ after the rotation. Thus B must go to D, and $\overline{BP}$ goes to $\overline{DP}$.

23. Given: Points P, Q, and R are the midpoints of the sides of $\triangle ABC$.

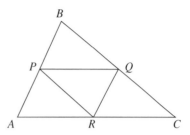

(a) Since P and R are midpoints of $\overline{AB}$ and $\overline{AC}$ respectively, the size transformation with center A and scale factor 2 takes P to B and R to C. Therefore $\triangle APR$ goes to $\triangle ABC$ under this size transformation.

(b) $\triangle BAC$

(c) The size transformation with center C and scale factor 2

25. In each move, the shaded region maintains the same area.

27. One path is shown. However there are other correct solutions. First reflect B to B_1, then B_1 to B_2. Aim A for B_2.

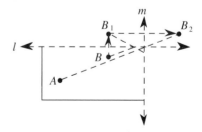

Answers for the Additional Problems

1. First draw a line of reflection (a vertical line, a horizontal line, or one of the two diagonal lines). Then shade pairs of squares and their images.

2. $R(-c, d)$, $S(-a, b)$ and $R(c, -d)$, $S(a, -b)$. Other answers are possible.

Chapter 9 Test

1. T

2. T

3. T

4. F–The reflection portion of the glide reflection does not necessarily take a line segment to a parallel line segment.

5. T

6. F–A reflection reverses the orientation of a triangle.

7. T

8. T

9. F–The isometry could be a rotation.

10. F–There is a similitude, but not necessarily a size transformation.

11. (a) A'(0, 1), B'(3, 4), C'(7, 3)
   (b) (a + 3, b + 2)

12.

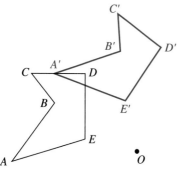

13.

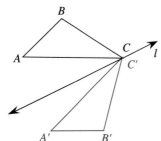

14.

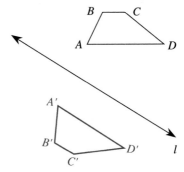

15.

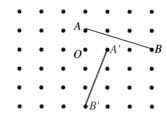

16.

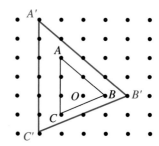

17. (a)

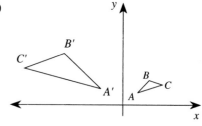

   (b) A'(−3, 6), B'(−12, 12), C'(−18, 9)
   (c) (−3a, 3b)

18. (a) Glide reflection

(b) Rotation

(c) Reflection

19. First, perform a size transformation with center D and scale factor $\frac{1}{2}$. Then reflect the resulting figure across $\overleftrightarrow{DC}$ and translate left 5 units. There are infinitely many other correct ways to describe this similitude as the composition of a size transformation followed by an isometry, depending on where the center of the size transformation is located.

20. The transformation could be the identity transformation (for example, a rotation of $0°$). The transformation could be a reflection with respect to a line l that contains $\overline{AB}$.

21. This can be accomplished by interchanging the translation and the reflection in the definition of the glide reflection.

22. Because $ABCD$ is a rhombus, the reflection with respect to $\overleftrightarrow{AC}$ takes $ABCD$ to itself, as does the reflection with respect to $\overleftrightarrow{DB}$. Because the diagonals of the rhombus are perpendicular, the combination of these two reflections will be a $180°$ rotation around E. Because such a rotation takes lines to parallel lines, $\overline{AB} \parallel \overline{CD}$ and $\overline{AD} \parallel \overline{BC}$. Thus $ABCD$ is a parallelogram.

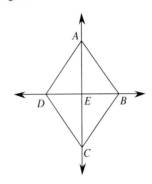

23. One successful path is shown.

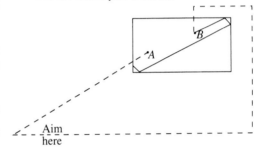

24. Place the logo at the origin and map a few key points using a size transformation of scale $\sqrt{2}$. This will have the effect of doubling the area of the logo.

Topic 1

1. (a), (b), (c), (e)

3. (a) 8   (b) 14   (c) -4

5. 73

7. (a) 7^6 (b) 30^3 (c) 4^{15} (d) 10^{10}

9. 2^4

11. (a) 9.7468   (b) -2.5198 (c) 5.3852

(d) 2.2985   (e) 11.4471

13. (a) $3\sqrt{5}$ (b) 18 (c) 4 (d) $\dfrac{5\sqrt{5}}{3}$

15. (a) $11x + 3y$ (b) $4\sqrt{5} - \sqrt{7}$ (c) $8\sqrt[3]{2}$

(d) $9\pi + 15$ (e) $\dfrac{5\pi}{2}$

17. (a) 46   (b) 18   (c) 82   (d) 4

19. (a) $48\pi \approx 150.8$ in^3 (b) $h = \dfrac{3V}{\pi r^2}$

21. $w = \dfrac{P - 2l}{2}$

23. (a) $x = 8$ (b) $x = 4$ (c) $x = 20$ (d) $x = 6$

25. $x = 3$ or $x = 4$

27. $x = \dfrac{-7 \pm \sqrt{17}}{2}$

29. (a) $x = \pm\dfrac{5}{2}$ (b) $x = 3$ or $x = -2$

(c) $x = \dfrac{-3 \pm \sqrt{5}}{2}$

31. $r = \sqrt{\dfrac{A}{\pi}}$

33. $x = 5$

35. $x \approx 6.167$

Topic 1 Test

1. F–A cube root could also be negative.

2. F–3.14 is rational because it is a terminating decimal.

3. T

4. F–$\sqrt{6} + \sqrt{5} \approx 2.45 + 2.34 = 4.79$, but $\sqrt{11} \approx 3.32$

5. T

6. (a) 40^3 (b) $\dfrac{1}{8^4}$ (c) 8^{18} (d) 2^{17}

7. (a) 4.8629   (b) 5.7236   (c) 0.7319

8. (a) $5\sqrt{3}$ (b) 10   (c) 6   (d) $\dfrac{\sqrt{21}}{2}$

9. (a) $16x + 5y$ (b) 5π

(c) $5\sqrt{2}$ (d) $\dfrac{11}{2}\sqrt{3} + 7$

10. (a) 43   (b) $\dfrac{29}{6}$ (c) 2.5

11. $S = 340$ in^2

12. $x = 2$

13. $x = \dfrac{5}{3}$ or $x = 7$

14. $x = \dfrac{11 \pm \sqrt{137}}{2}$

15. $x = \pm \dfrac{7}{\sqrt{5}}$

16. $r = 3\sqrt{\dfrac{3V}{2\pi}}$

Topic 2

1. (b) and (d)

3. (a) Roses are red, and the sky is blue.
(b) Roses are red, and the sky is blue or turtles are green.
(c) If the sky is blue, then both roses are red and turtles are green.
(d) If it is not the case that turtles are not green and turtles are green, then roses are not red.

5. $p \Rightarrow q$ and $\sim(p \wedge q)$

7. (a) T   (b) F   (c) T   (d) T
(e) T   (f) T   (g) T   (h) T

9.

p	q	$\sim p$	$\sim q$	$(\sim p) \vee (\sim q)$	$(\sim p) \vee q$	$(\sim p) \wedge (\sim q)$
T	T	F	F	F	T	F
T	F	F	T	T	F	F
F	T	T	F	T	T	F
F	F	T	T	T	T	T

p	q	$\sim p$	$\sim q$	$p \Rightarrow q$	$\sim(p \wedge q)$	$\sim(p \vee q)$
T	T	F	F	T	F	F
T	F	F	T	F	T	F
F	T	T	F	T	T	F
F	F	T	T	T	T	T

11. (a) Hypotheses: All football players are introverts, and Tony is a football player.
Conclusion: Tony is an introvert.
(b) Hypotheses: Bob is taller than Jim, and Jim is taller than Sue.
Conclusion: Bob is taller than Sue.

13. (a) Valid   (b) Invalid
(c) Invalid   (d) Valid
(e) Invalid   (f) Invalid
(g) Invalid   (h) Valid

15. (a) (i) only   (b) (ii) only
(c) (ii) only

17. (a) Syllogism   (b) Contraposition
(c) Syllogism

19. (a) T   (b) T
(c) Unknown   (d) T
(e) F   (f) Unknown
(g) T   (h) T
(i) Unknown   (j) F
(k) T   (l) Unknown
(m) Unknown   (n) T
(o) T

21. (a) $w \Rightarrow t$ and t, therefore w. Invalid.
      (b) $s \Rightarrow l$ and $\sim s$, therefore $\sim l$. Invalid.

23. Invalid.

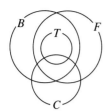

25.

p	q	$\sim p$	$\sim p \vee q$	$p \Rightarrow q$
T	T	F	T	T
T	F	F	F	F
F	T	T	T	T
F	F	T	T	T

The last two columns are the same.
Therefore, $\sim p \vee q$ is logically equivalent to $p \Rightarrow q$.

Topic 2 Test

1. T

2. F–$\sim q \Rightarrow \sim p$ is true.

3. F–You may be older than 20 and younger than 30.

4. T

5. F–It means q is necessary for p.

6. (a) $\sim q \Rightarrow p,\ \sim p \Rightarrow q,\ q \Rightarrow \sim p$
      (b) $q \Rightarrow \sim p,\ p \Rightarrow \sim q,\ \sim q \Rightarrow p$
      (c) $\sim p \Rightarrow \sim q,\ q \Rightarrow p,\ p \Rightarrow q$

7. (a) T   (b) F   (c) F
      (d) F   (e) T   (f) T

8. (a) Invalid   (b) Valid
      (c) Valid   (d) Invalid
      (e) Invalid

9. Row 1: T T T T T F F F T
      Row 2: F T F T T T F T T
      Row 3: F F F F T T T F T
      Row 4: T F F T F F T T F

10. One example is: Some Bs are A. All Cs are Bs. Therefore all Bs are Cs.

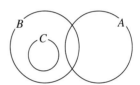

The Euler diagram represents the two hypotheses, but the conclusion is false.

11. All three

12. (a) A tetrahedron has edges.
      (b) No valid conclusion can be drawn.

13. (a) F   (b) F   (c) T

14. All three are true where the two statements are both true. The only time that the implication is false is when the hypothesis is true and the conclusion is false. In this case, the disjunction is true. Thus, the conjunction, disjunction, and implication are never all false at the same time.

Topic 3

1. (a) $x < 6$ (b) $x \leq -2.5$ (c) $x < -5$

3. (a) $x \geq 12$ (b) $x < -10$ (c) $x > -6$

5. Because the side opposite $\angle B$ is shorter than the side opposite $\angle C$.

7. (a) $59°$ (b) $\overline{PR}$ (c) $\overline{QR}$

9. (a) $163°$
      (b) $\overset{\frown}{AB}$ because $\angle AOB$ is larger than $\angle BOC$
      (c) $\overset{\frown}{AC}$ because $\angle AOC$ is larger than $\angle AOD$

11. (a) $\overline{AC}$ because it is closest to the center
      (b) $\angle B$ because $\overline{AC}$ is the longest chord
      (c) $\overset{\frown}{AB}$ because AB is the shortest chord

13. $BD < BC$, $BE < BC$, and $AB < BE$. Other correct answers are possible.

15. $11 < BC < 15$. That is, BC must be between $11''$ and $15''$.

17. Because the hypotenuse is opposite the 90° angle and the 90° angle is the largest angle in a right triangle, the hypotenuse must be longer than the other two legs by Theorem T3.3.

19. First, $\angle BDC > \angle A$ because $\angle BDC$ is an exterior angle of $\triangle ABD$. Because $\triangle ABC$ is isosceles, $\angle A = \angle C$. Therefore $\angle BDC > \angle C$. By Theorem T3.3, $BC > BD$. Because $AB = BC$, we have $AB > BD$.

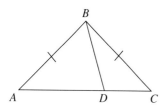

21. Method 1: Use the Pythagorean Theorem: Because $\triangle ACD$ is a right triangle, $(AD)^2 + (CD)^2 = (AC)^2$. Therefore, $(AD)^2 < (AC)^2$ and $(CD)^2 < (AC)^2$, or $AD < AC$ and $CD < AC$.

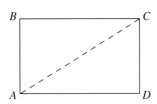

Method 2: Use Theorem T3.3: Because $\angle DCA < \angle D$ and $\angle DAC < \angle D$, by Theorem T3.2 we have that $AD < AC$ and $CD < AC$.

23. In $\triangle ACD$, (i) $\angle CAD < \angle ACD$ because $CD < AD$. In $\triangle ABC$, (ii) $\angle BAC = \angle BCA$ because they are opposite congruent sides. Adding the left sides and

the right sides of (i) and (ii), we have $\angle BAC + \angle CAD < \angle BCA + \angle ACD$. Thus $\angle BAD < \angle BCD$, or equivalently, $\angle A < \angle C$.

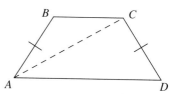

25. The area of $\triangle AOB$ is $\frac{1}{2}(OB)(OA)$. Extend $\overline{AO}$ through O to point E on $\overline{DC}$. Because $\overline{DC}$ is in the interior of the circle, $OE < OA$. Therefore, the area of $\triangle ODE$ is $\frac{1}{2}(OD)(OE) < \frac{1}{2}(OB)(OA)$, which is the area of $\triangle ABO$.

27. Let B be the point where ray $\overrightarrow{OA}$ intersects the circle and P be another point on the circle. By the Triangle Inequality, $OP < OA + AP$. But $OB = OA + AB$. OB and OP are equal because they are radii. So $OA + AB < OA + AP$. Thus, $AB < AP$.

29. If $a < b$ and $b < c$, then there are positive real numbers p and q such that $a + p = b$ and $b + q = c$. Substituting $a + p$ in the second equation for b yields $a + p + q = c$ for the positive real number $p + q$. Thus, $a < c$.

Topic 3 Test

1. T

2. F–since c and d may be negative

3. F–since both $\angle A$ and $\angle B$ may be less than 20°

4. T

5. T

6. $x < -1$

7. $x > -3$

8. (a) $\angle R$ because $\overline{ST}$ is the longest side
   (b) $\angle T$ because $\overline{RS}$ is the shortest side

9. (a) $\overline{XY}$ because Z is the largest angle
   (b) $\overline{YZ}$ because X is the smallest angle

10. (a) $\overparen{AC}$ (b) $\overline{BC}$

11. $11 < AC < 31$, or AC must be between $11''$ and $31''$.

12. (a) $\overline{DE}$ because $\angle EOD = 70°$ is larger than $\angle BOC = 50°$ and $\angle COD = 40°$

   (b) $\angle OED$. $\triangle BOC$, $\triangle COD$, and $\triangle DOE$ are isosceles, hence their base angles are congruent. Because $\triangle DOE$ contains the largest central angle, its base angles are the smallest.

13. (i) Because $\angle ADB$ is an exterior angle of $\triangle BCD$, $\angle ADB > \angle CBD$. But $\angle CBD = \angle ABD$, so $\angle ADB > \angle ABD$. Case (ii) is proven similarly.

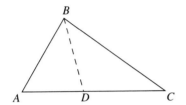

14. In pentagon $ABCDE$,
$$AC < AB + BC,$$
$$BD < BC + CD,$$
$$CE < CD + DE,$$
$$AD < DE + AE,$$
$$BE < AE + AB$$

Adding left sides and adding right sides, we have $AC + BD + CE + AD + BE$ is less than twice the perimeter.

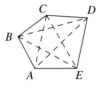

Topic 4

1. Lines are great circles and no two great circles are parallel. In fact, any two great circles intersect in two points that are endpoints of a diameter of the sphere.

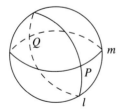

3. A "circle" on the sphere will look like a circle, but it will not necessarily be a great circle. For example, a circle of points equidistant from point P is shown.

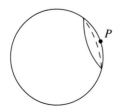

5. (a) $d_T(A, B) = 6$, $d_E(A, B) = \sqrt{18}$ or $3\sqrt{2}$
   (b) $d_T(A, B) = 10$, $d_E(A, B) = \sqrt{52}$ or $2\sqrt{13}$
   (c) $d_T(A, B) = 13$, $d_E(A, B) = \sqrt{85}$
   (d) Yes, it is true for (a)–(c). Yes, true for all pairs A and B.

7. (a) Area of the square is 16 square units. Area of the circle is 4π square units.
   (b) Area of the square varies depending on its placement on the axes. For example,

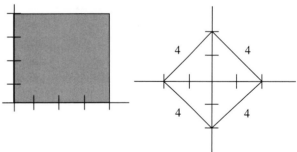

9. (a) B lies between A and C.
$$d_T(A, B) + d_T(B, C) = d_T(A, C)$$
   (b) B lies between A and C.
$$d_T(A, B) + d_T(B, C) = d_T(A, C)$$

11. Both of the following triangles have two angles of $45°$ and an included side of 2 units. Yet they are not congruent. Other examples are possible.

13. Lines in this geometry are the pairs of points AB, AC, AD, BC, BD, and CD. The lines that have a point in common are the pairs that share a letter. For example, AB and AC share point A. In every case, lines that have a point in common have exactly one point in common.

15. To verify this theorem, we list for each combination of a line and a point not on it, the one line in the set of six possible lines that does not intersect the given line.

Line AB and point C–line CD.
Line AB and point D–line CD.
Line BC and point A–line AD.
Line BC and point D–line AD.
Line CD and point A–line AB, and so on. In every case there is exactly one line parallel to a given line.

17. T–Each point occurs in exactly three lines. For example, A is on lines ABC, AED, and AGF.

19. T–For any given pair of points, there are exactly two lines that contain neither point. For example, for points B and G, the two lines that contain neither point are AED and CFD.

21. $\triangle ACD \cong \triangle DBA$ by LL, so $AC \cong BD$ by CP.

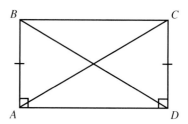

23. Assume $CD > AB$, and choose point E on $\overline{CD}$ such that $AB = DE$. Then $ABED$ is a Saccheri quadrilateral. Because $\angle BED$ is an exterior angle to $\triangle BCE$, $\angle BED > \angle C = 90°$. Thus, $\angle BED$ is obtuse. By problem 22, we know that the summit angles are congruent, so $\angle ABE$ is also obtuse. If $\angle ABE$ is obtuse, then $\angle ABC$ must also be obtuse. This results in a contradiction because the fourth angle in a Lambert quadrilateral is not obtuse. Therefore, our assumption that $CD > AB$ must be false, and we have that $CD \le AB$.

Topic 4 Test

1. F–There are infinitely many lines (great circles) through each pair of points.

2. T   **3.** T   **4.** T   **5.** T

6. No, this result does not hold in the geometry of the sphere. For example, in the figure shown, the exterior angle at C measures $90°$, but the two nonadjacent interior angles at A and B have a sum of $180°$.

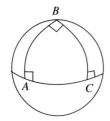

7. (a) Three lines: AB, AC, BC
   (b) One point. For example, AB and AC intersect in A.
   (c) No, because a line must contain two points. Given line AB, only point C remains.

8. (a)

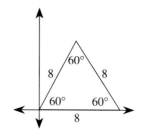

$$A = \frac{s^2\sqrt{3}}{4} = \frac{8^2\sqrt{3}}{4} = 16\sqrt{3} \text{ square units}$$

(b) Three possibilities are

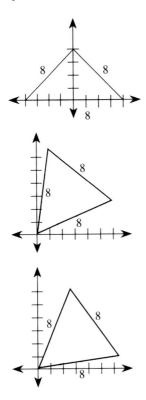

(c) In each case, the area is 16 square units.

9. The triangles are not necessarily similar. One counterexample is shown.

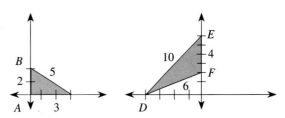

Each side of $\triangle FED$ is twice as long as a corresponding side of $\triangle ABC$, but $\triangle FED$ is not similar to $\triangle ABC$ because $\triangle ABC$ is a right triangle.

10. We have quadrilateral $ABCD$ in which $\angle A = \angle C = \angle D = 90°$ because $ABCD$ is a Lambert quadrilateral.

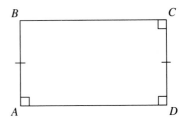

Also, $AB = CD$ because $ABCD$ is a Saccheri quadrilateral. By problem 19 of this section, the summit angles of a Saccheri quadrilateral are congruent. Thus $\angle B = 90°$, and so all four angles of $ABCD$ are right angles. Therefore, $ABCD$ is a rectangle.

Index

APPLICATIONS INDEX

Continued on page 592